Liaison Functions 2014

Core Programming Area at the 2014 AIChE Annual Meeting

Atlanta, Georgia, USA
16-21 November 2014

ISBN: 978-1-5108-1262-8

Printed from e-media with permission by:

Curran Associates, Inc.
57 Morehouse Lane
Red Hook, NY 12571

Printed by Curran Associates, Inc. (2015)

For permission requests, please contact AIChE
at the address below.

AIChE
120 Wall Street, FL 23
New York, NY 10005-4020

Phone: (800) 242-4363
Fax: (203) 775-5177

www.aiche.org

Additional copies of this publication are available from:

Curran Associates, Inc.
57 Morehouse Lane
Red Hook, NY 12571 USA
Phone: 845-758-0400
Fax: 845-758-2634
Email: curran@proceedings.com
Web: www.proceedings.com

TABLE OF CONTENTS

392003 Control and Automation of Fluid Flow, Mass Transfer and Chemical Reactions in Microscale Segmented Flow

Abstract:

Owing to the constant increase of the worldwide energy demand due to the growth of populations and development of countries, the emissions of greenhouse gases (mostly carbon dioxide (CO_2)) from fossil fuels will constantly increase. Currently, carbon capture and sequestration is one of the most efficient solutions to control and reduce the concentration of the released greenhouse gases in the atmosphere. However, the underground storage of the captured CO_2 is expensive and limited to some specific locations. On the other hand, the current rate of energy consumption along with the inevitable global warming makes energy as one of the crucial worldwide issues of the near future. Therefore, there will be a huge demand for new sets of sustainable energy substitutes. In 2008, it was found that CO_2 in the presence of a frustrated Lewis pair (FLP) and hydrogen can be quickly converted to methanol and water. It has been proposed that CO_2 fuel conversion using the captured CO_2 from power plants (or atmosphere) has the potential to be one of the key solutions to address the energy demand and global warming problems. However, measurement and characterization of the kinetics of CO_2 conversion reactions (i.e., to methanol, ethanol or CO) in the presence of metal-free catalysts (FLPs) are currently challenging or in some cases impossible, owing to the mass transfer and time-resolution limitation of the conventional batch scale characterization tools.

During the past decade, microfluidic (MF)-based technologies have shown promising solutions as a high throughput characterization tool for investigation, screening, and optimization of gas-liquid reactions.

In my PhD, I was working on a multidisciplinary project, collaborating between Mechanical Engineering and Chemistry Departments of University of Toronto. During this time I have made significant contributions in flow chemistry and microfluidic (MF) research fields specifically related to screening gas solubilities in liquids, measuring thermodynamic characteristics of gas-liquid reactions associated with CO_2 sequestration, characterizing CO_2-governed liquid-liquid phase separation process (switchable water) and on-demand preparation of high quality colloidal nanomaterials (e.g. CdSe/CdS).

During this time, I designed and developed the first automated MF platform that allowed the routine screening of solubility data for a wide range of CO_2-physical solvent mixtures. The platform allowed screening of approximately 300 different conditions (temperature and pressure) during one day. In the next step, the developed automated MF strategy was utilized to measure, for the first time, thermodynamic characteristics of CO_2-FLP reaction. I have also designed a two-phase MF platform for high temperature (240-330 C) synthesis of high quality (Fwhm ~30 nm) nanocrystals such as CdSe, CdS and ZnS. The silicon-based microreactor could produce high quality core or core/shell nanocrystals with desired spectral characteristics at a throughput of 0.6 ml/min.

As an independent scientist, I will focus on the development and utilization of MF platforms for the carbon capture in the presence of nanoparticles along with fuel conversion process. I will also study the underlying transport mechanisms in the presence of chemical reactions with experimental and numerical methods.

389363 Mechanistic Correlations Between Material Properties and the Dynamics of Catalytic and Photocatalytic Redox Processes

Abstract:

Chemical reactions catalyzed by materials with electronic band gap (e.g., metal oxides) proceed via redox cycles involving kinetic coupling between elementary steps that reduce and re-oxidize the catalyst materials. Such redox cycles involve transfer of electrons between the catalyst and substrate molecules, with or without photo-excitation of electrons within the material to facilitate the electron transfer, and their dynamics depend on positions of filled and empty electronic bands. Here, I will discuss the use of kinetic, isotopic and spectroscopic methods in combination with quantum mechanical calculations to understand the mechanism of such reactions and to derive rigorous relations between material properties and their catalytic functions.

389825 Understanding of Polymers in Confined Thin Films and Bulk Membranes: Fluorescence Based Approach

Abstract:

My research interest is to develop core understanding of material properties of functional polymers in thin film and bulk membrane format using fluorescence and other complementary surface characterization techniques (spectroscopic ellipsometry, quartz crystal microbalance, neutron reflectometry).

Bulk membranes have potential applications in bioseparation, water purification, fuel cell applications and so on. Virus filtration membranes work based on size exclusion mechanism and are used to remove viral contaminants during the processing and commercial production of therapeutic proteins. The biggest challenge in these processes is lack of sufficient understanding of virus retention mechanism. This knowledge is pivotal to understand the trends in flux and log reduction values (LRVs) of virus filtration membranes with increased volumetric throughput. The application of fluorescence confocal microscope to visualize the virus retention zones along the thickness of the membranes is a very exciting and new concept. This has been taken one step forward by adopting a multiple dye approach to track model bacteriophage retained inside virus filtration membranes (DV20, Vireolsve pro and Viresolve NFP) at different stages of membrane operations with intermittent pressure releases (Post-doc at ChE, Penn State, PI: Prof. Andrew Zydney). Fluorescence based techniques offer unique opportunities to distinguish between virus captured before and after process interruptions. This, in turn, gives interesting insights about the effect of pore structure and morphology on virus retention.

Fluorescence, when used in conjunction with other surface characterization techniques, provides a powerful platform to characterize transport properties and mobility of polymer and water in bulk polymer membranes as well as thin films. Bulk free-standing and confined supported systems behave very differently at hydrated state in terms of material properties. When the film thickness approaches to several tens of nanometers, the optical, mechanical and transport properties are dominantly controlled by the interfaces. The mobility of the polymer chains is constrained due to severe confinement effect. Although a great deal of work has been done so far to explore the material properties of ionomers in bulk membranes, the mobility, hydration and transport properties of functional ionomers in nm-thick films are relatively unexplored. The challenges in characterizing thin films start when the conventional bulk characterization techniques fail to predict the nanoscale changes in materials properties. Fluorescence based techniques can overcome many of these issues. Fluorescent probes sensitive to local viscosity and proton concentration were incorporated into ionomer thin films and membranes and the changes in fluorescence properties of the dyes as a function of film thickness and hydration were studied (Post-doc at MatSE, Penn State, PI: Prof. Michael Hickner). The changes in fluorescence offered valuable insight into stiffness and proton mobility of hydrated bulk and confined systems. The work clearly demonstrated the differences between confined and bulk systems. The study has prime importance in nanoscale understanding and advanced level designing of polymer-catalyst interface in fuel cells.

In addition, fluorescent p-conjugated oligoelectrolytes were synthesized and characterized in solution and solid state utilizing their light-harvesting properties. This was a part of my PhD work. Layer-by-layer self-assembled thin films of fluorescent oligoelectrolytes were investigated to develop fluorescence based bioassay platforms.

The multi-disciplinary work experience enabled me to innovate effective techniques to understand functional materials more deeply for a range of applications. As a future faculty, I would like to perform independent as well as collaborative research focusing on structure-property relationships and material development for (1) energy applications and (2) chemobiosensing. The application of fluorescence based techniques in fuel cell research is quite new, has already started offering interesting insights and has a lot of scope of work. An array of confined and bulk ionomeric systems will be probed using fluorescent functional sensory molecules and water-polymer-substrate interaction will be further explored. Other surface characterization techniques can help to validate the hypothesis from fluorescence based analysis and strengthen the arguments about unknown scenerios in confined systems. This will solve a lot of mysteries hidden inside these geometries. Techniques will be developed to probe zone-specific density, water/proton transport and mobility. This will be of specific interest and the knowledge earned from materials studies will be employed to develop new materials with desired hydration and transport properties. The study will be extended to light harvesting molecules in confined systems with varied geometry and microenvironment. Typical light induced energy and charge transfer phenomena will be tuned for better and cheaper solid state biodetection platforms with improved sensitivity and selectivity.

388999 Protein-Protein Interactions in Disease and Therapy

Abstract:

Protein-protein interactions mediate a wide range of biological processes including receptor-ligand binding, protein signaling cascades, antibody binding, cell adhesion and metabolic regulation. Dysfunction of protein-protein interactions can lead to a variety of diseases including cancers and neurodegenerative diseases. Antibody therapeutics also make use of highly specific protein interactions to provide targeted intervention. The ability to generate, characterize and evolve protein interactions provides a basis to understand disease pathogenesis and to develop potential therapies.

I am interested in how the arrangement of different amino acids leads to highly specific interactions between two proteins and to the overall disease state of an organism. In my work with David Colby at the University of Delaware we have investigated prion pathogenesis and therapy using high throughput technologies such as fluorescence activated cells sorting and deep sequencing and at multiple scales from recombinant DNA techniques to animal models. Prion diseases are a group of rapidly progressing, lethal neurodegenerative diseases that are propagated by the conversion of the native form of the prion protein (PrP^{C}) to an alternatively folded, disease causing form (PrP^{Sc}) catalyzed by the interaction of PrP^{Sc} with PrP^{C}. We used mutagenesis and directed evolution approaches to generate high affinity antibodies to disrupt prion pathogenesis and to characterize prion protein self-interactions. Future applications of the methodologies developed include investigation of allosteric enzyme sites and antibody interactions with non-antigen targets relevant to the biopharmaceutical industry. I am also interested in engineering education and have pursued several opportunities to lead students in classroom and laboratory settings.

Doped Metal Oxides – Increased Functionality for Energy Applications

James Dorman- University of Konstanz, Konstanz, Germany

Metal oxides are commonly used in current technologies due to the numerous properties that can be achieved with slight variations of the metal ion(s). Furthermore, the properties of these compounds can be engineered to increase their applicability through the incorporation of low concentrations of substitute ions into the crystal lattice or the addition of thin shell layers. Such doped metal oxide structures are implemented into state-of-the-art technology, such as catalysts, optical displays, semiconductor devices, etc., due to the following properties:

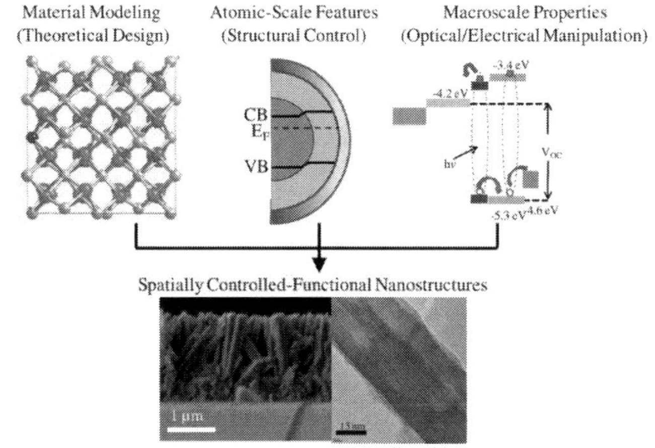

The schematic shows how material modeling, spatially control, and manipulation of the physical properties will be implemented for high functionality nanostructures.

- Predictable luminescence due to allowed transitions after crystal field splitting,
- Charge mobility control proportional to the electron/hole density,
- Electronic state addition around the conduction and valance bands,
- Increased chemical stability in the presence of harsh conditions, and
- Control over stress/strain of the system during chemical reactions.

However, the combination of host and dopant structures and their associated applications are far from optimized or used to their full potential. My research interests focus on the doping of metal oxides to control their physical and electronic properties with a specific focus on energy applications. Specifically, my research focuses on the engineering of spatially controlled doped metal oxide nanostructures for optical/optical (lighting and displays)[1,2], optical/electrical (photovoltaics)[3-5], and electrical/chemical (energy storage) energy conversion. The schematic highlights the implementation of these material modeling, spatial control, and the macroscale properties for the production of functional nanostructures for energy applications. The role of the dopant on the material properties can be confirmed through a combination of theoretical modeling[6] and experimentally[7] as discussed in my pervious publications. Specifically, the position and concentration of the dopant plays a critical role in both lighting and photovoltaic applications.[3,6] Furthermore, the process can be precisely controlled using a combination of wet chemical synthesis methods, i.e. hydrothermal growth and sol-gel chemistry, and low pressure deposition methods, such as atomic layer deposition.

(1) Dorman, J. A.; Choi, J. H.; Kuzmanich, G.; Chang, J. P. *The Journal of Physical Chemistry C* **2012**, *116*, 10333.
(2) Dorman, J. A.; Choi, J. H.; Kuzmanich, G.; Chang, J. P. *The Journal of Physical Chemistry C* **2012**, *116*, 12854.
(3) Dorman, J. A.; Weickert, J.; Reindl, J. B.; Putnik, M.; Wisnet, A.; Noebels, M.; Scheu, C.; Schmidt-Mende, L. *The Journal of Physical Chemistry C* **2014**.
(4) Bretschneider, S. A.; Weickert, J.; Dorman, J. A.; Schmidt-Mende, L. *APL Materials* **2014**, *2*.
(5) Weickert, J.; Zimmermann, E.; Reindl, J. B.; Pfadler, T.; Dorman, J. A.; Petrozza, A.; Schmidt-Mende, L. *APL Materials* **2013**, *1*.
(6) Dorman, J. A.; Choi, J. H.; Kuzmanich, G.; Bargar, J. R.; Chang, J. P. *Journal of Applied Physics* **2012**, *111*, 083529.
(7) Dorman, J. A.; Mao, Y.; Bargar, J. R.; Chang, J. P. *The Journal of Physical Chemistry C* **2010**, *114*, 17422.

379648 Multimodal Nanomedicines for Solid Tumors: Rational Drug Combinations, Targeted Cancer Therapies, Photo-Activated Drugs, and Synthetic Lethal RNA Interference

Abstract:

Nanoscale drug delivery is a power full tool for the treatment of solid tumors. Although a range of approaches have been described over the past several years, the vast majority seek to improve cytotoxic chemotherapies that have been in use for decades. Nanoscale delivery strategies integrating rational drug combinations, synthetic lethal RNA interference, stimuli-responsive drugs, or so-called molecularly targeted cancer therapeutics are and remain underexplored. Here, we present work from our laboratory describing the development of a range of such nanoscale platforms for drug delivery to solid tumors. These include: (i) self-assembled peptide co-polymer nanoparticles for synthetic lethal RNA interference of DNA damage response in ovarian and lung cancers, (ii) multivalent anti-androgen and anti-estrogen gold nanoparticles for the treatment of hormone-refractory breast and prostate cancers, (iii) photo-responsive nanoparticles that deliver drugs or achieve photothermal ablation in response to laser exposure, (iv) nanoscale delivery platforms targeting cancer stem cells, the hypoxic tumor microenvironment, or the tumor stroma, and (iv) Layer-by-Layer (LbL) nanoparticles for horizontal signal blockade in breast carcinoma.

383097 Optical Methods to Probe Biological Questions

Abstract:

The convergence of technical advances in optical imaging, biology (specifically genetic modification), and chemical synthesis provide a wealth of avenues to explore complex biological systems with fluorescent microscopy. My research interests focus on applying novel optical, chemical, and/or biological approaches to better understand medically related processes at a fundamental engineering level in cells and *in vivo*. Using these approaches we can measure biological processes, medical intervention success or failures, and monitor disease state or progress. During my graduate work I developed novel optical nanoparticle sensors to fluorescently measure ions and small molecules. These sensors were used to measure sodium flux during action potentials in cardiomyocytes and interstitial concentrations as a tattoo like minimally invasive clinical monitor. During my postdoc I have focused on combining polarized fluorescence microscopy and fluorescently labeled anticancer drugs to measure target engagement and binding activity in cells and at the subcellular level in *in vivo* tumor models. This approach, discussed in detail in this poster, allows us to better understand small molecule behavior in cells *in vivo* and assess tumor heterogeneity during therapeutic treatment both spatially and temporally. Building on these pursuits, I hope to develop a research program that explores difficult to measure biological questions in complex systems using novel fluorescence imaging approaches.

380477 Discovery of Disease-Specific Antibody Biomarkers and Their Targets

Abstract:

Discovery of biologic molecules specific to a diseased state, or biomarkers, can lead to diagnostic development, therapeutic target identification, and improved understanding of disease pathogenesis. Antibodies remain an attractive class of biomarkers given their amplification by the immune system, stability, and current clinical use. While the antibody repertoire represents a rich source for biomarker discovery, it has been difficult to impartially identify which molecules from this repertoire are associated with disease. This work demonstrates three molecular discovery processes centered on the utility of bacterial-displayed peptide libraries and fluorescence activated cell sorting (FACS) for identifying novel antibody biomarkers. We applied these methods to identify candidate molecular diagnostics for pre-eclampsia (PE), a condition with unknown etiology that affects 5-8% of pregnancies. Previously characterized autoantibodies that bind the angiotensin II type 1 (AT1) receptor in PE patients [1][2] are difficult to detect, vary in prevalence amongst studies, and most importantly, lack specificity. Thus, we sought to identify additional PE-specific antibody biomarkers.

Applying three quantitative screening strategies against a set of PE and healthy outcome pregnancies (HOP) identified unique disease-specific peptides from a 15 amino acid peptide library. With a two-color screening method, we used antibody fractions enriched from plasma to isolate significantly PE cross-reactive and specific peptides distinct from the AT1 receptor epitope. We used a panel of these antibody-detecting peptides to train and validate an Adaptive Boosting classification algorithm that achieved 80% diagnostic accuracy [3]. In a second screen, the library underwent sequential enrichment for PE antibody (immunoglobulin G) binding followed by removal of HOP antibody binding peptides against unprocessed, diluted plasma. This screen more closely replicated the native environment for the antibody binding interaction and yielded a strong consensus motif. Directed evolution expanded this motif for which BLAST searches identified a number of environmental antigens and human proteins that shared similarities.

Importantly, by evaluating antibody binding activity of bacterial-displayed 15-mer fragments from the strongest candidates, we linked this motif to a region of a common viral antigen (VA) and a human G protein-coupled receptor (hGPCRa). Furthermore, we verified that the synthesized VA fragment efficiently blocked antibody binding to the hGPCRa fragment and to HEK293T cells expressing the full length protein. Thus, this method enabled unbiased identification of a disease-specific antibody and characterization of its targets. In a third method, we combined library screening with next-generation sequencing to profile the antibody repertoire of four individual PE patients and four HOP and identify motifs present amongst these PE.

This work demonstrates that we can now probe the antibody repertoire using unbiased screening of bacterial-displayed peptide libraries to identify new markers of disease. These disease-specific antibody-detecting peptide reagents can be used for molecular diagnostics and/or characterizing antibody target(s), potentially elucidating novel therapeutic targets. Building upon this work, we can investigate how the antibody repertoire changes with age or after exposure to a particular virus or environmental factor.

[1] Wallukat, G. *et al.* Patients with Preeclampsia Develop Agonistic Autoantibodies against the Angiotensin AT1 Receptor. *The Journal of Clinical Investigation* **103**,945–952 (1999).

[2] Zhou, C. C. *et al.* Angiotensin Receptor Agonistic Autoantibodies Induce Pre-eclampsia in Pregnant Mice. *Nature Medicine* **14**,855–862 (2008).

[3] **Elliott, S.E.** *et al.* Characterization of Antibody Specificities Associated with Preeclampsia. *Hypertension* **63**,1086-1093 (2014).

389947 Programmable Dynamic Surfaces As Information Carriers

Abstract:

Tuning surface chemistry to achieve selective material properties has been an object of research for decades. Subtle variations in chemical environment at the surface of particles and materials often result in drastic changes in surface reactions, self- assemblies, surface toxicity, and more. Fine-tuning surface chemistry of materials (and/or nano particles' surfaces) in response to environmental signals could be an efficient way to dynamic surfaces capable of adjusting their surface properties and also manipulating material objects resting on such surfaces. Change of the surface properties could be triggered by alteration or gradient in chemical potential, external fields such as electrical or magnetic fields, light, or other environmental stimuli. Potentially applications of such smart surfaces might include: transferring matter (at different scales), capture and release of ions, self-cleaning materials, "smart" delivery carriers, etc. As an example, nanoparticles functionalized by photoactive ligands might undergo conformational changes upon light irradiation ultimately altering surface properties. Careful selection of grafted functional groups at the correct ligand density could yield photoactive surfaces of the particles having dynamic surface charge density or hydrophobicity. Since the relaxation time of light-induced processes is fairly short, such photoactive particles could be applied to trap and release ions or other charged moieties. Yet another type of dynamic surfaces could be fabricated by applying charged ligands on flat substrates of macroscopic extent. Under the influence of externally applied electric fields, surface ligands could actively rearrange in regions of high-low filed. If the fields were then changing in time over the surface, they could create "waves" of surface properties that could alter effective surface properties and ultimately move around objects placed on such surfaces. My research on the design and, later, implementation of such surfaces can be facilitated by multiscale modeling with which one can gain further insights into the structures of the switchable molecules and the overall surface properties that these molecules give rise to. I am also interested in understanding such surface-dynamic phenomena at the level of fundamental thermodynamics of nonequilibrium actuation, characteristic response times, and energetic efficiencies.

379483 Improved Treatment of Fungal Infections through Targeted Drug Delivery: How Can Biomembranes Help?

Abstract:

Fungal infections are a significant cause of morbidity and mortality especially in immune-compromised individuals. Fungal diseases are hard to eradicate as fungal cells develop resistance to most available antifungal drugs and are hard to target inside mammalian cells. Several life threatening fungal infections such as cryptococcosis and histoplasmosis are contracted by inhalation of spores, which then proliferate inside the macrophages of the alveolar region and disseminate throughout the body. The macrophage-pathogen interaction is the first step in these diseases and provides a suitable target for early treatment of the disease. Recent evidence suggests that microdomains composed of saturated phospholipids and sterols (i.e. lipid rafts) play an important role in macrophage-pathogen interaction. Abolishing lipid rafts in fungal cells is shown to drastically reduce their virulence. Also, many pathogens use lipid rafts as the site of entry to the host cell, suggesting that rafts can be used for targeted treatment of fungal diseases. However, understanding the role of lipid rafts in modulating infections and utilizing them for targeted therapy of fungal infections is a complicated research problem requiring knowledge in various fields such as engineering, biophysics and microbiology.

My graduate and post-doctoral experiences in chemical engineering and microbiology have uniquely prepared me to investigate this challenging research problem. In my graduate work, I have studied the interactions between phospholipid microdomains and polymeric nanoparticles. I have examined the effects of particles on microdomain size and surface properties. I have also studied the interactions between aerosolized particles and lung surfactant interfaces to gain knowledge on the potential adverse effects of inhaled particles used for pulmonary delivery. In my post-doctoral research, I have studied the role of plasma membrane biophysics in the virulence properties of *Cryptococcus neoformans* (the causative agent of the disease cryptococcosis). Using a number of fluorescence spectroscopy methods, I have shown that the structure of lipids in the plasma membrane of *C. neoformans* plays an important role in its ability to form a rigid membrane and cause life threatening infections. I have also examined the relationship between lipid rafts in mammalian cells and their infectivity by *C. neoformans*.

My future research interests combine my expertise in characterizing plasma membrane physical properties and the interactions between lipid rafts and nanoparticles. Through modulating the physical properties of plasma membrane, I aim to design strategies to increase the uptake of drugs or drug containing nanoparticles in mammalian cells. Also, I aim to use particle engineering to target fungal cells inside macrophages and treat the disease at the site of infection. I plan to use the knowledge provided by these research directions to develop new and more effective inhalable drugs against fungal infections for pulmonary drug delivery.

365051 New Approaches to Interrogate and Engineer Complex Biocatalytic Networks

Abstract:

I hope to run a multidisciplinary research group that employs novel theoretical and experimental approaches to interrogate and engineer biochemical systems of relevance to human health, energy, and the environment. My research program will have three broad goals: (i) to develop methods for controlling, studying, and altering enzyme activities of metabolic relevance, (ii) to employ those methods to answer fundamental questions of human disease, cellular metabolism, and molecular recognition, and (iii) to apply those methods, and correspondingly evolved theories, to develop novel enzyme inhibitors and to engineer metabolic networks.

My lab will employ concepts and techniques from protein engineering, catalysis, nanotechnology, synthetic biology, and applied mathematics. Our work will range in scope from highly applied (e.g. the development of a new biochemical sensor or an improved metabolic pathway) to highly theoretical (e.g. the use of dynamical systems theory to identify patterns in social networks or ecosystem dynamics that apply to the differentiation of cells or to the behavior of systems of mutually interacting biocatalytic species).

With this presentation, I will describe a few proposed projects consistent with these plans. Examples include (i) optical systems for spatiotemporal control of enzyme activity *in vivo*, (ii) metamaterials-based strategies for evolving enzymes for synthetic metabolic networks, and (iii) enzymatic evolution of small molecules to inhibit pathologically relevant enzymes.

378090 Multiscale Modelling and Process Simulation

Abstract:

Reliable modelling of chemical processes depends on the accuracy of the physical properties used to describe each aspect of the combined system. However, most systems of interest are sufficiently complex that normal operations are data poor. Optimization increases the need for data and compounds the challenge. Using numerical simulation to estimate both the importance of physical properties and the properties themselves provides a way to meet that challenge. Connecting atomic and microscale behavior accessible through electronic structure calculations (commonly with intermediate descriptions using force fields) to continuum, macroscopic behavior requires a familiarity with multiple approaches and computational techniques. One of the most promising avenues of investigation is using molecular simulation techniques, either molecular dynamics or Monte Carlo, to calculate excess energy data in both stable and unstable regions. Correlating these data provides physical property relations analogous to traditional constitutive models that can be implemented at minimal cost to provide atomic scale insight for system level implementation.

Design of heterogeneous catalysts for the conversion of biomass-derived compounds into fuels and chemicals

Jean Marcel R. Gallo, Department of Chemistry at the Federal University of São Carlos, Brazil.
email: jean@ufscar.br

There has been an increasing interest in renewable sources of carbon for the production of transportation fuels and chemicals due to the diminishing petroleum resources. Cellulose and hemicellulose, components of biomass, can be depolymerized into the monossacharides glucose and xylose, which yields a variety of important platform molecules for the production of fuels and chemicals. For the economical viability and sustainability of biomass conversion processes, development of selective and stable solid catalysts is necessary. My future research studies will focus on the context of biomass processing, such as preparing novel heterogeneous catalysts to address current process problems (such as low stability, low yields, and leaching), devising pathways for biomass conversion into chemicals, and studying reaction kinetics and mechanisms. Between these topics I can mention four examples: (1) bioethanol conversion into valuable chemicals; (2) hydrothermally stable catalysts for the conversion of monosaccharides into platform molecules; (3) conversion of furfuryl alcohol into levulinic acid; and (4) upgrading of crude glycerol obtained as biodiesel byproduct. I believe that my research interests lie on strategical areas for developing sustainable and economically viable biomass conversion processes that will be well received by the scientific community and attract fundings.

I hold a Ph.D. in Chemistry (in a joint program between the University of Eastern Piedmont in Italy and the State University of Campinas in Brazil under the guidance of Prof. Leonardo Marchese and Prof. Heloise O. Pastore) followed by three years of postdoctoral research in Prof. James A. Dumesic's group from the Department of Chemical and Biological Engineering at the University of Wisconsin-Madison. My research in Prof. Dumesic's group involved the design and synthesis of heterogeneous catalysts and reaction kinetic studies on the transformation of biomass-derived molecules to platform chemicals. This postdoctoral training has complemented my knowledge in the field of catalysis and reaction kinetics, while I have brought to the group a new field of expertise focused on materials chemistry. In September 2013, I left Prof. Dumesic Group and joined Prof. Jose Maria C. Bueno group at the Federal University of São Carlos, in Brazil. Since then I have been working in the conversion of ethanol into compounds of interest of the chemical industry.

JMR Gallo thanks FAPESP for the fellowship (process number 2013/15861-1).

387445 Climate Change, Air Quality and Public Health: An Uncertainty Analysis

Abstract:

The central goal of this research is to explore the connections between climate change, air quality and public health through numerical modeling, with an emphasis on uncertainty in simulations of human-climate interactions. Climate change is expected to significantly influence meteorology and impact future air quality. In addition, climate change mitigation strategies will also affect emissions and concentrations of air pollutants. Earth system models are vital to advance our understanding of atmospheric processes and are used as key tools to steer environmental policies with enormous economic and social implications. However, there are large uncertainties associated with climate modeling that further propagate into estimates of future air quality, as well as projections of associated health and economic impacts. These uncertainties must be identified and quantified to adequately assess the changes brought about by different climate policies. Furthermore, characterizing uncertainty across the complete climate change problem is essential to target the largest potential reductions to uncertainty in existing modeling systems and allow for improved impact and adaptation analyses.

Towards this end, a modeling framework to simulate air quality and associated health impacts under different climate scenarios has been developed using global earth systems modeling with atmospheric chemistry capabilities. Future pollutant concentrations are simulated with the Community Earth System Model (CESM), while the MIT Integrated Global System Model is used to predict climate change during the 21st century and beyond. The framework can be used to explore the most important sources of uncertainty in climate projections: emissions of climate forcers and conventional air pollutants under changing policy scenarios; model response due to structural uncertainty; and natural climate variability. By using this modeling framework to simulate changes in air quality for ensemble simulations of 21st century climate change, the propagation of climate uncertainties into atmospheric pollutant concentrations predictions, as well as associated health and economic impacts can be evaluated through probabilistic analyses. This information will assist in developing more meaningful climate impacts and risk assessments and better guide decision-making.

Dielectrophoresis (DEP) is an electrokinetic phenomenon used widely in microfluidic devices to manipulate, concentrate and separate particles. A polarizable particle in a nonuniform electric field experiences a physical force called the DEP force. In one format, called 'insulator-based DEP" (iDEP), particles flow through a microchannel which has zones where the cross section changes due to insulating obstacles, channel constrictions or other structures, while a DC electric field is applied across the channel. When an electric field is applied across the microfluidic channel, these structures generate regions of strong electric field gradients near themselves, which causes polarizable particles to experience a DEP force. iDEP has been used for more than a decade, and its advantages include relatively simple fabrication, and virtually eliminated electrode fouling and electrode reaction problems. Manipulation of a particular set of analytes in iDEP is dependent on a combination of the applied electric signal, channel features (design of insulating obstacles etc), suspending medium composition and microfluidic channel material [1]. We have investigated iDEP with <100 Hz AC signals instead of DC signals, and found that a given particle could be trapped under identical conditions in the same microchannel with 100 V_{RMS} weaker electric fields compared to DC signals [2]. Later, we compared several insulating obstacle geometries and found that under otherwise identical conditions, the DC voltage required for trapping the same particle could vary by 300 V between "long" circle shaped and "short" diamond shaped posts. By changing the electric signal or insulating obstacles it is possible to reduce the power requirements of iDEP devices without changing the sample and at no additional device complexity or cost. These studies are expected to pioneer more rationally designed iDEP devices [3].

1. Gencoglu, A., et al., *Particle Manipulation in Insulator Based Dielectrophoretic Devices.* Journal of Nanotechnology in Engineering and Medicine, 2013. **4**(2): p. 021002.1-7.
2. Gencoglu, A., et al., *Dynamic microparticle manipulation with an electroosmotic flow gradient with low frequency alternating current dielectrophoresis.* Electrophoresis, 2014. **35**(2-3): p. 363-373.
3. LaLonde, A., et al., *Effect of insulating posts geometry on particle manipulation in insulator based dielectrophoretic devices.* Journal of Chromatography A, 2014.

365893 Crystal Polymorphism and Texture Control during Solution Growth for Organic Electronics and Pharmaceutical Flow Chemistry

Abstract:

Solution deposition of organic semiconductors (OSC) is a leading contender for producing large-area, inexpensive, and flexible organic electronics. Using a new solution processing technique called solution shearing, we modify the molecular packing of a wide range of OSCs through processing alone. We quantitatively measure the molecular ordering resulting from different deposition conditions and show effective tuning of molecular packing that yields higher charge carrier performance without changing the underlying chemical structure. We investigate how solution flow and evaporation affects the morphological features and their impact on charge carrier transport. We also develop a novel high speed in-situ technique that uses X-ray diffraction to study how the polymorphs grow as a function of solution shearing parameters. This in-situ technique shows that the confined growth of OSC in a thin film promotes formation of a metastable crystal polymorph. We pattern high performance organic electronics through the use of surface selective patterning and nucleation control. The concept of nucleation control and crystal growth is also utilized to perform and extend microfluidic flow chemistry for pharmaceutically relevant reactions. Inorganic and organic solid formation can be processed without clogging the reactor by understanding the characteristics of crystals in flow. My proposed research will expand on the concept of metastable crystallization of OSCs and other organic molecules in order to reliably generate new polymorphs while controlling crystal morphology.

386728 Computational Design of Crystalline Materials Toward Storage and Efficient Use of Energy

Abstract:

One of the greatest challenges that mankind is facing is to continue the development of technologies that improve our quality of life without depleting our energy resources. At the center of addressing this challenge is the need to discover novel materials that allow us to develop processes and systems that make efficient use of energy. The advent of nanotechnology has given us unprecedented access to the atomistic structure of matter and enabled the rational design of materials by precisely controlling their nanostructure. My primary research interest is to use quantum and classical mechanics molecular simulations to (help) design the materials that will play pivotal roles in overcoming the bottlenecks related to energy consumption and/or storage in emerging technologies in areas as diverse as transportation, the chemical industry, and biomedicine. The goal is to capitalize on the link between properties/performance and nanoscale features.

This poster will provide an overview of my computational research work for the design of materials towards energy storage and/or efficient use of energy. During my doctoral work, I sought to design a catalyst/support system that could selectively produce single-walled carbon nanotubes (SWCNTs) of the same chirality. SWCNTs are tubular crystalline materials with outstanding chirality-dependent properties. Therefore, chiral-selective nanotube synthesis is eagerly pursued to avoid energy-intensive separations methods to obtain the desired chirality for applications such as biomedical imaging and photovoltaics. I investigated the nanotube growth mechanism to elucidate and discover ways that chirality could be controlled during synthesis, using reactive molecular dynamics (RMD) and density functional theory (DFT) methods. I investigated the feasibility and relevant conditions under which one could take advantage of the catalyst nanoparticle structure to act as a template to the nascent nanotube, and thus provided useful synthesis guidelines to experts in nanotube synthesis.

During my first postdoctoral project, I sought to design metal-organic frameworks (MOFs) that could be used in vehicular tanks to store fuels such as natural gas and hydrogen in high enough quantities to allow practical driving ranges. These gaseous fuels have specific energy contents (MJ/kg) superior to gasoline, but significantly lower energy density (MJ/L) at standard conditions, and thus their use currently requires energy-intensive compression. However, natural gas and hydrogen represent promising cleaner alternatives to reduce dependence on oil, so their adsorption on porous materials is being actively explored. MOFs are novel, remarkable, crystalline materials whose nanopore structure can be finely tuned by appropriate selection of essentially limitless possible combinations of inorganic "nodes" and organic "linkers." I used a variety of strategies for generating "crystal structures" and Monte Carlo simulations to calculate their adsorption properties. The goal was to establish performance limits (through data mining) for this class of materials and identify the best materials within these boundaries. Through close collaboration with synthetic chemists, some of the best MOFs identified via simulation have been synthesized, and their promising adsorption characteristics have been experimentally confirmed.

My current postdoctoral research focuses on the computational component of a collaborative project with experts in synthetic chemistry and catalysis. We aim to design hybrid MOF/nanoparticle catalysts for regioselective oxidation and/or hydrogenation toward the synthesis of commodity chemicals. Designing highly selective catalysts can result in "greener" reactions by reducing the need for post-reaction separation processes, which typically account for the bulk of the energy cost of operating a chemical plant. Mimicking the working mechanism of enzymes, the nanopores of the MOF component are expected to provide control on how the molecule "attacks" the surface of the nanoparticle component. In this research, I integrate insights obtained from classical molecular dynamics (MD) about the dynamics of the reactants inside MOF nanopores with insights from DFT calculations about the reaction energetics for the transformation of reactants into products on the catalyst surface. These calculations stand out from other DFT calculations in the literature in that the reactants are exposed to steric constraints using a novel faux-pore approach.

Building on my expertise in materials science, catalysis, computational methods, and data mining, I plan to mainly focus my academic career on the computational design and discovery of materials and catalytic systems that will either enable us to reduce energy consumption through highly active and selective catalysts (likely drawing inspiration from nature's enzymes), or directly use the required energy to convert reactants to products from abundant renewable sources such as in metal-oxide photocatalysts.

Doctoral Advisor: Professor Perla B. Balbuena, Texas A&M University, College Station TX

Postdoctoral Advisor: Professor Randall Q. Snurr, Northwestern University, Evanston IL

368013 Optimization of Valuable Intermediates By 11 Alpha-Hydroxylation of Steroid DHEA By Solvent-Enhanced Biocatalyst Beauveria Bassiana

Abstract:

This research evaluates the oxidative capacity of fungus *Beauveria bassiana*; a versatile whole-cell biocatalyst used in the biotransformation of chemicals. The goal of this research is to standardize, control and optimize the biocatalysis capacity of *B. bassiana* in the biotransformation of DHEA. To extend use of *B. bassiana* to commercial applications, the optimization of reaction conditions and accurate prediction of biotransformation products are necessary. This work enhances the selective hydroxylation capacity of Strain ATCC 7159, resulting in a cost effective and eco-friendly process for the synthesis of valuable 11α-hydroxy steroids. Our work establishes the metabolic pathway of DHEA and defines the optimum reactor arrangement, substrate concentration, reaction temperature and pH. NMR and LC-MS analysis were performed to characterize reaction products. Higher substrate conversion, selectivity and yield of desired product was achieved in a "Resting Cells" arrangement. As well, higher volumes of growing medium compared to reaction buffer, increase the activity of the biocatalyst. The apparent rate of reaction fits a Michaelis-Menten kinetics model with a maximum reaction rate of 4.45mM-day. Interestingly, when a diluted amount of substrate is used, a higher yield of 11α-hydroxy steroids is achieved. Also, reactions at 26°C with pH ranges between 6.0 and 7.0 resulted in the highest conversion (70%) and the higher product yield (45.8%). *B. bassiana* has the capacity to metabolize DHEA and similar steroids in different set-ups and has a promising future as biocatalyst to be used in the production of drug metabolites.

Porous Materials for Energy Applications: Shale Gas Recovery, CO_2 Sequestration, and Lithium-Ion Batteries

Gennady Gor
Department of Civil and Environmental Engineering, Princeton University, Princeton, NJ

As a faculty member, I will establish a world leading group specializing in porous materials for energy applications. My unique expertise is a combination of modeling chemical and mechanical processes in porous materials, with my toolkit spanning various computational techniques that cover both micro and macro scales. I am particularly interested in phenomena where fluids affect the mechanical properties and behavior of solid or soft porous materials. My areas of practical interest are shale gas recovery, CO_2 sequestration and lithium-ion batteries.

St. Petersburg State University: I received fundamental education in theoretical physics at St. Petersburg State University, Russia. My research was in the field of first-order phase transitions. Particularly, I studied how a system consisting of liquid droplets and vapor evolves in time. In my Ph.D. thesis, supervised by Prof. A. P. Grinin and Prof. F. M. Kuni, I developed a novel approach to the kinetic theory of non-isothermal nucleation using methods of thermodynamics, mathematical statistics, and transport phenomena.

Rutgers University: During the following two years I worked as a postdoctoral researcher at the Department of Chemical and Biochemical Engineering at Rutgers University (group of Prof. Alexander Neimark). My main focus was on modeling adsorption in nanoporous materials. I achieved the following results: I developed macroscopic and molecular theories of the deformation of nanoporous materials during adsorption. By means of Classical Density Functional Theory and Monte Carlo simulations I modeled nitrogen adsorption in novel types of nanoporous carbons and developed methods for characterization of these materials. These methods have been further successfully implemented in commercial instruments by Quantachrome Inc. I also modeled adsorption and transport of chemical warfare agents in polyelectrolyte membranes.

Princeton University: Currently I work at Princeton University with Prof. Jean Prevost (Department of Civil and Environmental Engineering) and Prof. Craig Arnold (Department of Mechanical and Aerospace Engineering). My first research project at Princeton was on modeling CO_2 leaks from geological storage. It involved the development, implementation, and application of methods of thermodynamics of fluid phases and mechanics for modeling the compositional flow of CO_2-brine mixtures. Then switch my focus to developing a multi-physics model for lithium-ion batteries to study the mechanical reasons of batteries aging and degradation.

Apart from research activities, I have always enjoyed teaching. I have taught a number of courses to a variety of different audiences: from high-school students to Princeton graduate students.

At the poster session, besides giving an overview of my past and current research, I will discuss my plans in research and teaching as a prospective faculty.

Selected publications (from 26 total):

[1] Cao Q., **Gor G. Y.***, Krogh-Jespersen K., Khriachtchev L. Non-Covalent Interactions of Nitrous Oxide with Aromatic Compounds: Spectroscopic and Computational Evidence for the Formation of 1:1 Complexes ***J. Chem. Phys.*** 2014, 140, 144304
[2] **Gor G. Y.***, Stone H. A., Prevost J. H. Fracture Propagation Driven by Fluid Outflow from a Low-permeability Aquifer ***Transport in Porous Media*** 2013, 100 (1), 69-92.
[3] **Gor G. Y.***, Paris O., Prass J., Russo P. A., Ribeiro Carrott M. M. L., Neimark A. V. Adsorption of n-Pentane on Mesoporous Silica and Adsorbent Deformation ***Langmuir*** 2013, 29 (27), p. 8601-8608
[4] Landers J., **Gor G. Y.**, Neimark A. V. Density Functional Theory Methods for Characterization of Porous Materials (Review) ***Colloids and Surfaces A*** 2013, 437, p. 3-32.
[5] **Gor G. Y.**, Elliot T. R., Prevost J. H. Effects of Thermal Stresses on Caprock Integrity During CO_2 Storage ***International Journal of Greenhouse Gas Control*** 2013, 12, p. 300-309
[6] **Gor G. Y.**, Thommes M., Cychosz K., Neimark A. V. Quenched Solid Density Functional Theory Method for Characterization of Mesoporous Carbons by Nitrogen Adsorption ***Carbon*** 2012, Vol. 50 (4), p. 1583-1590
[7] **Gor G. Y.**, Neimark A. V. Adsorption-Induced Deformation of Mesoporous Solids ***Langmuir*** 2010, Vol. 26 (16), p. 13021-13027

362104 Research and STEM Outreach at a Primarily Undergraduate Institution

Abstract:

There is a need for novel multifunctional nanoscale materials in the area of medicine and environment. My goal is to develop a successful research program at an undergraduate institute to focus on these needs by establishing an interdisciplinary well-equipped laboratory that encourages women and diverse groups in science and engineering. My research interests are, but not limited to, i) Pickering emulsions (PE) for biomedical and environmental applications, ii) Synthesis of spider-web peptides for biomimetic materials.

In modern undergraduate STEM education research has become an integral tool that provides training in innovation and collaboration in addition to preparing students for advanced degrees and careers in science and engineering. One of the challenges in performing research at an undergraduate institution involves designing simple projects that can be conducted with limited resources and involves undergraduates at all stages. The above proposed research activity intends to improve and increase research opportunities in Chemical Engineering Department at PUIs. Typically undergraduates are available for short durations - one semester or sometimes during the summer. This will be addressed by designing short, cumulative projects from the proposed work that can be completed quickly and the results generated will contribute towards answering the research questions.

In addition to scientific research, I am also interested in developing K-12 STEM outreach activities that involves active participation of the high school schools students, pre-service and K-12 educators. Some of my recent initiatives include: 1) a Chemical Engineering based workshop for pre-service K-12 teachers, 2) acting as a faculty advisor to local high schools for the Performance Standards Project Science Fair, 3) hosting a STEM day workshop for 7th grade and high school girls

380715 Molecular Engineering of Crystallization in Pursuit of More Potent Functional Materials, Cleaner Sources of Energy and Better Climate

Abstract:

Crystallization is a versatile phenomena in nature. For millennia, it has been widely used for extracting salt and other minerals, and is perhaps the oldest of the traditional chemical engineering separation processes. The importance of crystallization is however not confined to purification only. Crystallization is a vital component of many important natural phenomena, such as precipitation. It can also be employed for making functional materials, as many of these materials have crystalline, quasi-crystalline, or liquid-crystalline long-range order, and their functional properties usually emanate from their underlying long-range order. For instance, optical band gaps can only be realized in certain crystalline and quasi-crystalline arrangements of atoms, molecules or colloidal building blocks. Consequently, studying crystallization processes with the aim of discovering new crystallization routes, or engineering the existing crystallization processes, is of broad scientific and technological interest. Molecular engineering of crystallization processes is however not possible without having a thorough understanding of the kinetics and the mechanisms of the underlying processes. Computer simulations are invaluable tools in this endeavor since the existing experimental techniques lack the necessary spatiotemporal resolution for studying molecular mechanism(s) of crystallization. Typically, crystallization proceeds through a nucleation and growth mechanism, with the nucleation stage being the rate-limiting step. Historically, several advanced simulation techniques have been developed for studying rare event phenomena such as nucleation. However, there are various technical difficulties in applying these generic algorithms to crystallization processes.

One such difficulty arises in studying crystallization in systems in which structural relaxation is slow. In such circumstances, developing more efficient computer software can assist in expanding the range of accessible timescales. However, the existing advanced simulation techniques must also be modified to cope with the issue of correlated configurations/trajectories. A notable example is the supercooled water. Water transforms into hexagonal ice at ambient pressures. However, its crystallization can be suppressed for temperatures as low as 40 K below its melting point. Therefore, the ability to predict the extent and the rate of crystallization is key in understanding the behavior of supercooled aqueous systems. For instance, the fraction of icy droplets in a cloud is a very important parameter to many climatological models, and determines the light-absorbing properties and the precipitation propensity of a cloud, but the extent of freezing in atmospheric droplets and aerosols can not be determined from thermodynamics only. However, computational studies of ice nucleation are very challenging and the problem of calculating the rate of homogeneous nucleation of ice in realistic models of water has been dubbed as one of the most challenging open problems in computational statistical physics, besting the efforts of generations of computational scientists. Another aqueous system that is of utmost interest in the context of clean sources of energy is the water-gas mixtures that can transform into clathrate hydrates. Clathrates are crystalline solids in which water molecules organize into polyhedral cages that encompass small guest molecules such as hydrocarbons, carbon dioxide and nitrogen. Methane hydrates are present in the ocean bed and in the permafrost and it is believed that the methane trapped in the form of hydrates in those environments is orders of magnitude larger in quantity than the methane in known oil and natural gas resources. Methane hydrates can become sources of clean energy in the future. However, the mechanism and the kinetics of hydrate formation are not fully understood, due to the very same reasons that make computational studies of ice so challenging. In my postdoctoral years in Princeton, I developed a coarse-grained variant of the forward-flux sampling algorithm, and successfully used it for calculating nucleation rates in a fully atomistic model of water. As an assistant professor, I intend to expand on my postdoctoral work and study the problem of ice nucleation under atmospherically relevant circumstances, e.g. in droplets and films of supercooled water. There are several interesting questions to be asked in this regard, such as the effect of vapor-liquid interfaces on crystallization, or the role that inorganic impurities play in modifying the rate and the mechanism of crystallization. The findings of this investigation will be pivotal to the area of atmospheric and environmental sciences and can help us gain a better understanding of phenomena such as global warming and climate change. It can also assist us in devising guidelines for engineering our climate, e.g. by engineering the radiation budget and the precipitation patterns across the biosphere via changing the properties of clouds. I am also interested in the problem of hydrate nucleation, mostly at realistic supersaturations that have been mostly overlooked in existing computational studies of hydrates. Understanding the mechanisms of nucleation in those systems can assist us in developing better technologies for energy storage and carbon sequestration. It can also help us prevent the problems that hydrate formation can cause in production and transportation facilities in the petroleum industry.

Another difficulty in studying crystallization is the issue of nonconventional symmetries that can make the characterization of the disordered and the crystalline states a nontrivial task. The most notable examples are the complex structures that assemble from anisotropic nano- and colloidal building blocks. Many of these nonconventional symmetries can result in interesting functional properties, such as photonic band gaps, negative refractive index. During my doctoral years at the University of Michigan, we established the striking capably of anisotropic building blocks to assemble into structures with complex crystalline and quasicrystalline order. For instance, we showed that hard tetrahedra can assemble into a dodecagonal quasicrystal, a discovery that was followed by further discoveries of other intriguing disorder-order transitions in other hard and soft particle systems. Today, it is relatively straightforward to study the thermodynamics and the self-assembly propensities of different building blocks. However, studying kinetics of such transformations are less straightforward. There are two types of difficulties in applying the existing advanced simulation techniques to these crystallization processes. First of all, many of these structures have nonconventional symmetries, and can thus not be properly characterized by the existing order parameters that are widely used for simple atomic and molecular systems. It is necessary to emphasize that the problem of identifying a good oder parameter for these advanced simulation techniques is not just an exercise in geometry, but instead requires a large deal of experience and a lot of effort. The second difficulty is that no efficient algorithms exist for following the dynamics of systems that are comprised of building blocks with singular surfaces e.g. hard polyhedra. Combining my graduate experience in studying such nontrivial symmetries, and my postdoctoral experience in using and modifying advanced sampling techniques, I intend to tackle these two problems to obtain a mechanistic understanding of the processes that give rise to such nontrivial structures.

HIGHLIGHTS [AS OF MAY 2014]:

-Five first-author publications.

-One submitted manuscript.

-Five to be submitted within the next few months.

Multifunctional Polymer Nanocomposites towards Electromagnetic Interference Shielding and Flame Retardation

Qingliang He

Integrated Composites Laboratory (ICL), Dan F. Smith Department of Chemical Engineering, Lamar University, Beaumont, TX 77710 USA

ABSTRACT: Electromagnetic pollution and fire hazard are two major problems that compromise the sustainable development of high technologies in our modern society. Although extensive efforts had been made trying to solve both the two problems separately, the tentative for exploring materials which can solve both the two aforementioned problems simultaneously has rarely been reported. Polymer matrix nanocomposites reinforced with nanofillers possessing unique filler architectures with easily tuned composition are ideal for targeting desirable functionalities. However, the major challenge by using polymer matrix nanocomposites for both the electromagnetic interference (EMI) shielding and flame retardation is the unknown relationship between each other. Hence, the effective improvement for both the two functions by one material is hard to design.

Based on the mechanisms of EMI shielding (magnetic and dielectric materials, which may consist of elements possessing catalytic function for flame retardancy) and flame retardation (carbon and ceramic nanomaterials, which may contribute as dielectric components for EMI shielding), multifunctional polymer matrix nanocomposites with high efficiency in both EMI shielding and flame retardation have been developed. To be specific, polypropylene nanocomposites reinforced with nanofillers consists of magnetic (iron, cobalt and its core shell metal @metal oxide nanoparticles), dielectric materials (carbon nanomaterials including carbon nanotubes and graphene nanosheets, ceramic fillers or coating such as SiO_2) were synthesized using a facile one-pot wet chemistry approach.

Regarding the structure design, hierarchical 3-D filler systems with different dimensionalities (0D, 1D, 2D and 3D) were integrated to achieve unique architecture in polymer matrix; and surfactant with unique functional groups will also be introduced in order to improve the polymer-filler interfacial bonding strength. Meanwhile, how the filler composition, dispersion quality and assembly patterns in polymer matrix will affect the physicochemical properties and the corresponding functionalities for the as-prepared nanocomposites will be studied and the structure-property-performance relationship will be revealed.

SELECTED PEER-REVIEWED PUBLICATIONS (12 out of 40):

1. **Q. He**, T. Yuan, S. Wei, N. Haldolaarachchige, Z. Luo, D. P. Young, A. Khasanov, Z Guo, Morphology and phase controlled iron oxide nanoparticles stabilized with maleic anhydride grafted polypropylene, ***Angewandte Chemie International Edition*** 51(**2012**): 8842 -8845
2. **Q. He**, T. Yuan, J. Zhu, Z. Luo, N. Haldolaarachchige, L. Sun, A. Khasanov, Y. Li, D. P. Young, S. Wei, Z. Guo, Magnetic high density polyethylene nanocomposites reinforced with in-situ synthesized Fe@FeO core-shell nanoparticles, ***Polymer*** 53 (**2012**): 3642-3652
3. **Q. He,** T. Yuan, X. Zhang, Z. Luo, N. Haldolaarachchige, L. Sun, D. P. Young, S. Wei, Z. Guo, Magnetically Soft and Hard Polypropylene/Cobalt Nanocomposites: Role of Maleic Anhydride Grafted Polypropylene, ***Macromolecules*** 46 (**2013**): 2357-2368

4. **Q. He**, T. Yuan, Z. Luo, N. Haldolaarachchige, D. P Young, S. Wei, Z. Guo, Morphology and phase controlled cobalt nanostructures in magnetic polypropylene nanocomposites: the role of alkyl chain-length in maleic anhydride grafted polypropylene, ***Chemical Communications*** 49 (**2013**): 2679-2681
5. **Q. He**, T. Yuan, S. Wei, Z. Guo, Catalytic and Synergistic Effects on Thermal Stability and Combustion Behavior of Polypropylene: Influence of Maleic Anhydride Grafted Polypropylene Stabilized Cobalt Nanoparticles ***Journal of Materials Chemistry A***, 1 (**2013**): 13064–13075
6. **Q. He**, T. Yuan, X. Yan, Z. Luo, N. Haldolaarachchige, D. P. Young, S. Wei, Z. Guo, One-pot Synthesis of Size- and Morphology- Controlled 1-D Iron Oxide Nanochains with Manipulated Magnetic Properties , ***Chemical Communications***, 50 (**2014**): 201-203
7. **Q. He**, T. Yuan, X. Yan, D. Ding, Q. Wang, Z. Luo, T. D. Shen, S. Wei, D. Cao, Z. Guo, Flame-Retardant Polypropylene/Multiwall Carbon Nanotube Nanocomposites: Effects of Surface Functionalization and Surfactant Molecular Weight, ***Macromolecular Chemistry and Physics***, 215 (**2014**): 327–340
8. **Q. He**, T. Yuan, S. Wei, Z. Guo, Unusual Function of Modified Polyolefins for Manipulating Magnetic Nanostructures, ***JOM*** (Invited review contribution), 66 (**2014**): 655-659
9. J. Zhu, **Q. He**, Z. Luo, A. Khasanov, Y. Li, L. Sun, Q. Wang, S. Wei, Z. Guo, Property manipulated polypropylene-iron nanocomposites with maleic anhydride polypropylene, ***Journal of Materials Chemistry*** 22 (**2012**): 15928-15938
10. X. Zhang, **Q. He**, H. Gu, H. S. Wei, Z. Guo, Polyaniline Stabilized Barium Titanate Nanoparticles Reinforced Epoxy Nanocomposites with High Dielectric Permittivity and Reduced Flammability, ***Journal of Materials Chemistry C*** 1 (**2013**): 2886-2899
11. X. Zhang, **Q. He**, H. Gu, H. A. Colorado, S. Wei, Z. Guo, Flame retardant electrical conductive nano-polymers based on bisphenol F epoxy resin reinforced with nano polyanilines, ***ACS Applied Materials & Interfaces*** 5 (**2013**): 898-910
12. H. Gu, J. Guo, **Q. He**, Y. Jiang, Y. Huang, N. Haldolaarachchige, Z. Luo, D. P. Young, S.Wei, Z. Guo, Magnetoresistive Polyaniline/Multi-Walled Carbon Nanotubes Nanocomposites with Negative Permittivity, ***Nanoscale***, 6 (**2014**): 181-189

AWARDS/HONORS:

1. 1st Place Award, *International Polyolefins Conference 2012*, Houston, TX, USA, 2012
2. Scholarship Award, *Society of Plastics Engineers*, *Lamar University Chapter*, Beaumont, TX, USA, 2012
3. Scholarship Award, *South Texas Section, Society of Plastics Engineers*, *College of Engineering, Lamar University*, Beaumont, TX, USA, 2012
4. 2nd Place Research Excellence Award in the *2012 Chinese-American Chemical Society Southwest Chapter (CACS-SW) Annual Poster Competition*, Houston, TX, USA, 2012
5. Scholarship Award, Polymer Modifiers and Additives Division (PMAD) of the *Society of Plastics Engineers*, Houston, TX, USA, 2013
6. 2nd Place Research Excellence Award in the *2013 Chinese-American Chemical Society Southwest Chapter (CACS-SW) Annual Poster Competition*, Houston, TX, USA, 2013

Understanding and Predicting Chemical Reactivity in Supported Metal Catalysis

David Hibbitts

Metal-catalyzed reactions occur on polycrystalline particles, often at high coverages of reaction intermediates and (in some instances) with H_2O present. These complicating factors, however, have only recently been addressed by theoretical treatments of chemical reactions, due to the methodological and computational challenges they present.

Aerobic oxidation of alcohols (including diols) on Pd, Pt, and Au catalysts is active for the formation of carboxylic acids with H_2O and hydroxide present. Density functional theory (DFT) calculations performed with discrete H_2O molecules found that alcohol oxidation is a base- and metal-catalyzed reaction in which hydroxide acts as a Brønsted base in solution and at the H_2O-metal interface to deprotonate alcohols and nucleophilically attack aldehyde intermediates to form carboxylic acids. These conclusions were consistent with kinetic data, which showed that rates are proportional to the pH and pKa of the alcohol and isotopic data, which demonstrated that the additional O-atoms in the product acids originated from the H_2O solvent rather than from the O_2-coreactant. This oxidation process consumes hydroxide and produces electrons, a process balanced by oxygen reduction reactions (ORR) in which O_2 abstracts H-atoms from H_2O (to replenish hydroxide) in subsequent steps to form OOH* and *HOOH* species prior to O-O cleavage. These reactions combine to form a short-circuited electrochemical cell, permitting insights into the two half-reactions to be applied to electrochemical applications.

H_2O also plays a role during gas-phase Fischer-Tropsch synthesis (FTS), which forms large hydrocarbons and H_2O from synthesis gas (CO-H_2) derived from biomass, coal, or natural gas. Kinetic and isotopic data coupled with DFT calculations suggest that CO activates following initial reactions with chemisorbed H-atoms (H*) to form *HCOH* preceding C-O cleavage rather than via direct CO* dissociation, which has been shown to have much larger barriers on both low-index terraces and defect sites of 201- and 586-atom Ru and Co nanoparticles modeled at high CO* coverages. H_2O increases turnover rates on Co and Ru catalysts and DFT calculations on such models show that H_2O acts as a H-shuttle in which H* reacts with nearby H_2O to form a short-lived H_3O^+ intermediate, which then protonates CO* (to form COH*) or HCO* (to form *HCOH*) both of which lead to reduced barriers for *HCOH* formation, increasing the rate of CO activation.

The H*-assisted mechanism of CO activation via *HCOH* intermediates on Co and Ru is reminiscent of O_2 activations via OOH* and *HOOH* species on Pd, Pt, and Au described above and was also observed during NO-H_2 reactions in which NO reacts with H* to form *HNOH* species prior to N-O cleavage on Pt. Kinetic, isotopic, and theoretical studies at high NO* coverage of NO-H_2 reactions on Pt indicated that NO activates through that H*-assisted mechanism to form *HNOH*, which then dissociates into NH* and OH* intermediates that undergo fast reactions to give N_2O, NH_3, N_2, and H_2O products.

This work has indicated the necessity of rigorous kinetic studies and isotopic data to be combined with DFT calculations on complex catalyst models, in which the effects of H_2O, high coverages of co-adsorbates, bimolecular reactions, and polycrystalline surfaces may be required to correctly identify reaction mechanisms. Future plans are to continue such comprehensive studies through research that combines gaseous and aqueous-phase kinetic measurements with isotopic investigations and DFT calculations to understand the factors that affect metal catalyst reactivity and selectivity.

387872 Bridging the Gap Between Academia and Industry: Designing Curricula to Prepare Students for the Profession

Abstract:

One purpose of an undergraduate engineering education is to prepare students to become practicing engineers. However, in recent years, industry has expressed concern that while graduating students are technically competent, they are less prepared for other aspects of engineering jobs. Curricula are well-designed to convey technical content, but focus less on skills such as communication, project management, creativity, and teamwork. Although capstone design courses emphasize the importance of working in teams and communicating progress, students would be better prepared for industry if this emphasis were woven throughout the entire undergraduate engineering curriculum. My research interests lie in developing innovative ways to equip students with these skills early on and throughout their entire undergraduate education, preparing them to attack open-ended problems, develop innovative solutions, communicate ideas, interact with peers, and manage their projects and time.

My doctoral work at Purdue University involved working closely with industry partners to develop an automation and control system for a pharmaceutical manufacturing process that utilized inkjet printing technology. This project not only gave me a solid foundation in chemical engineering fundamentals, but also convinced me of the importance of closely linking curricula with industry principles in order to better prepare students to serve the needs of the profession. During my graduate studies, I sought out many leadership opportunities, including being president of our Graduate Student Organization and heading Future Faculty seminars for interested graduate students. I also devoted a large portion of my time to teaching, receiving two teaching awards for my semesters as a teaching assistant. Through my research, leadership and teaching experiences, I realized that there was an opportunity to improve curricula and pedagogy to better prepare future engineers, leading to my current focus on engineering education research.

My current post-doctoral research at Oregon State University focuses on developing technology to instill conceptual understanding in students through different pedagogical approaches. Through the use of tools such as the AIChE Concept Warehouse and the Industrially-Situated Virtual Laboratory Project, we focus on methods that challenge students to grasp fundamental concepts, as opposed to teaching students to memorize content and algorithmically solve problems. These approaches help to improve knowledge retention, conceptual understanding and transfer and also help to better prepare students to work with the open-ended problems they will encounter as practicing engineers. As a professor, I plan to continue in this area of research by investigating student learning and creating ways to bridge the gap between education and industry. I want to help students develop the skills vital to a career in engineering throughout their entire course of study, better preparing them to work in a challenging professional environment. I look forward to transforming engineering curricula while educating and developing future engineers.

387005 Engineering Biomimetic Cues to Restore Musculoskeletal Tissue Function

Abstract:

Musculoskeletal injuries and diseases, which include meniscal tears and degenerative disc disease, are a significant health concern in the United States. Current treatments, however, typically rely on donor tissues (either allo- or autografts) and suffer from poor availability. Furthermore, donor tissue does not always adequately restore function, integrates poorly with surrounding tissue, and can have a high morbidity. Tissue engineering aimed at replicating and/or restoring the biological and mechanical cues required for tissue function offers an advantage to current treatments and prevents further musculoskeletal degeneration. My research has focused on the intelligent design of hydrogels towards this aim.

During my Ph.D., under the advisement of Anthony Lowman and Giuseppe Palmese, I developed and characterized a fiber-reinforced hydrogel composite for meniscal replacement aimed at replicating the anisotropic mechanical properties of the native meniscus. The hydrogel composite was characterized in compression (0.1 - 0.8 MPa) and tension (0.1 - 250 MPa) for a number of formulations demonstrating fine control over mechanical properties within the range of the human meniscus. Furthermore, the fiber-hydrogel interface was tailored for optimal adhesion using a novel biocompatible grafting technique to form a direct covalent linkage between the fiber and hydrogel, where grafting resulted in a 20-fold improvement to interfacial shear strength. Following implantation into human cadaveric knees, ex vivo gait simulations showed improved contact areas and stresses along the tibial plateau for the hydrogel composites in comparison to a meniscectomy.

Currently, I am a postdoctoral fellow in Jason Burdick's laboratory researching synergistic growth factor delivery for improved osteogenesis using proteolytically degradable hydrogels. A critical-sized calvarial defect was used to determine the effect of combined SDF-1α, a known chemokine, and BMP-2 delivery on bone formation in vivo. The treatment group with combined delivery showed significantly higher bone formation when compared to hydrogels loaded with either growth factor alone or the empty defect group. More importantly, SDF-1α, when delivered in combination with BMP-2, effectively reduced the required BMP-2 dose for similar bone formation by increasing cellular invasion into the defect site. Ongoing work evaluating the activation of other osteogenic signaling pathways is being investigated to further improve BMP-induced osteogenesis.

In the future as a faculty member, I will use my experience with and knowledge of polymer science, biomechanics, and growth factor signaling to develop finely tunable materials that replicate and restore the mechanical, structural, and biological signals needed for function in musculoskeletal tissues.

384727 Developing Novel Biocatalysts Using Synthetic Biology

Abstract:

Synthetic biology is an emerging field to develop new biological systems that perform novel functions for useful purposes. The field has been rapidly growing by designing synthetic genetic circuits, demonstrating biological principles, and developing artificial cells. Advances made have inspired a variety of new applications in industrial and clinical processing to address current needs in sustainable energy and food supply as well as novel therapeutics development. However, it is still early in the development of real-world applications of synthetic biology, and progress is limited by the complexity and the unexpected responses of biological systems. My faculty research program seeks to address this limitation. We will develop biocatalysts for a variety of applications by using interdisciplinary approaches that analyze, understand, and control cellular processes on a molecular and systems level. Specifically, I am interested in three areas: i) engineering biofilms for disease treatment and therapeutics production ii) cell-free synthetic biology for developing artificial metabolic pathways that are impossible, or otherwise difficult, to generate using cell-based systems, and iii) microbial electrosynthesis using engineered biofilms and cell-free systems to convert electricity to biochemical, therapeutics, and fuels. I will apply a unique and diverse set of skills in synthetic biology, protein engineering, genome engineering, and cell-free engineering for the development of novel biocatalysts and biosynthesis pathways for clinical and industrial uses. My work will allow me to build a leading research program for compelling applications in medicine, basic science and quantitative applied biotechnology.

377619 Engineering the Flow Behavior of Colloidal Materials through Surface Modification and Shape Anisotropy

Abstract:

The idea that colloidal materials can be designed with enhanced rheological properties by incorporating shape and surface anisotropy is motivated by our finding that structural rigidity can be used to predict the nonlinear elasticity in gels that have undergone yielding. We demonstrate that the predictive power of microscopic theories can be improved when both the microstructural rigidity and the dynamical heterogeneity induced by yielding are taken into account. Because thermal rupture forces play a critical role in yielding, we develop a model gel system in which rheological measurements can be carried out in conjunction with microscopy experiments and direct force measurements using optical tweezers. The validity of structural rigidity is also tested by synthesizing colloidal spheroids and roughened particles. We find that colloidal oblate spheroids self-assemble into tilted structures in a specific range of volume fractions and attraction strengths. These structures could pave the way for maintaining the elasticity of colloidal gels even at exceedingly low particle loading. Finally, we observe a significant increase in the shear thickening behavior of rough colloids in concentrated suspensions. These results collectively provide experimental support for the applicability of structural rigidity as a guiding principle in engineering the flow properties of a broad range of soft matter.

390384 Microfluidic Biosensors to Monitor Health

Abstract:

Many deaths in the U.S. are attributed to chronic diseases. Diabetes, heart disease, and stroke are all examples of preventable and/or manageable chronic diseases that affect many people under the age of 60. This means that on a yearly basis large amounts of money are spent on health care for treatment. Along with this expense comes compromised quality of life (in some cases disability), and escalating health care costs. Healthcare cost can be alleviated if there is technology available that allows early disease detection, so that diet and exercise can be implemented as a treatment option. Microfluidics and dielectrophoresis (DEP) are technologies applicable to disease detection. Specifically microfluidics manipulates small volumes of fluid (μL) and DEP uses electric fields to detect differences and separate biological cells. Together DEP and microfluidic devices are useful for characterizing biological cell properties and distinguishing between healthy and unhealthy cells. Microfluidic devices are fabricated at a low cost making them suitable in healthcare.

The theme of my research will be improved healthcare using microfluidic biosensors with a focus on (1) health and (2) exercise assessments. I would like to use my knowledge in DEP to achieve these assessments in microdevices. Health care assessments such as blood screens completed in microfluidic biosensors can give indication of disease on a shorter time scale. Wearable microfluidic biosensors that monitor a person's health during exercise are also important in healthcare especially for person's overweight. I have experience using DEP technology to study red blood cells, polystyrene beads, and human mesenchymal stem cells (hMSCs). The focus of my doctoral work was on the quantification of hMSCs dielectric properties (membrane permittivity and capacitance) for the design of a continuous cell sorting microfluidic device. I've also worked on the development of a new DEP experimental technique that rapidly determines the DEP response spectra of biological particles via frequency sweep rates. This work was optimized using polystyrene beads and extended to red blood cells. My plan is to continue research on DEP and biological cells with expansion into the area of exercise science to help improve healthcare.

388149 Hierarchical Hybrid Assembly of Functional Nanomaterials

Abstract:

Functional nanomaterials will play critical role in many future technologies related to energy, environment and health. Next generation hybrid or plug-in-hybrid vehicles would require high-power, high-energy density energy storage devices with lower emission for sustainability. To achieve these goals, it is essential to design, identify and sythesize materials that can accommodate efficient transport of ions and electrons in addition to high surface area. Energy storage devices based on nanoscale materials (nanoparticles, nanotubes or nanofibers of electroactive polymers, metal oxides and carbon nanotubes) recently gained significant interest to afford high-power, high-energy density without compromising cycle life. During my postdoctoral research at MIT, I've developed expertise in the design of novel electrode assembly for energy storage devices. I focused on developing hybrid electrodes based on conjugated polymers (Polyaniline, PANi), metal oxides (TiO_2), graphene oxide (GO) and multi-walled carbon nanotubes (MWNT); these nanostructured hybrid electrodes performed exceptionally well (compared to graphitic electrodes) delivering high-power, high-capacity simultaneously for electrochemical pseudocapacitors that suitable for electric vehicles or plug-in electric vehicles. Along with conventional electrode development, I designed alternative layer-by-layer assembled electrodes based on electrostatic interaction of nanomaterials. The film architecture, assembly technique and post treatment were optimized to create high performance electrodes. These electrodes were engineered to achieve high electronic conductivity and high porosity for improved ion transport of aqueous or organic electrolytes rendering higher capacitance than graphite electrodes with excellent mechanical and electrochemical stability over thousands of cycles. The development of hybrid electrodes of TiO_2, MWNTs and PANi nanofibers with controlled aspect ratio and morphology exhibits attractive potential in terms of improved electronic conductivity and porosity that are competitive with the expensive state-of-the-art RuO_2. However, these multilayer electrodes of nanoscale materials would reach a performance limit unless we optimize the selection of electrolytes which affect ion conduction and thereby overall electrochemical performance. Polymer-based gel electrolyte could be a solution to address this challenge that could be part of the separator membrane in the overall electrochemical cell design. In addition, my Ph.D research experiences on polymeric membrane design and fabrication would allow me to develop such novel separator membranes. In future, as an independent researcher, I am interested in the design and facile synthesis of nanomaterials and their assembly for hierarchical electrodes and electrolytes. Using thses electrodes, my lab will focus on fundamental aspects of charge storages, ion transport which could be ultimately translated to manufacturing. These nanostructured materials along with polymeric/ionic gel electrolyte would have tremendous potential to impact the overall electrode design and performance for improve energy storage devices like metal-air batteries and hybrid supercapacitors.

Searching for epigenetic patterns as biomarkers for cancer diagnosis

Isabel Jimenez-useche[1,2]

[1] School of Chemical Engineering, Purdue University, West Lafayette, IN, USA

[2] School of Engineering Education, Purdue University, West Lafayette, IN, USA

DNA methylation plays an essential role in various biological processes such as stem cell differentiation, imprinting, X-chromosome inactivation, etc. Increased DNA methylation levels have been associated with chromatin compaction leading to gene silencing. For example, abnormal DNA methylation is associated with silencing of tumor suppressor genes and is observed in the onset of tumorigenesis. There is evidence suggesting that not all methylation events are relevant in chromatin compaction and the initiation of cancer. It seems that methylation at certain locations of the DNA might be key to start chromatin compaction and gene silencing, but the location of this methylation sites is still unknown. In order to identify DNA methylation locations that could potentially be involved in chromatin compaction and gene silencing, my research focus has been on studying the effects of different DNA methylation patterns in the modulation of chromatin compaction.

Using multiple molecular biology techniques, I constructed chromatin-like molecules representing three levels of chromatin organization: 1) Naked DNA, 2) Nucleosomes and 3) Nucleosome arrays. These *in vitro* systems allowed the evaluation of the direct effect of DNA methylation on the conformation and dynamics of chromatin without the influence of other proteins found in the cell nucleus. In addition, I have introduced different DNA methylation patterns to my *in vitro* systems to study the effects of the position of the methylation sites in the chromatin structure.

My results showed that compactness of chromatin-like molecules with a stretch of CG dinucleotides in the central dyad of the nucleosomes or at the major groove locations showed a dependence on their methylation status. Specifically, methylation of a stretch of dinucleotides decreased the relative compactness of nucleosomes and increased tetranucleosome compaction. The opposite effect was observed for nucleosomes and tetranucleosomes with CG dinucleotides located at the major grooves. These findings confirm that the presence of methylation in certain locations within chromatin lead to distinctive effects on the compactness of chromatin-like molecules. These results suggest that these two methylation patterns could potentially be biologically relevant for gene expression regulation. Although the biological significance of these methylation locations is still to be determined, the results of this research are instrumental in understanding the mechanism of chromatin compaction by DNA methylation and could be applied in the identification of new and more accurate DNA methylation biomarkers for early detection of cancer.

In addition to my work with epigenetic modifications, I have always been fascinated by the ability of instructors to facilitate the process of learning, in particular in engineering disciplines. This interest lead me to complement my previous research experience with a new research focus in the area of Engineering Education. The goal of my current research is to assess the experience of first-year international undergraduate students in engineering programs in American universities. I have been able to identify four different challenges that international students face when coming to American Universities. These challenges are: 1) cultural differences, 2) communication challenges, 3) personal challenges and 4) academic challenges. Each of these challenges has different implication in the students' academic performance. I intend to assess the impact of these challenges using multiple assessment instruments such as surveys, questionnaires, interviews, focus groups, students' grades and other kind of evaluations. The results of this research will be instrumental in identifying relevant factors that might be addressed to improve the experience of international students in American universities.

393664 Active colloids at the oil-water interface: The biophysics of microbial oil degradation

Abstract:

In natural habitats, microbes regularly encounter a wide variety of interfaces that influence their behavior and activity. Bacteria often accumulate near interfaces and will eventually attach to them. In favorable conditions, bacterial will grow and divide leading to colonization of the interface. Dense colonies formed at the liquid-solid interface are referred to as biofilms while those formed at the liquid-gas interface are referred to as pellicles. Bacterial colonies also form at the liquid-liquid interface, for example at the oil-water interface, although the process is not well understood. Biofilms can be undesirable and destructive causing infection and biofouling, however, they can be beneficial in the production of biofuels and environmental processes such as waste water treatment. Bacterial attachment to an interface is the first step in colonization, therefore understanding the mechanisms of bacterial attachment will provide controlled methods to prevent or enhance biofilm formation.

My research interests are broadly in the field of experimental soft condensed matter physics. My doctoral research (with Richard Lueptow and Julio Ottino, Northwestern University) focused on mixing and segregation dynamics of granular mixtures in rotating tumblers. My postdoctoral research (with Paulo Arratia, UPenn) focused on establishing a link between the molecular microstructure and the macroscopic rheological properties of complex polymeric fluids. My current research (with Roman Stocker, MIT) focuses on understanding the fundamental interactions between microbes and liquid-liquid interfaces, specifically the biodegradation of crude oil by marine bacteria. The degradation of oil in the ocean is a process of immense societal and ecological importance whose efficiency ultimately hinges on fundamental problems in colloidal and interface physics because degradation occurs when individual bacteria attach to individual oil drops.

I will discuss experimental and theoretical results on the effect of oil drop diameter on the attachment of marine bacteria to the oil-water interface. Using phase-contrast and fluorescence microscopy, we directly observed the dynamics of marine bacteria in the presence of oil droplets of varying diameter within microfluidic devices. We find that microbial growth and degradation become strongly limited for drops smaller than a critical diameter. A theoretical model of attachment supports this finding, suggesting that rendering oil droplets too small by the excessive addition of dispersants, as often done in oil spills, might be a counterproductive strategy. This research aims to pursue a deeper understanding of the fundamental oil-microbe interactions and promises to open the door to a broad set of other problems in interfacial physics, hydrocarbon microbiology, and other oil spill management related questions.

381161 Design of Multifunctional Materials for Energy Related Applications

Abstract:

The materials design by first-principles modeling provides an opportunity to study new materials for energy related applications. Today, the importance of this approach rapidly increases in chemical engineering and other disciplines. The atomic level of information, such as the properties of matter, will be obtained by the computational methods like density functional theory (DFT). The specific systems we will study are (a) crystalline membranes for gas purification (b) proton-conducting perovskites for fuel cells (c) Li-ion electrolytes, and (d) ferroelectric and multiferroic materials.

Post-doctoral advisor: Professor Craig J. Fennie, School of Applied & Engineering Physics, Cornell University

Doctoral thesis advisor: Professor David S. Sholl, School of Chemical and Biomolecular Engineering, Georgia Institute of Technology

Design of catalysts for energy conversion and storage

Yijin Kang, Materials Science Division, Argonne National Laboratory

My research interests and achievements focus on fundamental understanding of catalytic processes and development of highly active (or selective) and durable catalysts for the reactions of industrial importance, especially for energy conversion and storage. I establish a research system that connects fundamental investigation on well-defined extended surfaces (e.g. single crystal surfaces), extrapolation onto nanocrystals with highly controlled shape and size, exploration of interfacial interaction using novel nanocrystal superlattices as platform, and finally design of high performance catalysts in which all the possible beneficial properties from complex functional structures are implemented.

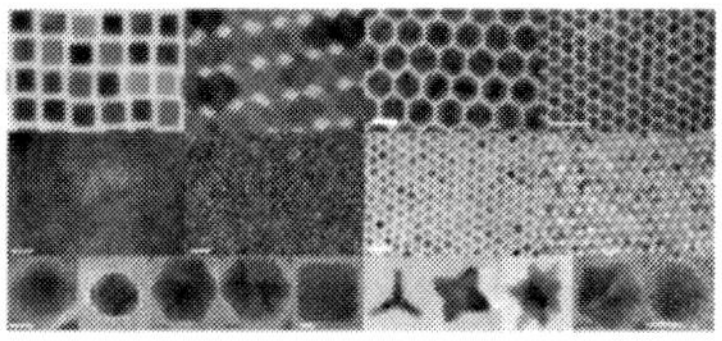

Figure 1. High quality Pt nanocrystals with various shapes. The size is tunable in the range of 2-30nm.[1]

The fundamental understanding of correlationship between materials structure and catalytic properties is the key to design new catalysts with desired properties. The very first step would be to understand the most simplified surface processes using well-defined extended surfaces and computer-powered theoretical simulations. However, the real catalysts used in industrial processes are mainly nanomaterials that take advantage of their high surface-to-volume ratio. It is necessary to confirm that the desirable properties found in simplified cases are actually functioning in the nanomaterials. Therefore, well-defined nanocrystals with exposed surfaces mimicking the simplified surfaces which are used in theoretical studies are highly valuable and appreciated. My work at this level focuses on providing high quality model materials at nanoscale. For instance, in order to study the surface sensitive reactions on Pt surface at nanoscale, I prepared high quality Pt nanocrystals with various controlled shapes[1,2], exposing selective surfaces (e.g. {111} on octahedra and icosahedra, {100} on cubes, and {110} on rhombic dodecahedra). The library of such highly controlled nanocrystals with desired surface is not limited to single element metals[3], but also include alloys[4,5] and oxides[6].

Besides desired surfaces, the interactions among the components within the catalysts are also important. For example, in the Pt-Ni alloy Ni atoms modify the electronic structure of Pt, weakening the adsorption of OH_{ad} spectator species on Pt surfaces, thus enhancing the oxygen reduction reaction (ORR) activity. Because of certain desired metal-oxide interfacial interactions that promote catalysis or introduce synergistic effect, metal catalysts supported on oxide supports are the other class of important catalysts that exploit beneficial interactions. In additional to well-controlled size and shape, the structures rendering the desired interactions are needed to carry out the investigation of interfacial effects. I established a platform using binary nanocrystal superlattices (BNSLs) to construct tunable interactions[7,8]. For example, by constructing Au-FeO_x BNSLs with various structures (AB, AB_2, AB_{13}), I was able to control the population of Au-FeO_x contacts. In the investigation of CO oxidation, a linear correlation between population of contacts and activity was found. Such correlation provided strong evidence that the interface of Au-FeO_x contacts is responsible for the extraordinary activity toward CO oxidation, rather than small sized Au[7].

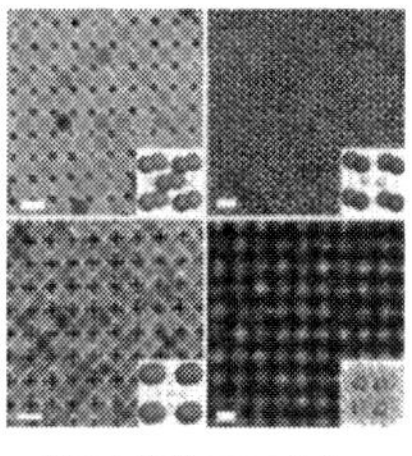

Figure 2. Nanocrystal superlattices, allowing programmable inter-particle contacts.[7]

Finally, with all the knowledge learned from theoretical studies and investigation on model materials, I am able to design catalyst that incorporates the beneficial properties in a single nanostructure. For example, my Berkeley collaborators observed a shape evolution from solid $PtNi_3$ nanocrystal to hollow Pt_3Ni nanoframes[9]. I took advantage of the hollow structure and Pt_3Ni composition of the nanoframes. I treated the surface of nanoframes to promote the surface segregation, obtaining the surface structure mimicking the optimal $Pt_3Ni(111)$-Pt-Skin for ORR. The hollow structure of nanoframes not only substantially increases the active surface area, but also provides a capillary force to hold a specific ionic liquid in which oxygen solubility is much higher than that in regular acid electrolyte. Therefore, combining all these beneficial properties for ORR in surface segregated Pt_3Ni nanoframes, I was able to achieve a 36-fold enhancement in activity in comparison to state-of-the-art Pt/C catalysts[9]. The similar approach is applicable to other catalyst systems and for other important reactions. For instance, I made Pt-Pb core-shell structure with desirable Pt_3Pb intermetallic structure, in order to overcome the poisoning problem thus to enhance the overall activity for formic acid oxidation[10].

Figure 3. Pt_3Ni catalysts for ORR (oxygen electrode of fuel cells) that incorporated all beneficial properties in one nanostructure.[9]

In summary, I have established an effective catalyst designing route that extracts knowledge from model study and theoretical computation, fabricates desirable surfaces and constructs beneficial structures, and ultimately directs the production of real catalysts of industrial interests. I am seeking for a tenure-track faculty position at R1 universities at the cycle of 2014-2015. I would like to focus my future research on energy conversion and storage, especially on solar-to-fuel conversion and advanced battery technologies.

Notes and references:

1. Kang *et al.* ***J. Am. Chem. Soc.*, 2013**, 135 (7), 2741
2. Kang *et al.* ***ACS Nano*, 2013**, 7(1), 645
3. Kang *et al.* ***Angew. Chem. Int. Ed.*, 2010**, 49(35), 6156
4. Kang *et al.* ***J. Am. Chem. Soc.*, 2010**, 132(22), 7568
5. Kang *et al.* ***ACS Nano*, 2012**, *6* (6), 5642
6. Kang *et al.* ***Angew. Chem. Int. Ed.*, 2011**, 50(19), 4378
7. Kang *et al.* ***J. Am. Chem. Soc.*, 2013**, 135 (4), 1499
8. Kang *et al.* ***J. Am. Chem. Soc.*, 2013**, 135 (1), 42
9. Kang *et al.* ***Science*, 2014**, *DOI: 10.1126/science.1249061*
10. Kang *et al.* ***ACS Nano*, 2012**, 6 (3), 2818

393556 Investigating the Selective Removal of Anions and Cations in Electrodialysis/ Electrodialysis Reversal (ED/EDR)

Abstract:

Fresh water resources are inherently limited by their quantity, and it has drawn the world's attention. Electrodialysis and electrodialysis reversal (ED/EDR) are two closely-related water purification processes which use an electrical field to propel ions through ion exchange membranes. The most effective use of these technologies has proven to be the production of potable water from brackish water resources. In brackish water resources as well as industrial water streams, water compositions differ, and removal of particular monovalent or divalent ions may be more desired than others. Therefore, removing ions at different rates under more controllable conditions is valuable. Selective removal of ions in an ED / EDR process can decrease the energy due to removing the target ions rather than unwanted ones. This experimental study, which investigates the nature of ion removal and the effects of operating parameters on the process, was conducted to explore the dominant empirical model on the selective removal of monovalent ions and divalent ions in the laboratory-scale ED process.

During my PhD studies, I have conducted the experiments in pilot scale EDR at the Brackish Groundwater National Desalination Research Facility (BGNDRF) in Alamogordo, NM. This set of experiments provided me a general idea regarding the effect of operating parameters on the removal of different ions in the EDR process. Utilizing this data, I designed, and built a complete laboratory scale ED system equipped with a complete data acquisition system. The reason that the data acquisition system was designed and used in this system was to provide continuous and accurate monitoring and recording the operating conditions during the entire experiments. The collected data was used in developing an empirical model to predict the removal of monovalent and divalent ions in the ED/EDR process at various operating conditions. The developed model can be widely used for full-scale electrodialysis operations for which removal of some of the specific ions is desired.

As an independent scientist, I will focus on improving the ED/EDR as well as other membrane-based desalination technologies in order to reduce the energy and cost of the process without impacting performance of these processes. Additionally, I will work on using renewable energies, specifically solar energy, to provide off-grid desalination systems for rural, remote off-grid areas to provide water for people who are suffering from fresh water scarcity.

390958 Combining Computation and Experiment to Uncover Environment Friendly Solutions to Energy Problems

Abstract:

The world's dependence on fossil fuels has lead to the need for alternate sources of energy as supplies dwindle, as well as a growing need to remove harmful compounds from the air. Solar energy and hydrogen energy are promising candidates for supplanting fossil fuels while novel adsorbents like metal-organic frameworks (MOFs) are promising materials for removing harmful gases. To date, my research has been concerned with both aspects of the fossil fuel problem. Specifically, my M.S. and Ph.D studies were concerned with solar energy and hydrogen energy. In order for solar cells to be a feasible alternative to fossil fuels, they need to have a competitive power conversion efficiency. By adding partially sulfonated ionomer to the active layer of the solar cell, my research led to an increase in power conversion efficiency. Regarding hydrogen energy, my research was focused on understanding the thermodynamics of metal hydride reactions for hydrogen storage applications. Specifically, the goal was to screen thermodynamically promising metal hydride reactions from a full database of metal hydride mixtures using first-principles calculations. The large-scale screening ultimately provided a number of promising single-step or multi-step metal hydride reactions. My postdoctoral research at Northwestern University was related to investigating MOFs for the removal and separation of harmful gases. I used computational methods to screen and assess functional groups that could be incorporated into MOF ligands to preferentially adsorb harmful gases under humid conditions. In addition, I was also involved in assessing new methods for characterization of MOFs, as well as assessing MOFs for various gas separations. Currently, I am interested in lithium ion battery issues as a research director and postdoctoral fellow at Georgia Institute of Technology. Inspired by my research background, my future research will be focused on producing energy sources, such as methane or methanol, through carbon dioxide reduction on nanosized clusters or other shapes of metals and metal-based compounds. The nanoscale particles will have different chemical properties from bulk-based surfaces, and thus understanding the effect of the particle size, shape, and type on the carbon dioxide reduction would be an interesting topic for environment-friendly energy applications. Specifically, I have a plan to examine the changes in the mechanisms and activation barriers of the carbon dioxide reduction, increasing the particle size or changing the particle shape and the type of metals.

389054 Design of Biopolymer Building Blocks for Novel Self-Assembled Materials

Abstract:

Bioinspired and biomimetic materials, in particular hydrogels, are currently one of the most interesting areas in academic and industrial research, encompassing both biomedical (drug delivery, tissue engineering) and industrial applications (filters, sensors, and actuators). To meet these emerging demands, studies have focused on tailoring a material's mechanical properties (e.g. mechanical strength, extensibility, toughness, injectability, self-healing, and lifetime) and mimicking the physical/chemical properties of natural materials (e.g. extracellular matrix and muscle). Biopolymers, in particular natural proteins and synthetic polypeptides, have received a great deal of attention as elements in these materials. Amino acid sequences of polypeptides can be precisely controlled and easily modified through recombinant DNA technology while large batches of high purity and monodisperse polypeptides can be cost-effectively biosynthesized. More importantly, different sequences can produce unique folded structures that result in distinct biophysical and biochemical properties at the molecular level including mechanical properties, molecular recognition abilities, and intelligent responses to external stimuli. If these exceptional molecular properties of biopolymers are incorporated into macroscopic materials, multifunctional biopolymer-based materials that exceed those attainable from purely synthetic polymer-based materials can be designed.

Mechanical behaviors of muscle can be potentially mimicked by engineered hyperelastic materials/hydrogels constructed by crosslinking folded protein structures. Using a custom-designed atomic force microscope (AFM) for single-molecule force spectroscopy, structure-mechanical property relationships of α-helical repeat proteins and globular proteins have been characterized, demonstrating that similarly structured proteins can be used as building blocks for hyperelastic materials due to their reproducible large extensibility and energy dissipation, and fast response to external forces. In addition, to maintain proper folding and to preserve the desired mechanical properties in macroscopic materials, end-crosslinking of the proteins was investigated. Simple physical crosslinking was performed by fusing subunits of protein complexes to their termini and it was confirmed to successfully construct the mechanical protein-based supramolecular self-assemblies as designed. A new technique was also developed to directly quantify the interfacial rupture strength of the subunits in their individual assemblies, permitting the investigation of various protein complexes that can be used as molecular "screws" or "bolts" in novel biopolymeric materials.

Protein (nucleoporin)-based biological hydrogels with selective filtering properties have been mimicked by constructing biosynthetic hydrogels built from newly designed nucleoporin-like polypeptide (NLP) building blocks. The biological hydrogels in the nuclear pore have been attractive due to their capability of selectively filtering $< 0.1\%$ of all proteins in a eukaryotic cell while translocating over 1,000 molecules per second. Development of similarly behaving biopolymer gels can potentially impact separation technologies and defense applications but is limited by the low synthesis yield of proteins and long gel processing time. Artificially engineered high-yield NLP building blocks that can mimic the structural properties of nucleoporins have been successfully developed, and rapid gel construction is achieved. Using a combination of selective permeability fluorescence assays, shear rheology, and Raman spectroscopy, the mechanism of artificially engineered NLP hydrogel function is studied.

380028 Development of Microporous Membranes and Thin Films for Energy and Environmental Systems

Abstract:

My research interest is in the areas of inorganic membranes and thin films of nanostructure, particularly the nanoporous zeolites and zeolite membranes. I have focused on developing methodologies for synthesizing materials of desired microstructure and chemical composition and understanding the materials' structure-property relationships. My research has been particularly concentrated on the ceramic membranes for chemical separations and catalytic reactions with direct applications in the emerging clean energy technologies and environmental management systems.

During my PhD program, I have been involved to develop a new method of controlled modification of zeolite pore structure that can make the membrane for simultaneously achieving high selectivity and high permeance in the separation of similarly-sized small molecules such as H_2/CO_2. The modified zeolite membrane was able to realize the long-anticipated zeolite membrane reactor for high-temperature catalytic reactions for the pre-combustion CO_2 capture and H_2 production from fossil fuels and biomass. I have successfully established and validated a one-dimension model to simulate the membrane reaction performance beyond the experimental conditions. In my post-doctoral study, I have continued to work on the development of gas separation membranes and membrane reactor systems for major commercial applications in catalysis, adsorption, and ion exchange. In addition, I have been involved to validate the modeling works by obtaining experimental data via synthesis, characterization and evaluation of nanoporous ceramic membranes.

Based on my knowledge of membranes/catalysts, instrumentation design, and reaction modeling, my future research will focus on the fundamental issues including synthesis of novel materials and membranes and understanding their structure-property-performance relationships. Also, by developing innovative materials/membranes associated with clean and sustainable energy technologies, I am going to establish strong research for studying molecular transport, diffusion, gas-solid reaction mechanisms in the nanostructured materials and thin films.

393444 Solid-State Self-Assembly : Fundamentals and Applications

Abstract:

Nanoscale science and technologies has been developed tremendously during the last two decades, introducing a variety of nanomaterials with unique properties. However, incorporation of the properties into macroscale functional applications has been limited. An essential challenge is the integration of such unique properties into assemblies for micro- and macroscale devices and discovering appropriate conditions under which this can be accomplished. Here I explore the self-assembly of nanoparticles (NPs) in solid-state under external stress for discovering fundamental understandings of mechanisms and dynamics for various engineering applications. 1. Yoonseob Kim et al. ***Nature***, 2013, 500, 59-63 first demonstrated an example of excellent stretchable conductors from self-assembly of spherical NPs in solid-state. Free-standing stretchable conductors were assembled by layer-by-layer assembly. High conductivity and stretchability were observed and the properties originated from dynamic self-organization of NPs under stress. Modified percolation theory to incorporate the self-assembly of NPs gave excellent match with experimental data. 2. The current study demonstrated the chiroptical activities of NPs in solid-state for the applications of biosensing devices and optoelectronics. Unconventional transfer of macroscale stresses shaped nanoscale assemblies with handedness. Chiroptical responses were reversibly tunable by controlled stresses. Computational simulation supported the chiroptical properties originated from self-assembly of NPs in solid-state.

Based on my past experience, I will develop a research plan that includes, but are not limited to, more fundamental studies about 1) initiation force and main driving force of self-assembly, 2) relation between nanoparticles and external stresses, 3) form of probable assemblies not only in chain and cell but for higher dimensionality, and 4) optimization to the commercially available materials. Understanding the fundamentals has great potentials to lead to improved material designs for advanced applications. Specific objectives include the following: 1) stretchable semiconducting nanoparticle composites for reversibly tunable photocurrent with discovery of unique pathways, 2) electro-tunable mechanical properties of angioplasty balloons with enhanced performance including in-vivo experiments, 3) reversibly stretchable Li-ion batteries with high performance during many repeating cycles of huge mechanical stress.

The works were funded, in part, by Rackham Predoctoral Fellowship, University of Michigan, STX foundation, Seoul, Korea, and US Air Force Office of Scientific Research.

Revealing Molecular Mechanisms by Single Molecule Fluorescence Microscopy

Younghoon Kim

[1]*Department of Bioengineering, University of Illinois at Urbana-Champaign, Urbana, IL 61801*

Research Interests:
Single Molecule Fluorescence Microscopy is an emerging multi-disciplinary field that currently finds numerous applications in many diverse industries ranging from Biological Diagnostics to Traditional Reaction Engineering. Single Molecule Fluorescence Resonance Energy Transfer (smFRET) has rapidly become an integral part of nanobiotechnology, and enables researchers to gain insights into systems that otherwise would have been inaccessible. Direct monitoring proteins, nucleic acid, and individual movement of biomolecules provide rapid, deep understanding of molecular mechanisms, that cannot be seen in mass or ensemble state. For example, using smFRET, nucleic acid processing enzyme, helicase motion was monitored to understand the liquid-gas like transition state in sub-cellular level. smFRET allowed to observe milli-second to minute time scale molecular motions. Perhaps the main advantage of this experimental approach is that the unmatched detection level and real-time motion in simple robust manner. Also it gives direct, intuitive results that does not need any over-simplifications or complicated mathematical process and direct monitoring of heterogeneous populations.

Postdoctoral Project: "*DEAD-box helicase in the processing body*" and "*Bright and Photostable Fluorescent Labels for Single Molecule Microscopy*" under supervision of Sua Myong, Bioengineering and Charles M. Schroeder III, Chemical and Biomolecular Engineering, University of Illinois, Urbana-Champaign
PhD Dissertation: "*Cure for Muscular Dystrophy by Self-Assembled Polymer Vesicles and Worm-Micelles*" Under supervision of Dennis E. Discher, Chemical and Biomolecular Engineering, University of Pennsylvania

Research Experience:
My academic career path has been a blend of many fields of biology, biotechnology and chemical engineering. My formal training was in biophysics, which I have learned from many different experimentalist by the projects that I have worked on. I was always close collaborations with other engineers and scientists in the fields of rheology, medicinal chemistry, drug delivery, material science and molecular biologist. As a result of these collaborations I have acquired invaluable and versatile experience in many research areas: biophysics and biochemistry, biomedical imaging, material science, biology, and engineering design. Moreover, as a result of engineering mind-set, I have published two patents covering world and US, which has been actively discussed to commercialize the product.

Industry Experience:
It would be unique experience for a chemical engineer to be a traditional process engineer and later to be a biotech engineer. After bachelor's degree, I have worked at the traditional cement plant as a senior process engineer, and gained invaluable experience. After PhD degree, flourishing environment of biotech company enticed me and realized that how the new technologies can nurture small company to mid-scale or big one. I would like to pass my experience to the next generation chemical engineers how two industries have common points and what are our strengths.

Teaching Experience:

Aside from my research career, I also have extensive teaching experience. I spent two semesters teaching basic math and science to middle school students as a part of outreach program; I TAed and undergraduate lecture and laboratory courses in Chemical Engineering at the University of Pennsylvania; and, lastly, I have mentored high school student, high school teachers, undergrad students and graduate students, resulting a national-wide award and fellowships from the research I had with.

Towards the Next Generation of Chemical Analysis and Semiconductor Devices: Harnessing Light and Spin

Jonathan King
Department of Chemistry, University of California, Berkeley
Materials Sciences Division, Lawrence Berkeley National Laboratory

Nuclear magnetic resonance spectroscopy (NMR) and imaging (MRI) are two of the most powerful analytical techniques available to engineers, scientists, and medical professionals, enabling a wide range of chemical and structural analysis. However, they require costly infrastructure and are plagued by low sensitivity resulting from small nuclear spin magnetization. My laboratory will harness the highly polarized spin states of photons and electrons to control nuclear magnetization, resulting in greatly enhanced signals from magnetic resonance techniques. We will build upon my previous work with nitrogen vacancy (NV) centers in diamond [1] to develop methods of coupling spin-polarized electrons to nuclear spins in target molecules. This "hyperpolarization" gives the potential for more than 10,000-fold enhancement of NMR and MRI signals, greatly lowering sensitivity limits and reducing experimental timescales. Such enhancements will be transformative, allowing the analysis of dilute biomolecules and microscopic samples over rapid timescales, and generating technology for portable, low-cost magnetic resonance instruments.

While traditional NMR spectroscopy yields a great deal of chemical information, it cannot directly probe the chirality of molecules. We will make use of optically detected methods of magnetic resonance, such as atomic magnetometry, to enable zero-field NMR, where the geometry of spin-spin couplings may be fully resolved. These couplings include the traditionally ignored antisymmetric J-coupling, which is linked to the chirality of the analyte molecule. This chiral recognition, in combination with the chemical specificity of NMR, will provide a powerful new technique for the analysis of pharmaceuticals and biomolecules, where chirality is often a dominant factor in the biological function of a molecule.

Semiconductor devices that harness "spin current" can be faster and consume far less energy than traditional devices using charge current. Magnetized nuclei may be used to manipulate spin currents in semiconductors, enabling a host of "spintronic" devices. We will use a combination of optical and magnetic manipulation of nuclear spins in gallium arsenide [2,3,4] to generate re-writable, 3-dimensional patterns of magnetization without the need for microfabrication or lithographic patterning. These structured regions of magnetism will serve as the architecture for spintronic devices integrated into bulk semiconductors.

1. **J. P. King**, P. J. Coles, J. A. Reimer, "Optical polarization of ^{13}C nuclei in diamond through nitrogen vacancy centers." *Physical Review B*, 81, 073201 (2010).

2. **J. P. King,** Y. Li, C. A. Meriles, J. A. Reimer, "Optically Re-Writable Patterns of Nuclear Magnetization in Gallium Arsenide." *Nature Communications*, 3, 918 (2012).

3. Y. Li, **J. P. King**, L. Peng, M.C. Tamargo, J. A. Reimer, C. A. Meriles, "Helicity independent optically-pumped nuclear magnetic resonance in gallium arsenide." *Applied Physics Letters*, 98, 112101 (2011).

4. Y. Li, **J. P. King,** J. A. Reimer, C. A. Meriles, "Near-band-gap photo-induced nuclear spin dynamics in semi-insulating GaAs: Hyperfine- and quadrupolar-driven relaxation." *Physical Review B*, 88, 235211 (2013).

381902 Optimal Control of Neural and Small Length Scale Dynamical Systems

Abstract:

In the last decade, dramatic advancements in experimental and computational techniques have flushed tremendous opportunities in the study of fundamental questions of science and engineering by taking the approach of stochastic modeling and control of dynamical systems. In parallel, emerging applications through these advances have ignited the development of new technologies by integrating advanced control strategies with stochastic dynamical models. One such example is the brain-machine interface. Over the course of my doctoral and postdoctoral studies, my research interests have mainly been focused on developing optimal control strategies for applications in closed-loop neural prostheses, deep brain stimulations, and closed-loop drug delivery systems. During this poster session, I will highlight examples from both my graduate research and postdoctoral research which demonstrate my capability in advancing these subjects by integrating probabilistic tools with optimal control policies. Building on these results, I will describe my future academic plans to commence a distinct, interdisciplinary and independent research program at the interface of engineering and computational neuroscience. I will explain how my interdisciplinary expertise in the theory of stochastic processes, neuroscience, optimization, dynamical systems and control uniquely positions me to capitalize on emerging opportunities at this area.

I am currently working as a postdoctoral research associate in the department of Electrical and Systems Engineering at Washington University in St. Louis. The major theme of my studies is centered on understanding the effect of system dynamics on the emerged system behaviors such as synchronization in complex dynamical systems, in particular pathological synchronization in Parkinson's disease, and developing optimal control strategies to achieve the desired system behavior in such systems. During the course of my doctoral studies at Lehigh University, I developed optimal control-theoretic framework for investigating the design of closed-loop neural prostheses. In parallel, I developed optimal stochastic control strategies for regulating behaviors of particles driven by Brownian motion.

390372 Optimization of Nonsmooth Chemical Process Models

Abstract:

Nonsmoothness can enter chemical process models through sources such as transitions in phase or flow regime, activation of safety mechanisms, and switches between discrete operating modes. Established numerical methods for simulation, sensitivity analysis, and optimization typically require derivative information, and have limited applicability to nonsmooth systems. Specialized numerical methods exist for nonsmooth problems, such as bundle methods for local optimization, and semismooth Newton methods for equation solving. These nonsmooth methods require local sensitivity information in the form of the generalized derivatives developed by Clarke and Nesterov. However, established methods for evaluation of these generalized derivatives are limited either in scope or in the quality of the obtained derivatives. Nonsmooth dynamic systems introduce particular difficulties, since little is known *a priori* about the behavior of state variables as functions of system parameters, and since even theoretical descriptions of generalized derivatives are lacking for dynamic systems.

For my PhD work with Paul Barton in the Department of Chemical Engineering at the Massachusetts Institute of Technology, I have focused on describing and evaluating generalized derivatives for nonsmooth systems. I developed and implemented a variant of automatic differentiation that evaluates generalized derivatives for composite nonsmooth functions. This method is the first computationally tractable numerical method for accurate generalized derivative evaluation, and is fully automatable.

Several new results were also obtained for nonsmooth dynamic systems that are represented as systems of parametric ordinary differential equations (ODEs) with nonsmooth right-hand side functions. Firstly, I obtained sufficient conditions under which the state variables of such a nonsmooth dynamic system are in fact smooth functions of system parameters. These conditions are computationally inexpensive and simple to verify *a posteriori* during a simulation run, and can sometimes be shown to hold *a priori*. Secondly, I provided the first theoretical description of a useful generalized derivative for a general nonsmooth dynamic system. As in classical sensitivity analysis approaches for smooth dynamic systems, this generalized derivative is expressed in terms of an auxiliary dynamic system. These nonsmooth dynamic sensitivity results were generalized to certain hybrid discrete/continuous systems, which can be used to model systems with discontinuous jumps.

363150 Synthetic Approaches to Control Cell Fate and Function

Abstract:

Cells are complex and transformable machines. In addition to their homeostatic functions, they are capable of sensing and processing diverse biochemical and biophysical signals. Furthermore, isogenic cells are capable of transforming into a multitude of states, including diseased ones, each with distinct sets of active regulatory programs. To decode these complex cellular programs, synthetic approaches provide exquisite quantitative control of the spatiotemporal, physical, and chemical properties of both extra- and intra-cellular regulatory components. My research focuses on two complementary experimental approaches: 1) Engineering synthetic cellular microenvironments with defined spatiotemporal, physical, and chemical properties; 2) Engineering synthetic intracellular signal-transducing machineries that have unique input-output functions and can interface with complex, native regulatory systems such as chromatin and the actin cytoskeleton. The main applications of these synthetic approaches are to: 1) Quantitatively interrogate and control the cellular programs and machines that define distinct stem cell and disease states and their developmental or oncogenic transformations; 2) Elucidate molecular mechanisms underlying cell regulation, including chromatin dynamics and cellular mechanotransduction. My research takes advantage of methods from and collaborations with the fields of synthetic biology, systems biology, materials science and engineering, stem cell and cancer biology, and biophysics. Together this work may be applied to diverse areas in regenerative medicine, cancer biology, and eukaryotic cellular biotechnology.

Past and current work in these areas include: 1) Organic-inorganic surface chemistry (Keung et al 2005, 2007); 2) Biophysical control of neural and embryonic stem cell differentiation and proliferation (Keung et al 2009, 2010, 2011, 2012, 2013); 3) Synthetic, site-specific genome-targeting of chromatin factors to engineer cellular logic, spatial transcriptional regulation, and memory (Keung et al 2014).

390749 Towards a Sustainable Energy Future: Modest – Model Aided Optimization and Design of Energy Systems' Toolkit

Abstract:

Just over a decade into the 21st century and new millennium, it is becoming increasingly evident that energy security and climate change are two of the most defining challenges of our time and generations to come. Unfortunately, there is no *silver bullet* and we need to nurture a range of technologies to address these challenges. To accomplish this, motivated by my own diverse research experiences, several ***clean and green technologies*** need to be developed further. These include: .

- *Bioenergy (including advanced biofuels)*
- *Carbon capture, transport and sequestration (technologies and infrastructure)*
- *Hydrogen driven transportation (technologies and infrastructure)*
- *Clean coal technologies (including integrated gasification combined cycle)*
- *Renewable energy sources (including solar and wind)*
- *Energy systems of the future (e.g., urban & rural energy systems, zero-carbon homes)*
- *Water and healthcare (technologies and infrastructure)*

Together, all these technologies have the potential to alleviate both the energy security and climate change concerns. This presentation will discuss effective use of various *Systems Engineering* tools in a comprehensive framework **MODEST (Model aided Optimization, Design of Energy Systems' Toolkit)** to *foster scientific innovation and facilitate the acceleration of commercial deployment of these technologies. Special emphasis will be on large-scale real world energy systems.*

388324 The Physical Cell: Impact of Mechanics and Rheology on Cellular Function

Abstract:

The internal microenvironment of a living cell comprises an intricate choreography of molecules that must be transported from one location to another, elastic forces that must be overcome or harnessed into useful work, and interdependent chemical reactions whose rates must be carefully controlled. My research focuses on examining the basic physical phenomena that underly biological processes at the molecular and cellular level. I develop multiscale models, employing both analytical and computational techniques, that interface directly with experimental data. My work draws on concepts from continuum mechanics, fluid dynamics, and statistical physics to address the interplay of structure, transport, and biological function.

As a doctoral student with Prof. Andrew Spakowitz, I focused on studying how the mechanical properties of DNA impact the packaging and accessibility of the genome. I demonstrated that the elasticity of the DNA molecule can give rise to tension-mediated cooperative binding between DNA-bending proteins, allowing them to interact across a wide range of length scales without direct protein-protein contact. I also investigated the role of this elasticity in the packing of DNA into compact chromatin fibers, and the connection between packaging fluctuations and force-dependent kinetics of enzymes that access the DNA. My work on the statistics of large DNA-protein complexes spurred the development of a generalized approach for coarse-grained modeling of polymer systems by mapping onto effective elastic chains. This methodology enables simulations of polymer dynamics and statistics on much larger length scales than was previously possible.

As a postdoctoral scholar with Prof. Julie Theriot, I investigate cytoplasmic mechanics in motile neutrophils. These cells exhibit rapid motion accompanied by dynamic changes in shape that arise from large-scale fluctuations at the cell boundary, driven by active polymerization processes. Using a combination of analytical theory and simulations, I study the motion of particles near an actively fluctuating membrane. I apply this model directly to cytoplasmic dynamics in motile cells by treating the cell interior as a viscous fluid driven by boundary fluctuations that are tracked using phase microscopy. Fluorescent tracking of intracellular particles indicates that their relative motion is consistent with the predictions of this simple model. In particular, I demonstrate that fluid flows arising from the cell shape dynamics can contribute significantly to mixing and transport of large complexes and organelles within the cell.

My future research goals center on studying the physical processes underlying the propagation of chemical and mechanical signals within a cell. I intend to use my background in multiscale biophysical modeling as a springboard for addressing fundamental problems of how transport phenomena, active mechanical fluctuations, and intracellular spatial organization impact the structure and kinetics of biomolecular networks. Hijacking cellular machinery to engineer novel circuits or to induce therapeutic effects necessitates an ability to harness the kinetic properties of these reaction networks. The role of cytoplasmic rheology and mechanics in modulating biomolecular processes within the cell thus presents a rich set of questions of direct relevance to cellular engineering.

370573 Controlling the Electronic Properties of Surfaces and Interfaces for Energy Applications

Abstract:

The goal of my research is to develop nanostructured materials for energy conversion in which to investigate the fundamental surface and interface interactions. This poster gives an overview of my past and current research efforts on the design and development of nanostructured materials for various applications, including (i) fundamental research on ferroelectric perovskites for nanodevice applications; (ii) fundamental processing-structure-property relationships that govern organic-inorganic nano-interfaces for organic transistors and solar cells (iii) materials synthesis of transition metal oxides for heterogeneous catalytic systems. I will establish an interdisciplinary research program that lies at the interface of engineering, physics, and materials science. My laboratory will study the fundamental properties of nanostructured materials and focus on developing model systems to design efficient and cost-effective materials for energy applications.

370684 Nanotechnology-Based Breakthroughs in Biology and Energy

Abstract:

In my independent faculty career, my research will focus on nanotechnology-based approaches to understanding and controlling fundamental electronic and mechanical properties of biological materials for advancing both biomedical and energy sciences. I am particularly interested in interfacing biology with functional electronics to yield breakthroughs in drug development, to enable new avenues of powering biomedical devices, and to probe biological changes in a truly non-invasive way. For my doctoral research, I developed a novel method for understanding charge transport at nanoscale metal-oxide interfaces with the use of scanning probe techniques and extended that knowledge to understanding and controlling polarization switching in ferroelectric devices. Ferroelectrics are highly efficient mechanical couplers and are ideal candidates for interfacing with biology for a number of applications. As a postdoctoral fellow, I have developed new techniques to measure tight junction adhesion forces at piconewton resolution. With innovative electrode designs, I have developed a novel bioimpedance system to track changes in cellular dynamics – in-real time – while modulating ion transport gateways. With my background and skills, I am uniquely positioned to launch an interdisciplinary, collaborative Nanotechnology in Biology and Energy program that incorporates the expertise of both scientists and engineers.

388936 Polymers, Colloids and Composites: In the Service of Chromatography with New Porous Materials

Abstract:

Polymer Science, Colloid Science and Chromatography can be interfaced during the development of functional materials for particular applications.

During my doctoral studies in ETH Zurich, I developed a new procedure towards macroporous polymeric microparticles. The major novelty of this procedure is that the porous structure is created and controlled without a template, through the controlled aggregation and coalescence of latex nanoparticles.

These macroporous particles were packed into chromatographic columns and functionalized via controlled radical polymerization with polymer chains of variable composition. Thus, different chromatographic modes were addressed, namely ion-exchange, as well as temperature-responsive chromatography, which relies on an unconventional separation mechanism. Several classes of compounds have been successfully separated, such as small pharmaceuticals, peptides, proteins and monoclonal antibodies. This whole procedure, from the preparation of the base macroporous material, to functionalization, to chromatographic application, has been recently secured by two patents.

Diverging from this project, I also became interested in different approaches for controlling the stability latex nanoparticles. To this end, I synthesized a set of block-copolymer surfactants via controlled polymerization and studied first their self-assembly behavior in an aqueous solution. Subsequently, these surfactants were applied in order to control the shear-induced aggregation of commercial latexes, which is an important issue in the recovery of polymers from a dispersed state.

In my postdoctoral research in Lawrence Berkeley National Lab, I developed porous polymeric monoliths in a thin-layer format for the efficient thin layer chromatographic (TLC) separation of proteins, peptides and fatty acids. Their pore surface was modified either with an in-house synthesized polymer or with metal nanoparticles of different noble metals, namely gold and silver, starting from the corresponding colloidal dispersion. All applied modifications provide a dual function: appropriate modulation of chromatographic interactions, coupled with the assistance in the desorption/ ionization of the separated compounds by a laser beam. The latter allowed the analytes detection by MALDI Mass Spectroscopy immediately after separation, without the use of an additional "matrix" substance, which needs to be conventionally applied.

Recently I have initiated a project for the preparation of a composite porous monolith in a capillary format, modified with a layer of a metal-organic framework (MOF), aiming to the liquid-phase separation of small molecules. MOF are an advanced class of porous materials, composed of metal centers interconnected by organic linkers. They are currently intensively investigated, owing to their unique combination of high porosity, ultra-high surface area and tailorable structures. Thanks to their versatility, a range of different interactions between MOF and molecules that can enter their porous network can be potentially exploited for multiple applications, including separations. Nevertheless, little attention has been devoted so far to liquid-phase separations.

Apparently, supports with the appropriate porous structure are necessary, where the MOF can reside. Furthermore, procedures for the controlled formation of the MOF in the confined interior of such a porous support need to be developed, ensuring that it is properly anchored and it also exhibits structural integrity. After resolving those issues, analytical-scale separations can be addressed. As soon as the MOF scalability is tackled, preparative-scale applications could be feasible as well. Since this is an emerging field, I anticipate this type of studies to extent into the future and I hope to be able to undertake them as an independent investigator.

I therefore wish to create a research program, encompassing the following overlapping areas: (i) Developing porous polymeric or composite materials with defined pore structure and surface chemistry that can perform efficient chromatographic separations of biologically relevant compounds, (ii) Understanding and controlling the stability of colloidal systems, such as dispersions of polymer and metal nanoparticles, for both fundamental and applied purposes, (iii) Studying the self-association of polymers with tailored composition into structures, which can be utilized for multiple applications, such as structure templating and encapsulation.

Concerning my teaching and mentoring experience and vision, I have been a teaching assistant in undergraduate chemistry laboratory courses for three years. I have also supervised more than 10 undergraduate interns, assisting my research work. I have indeed learned a lot by working with younger colleagues; I enjoy the challenge of instilling motivation and inspiration for science in people with different personalities. As my scientific background is broad, I feel confident with undertaking teaching duties in several areas, within an engineering or a natural sciences curriculum. Such lectures could include Polymer Science & Engineering, Colloid & Interface Science, Analytical Chemistry, Separation Science & Technology and related subjects.

Single-Molecule Visualization of Corona Phase Molecular Recognition

Molecular recognition is central to the design of therapeutics, chemical catalysis and sensor platforms, with the most common mechanisms involving biological structures such as protein antibodies. However, identifying and isolating an antibody for a particular biological molecule is often costly and time-consuming, and their use is limited to physiological conditions in which protein antibodies remain stable. This work reports the development of a synthesis platform for the design of synthetic sensors, based on the principle of carbon nanotube Corona Phase Molecular Recognition (CoPhMoRe). The design of sensors with CoPhMoRe is based on the adsorptive properties of engineered heteropolymers to carbon nanotubes, which create a unique corona with molecular recognition properties for a target analyte. This phenomenon is shown to be generic, with new recognition complexes demonstrated for riboflavin, l-thyroxine, and estradiol, and confirmed with single-molecule imaging of the nanotube and the corona phase. The complexes can be used as new types of sensors based on modulation of carbon nanotube optical fluorescence, as demonstrated using a synthetic antibody for real time spatio-temporal detection of riboflavin in live murine macrophages. The analyte library for these synthetic sensors could be amplified to identify synthetic sensors for a variety of different molecules for which there currently exists no detection method.

Biofilm and swarming genesis: rotary flagellar motors as mechanosensors

Pushkar Lele
Department of Molecular and Cellular Biology, Harvard University
Cambridge, MA 02138

Mechanosensing in bacterial cells leads to the initiation of biofilms and swarming in multiple species. The mechanisms of bacterial sense of touch, however, remain unknown. I am interested in determining the molecular mechanisms by which bacterial rotary motors sense surfaces and trigger biofilm formation and swarming.

The proposed research builds on recent discoveries of stimuli-dependent changes in motor self-assembly [1, 2]. I developed a novel optical-tweezer based assay to mechanically stimulate single motors, which enabled me to identify the motor components that receive mechanical signals and subsequently regulate function. These findings provide a handle on an important developmental problem: how do mechanical forces initiate cell development and biochemical signaling? My research program will focus on the mechanisms of self-assembly in biological macromolecules, signaling networks, mechanotransduction, and physics of motility. The ultimate objectives of this research will be to determine important adaptation mechanisms in micro-organisms, to identify targets for drug discovery and to develop effective strategies to combat the growing menace of antibiotic resistance.

[1] P. P. Lele, Basarab G. Hosu, H. C. Berg, "Dynamics of mechanosensing in the bacterial flagellar motor", under revision, Proc. Natl. Acad. Sci. USA.
[2] P. P. Lele, R. W. Branch, V. Nathan, H. C. Berg, "Mechanism for adaptive remodeling of the bacterial flagellar switch", Proc. Natl. Acad. Sci. USA (2012), 109 (49), 20018-20022.

385536 Engineering Biomimetic Self-Assembled Materials

Abstract:

My research interests involve the design, characterization, and application of novel self-assembling materials. In particular, I am interested in biomimetic systems in which carefully tuned molecular interactions create functional and responsive materials for applications ranging from nanoscale electronics to drug delivery. Electrostatic interactions play a fundamental role in both creating spatially patterned interfaces and also creating separate polymer rich phases (complex coacervation) for both equilibrium and non-equilibrium processes, and yet are still largely misunderstood. Using rational design of peptides and polymers I intend to explore the role that these interactions play in compartmentalized delivery systems and as a dynamic templates to create hierarchical hybrid materials composed of organic and inorganic constructs. My goal is to thoroughly understand the thermodynamic and kinetic processes driving the self-assembly of these materials and take them from the bench to their functional application.

My postdoctoral work with Professor Matthew Tirrell (The Institute for Molecular Engineering, University of Chicago &Argonne National Laboratory) involves the design, characterization and application of polyelectrolyte complex micellar systems. We have designed polypeptide-based micelles that can self-assemble to have either liquid or solid cores depending on their secondary structure, which provides unique self-assembly and degradation properties. After carefully characterizing model systems, we designed polyelectrolyte complex micelles that contain therapeutically relevant charged molecules such as miRNA and peptides, specifically for the treatment of atherosclerotic lesions and cancer. The modular nature of these assemblies enables the addition of a targeting ligand to increase the efficacy of delivering miRNA, creating polyelectrolyte complex micelles with a targeting ligand outside the corona of the micelle. Our rationally designed micelles have successfully delivered miRNA *in vitro* and are currently being tested *in vivo.*

My graduate work with Professor Raymond Tu (Chemical Engineering Department, The City College of New York) involved the interfacial templating of inorganic materials using self-assembled peptides. In nature, biological molecules form interfaces that assemble patterns of chemical functionality with exceptional precision. The role of dynamics during the assembly of biological molecules appears to be important for mineralization processes. Our work applied model sheet-forming peptides at interfaces to explore the dynamics of assembly in order to template mineral growth. Thermodynamic analysis of structure formation with increasing surface pressure allowed us to understand the nature of self-assembly with iterative changes in the peptide sequence. Additionally, we investigated the dynamics of the self-assembled state, where the organic phase switches between short- and long-range order as a function of surface pressure. This model system allowed us to explore the influence of electrostatic interactions on self-assembly and additionally the influence of short- and long-range order on the nucleation and growth of inorganic material. This is in contrast to a system that starts with a well-ordered preformed template that defines the epitaxial growth of the mineral phase. Versions of our model peptides are modified to include histidine in order to nucleate Au nanocrystals in both the short and long range ordered organic matrix showing that the phase behavior of the peptide influences the crystallinity and shape of the resultant nanocrystals.

Mathematical Modeling and Global Optimization for Planning and Scheduling of Petrochemical Processes, and Natural Gas to Liquid Fuel Energy Systems

Jie Li
Department of Chemical and Biological Engineering
Princeton University
Princeton, NJ 08544-5263

In the last twenty years, the petrochemical industry has succeeded by creating markets and supplying them with suitable products used to create goods such as plastics, cosmetics, lubricants, and paints. Petrochemical production begins in a refinery that separates crude oils mainly into lighter components such as naphtha, light naphtha, top oil, and liquid fuels including gasoline, diesel and jet fuel. The lighter components are further processed into various petrochemicals such as ethylene, propylene, butadiene, benzene, toluene, xylol and some other high-valued products via cracking, butadiene extraction, hydrotreating, etherification, and polymerization processes. Nowadays, tighter competition, strict environmental regulations, and lower-margin profits, drive the petrochemical industry to apply new technologies to improve their planning operations.

In this presentation, detailed nonlinear models are first proposed to predict product yields and properties in production units including crude distillation unit, vacuum distillation unit, hydrocracking units, catalytic cracking units, ethylene-cracking units and other processing units present in a petrochemical plant. The yield and property prediction models for crude distillation and vacuum distillation units are developed using swing cut theory based on crude assay data. Empirical non-linear models are developed for other processing units, including bilinear, trilinear, quadratic, polynomial, and exponential terms. Parameter estimation for all of the developed models is performed by globally minimizing the least squares error between the parametric models and real production data. Moreover, property indices in blending units are linearly additive and calculated on weight or volume basis, which introduce bilinear and trilinear terms. We also introduce binary variables to denote different operation modes for several production units.

The entire planning model is a non-convex mixed integer nonlinear optimization model (MINLP). To solve this large-scale nonlinear model, we propose an optimization-based procedure to obtain the tightest lower and upper bounds for variables especially the variables involving nonlinear terms. Then, we incorporate those tightest lower and upper bounds into commercial solver ANTIGONE to obtain ε-global optimality. Finally, a user-friendly platform is developed to allow the user to modify the planning model by updating model parameters when new data is available, product demands and specifications, cost parameters and many more. Several large-scale industrial examples are solved to illustrate the efficiency of our proposed model and global optimization approach.

381981 Sustainable Fuels and Chemicals from Catalytic Conversion of Natural Gas and Biomass

Abstract:

As petroleum reserves decrease, price increases and the impact of rising atmospheric CO_2 is exacerbated, world production of fuels and chemicals must shift to low-carbon and sustainable resources, such as biomass and natural gas. Conversion of these sustainable resources into fuels and valuable chemicals is very attractive, but also very challenging. Innovative technologies for this conversion could help break the technical barriers and lower the costs for commercial applications. I have been working on developing new technologies and new catalysts for biomass and natural gas conversion since my Ph.D. work.

My Ph.D. study was on fast pyrolysis of biomass and catalytic upgrading to liquid fuels and valuable chemicals. One of the most important contributions is the development of a new electrocatalytic method with a ruthenium electrocatalyst for biomass conversion, converting electrical energy into carbon-neutral drop-in replacement fuels and valuable chemicals. This method provides substantial advantages, including no need of hydrogen gas, operation at mild conditions to reduce coke formation and catalyst deactivation, and a great way to store intermittent renewable electricity as chemical energy in liquid fuels. The other significant contribution is the development of a new screw-conveyor pyrolysis reactor with compact design, enabling flexible application in regional biomass processing depots, showing great potential to solve biomass supply issue.

To deepen and expand my knowledge in catalysis and sustainable energy, I continued my postdoc research with Prof. Fabio H. Ribeiro and Prof. W. Nicholas Delgass at Purdue University. One of my projects is to understand the catalytic active sites on precious metal catalysts (Au on metal oxides) for water-gas shift reaction from the fundamental and atomic perspectives by studying kinetics and applying various detailed catalyst characterization methods. Another important aspect of my work involves synthesis and characterization of heterogeneous lewis acid zeolite catalysts, such as Sn-, Ti- and Zr-zeolites. These catalysts are being tested for condensation of glycerol with aldehydes or ketones to produce fuel additives for diesel fuel, and also to probe the active sites and reaction mechanism. As an expansion on sustainable energy study, I have been studying methane activation via oxidative coupling for ethylene formation using $Mn/Na_2WO_4/SiO_2$. The major work is on understanding of the active sites for methane activation by doing kinetic study and catalyst characterization and using the information gained to design better catalysts.

My future research interests are applying the knowledge of catalyst synthesis, active sites characterization and kinetic studies in biomass and natural gas conversion, such as bio-oil upgrading, catalytic pyrolysis, ethanol and glycerol conversion, and syngas to liquid fuels. Another interest lies in electrocatalytic conversion of biomass and natural gas using renewable electricity (such as photovoltaic and wind energy) and simultaneously storing intermittent renewable electricity as chemical energy in the form of liquid fuels and valuable chemicals. I will focus on designing bimetallic and bifunctional electrocatalysts, developing high temperature electrochemical systems and integrating renewable electricity techniques with electrocatalytic method for biomass and natural gas conversion.

388624 Two-Dimensional Molecular Sieves for Selective Transport

Abstract:

Two-dimensional molecular sieves are highly desired building blocks for the fabrication of molecular selective membranes. Miniaturization of the third dimension opens up opportunities for the synthesis of membranes with a thickness of just a few atoms. Due to remarkably short transport path of the molecules, these membranes will have profound impact in the field of separation technology. Apart from membranes, these two-dimensional porous sheets will find applications in catalysis, sensors, low-dielectric constant materials, barriers, etc.

Zeolite membranes are attractive due to their high performance (permeance and separation factor) and robust chemical and thermal stability. Attempts to synthesize segregated, two-dimensional zeolite sheets have often led to an intergrown or a layered zeolite. For the first time, a purified, dispersed suspension of highly crystalline 1.5 unit-cell-thick zeolite nanosheets was prepared by the exfoliation of layered zeolites followed by solution processing. Compared to time-consuming and expensive coating techniques involving modification of supports by masking and/or functionalization, a simple scalable filter coating technique was developed to fabricate a sub-100 nm-thick, oriented and compact coating of nanosheets. Wetting properties of the nanosheets were tuned, allowing it to deposit on the air-water interface. This allowed a monolayer coating of nanosheets on non-porous supports, opening opportunities for applications in sensors, low-dielectric constant materials etc. Thin, zeolite films prepared by hydrothermal treatment of the nanosheet film showed exceptional performance in the separation of xylene isomers, sparking industrial interest in deploying these membranes for the large scale purification of *p*-xylene. A mathematical model based on adsorption and diffusion kinetics was used to gain insights into the parameters affecting membrane performance at industrial process conditions.

In my future research, I aspire to expand the list of two-dimensional materials for molecular separations. Especially, one atom thick graphene sheets, MoS_2 sheets, aluminophosphate sheets, etc. hold exciting potential towards the fabrication of the thinnest membrane. Strategies for the generation of pores with tunable size in the non-porous atomic sheets will be explored. The transport of molecules with varying size and physical properties will be modeled to explore the full potential of two-dimensional membranes.

Design and Development of Advanced Materials for Energy Storage and Conversion

Qi Lu, Chemical Engineering, Columbia University/University of Delaware

It is now a common realization that our society is moving towards renewable and sustainable energy sources. In witness we are observing an increase in renewable energy production from sun and wind. Due to their intermittent availability however, energy storage and energy conversion techniques are starting to play a more and more important part in our lives.

My research focuses on the understandings of fundamental principles that govern the materials performance using well-defined model systems and first-principles-based calculations, based on which the rational design of high-performance materials for energy storage and conversion applications was carried out. For example, my recent work on the development of supercapacitor electrodes demonstrates my research philosophy: three key parameters that control the behavior of supercapacitor electrodes were identified through a series of mechanistic investigations on model systems[1]. Based on the gained fundamental insights, a novel nanotechnology was developed to simultaneously optimize those parameters and to achieve practical supercapacitor electrodes with both high-energy and power densities[2]. In addition, every fabrication procedure employed in this technology was designed to be cost-effective and industrially scalable. As a result, this technique has been adapted for commercialization in a startup company for grid-scale electricity storage.

Figure 1. A supercapacitor electrode with interconnected metal core and active outer layer.

Figure 2. A nanoporous Ag catalyst that converts CO_2 to CO with high-selectivity and efficiency.[3]

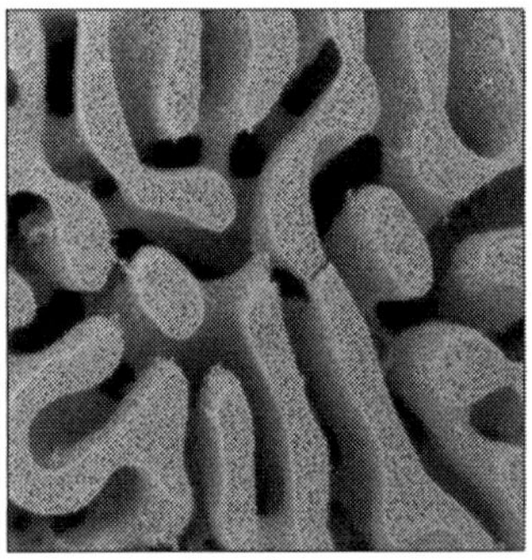

Figure 3. A Bimodal porous CuTi catalyst exhibits a Pt-like surface reaction sites for H_2 evolution reaction.[4]

In two of my most recent works, I developed highly active, non-precious metal-based catalysts for the electrochemical conversion of CO_2 and H_2O to CO and H_2, respectively. The design of this catalyst was investigated using density function theory-based calculations. Dealloy-based techniques were developed afterwards to synthesize practical catalysts with high surface area and density of active sites. The structure of the catalysts was further tailored to maximize the material utilization and electrode conductivity, achieving catalytic materials with optimal over-all performance for CO_2 conversion[3] and H_2 evolution reaction[4].

My future research will focus on the following aspects: 1) battery-capacitor hybrid electrochemical energy storage for solar- and wind- power grid applications; and 2) CO_2 and CH_4 conversion to value-added fuels and chemicals. The overarching goal of my research program is to provide the society advanced materials for future energy usage, which will be pursued via a combination of computational and experimental techniques.

References:

1. Lu Q. *et al.* "Nanostructured Electrodes for High-performance Pseudocapacitors" ***Angew. Chem. Int. Ed.***, 2013, 52 (7): 1882–1889

2. Lu Q. *et al.* "Supercapacitor Electrodes with High-Energy and Power Densities Prepared from Monolithic NiO/Ni Nanocomposites" ***Angew. Chem. Int. Ed.***, 2011, 50 (30): 6847–6850 (Covered by MaterialsViews, ChemistryViews, PhysOrg, *etc.*)

3. Lu Q. et al. "A Selective and Efficient Electrocatalyst for Carbon Dioxide Reduction" ***Nat. Commun.*** 2013, 5:3242 (Covered by Science Daily, R&D, Science World Report, etc.)

4. Lu Q. et al. unpublished results

390292 Next Generation Approaches to Biomolecular Engineering: Synthetic Biology Meets Directed Evolution

Abstract:

Rapidly advancing technologies like multiplex oligonucleotide synthesis and next-generation high-throughput sequencing have opened the door to tremendous progress in synthetic biology. My experience in both protein engineering and strain engineering puts me in an excellent position to forge the next generation of **synthetic biology tools for biomolecular engineering**.

Background

My research career started as an undergraduate at Johns Hopkins University under Prof. Marc Ostermeier, where I helped engineer **allosteric molecular switches via directed evolution**. For my M.S.E. research I created several modifications of chimeras of maltose binding protein and β-lactamase (BLA), the enzymatic activity of which could be allosterically regulated by the addition of maltose (published in *PNAS*). During my PhD with Prof. Matthew DeLisa (Cornell), I developed a **suite of tools for examining protein folding, protein-protein interactions, and post-translational modifications** in *E. coli*. I developed a protein folding reporter for the bacterial periplasm using a BLA-based selection. I showed that we could not only probe intrinsic protein folding, but **factors in the cellular environment** (e.g., chaperones, redox state) that affect protein folding (published in *Protein Science*). Later, I refactored the protein folding reporter to function as a detector for N-linked glycosylation in bacteria as a pipeline for **engineering post-translational modifications** (published in *Biotechnology Journal*).

Current Research

The postdoctoral work I am pursuing with Prof. Ryan Gill at CU-Boulder has provided me a broader view of both **metabolic engineering** and the design of directed evolution experiments. I am currently involved in three projects: (1) Performing selections on **genome-scale libraries** to discover genomic factors underlying **bacterial antibiotic resistance**. (2) Combining multiplex oligonucleotide synthesis and recombineering, I developed a platform for directed evolution of **orthogonal repressors for cellular circuits** on both the DNA and RNA level. (3) **Regulator engineering for ethanol tolerance** in *E. coli*. This work has given me a unique perspective on constructing targeted libraries on the genome scale, generating programmed diversity to complement my expertise in selection design. Specifically, leveraging the fast-developing technologies of multiplex oligo synthesis and next-generation sequencing will be key to my research as a principal investigator.

Future Research Directions

My lab will apply the latest synthetic biology technologies to tackling protein expression in bacteria with applications fpr human health, biofuel production, and synthetic circuit design. **Targeted libraries of expression elements** coupled with selections for protein folding will allow optimization of expression in difficult-to-produce proteins (e.g., membrane, heterologous proteins). **Genome mining** of heterologous proteins leverages the tremendous and ever-increasing amount of metagenomic data to inform and improve selection of enzyme activity. Sub-genome scale libraries of various families of proteins will provide in-depth exploration of cellular factors (and combinations thereof) that contribute to protein production. Finally, expanding into Gram-positive bacteria with **genome-scale libraries** on lactic acid bacteria will provide insight into protein folding and secretion in these clinically-relevant but understudied organisms.

Teaching

Beyond my research accomplishments, I have been fortunate to gain valuable teaching experience. In graduate school I served as a teaching assistant for four different courses, with topics ranging from process control and reaction kinetics (ChemE) to understanding wine and beer (Enology and Viticulture). Throughout my teaching pursuits I strove to implement **active and cooperative learning techniques** to keep students engaged and reinforce concepts learned in lecture. TA evaluations I have received consistently mention my **enthusiasm for the material and creative approaches to teaching** and exam review. I was also employed by Cornell Engineering Learning Initiatives, where I constructed and facilitated Engineering TA Development workshops in Assessment, Academic Integrity, Active and Cooperative Learning, Public Speaking, and Microteaching.

Inspiring students to pursue biomolecular engineering, whether in the classroom or the laboratory, is highly valuable. After all, Niels Bohr once supposedly said: "Biology is too important to be left to the biologists."

379325 Electrochemical Reduction of Carbon Dioxide with Highly Dispersed Metal Nanoparticles

Abstract:

The electrochemical conversion of carbon dioxide into hydrocarbons is an alternative route for synthesizing fuels and feedstocks that are typically derived from oil or natural gas. This route also represents a potential strategy to store electrical energy derived from intermittent sources of clean energy. Although metals in the form of polycrystalline foils are increasingly well-characterized as electrocatalysts for carbon dioxide reduction, their nanoscale counterparts remain poorly understood. Highly dispersed nanoparticle catalysts are ideal for practical electrolyzers due to their high surface area and because they can be integrated into gas diffusion layers in membrane electrode assemblies. The unique ensembles of surface atoms on gold and copper nanoparticle catalysts dramatically alter their activity and stability during carbon dioxide reduction compared to polycrystalline foils.

Our experiments reveal that the primary route by which gold nanoparticles lose surface area during electrochemical carbon dioxide reduction is by forming dendrites on the electrode surface. Random walk simulations coupled with fractal theory establish that the mechanism by which the dendrites form involves diffusion of entire nanoparticles along the support, followed by their collision and fusion to form dendritic structures. Although the mobility of nanoparticles along a support is typically thought to be limited at room temperature, our results demonstrate that nanoparticle catalysts readily move along the support during electrochemical carbon dioxide reduction. We show that an in-depth understanding of the polarization-dependent surface chemistry of gold nanoparticles can be used to limit the dendritic assembly process, thereby increasing surface area for catalysis. We also extend our understanding of surface area loss in gold particles to explain the instability of copper nanoparticles during catalysis.

Both gold and copper nanoparticles exhibit unique activities and selectivities for electrochemical carbon dioxide reduction compared to the polycrystalline foils. We find that there is a continuum in the selectivity for hydrocarbon products between copper nanoparticle and foil electrodes. Detailed electrochemical analysis reveals that nanoscale copper metal particles reduce carbon dioxide to hydrocarbons through a mechanism involving intermediates that are distinct from those observed on polycrystalline foils. Our insights into the structural and chemical factors that influence the activity and stability of copper and gold nanoparticles have enabled us to design advanced catalytic networks for electrochemical carbon dioxide reduction.

379868 Multiscale Modeling to Make Cellulosic Biofuels More Abundant and Affordable

Abstract:

As recognized by the 2013 Nobel Prize in Chemistry, "computer models mirroring real life have become crucial for most advances made in chemistry today."[1]The focus of my research is the application of theoretical and computational methods for uncovering reaction mechanisms that convert non-food biomass into renewable fuels and chemicals. Specifically, my PhD work examines two approaches to cellulose deconstruction: thermochemical and enzymatic processes. In the thermochemical portion of my work, I use quantum mechanics to investigate the elementary steps in cellulose depolymerization under fast pyrolysis conditions. For example, competing mechanisms, specifically homolytic vs. heterolytic, had been debated for many years. Experiments designed to test each hypothesis have been inconclusive. Using computational chemistry, specifically quantum mechanics, I modeled both pathways and discovered a more likely mechanism that better matches for experimental results. My latest work examines the effects of inorganic ions on reaction kinetics, to help understand how these naturally-occurring biomass components change product yield. Additionally, my work has revealed mechanisms that provide targets for catalyst design. The results from this part of my dissertation enable more accurate, mechanistic modeling of biomass pyrolysis that can be used to improve cellolosic biofuel process design and efficiency.

In contrast to biomass fast pyrolysis, which relies on high temperatures to break cellulose's strong bonds, enzymes manipulate cellulose to allow depolymerization at ambient temperatures. Many of the details of how enzymes perform these feats remain unclear. Structural biology has revealed the three-dimensional structures of many enzymes, with and without a substrate present. Molecular simulations allow us to investigate an enzyme's action in the fourth dimension, tracking catalysis over time. These efforts are computationally challenging due both to the system size and the high level of accuracy needed to accurately capture what can be profound stereoelectronic effects resulting from small perturbations in atomic and electronic structure.

Sustainable energy will remain the focus of my future research, and I will continue applying and expanding my expertise in multiscale modeling to make renewable energy more abundant and affordable.

---

[1] "The Nobel Prize in Chemistry 2013 - Press Release". *Nobelprize.org*. Nobel Media AB 2013. Web. 10 May 2014.

385713 Molecular and Mesoscopic Design and Understanding of Energy Materials

Abstract:

The need for novel materials for energy storage and conversion has become increasingly critical as society grapples with limited fossil fuel reserves and the threat of climate change. The challenges and opportunities associated with the discovery and optimization of new energy materials are rooted in a fundamental understanding, and the subsequent control, of how material properties are related over multiple length and time scales. My future research group will seek to design, synthesize, and characterize novel materials for energy applications, with an emphasis on measuring, understanding, and controlling the crucial molecular- and meso-scale phenomena that govern their macroscopic functions. A core aspect of our research strategy will consist of detailed and quantitative characterization of relevant material properties up from the atomic length scale, pushing the limits of current methods when necessary. Among other techniques, novel solid-state nuclear magnetic resonance (NMR) methods will be a cornerstone of such investigations, which enable atom-selective measurements of molecular-level compositions, structures, dynamics, and interactions, as well as transport processes. Multi-scale understanding, combined with a synthesis strategy emphasizing simultaneous control of chemical compositions, material structures, and interfaces, will enable us to more efficiently explore new energy materials composed of low-cost, earth-abundant elements. Initial efforts will focus on materials and devices for electrochemical energy storage, with intentions to also expand to systems for energy conversion. Thematically, we expect to contribute to a broad range of interesting and technologically important problems in chemical engineering, materials science, physical chemistry, and energy.

Within this context, I will highlight my research experiences as a Ph.D. student at the University of California, Santa Barbara, and Marie Curie postdoctoral researcher at the CNRS, France, as well as discuss my future faculty research plans. During my Ph.D., with Profs. Brad Chmelka and Todd Squires, my research broadly focused on elucidating the molecular-level origins of meso- and macro-scale structures and properties in self-assembled and/or hierarchically structured solids, such as mesostructured zeolites, organosilica nanocomposites, and elastomeric organosiloxane foams. In parallel, I modeled electrokinetic flows over surfaces with meso-scale roughness and mass transport to surface-based microfluidic sensors, establishing the dominant physics that influences the transport of ions, analyte, and fluid in distinct asymptotic regimes. At the CNRS, and with the French Network for Electrochemical Energy Storage, my research has focused on measuring disorder and transport properties in novel crystalline electrode and polymer electrolyte materials for lithium-ion batteries, including the development and application of novel solid-state NMR experiments. Collectively, these experiences provide a unique foundation for my proposed faculty research program encompassing materials for energy.

Nanoscale Chirality and Sustainability Laboratory

Kevin M. McPeak

Optical Materials Engineering Laboratory, Department of Mechanical and Process Engineering, ETH Zurich, 8092 Zurich, Switzerland,

The interactions between nanoscale chiral (a shape is said to be chiral if its mirror image is not superimposable on itself) inorganic material, chiral biomolecules and light is leading to a host of new devices with applications in chiral sensing, optical tweezers, optofluidics, nanomechanics, chiral separations and asymmetric catalysis. During my postdoctoral work at ETH Zürich I developed a new method to fabricate chiral nanoparticles and surfaces for visible and ultraviolet plasmonics. The chiral colloids display record molar circular dichroism in the visible and when fabricated from aluminum give a strong chiral optical response in the ultraviolet. This response is resonant with many chiral biomolecules opening up applications for enhanced structural chirality detection in biology, chiral sensing and asymmetric photocatalysis.

The late Nobel Laureate, Richard Smalley, declared "Energy is the single most important challenge facing humanity today." Solar energy is the only clean, renewable energy source available that is capable of supplying the 13 terawatts that are necessary to power the planet. At ETH Zürich I developed a low temperature, all solution-processed periodic nanostructured quantum dot solar cell. Furthermore, during my PhD I studied the chemical bath deposition of nanomaterials for low cost photovoltaic devices. I developed a combinatorial approach for depositing oxide and chalcogenide nanomaterials which enabled rapid and unprecedented understanding of the effect of growth conditions on optical, electronic, and structural properties of the material.

My laboratory will have two primary research areas: 1) fabricating novel inorganic chiral nanomaterials and 2) developing low cost, materials abundant photovoltaic and photocatalytic devices. We will study the interactions between light, inorganic chiral nanomaterials and chiral biomolecules. This fundamental work will allow us to develop new chiral sensors, enantiospecific separation techniques and new routes to asymmetric catalysis which would be beneficial to human health. Lastly, the development of low cost, materials abundant photovoltaic and photocatalytic devices is paramount to energy sustainability and human health in the world.

Selected publications:

Complex Chiral Colloids and Surfaces via High-Index Off-Cut Silicon
McPeak *et al.*, Nano letters ASAP
DOI: 10.1021/nl501032j

Microreactor Chemical Bath Deposition of Laterally Graded $Cd_{1-x}Zn_xS$ Thin Films: A Route to High-Throughput Optimization for Photovoltaic Buffer Layers
McPeak *et al.*, Chemistry of Materials 25 (3), 297-306
DOI: 10.1021/cm3023506

In situ x-ray absorption near-edge structure spectroscopy of ZnO nanowire growth during chemical bath deposition
McPeak *et al.*, Chemistry of Materials 22 (22), 6162-6170
DOI: 10.1021/cm102155m

Research Statement: My long-term vision is to contribute to the development of innovative, impactful biomaterials for fundamental and translational tissue engineering applications and to contribute to the development of the next generation of biomaterial scientists. In particular, as an independent researcher, I will lead a research team in the development and evaluation of cutting-edge biomaterials in the context of novel disease modifying agents as well as for organ replacement using Adult Stem Cells (ASCs). My doctoral and post-doctoral studies have given me strong training in polymeric biomaterial development and characterization, molecular biology techniques, and the culture of primary and stem cells (adult and embryonic). In addition, as a young researcher I understand the relevance of interdisciplinary collaboration, which in my case has resulted in 20 peer-reviewed manuscripts in the area of tissue engineering, biomaterials and regenerative medicine. This expertise will serve as a basis for me to build my own research program. My research will initially be focused on the design of mechanically tunable substrates of varying geometries, chemical and biochemical composition capable of supporting prolonged *in vitro* expansion of ASCs while maintaining their multipotency and therapeutic capabilities. This initial research work will contribute to our understanding of the cellular mechanisms that maintain "stemness" while directly contributing to the improved design of *in vitro* culture environments that enhance the clinical utility of ASCs.

Education:

- PhD, Chemical Engineering; Texas A&M University, College Station TX, May 2011. Advisor: Dr. Mariah Hahn, PhD
- BS, Chemical Engineering; Universidad Industrial de Santander, Bucaramanga Colombia, May 2005.

Research/Work Experience:

- Postdoctoral Research Associate, 6/11-present
  Department of Biomedical Engineering, Rensselaer Polytechnic Institute, Advisor: Dr. Mariah Hahn
- Graduate Research Assistant, 3/06-05/11
  Department of Chemical Engineering, Texas A&M University, College Station, TX
- Lecturer, 10/04-01/06
  Instituto Marco, Bucaramanga, Colombia
- Undergraduate Research Assistant, 01/04-04/05
  Centro de Investigación en Catálisis (CICAT), Bucaramanga, Colombia

Selected Publications (from a total of 20 peer-reviewed publications):

- A.C Jimenez-Vergara, **D.J. Munoz-Pinto**, M.S Hahn. Influence of pressurized cyclic stretch and endothelial cell presence on multipotent stem cell osteogenic commitment. Integr Biol. 5 (7): 983-92, 2013.
- **D. Munoz-Pinto**, X. Qu, L. Bansal, HN. Hayenga, J. Hahn J, MS. Hahn. Relative impact of form-induced stress vs. uniaxial alignment on multipotent stem cell myogenesis. Acta Biomateralia. 8 (11): 3974-81, 2012.

- **D. Munoz-Pinto**, A.C. Jimenez-Vergara, Y. Hou, H.N. Hayenga, A. Rivas, M. Grunlan, M.S. Hahn. Osteogenic potential of poly(ethylene glycol)-poly(dimethylsiloxane) hybrid hydrogels. Tissue Eng A, 18(15-16):1710-9, 2012.
- **D. Munoz-Pinto**, B. Grigoryan, J. Long, M. Grunlan, M.S. Hahn. An Approach for Assessing Hydrogel Hydrophobicity. J Biomed Mater Res A. 100(10):2855-60, 2012.
- **D. Munoz-Pinto**, R. McMahon, M. Kanzelberger, A.C. Jimenez-Vergara, Y. Hou, M. Grunlan, M.S. Hahn. Inorganic-organic hybrid scaffolds for osteochondral regeneration. J of Biomed Mater Res A. 94 (1): 112-21, 2010.

380676 Understanding Diffusion of Small Molecules in Polymer Glasses

Abstract:

The Free Volume Model (FVM) proposed by Cohen and Turnbull over 50 years ago had provided a simple picture that relates the structural properties of amorphous glasses to the diffusing behaviors of small molecules in them. Albeit the semi-phenomenological nature of the model, FVM has met with great success in matching experimental results and providing guidance in designing new materials for membrane separations. However, accumulating experimental data on diffusion of gases through polymer membranes have increasingly suggested inconsistencies of the model. In attempts to correct these deviations within the framework of FVM, effects of activation energy, cavity connectivity and accessible free volume had been introduced. These modifications are largely phenomenological due to the lack of a systematic understanding of the diffusion process in glassy polymers. In this study, using molecular dynamics simulations we aim to provide a coherent picture of small molecules diffusing in polymer glasses. Our results suggest that there are three distinct diffusion regimes depending on the relative size of the diffusant and "cavities". The dominant mechanism that controls diffusion in each regime is identified respectively. Based on the identified mechanisms, deviations from the FVM can now be systematically understood and predicted. Our results provide a basis on which further development of more comprehensive diffusion models will be constructed, a work that is undergoing.

384120 Understanding Artificial Photosynthesis Systems from the Nanoscale to the Device Level

Abstract:

The world's development and the implementation of a sustainable lifestyle depend on significant advances in the alternative energy sector. These advances will require basic scientific developments in energy generation, capture and storage, as well as large scale economic policy drivers to motivate the necessary changes to our current energy infrastructure. I have actively participated in the search for solutions to energy problems with deep fundamental and technological implications in the field of artificial photosynthesis. Artificial photosynthesis devices are complex energy conversion systems that need to absorb sunlight, create and transport charges, and electrochemically generate energy rich molecules such as hydrogen that can be used as a fuel. Given the complexity of the process, integrated systems need to be designed to efficiently carry-out all of these processes in parallel. My research has tackled all of these design aspects ranging from the materials nanostructure level to the overall device design structure, with particular emphasis on the role of hybrid polymeric materials for solar-fuel generation. Here, I will highlight research insights from the basic understanding of functional nanoparticle self-assembly in polymer membranes for solar-energy harvesting, to the structural and transport characteristics of ion-conducting polymers at inorganic interphases for electrochemical water splitting. Lastly, broader aspects on the implementation and device design of cost-effective solar-fuels technologies will be discussed.

Building on the systematic approaches applied on solar-fuel devices, my research program will focus on understanding multifunctional polymer composite systems for energy applications. Specifically, we will first design materials using basic physical models to understand the drivers for performance and potential for a particular application, as well as devise the required morphologies to achieve the desired properties. Then, we will devote the bulk of our research efforts to understanding the self-assembly aspects of model composite systems that lead to the required morphologies by using a wide range of polymer processing techniques and structural characterization tools. Lastly, understanding the effects of composites' nanostructures on their properties (e.g. electrical, electrochemical and physical) will allow us to achieve optimized hybrid materials that can be implemented into devices and will also point out towards directions for improvements in their design. Examples of functional energy systems based on hybrid nanostructured materials include electrochemical catalyst layers, supercapacitor and battery electrodes where the structure and characteristics of the electrochemical interfaces are critical for the device performance.

The Design of Catalytic Materials at the Molecular Level for Sustainable Product Synthesis

Eric G. Moschetta

School of Chemical & Biomolecular Engineering

Georgia Institute of Technology, 311 Ferst Drive NW, Atlanta, GA 30332

My research interests involve the design and implementation of homogeneous and heterogeneous molecular catalysts for production of synthetically useful materials. As a PhD student at The Pennsylvania State University under the guidance of Prof. Robert M. Rioux, I explored the thermodynamics of metal-ligand associations in solution using microcalorimetry for a series of ligands and Pd metal centers. My work showed that ligand binding was enthalpy-driven, that the chelate effect for binding bidentate ligands was enthalpy-driven and not entropy-driven, and that solvent reorganization contributed significantly to the stability of chelated complexes via enthalpy-entropy compensation analysis. I also investigated hydrosulfonation and hydrothiolation reactions using homogeneous and heterogeneous Rh catalysts and examined the structure-function relationships between the active species and their activity and selectivity. This study was the first to use Rh as a hydrosulfonation catalyst and showed that lowering the concentration of the exogenous phosphine ligand that formed the Rh catalyst *in situ* changed the chemo- and regioselectivity in favor of the desired Markovnikov vinyl sulfonate esters. My PhD work allowed me to develop my problem solving skills from the perspective of a chemist as well as the perspective of an engineer.

I am currently a postdoctoral fellow at the Georgia Institute of Technology under the guidance of Prof. Christopher W. Jones where I am expanding my repertoire of catalysis knowledge while delving into new areas such as carbon capture. Whereas my PhD experience imparted a deep knowledge of bonding and catalysis with molecular catalysts, my post-doctoral work emphasizes material design and synthesis. I am probing the impact of Lewis and Bronsted acidic heteroatoms incorporated into amine-functionalized silica materials that act as bifunctional, or cooperative, acid-base catalysts. I am also studying the effects of active sites accessibility and active site dynamics, for example by tuning the alkyl linker length of amine-functionalized silica catalysts. Additionally, I am developing a new procedure for synthesizing amine-functionalized silica adsorbents for applications in catalysis and CO_2 capture. Finally, I am also part of a collaboration through the Center for Selective C-H Functionalization (CCHF, an NSF Center for Chemical Innovation) where I explore enantioselective C-H activation catalysis in flow using a unique hollow fiber reactor that allows for high product yield and enantioselectivity. This collaboration involves organic and organometallic chemists as well as chemical engineers and has been an excellent opportunity for me to expand my areas of research interest.

I plan to apply my synthetic, analytic, and material characterization skills to my future research interests, which include green chemistry, developing sustainable chemical syntheses, and catalyst design. I will emphasize the design of reusable, inexpensive, and environmentally benign catalytic materials for sustainable chemical transformations. My research will focus on the atom efficiency of these chemical transformations and how the structure of each catalyst dictates its activity and selectivity. My goal is to combine the philosophy of a chemist, in terms of materials synthesis and catalyst design, with the philosophy of a chemical engineer, in terms widespread application and sustainability. Ultimately, I will distinguish myself in the catalytic world by considering more complicated organic reactions for heterogeneously-catalyzed systems and applying my current knowledge and skills, such as silica synthesis/functionalization and calorimetry, to other fields such as drug delivery where my background in catalysis will provide a unique perspective in problem solving.

Publications

1. E. G. Moschetta, K. M. Gans, R. M. Rioux, "Elucidating the Roles of Enthalpy, Entropy, and Donor Atom in the Chelate Effect for Binding Different Bidentate Ligands on the Same Metal Center" *J. Catal.* **2014**, *309*, 11-20.

2. E. G. Moschetta, K. M. Gans, R. M. Rioux, "Characterization of Sites of Different Thermodynamic Affinities on the Same Metal Center via Isothermal Titration Calorimetry" *J. Catal.* **2013**, *302*, 1-9.

3. Y. Yang, E. G. Moschetta, R. M. Rioux, "Addition of Sulfonic Acids to Terminal Alkynes Catalyzed by a Rhodium Complex: Ligand Concentration-Controlled Reaction Selectivity" *ChemCatChem* **2013**, *5*, 3005-3013.

4. Y. S. Choi, E. G. Moschetta, J. T. Miller, M. Fasulo, M. J. McMurdo, R. M. Rioux, T. D. Tilley, "Highly Dispersed Pd-SBA15 Materials from Tris(*tert*-butoxy)siloxy Complexes of Pd(II)" *ACS Catal.* **2011**, 1166-1177.

5. E. G. Moschetta, N. A. Brunelli, C. W. Jones, "Reaction-Dependent Heteroatom Modification of Acid-Base Catalytic Cooperativity in Aminosilica Composite Materials" **2014**, *In Press*.

6. E. G. Moschetta, N. A. Brunelli, V. E. Collier, G. I. Lindy, C. W. Jones, "The Effects of Alkyl Linker Length and Pore Size on the Cooperative Catalytic Activity of Amine-Functionalized Silica Materials" **2014**, *In Preparation*.

7. E. G. Moschetta, A. P. Holewinski, C. W. Jones, "Amination of Alkylhalide-Functionalized Silica for Applications in Catalysis, Adsorption and ^{15}N NMR Spectroscopy" **2014**, *In Preparation.*

8. E. G. Moschetta, K. M. Chepiga, N. A. Brunelli, Y. Feng, C. W. Jones, H. M. L. Davies, "Immobilization of Dirhodium Catalysts in Hollow Fibers for Applications of C-H Functionalization in a Flow Reactor" **2014**, *In Preparation.*

387032 In-Situ Polymerization of Functional Materials through CVD Pathway for Energy and Clean Water Resources

Abstract:

Polymers are versatile materials with diverse application in many different disciplines. However the current emerging technologies that rely on nanoscale materials fabrication pose a new challenge in polymer integration into devices. Solar cells, batteries, and capacitors with their mesoscopic dimension are good examples. The solvent intractability makes the integration of polymers into various devices with nanostructured electrodes very cumbersome. Here we rely on Initiated and oxidative chemical vapor deposition (iCVD/oCVD) as unique, solvent free method that enable polymerization and coating in one step to address the current challenges facing device fabrication at nanoscale metric regime.

Here we demonstrate how *in-situ* polymerization using chemical vapor deposition (CVD) can be utilized to integrate polymers within photoanode of a dye senstitzed solar cell. The importance of polymer chemistry investigated by integrating different polymer electrolytes such as poly (2 hydroxyethyl methacrylate) (PHEMA), poly (glycidyl methacrylate)(PGMA), and poly(4-vinylpyridine) (P4VP) into cell architecture. For all the three polymers deposited, we found that fill quality which is defined as the ratio of the volume occupied by polymer over the pore volume realizes an optimum as a function of monomer surface availability. The observed phenomena attributed to the difference in sticking coefficient and surface diffusivity of the monomers. The importance of the interfacial energetic studied using different side chain pendants group and the result compared with the simulation result of our first principal modeling of the cell as a pesudohomogenous medium. We further applied our unique oCVD process to enable step polymerization of conjugated polymer with exceptional and tunable properties. Highly non-soluble unsubstituted polythiophene was deposited with measured conductivity as high as 70 S/cm, which is an indication of low degree of defect. The charge storage capacity of the polymer deposited within nanostructed electrode was evaluated and proved to be enhanced due to the nanostructue effect and nanoconfinement. We used our finding and integrate polythiophene within conventional super-capacitor electrode. A 250% increase in volumetric capacitance was realized when the polymer loading to matrix mass was optimized. We further synthesize highly conductive Poly(3,4-ethylenedioxythiophene) (PEDOT) with near metallic conductivity. The importance of processing parameter on conductivity and the film composition was studied and the reason behind high conductivity of the as deposited film attributed to the overoxidation state of polymer that is stabilized with the oxidant complexes imbedded within the film. The highly conductive film was used as a hole transport media in a all solid state solar cell. A titania mesoporous electrode was deposited on a 50 nm compact layer of titanium dioxide which served as the electron transport layer and thin coating of dedoped oCVD polythiophene was used as absorber layer. The PEDOT with high conductivity was deposited on this assembly and the cell was fabricated using a gold contact on the top of the PEDOT layer. The measured photocurrent efficiency of 0.8% without any optimization was realized. To further benefit from novel iCVD process the potential application of the iCVD in fabricating microporous membrane for membrane distillation (MD) was examined. The exceptional coating ability of iCVD to deposit thin polymer layer on a hydrophilic microporous support and converting the surface to superhydrophic state proved the practicality of this approach for MD membrane fabrication application.

364305 Overcoming Barriers in Structural Biology Though Novel Method Development

Abstract:

The 3-dimensional structures of proteins and other biomolecules are critical to understanding the mechanisms of biological processes. By far the most successful method in structural biology for structure determination is X-ray crystallography, a fact highlighted by the over 88,000 protein and nucleic acid structures deposited in the Protein Data Bank. Despite its remarkable successes, X-ray crystallography suffers from two major bottlenecks: the relatively large quantity of biological material required to screen crystallization conditions, and the need to form large well-ordered crystals. This poster is a presentation of recent research on overcoming these two main roadblocks in order to facilitate the structural study of more difficult biological targets.

The first part of the poster focuses on the strategies used to address the large quantities of protein necessary for structural studies. The project specifically concentrated on improving the overexpression of membrane proteins in *Esherichia coli*. Membrane proteins are extremely important therapeutic targets and hold potential for practical applications in bionanotechnology. However, low yields in normally robust expression hosts such as *E. coli* have hindered progress in our understanding of their structure and function. By engineering the protein folding pathways in *E. coli*, we showed the expression levels of several membrane proteins could be improved 3 to 7 fold. Additionally, expression vectors were improved by directed evolution, resulting in reduced expression-induced toxicity and enhanced functional membrane protein yields in *E. coli*

The second part of the poster details the progress made on overcoming the other major roadblock of protein crystallography: the formation of large well-ordered crystals. For difficult proteins (e.g., membrane proteins and protein complexes), initial crystallization screens will often times only produce small microcrystals which, despite years of attempted optimization, will never grow to suitable sizes required by traditional X-ray crystallography. We demonstrated for the first time that it is possible to use electron crystallography to solve a protein structure from exceedingly small 3-dimensional microcrystals, orders of magnitude smaller than what is required for X-ray crystallography. The development of this new method, which we named 'MicroED', promises to have a large impact on structural biology by allowing structural data to be obtained from protein crystals previously thought to be unusable.

384927 Effect of Physical and Chemical Properties of Nanoparticles on Small Molecule Release from Liposomes

Abstract:

Drug nanocariers such as liposomes are often limited by slow and non-specific release. A novel strategy is to initiate drug release by an external agent such as near infra-red (NIR) pulsed laser light absorbed by encapsulated or tethered nanoparticles (NP). The absorbance of NIR light is desirable because it causes minimal thermal injury to normal tissues and light penetration of several centimeters. We can synthesize Hollow Gold Nanoparticles (HGN) and Sulfide Cupper Nanoparticles (CSN) to absorb NIR light over the range from 700-900 nm. NIR picosecond pulsed laser light absorbed by NP is rapidly converted into thermal energy; the photothermal heating of the NP results in large temperature gradients that leads to the formation of transient vapor nanobubbles in aqueous solution. The collapse of the nanobubbles induces the rupture of the liposome membrane and the subsequent drug release with minimal damage to the contents and/or surroundings. We find HGN show higher specific absorbance of the NIR light than CSN which leads to more efficiently delivery of liposome contents. We have systematically synthesized various sizes of monodsiperse HGN and CSN, and determined the relative efficacy of laser power and duration on the rate of contents release from 200 nm liposomes. A fluorescent dye was encapsulated within a liposome as a model agent to study dye release kinetics triggered by a picosecond pulsed NIR laser irradiation. Subsequently, an anticancer drug was encapsulated within the liposomes to deliver the drug to prostate cancer cell in vitro; near-complete cell killing was observed. The study leads to a better understanding of the capabilities of multi-functional nanomaterials for NIR laser-controlled drug release in order to facilitate the development of optimized more complex therapeutic systems.

390007 Design of Novel Hydrogen and Compressed Natural Gas Fueling Stations

Abstract:

During my PhD at UCLA under Prof. Manousiouthakis, the field of research I have been working on is hydrogen economy, specifically about the distribution of hydrogen fuel to the end user. One of the key challenges for the deployment of hydrogen fuel cell vehicles is the lack of fill-up practices that are time efficient and safe. Moreover, hydrogen fueling station operators do not employ an optimal mass flowrate policy that can satisfy the aforementioned constraints. A hydrogen vehicle user will like that the replenishment of fuel to his/her vehicle take the same time as a gasoline car. On the other hand, the type IV tanks, based on carbon fiber wrapping, have a maximum temperature limitation of 85 degree Celsius, so the fill-up process must ensure that this temperature is not surpassed. The research I have conducted has led to the development novel mathematical models and software that can simulate and optimize the fill-up process of gas fuel such as hydrogen, and even compressed natural gas (CNG). The models are based on a self-consistent thermodynamic model for real gases, transport phenomena principles, and conservations laws, which predict hydrogen's temperature and pressure time evolutions inside the vehicle tank, the station storage tank, the piping connecting the tanks, and the dispenser isenthalpic valve. The model results have proven to match, within 2%, corresponding experimental data from an actual fill-up process. Furthermore, in order to minimize the fill-up time and to determine an optimum mass flowrate policy, I formulated the fill-up process as a minimum time optimal control problem, which through the used of a novel decomposition of the problem into a process simulation problem, that checks for feasibility of the fill-up independent of time, and a simple minimum time optimal control problem, the global minimum fill-up time can be obtained analytically along with an optimal control policy on the mass flowrate. The optimization studies results lead to a fill-up time strategy that can improve current fill-up practices.

In addition to my core research, I have also performed as the safety officer for Prof. Manousiouthakis experimental lab, having to go to intensive safety training and yearly inspections. Additionally, I have participated in the construction of the UCLA hydrogen fuel cell go-kart, and I have served as hydrogen system lead, safety inspector, and project leader. I am a very active member of UCLA Center for Excellence in Engineering and Diversity efforts to motivate students coming from minority populations to attend the UCLA engineering school. I have been participating in the Mathematics Engineering Science Achievement (MESA) program, laboratory tours, panels, talks, tutoring, and volunteering and organizing events since 2006. My research interests are alternative energy sources for vehicles, production of "green" hydrogen, design of hydrogen fueling stations, hydrogen safety, and hydrogen fuel cell vehicles integration.

384783 Understanding the Influence of the Host Environment on Phenotypic Heterogeneity of Bacteria for Development of Effective Therapeutics

Abstract:

Phenotypic heterogeneity in bacterial populations is a common mechanism by which bacteria survive environmental stress. For example, bacterial persisters have the ability to tolerate extraordinary levels of antibiotics due to their phenotypic state. Unlike the resistant bacteria that arise from heritable mutations, persisters are genetically identical to their antibiotic non-tolerant progenies. Persisters are a significant public health problem because they are enriched in biofilms and have been hypothesized to underlie the proclivity of biofilm infections to relapse. Though the medical relevance of persisters is associated with infection sites in the human body, persistence has largely been studied *in vitro* using synthetic media. Pathogens associated with persistent bacterial infections such as tuberculosis, gastrisis, cystic fibrosis and urinary tract infections are developed in host environments including liver, lung and epithelial cells, which are far more complex environments than that attainable *in vitro*. Studies regarding the interaction between persisters and host environment are widely missing, and therefore, it is still unknown how host environments influence bacterial physiology with regard to persistence. Based on the expertise I have gained during my postdoctoral and graduate training, including both computational and experimental experiences for both mammalian and bacterial systems, I am planning to study the effects of host conditions on persister levels and physiology, and develop experimental models including mammalian systems to analyze the effects of host-pathogen interactions on persistence. After establishing experimental models, persister physiology will be analyzed by genetic, biochemical, and flow cytometric assays to elucidate molecular mechanisms responsible for persistent state, which can form the basis of novel therapeutics for the treatment of chronic infections.

384324 Expanding the Chemistry of Life for the Development of Novel Therapeutics and Synthetic Organisms

Abstract:

Synthetic biology has a fundamental scientific focus on developing biological parts to understand and manipulate biological systems. My multidisciplinary research program will focus on utilizing expansions in the chemistry of genetically encoded systems at both the protein and nucleic acids level for the development of engineered biological systems. Aim1: At the protein level, my work will focus on the co-translational incorporation of L-phosphoserine (Sep) into proteins. Our biochemical understanding of protein phosphorylation has been fundamentally limited by our technological inability to manufacture useful quantities of homogenous, site-specific phosphoserylated human proteins. The platform I have developed during my postdoctoral research now allows robust production of active human phosphoproteins. I will use this capacity to address previously intractable questions in cellular signaling associated with epigenetic regulation and disease progression. This work has the potential to add breadth and depth to our understanding as well as provide novel therapeutic windows. Aim 2: While engineered biological components display structural and functional orthogonally to their natural counterparts, they lack chemical orthogonality. To overcome this challenge, I will focus on expanding nucleic acid biochemistry by integrating 2 additional letters to the genetic code. My work will utilize synthetic nucleobases imidazoaminopyrimidone (letter: P) and nitroaminopyridone (letter: Z) in conjunction with G, A, T, and C to develop synthetic DNA and RNA 'parts' for biological engineering applications. P and Z form Watson-Crick pairing like G-C and A-T, do not significantly alter DNA stereochemistry, and are compatible with fundamental biochemical processes such as DNA replication and transcription. The methods I developed during my postdoctoral research allow for assembling P/Z containing synthetic genes. I will use this chemical expansion of the genetic code to develop modular synthetic gene components such as promoters, terminators, ribosome binding sites, riboswitches, as well as chemically-orthogonal translation systems for non-standard amino acid incorporation. In the long run, such biological 'parts' displaying chemical orthogonality provided by the P:Z pair could play an integral role in the bio-containment of synthetic organisms. These efforts will be sustained by my training in RNA-protein biochemistry and augmented by my experience in biological engineering. Furthermore, these aims will provide a framework for training Masters' and PhD level students, and will complement my interest in teaching undergraduate students.

385284 Integrating Computational and Experimental Methods to Discover Disease Causes and Design Protein Therapeutics

Abstract:

With the advent of genetic sequencing, there has been a clear trend in medicine to develop increasingly specialized products. Antibody therapeutics effective for only a specific gene mutation in cancers and individualized genome sequences are just two examples. As more data becomes available, there is a clear need for experimental and computational methods that can identify the most significant facts and utilize them to generate effective results. My vision is to have a combined computational and experimental group that identifies the causes of diseases and develops treatments for them. The graduate and postdoctoral research I have completed has created a strong foundation for achieving this goal.

At Penn State University, my graduate research was on the development of computational tools for protein engineering and *de novo* design, with a specific focus on antibodies. The Optimal Complementarity Determining Regions method (OptCDR) was developed to design the antigen-binding portions of antibodies, known as CDRs. Using Mixed Integer Linear Programming optimization, a rotamer library, and molecular mechanics force fields, libraries of any user-specified size of antibody CDRs against any specified antigen epitope can be generated. Other work focused on creating a database of modular antibody parts that can be accurately assembled to model antibody structures and using these parts in the Optimal Method of Antibody Variable region Engineering (OptMAVEn) to *de novo* design fully human antibodies. As computation speeds increase and force fields are refined, novel methods such as OptCDR and OptMAVEn will be essential tools to design protein therapeutics for specific objectives.

The postdoctoral research I have conducted at the University of California, Santa Barbara is experimental work to find environmental factors associated with diseases. Through a combination of magnetic selection and fluorescence activated cell sorting of a large peptide library and next-generation sequencing, it is possible to identify a large database of peptides bound by a patient's antibodies. In turn, computational analysis of the identified amino acid sequences from multiple disease and control patients can find motifs that are disease specific and cross-reactive. These motifs can be used to hypothesize the environmental factors that are contributing to or causing the disease.

As experimental methods improve, increasingly vast databases will become available. It is imperative that effective computational methods are available to analyze and utilize this data. However, computations alone are insufficient and they must be coupled with experimental techniques to evaluate the generated hypotheses. My future plans are to build upon what I have done and continue to develop computational and experimental methods for identifying causes of diseases and developing treatments for them. The computational and experimental research I have done so far is an excellent foundation for developing integrated methods to utilize the large amounts of data that are becoming available.

Candidate: **Natesh Parashurama MD PhD**

Title: Liver Regenerative Medicine and In Vivo Molecular Imaging for the Study of In Vivo Liver Organogenesis, Liver Disease and Development of New Diagnostics and Therapeutics

The emerging field of liver (hepatic) regenerative medicine encompasses understanding hepatic stem cell biology, organogenesis, tissue regeneration, and tissue homeostasis for the study of liver disease and development of new diagnostic and therapeutic strategies. The liver, the largest internal organ in the body, performs critical metabolic, endocrine, and synthetic physiological functions. As such, it displays elaborate cellular phenotypes and intricate tissue architecture, including hexagonal, lobular structure, complex vascular spaces with specialized endothelial cells, tissue specific fibroblasts and macrophages, complex blood supply (hepatic artery and portal vein) and heterogeneous billiary duct cells which line billiary duct drainage systems. With respect to capacity for organ regeneration, the liver is unmatched. For example, preclinical, rodent studies have indicated acute hepatocellular loss results in compensatory replication of neighboring adult hepatocytes. In contrast, chronic hepatocellular injury, which often leads to end-stage disease, is compensated for by small, liver stem/progenitor cells lying at the junction of intrahepatic bile ducts and the Canals of Hering. Furthermore, circulating and immune cells, like bone marrow mononuclear cells, mesenchymal stem cells, and tissue macrophages, all have major roles in orchestrating repair in liver injury. These repair mechanisms have been translated to established treatments, like human liver transplantation, and promising therapies, such as human hepatocyte, human bone marrow-derived, and human liver stem/progenitor cell transplantation. However, current and new therapies require harvesting of human tissue, underlying mechanisms of tissue homeostasis remain elusive and hard to decipher in vitro, in animals and humans, and new approaches and techniques are needed answer critical scientific questions and take steps towards new diagnostics and therapeutics aimed towards regeneration.

A deeper understanding of human, three-dimensional (3D) liver organogenesis would naturally advance liver regenerative medicine. Unfortunately, crucial steps of liver organogenesis take place in utero, and most research has been focused on making functional hepatocytes rather than re-creating liver organogenesis. Nevertheless, this would lead to more accurate studies of acute and chronic regenerative capacity, metabolic, detoxification, and synthetic functions, and disease modeling. A second point is the need for being able to assess 3D liver organogenesis while it is occurring. Current techniques rely on removing tissue, treating the tissue and analyzing it in a static fashion. However, techniques are also needed to assess dynamic molecular events in real time, in the intact living subject, during organogenesis or regeneration. These needs are addressed by techniques in the field of molecular imaging, which aims to study biological events and pathways in the intact living subjects. Thus the combination of studying 3D liver organogenesis, and applying current techniques and developing new techniques in molecular imaging, will likely further the field of hepatic regenerative medicine.

389491 Characterization of Conducting Polymers for Lithium Battery, Transistor, Thermoelectric Applications

Abstract:

My Ph.D. research at UC Berkeley under the advisement of Prof. Nitash Balsara involved the synthesis and characterization of simultaneous electronic and ionic conducting block copolymers (P3HT-*block*-PEO) for lithium battery electrodes. Block copolymers can self-assemble and form co-continuous nanoscale domains, which are necessary for enabling redox reactions of the active material ($LiFePO_4$). In addition, the block copolymer serves as a binder to hold the redox-active material particles together. This simplifies the electrode design as one material serve both binding and charge transport functions. The application of this material in a lithium battery with $LiFePO_4$ showed specific capacity reaching the theoretical limit and with minimal capacity fade after ten cycles, thus demonstrating feasibility of P3HT-*b*-PEO as a conductive binder material. Furthermore, the ability of the conductive binder to switch between electronically conducting and insulating states in the positive electrode provides an unprecedented route for automatic overdischarge protection within the battery. This is in stark contrast to traditional lithium ion batteries where external electronics is used to provide overdischarge protection.

My current research as postdoctoral researcher at UC Santa Barbara (since May 2013) focuses on structure-property relationships of semiconducting polymers for organic field-effect transistors (OFETs) and organic thermoelectrics. One of the research directions under the advisement of Prof. Ed Kramer has been the structural characterization of aligned semiconducting polymers where the measured mobility has been greater than 20 cm^2/Vs. A combination X-ray scattering and electron microscopy techniques provides evidence for exceptional alignment of the polymer chains. This allows for an efficient intrachain charge transport between source/drain contacts, thus resulting in the measured high mobility. Another research direction under the advisement of Prof. Michael Chabinyc has focused on how various processing conditions effect the thermoelectric properties (Seebeck coefficient, electronic conductivity, and thermal conductivity) of highly doped semiconducting polymers.

My Ph.D. and postdoctoral research provides a unique background in electrochemistry, polymer physics, and solid-state physics which have been applied to lithium batteries, organic transistors, and organic thermoelectrics. My proposed research will build upon my diverse Ph.D. and postdoctoral research background. The foundation of my research group will focus on the versatility of conductive polymers for not only energy storage and conversion applications, but topics branching away from previous research applications.

393540 Syngas Mass Transfer Delivers Fermentation Potential

Abstract:

Bacterial conversion of CO and H_2 from synthesis gas opens a potential supply of energy and chemicals from renewable biomass and waste feedstocks. This potential is being tested in commercial scale plants that use vegetative wastes and industrial waste gases for production of ethanol. Commercial success will hinge on efficiency, selectivity, productivity and stability of the process deployed. The microorganisms that perform the syngas fermentation use the Wood-Ljungdahl biochemical pathway to obtain energy as electrons and protons from the oxidation of CO and H_2 on enzymes inside the cell. The transfer of the substrates CO and H_2 from the gas phase into the cell determines the electrochemical potential inside the cell, and thus the reaction kinetics and the thermodynamic ratio of products. As such, the application of gas/liquid mass transfer is the most important parameter for control of the syngas fermentation process. Analysis of syngas fermentation that employs mass transfer, reaction kinetics, thermodynamics and mass balance derived from the pathway provides a basis for successful control and equipment design in syngas fermentation. Control of mass transfer can be used to effect high energy conservation, high selectivity for desired product and high productivity in stable continuous fermentation.

392845 Establishing Governing Equations for 3D Cell Culture in Perfusion Bioreactors

Abstract:

My research interests are focused on developing new biomedical products such as biomaterials and bioreactor systems that can be used for drug discovery and testing, 3D cell culture and large tissue regeneration. Culturing cells and regenerating tissues *in vitro* on 3D scaffolds involves several challenges, such as efficient nutrient transportation, uniform stress distribution, and the removal of wastes. Perfusion bioreactors not only allow reproducibility but also provide a controlled environment for production of tissues. My goal is to establish fundamental governing equations for the design of tissue engineering bioreactors and scaffolds. My doctoral research involved testing governing equations related to nutrient permeability, mechanical and structural properties of the scaffolds, as well as nutrient consumption kinetics. The strategy was to use a combination of computational modeling and experimental analysis. Large scaffolds with a high aspect ratio were utilized so that the obtained experimental measurements have high signal-to-noise ratio. This allowed the validation of the governing equations used in the computational models with high fidelity. Three different scaffold preparation techniques, freeze drying, salt leaching, and electrospinning were used to fabricate scaffolds with different microarchitecture. Chitosan, gelatin, and polycaprolactone polymers were used to prepare scaffolds. Two types of bioreactor configurations, flow-through and axial-flow, were used in this study. Both were designed to hold same sized scaffolds, but differ in flow configuration, which made them suitable for evaluating and validating the equations. Bioreactors of appropriate flow configuration were constructed in-house for experimental analysis. Computational Fluid Dynamics (CFD) simulations were performed to predict pressure drop, shear stress, deformation, nutrient distribution profile and exit concentration at various operating conditions. These insights help monitor in vitro tissue regeneration, understand the effect of mechanical stimulus on 3D cell culture, and improve quality of the regenerated tissue.

My future research plan will involve developing *a priori* charts for designing systems involving porous media can help simplify the design process for bioreactor systems. The design variables such as flow rate, porosity and permeability can be transformed into dimensionless variables and characteristic curves can be developed to relate these dimensionless variables. This process will significantly reduce the amount of effort required to improve existing bioreactors and develop new ones. Further, I plan to establish other governing equations such as nutrient diffusivity in porous media, kinetic parameter, and non-viscoelastic nature of bioscaffolds to help improve our understanding of the tissue regeneration process and better mimic the native tissue growth phenomenon. Additionally, I would apply my skills and experience from developing the OSU AppCenter at Oklahoma State University in finding ways to use mobile apps to improve productivity of research and development activities. The poster will include examples of my research projects that constitute my research ideas. During the session I will give an overview of the potential and challenges associated with my research plan.

Development of catalysts for energy and environmental applications

Marc D. Porosoff and Jingguang G. Chen

Over the next 30 years, the global energy demand is expected to increase by 60% to maintain pace with current projections of population and economic growth.[1] Fossil fuels will continue to provide a significant fraction of the energy supply; however, cleaner technologies, including solar, wind and nuclear will become critical industries to meet the required demand. As the burning of fossil fuels continues into the foreseeable future, it is necessary to develop more efficient processes to decrease energy usage and the associated CO_2 emissions. With the threat of climate change and ocean acidification becoming more apparent each year,[2] efforts must be put forth to not only develop new forms of clean energy, but also capture and utilize CO_2 as a commodity. These large-scale CO_2 conversion processes require the discovery of effective and low-cost catalysts.

Development of catalysts for CO_2 conversion can be divided into three areas: synthesis, characterization, and discovery, with significant research opportunities available in each area. In catalyst synthesis, the most promising methods include: atomic layer deposition, colloidal synthesis, and incipient wetness impregnation. Although innovative synthesis methods, such as core-shell nanoparticles, can be extremely effective, the complexity of synthesis does not always translate to low-cost catalysts for CO_2 conversion. For example, although relatively simple, uniform bimetallic particles can be synthesized via incipient wetness impregnation. Catalyst characterization requires the utilization of a wide range of techniques. A previous study on CO_2 conversion used extended X-ray absorption fine structure (EXAFS), CO chemisorption, and transmission electron microscopy (TEM) to confirm the formation of well-distributed, bimetallic particles synthesized by incipient wetness impregnation.[3]

Catalyst discovery is the most general area of catalyst development, but perhaps the most important. Previous studies demonstrate that reducible oxide supports, such as CeO_2, exhibit high activity in CO_2 conversion experiments because of their ability to exchange oxygen with CO_2.[3] Molybdenum carbide (Mo_2C), a non-precious metal material, also shows promise as an active and selective catalyst for CO_2 conversion.[4] These materials are strikingly different, but have a fundamental similarity, which is they are each able to uptake and exchange oxygen with CO_2. Because of the importance of oxygen exchange in reactions with CO_2, oxygen binding energy could be a valuable descriptor when searching for CO_2 conversion catalysts. The discovery of such descriptors is an extremely powerful tool that can be applied beyond CO_2 conversion and into other applications of environmental catalysis.

References

[1] U. E. I. Administration, (Ed.: DOE), Washington DC, **2013**.

[2] T. R. Knutson, R. E. Tuleya, *J. Clim.* **2004**, *17*, 3477-3495.

[3] M. D. Porosoff, J. G. Chen, *J. Catal.* **2013**, *301*, 30-37.

[4] M. D. Porosoff, X. Yang, J. A. Boscoboinik, J. G. Chen, *Angew. Chem. Int. Edit.* **2014**, *Accepted*.

367996 Molecular Engineering Strategies for Tunable Assemblies and Hybrid Materials

Abstract:

The intelligent use of processes such as self-assembly, combined with the ability to manipulate the chemical structure of polymers, can lead to a wide array of new materials. Such functional materials can be the solution to many 21st century challenges in fields such as chemistry or bioengineering. Examples of polymer-based materials that combine these two elements will be presented. In the first example, complex coacervation (i.e. a liquid-liquid phase separation phenomenon) is used as a platform for soft material design. To understand the different aspects of complex coacervation (e.g. thermodynamics, affecting parameters, rheological and interfacial properties) polypeptides are used as a model system. More complex molecular design can be utilized wherein polyelectrolyte domains are connected to neutral polymer blocks. Use of such co-polymers can result in the formation of a variety of self-assembly structures with coacervate core domains including nanometer-sized micelles or tunable hydrogels. The second example includes a technique for polymer functionalization of carbon nanotubes (CNTs) that helps circumvent their inherent insolubility, and considerably widens the scope of nanocomposite materials that can be produced. The strategy involves attachment of substituted polymerization initiators onto a CNT surface. With a judicious choice of substitution, initiators of most popular polymerization techniques can be attached. Complete control over grafting percentage of initiator and surface-initiated polymerizations allows synthesis of nanocomposite materials with desired compositions, which is essential for any application. The resulting nanocomposite materials exhibit improved mechanical and thermal properties, compared to pure polymers.

390550 Orthogonal Engineering of Block Copolymers: Tools, Techniques and Applications

Abstract:

Application-directed chemical modification of synthetic and natural polymers can generate bio-functional analogues with diverse topology. Synthetic block copolymers with ordered structural hierarchy are powerful fabrication tools for creating self-assembled systems with applied functionality. Robust and sterically stabilized nanostructures, such as conjugates, micelles and vesicles can be engineered from these synthetic block copolymers to design effective and targeted drug delivery vehicles. However, limited availability of functional monomers, synthetic complexity of polymerization, and lack of orthogonality during post-polymerization processes require efficient and modular synthesis of the polymer backbone. Intelligent application of functional-group specific chemistry, multivalent amplification of linear polymers using dendritic building blocks, and hybridization of natural polymers with synthetic macromolecules can open up a wide range of possibilities for the development of bio-inspired materials with desired form and function.

387753 Catalytic Processes: From Molecules to Complex Reaction Networks

Abstract:

Designing "optimal" processes in terms of yields, selectivity, reaction rates, economics, and environmental impact is an open challenge in the field of heterogeneous catalysis. This can involve multi-dimensional decision-making, including the choice of chemistry, catalyst, and the reactors. My research aims to apply state-of-the-art computational tools and developing new ones for detailed modeling and design of heterogeneous catalytic processes. To this end, I propose to build and expand on my PhD experience at University of Minnesota under the co-advisorship of Profs. Daoutidis and Bhan on developing methods and software for elucidating complex reaction networks and my current research co-advised by Prof. Mavrikakis and Maravelias at University of Wisconsin focusing on using DFT and mathematical optimization for a first-principles-based approach to rigorously model and design catalytic systems.

In this poster, I will present a multi-pronged approach involving computational chemistry, microkinetic modeling, numerical optimization, cheminformatics, graph theory, and network analysis that allows for elucidating molecular events such as surface reactions and relating them to reactor-scale phenomena for systems spanning few tens of species and reactions to several thousands. Specifically, I will discuss: (a) identification of molecular structure-property relationships in competitive adsorption of organonitrogen compounds in gas oil hydrotreating using DFT and vdW-DF methods, (b) leveraging nonlinear programming methods for rational catalyst design, design of experiments, and parameter estimation, and (c) development of a new code, RING, for automated rule-based generation, mechanism analysis, and modeling of complex reaction networks.

Transforming the art of catalyst preparation into a science

Ana C. Alba-Rubio

Department of Chemical and Biological Engineering, University of Wisconsin, Madison, Wisconsin 53706, United States.

Multi-metallic nanoparticles constitute a class of materials offering the opportunity to tune their properties via the composition, atomic ordering and size. In particular, supported bimetallic nanoparticles have generated intense interest in catalysis. However, traditional synthesis methods often lack precise control, yielding a mixture of nanoparticles with varying particle size and composition. This heterogeneity complicates studies to link catalytic properties to structural and/or compositional changes in the catalyst. My experience in both catalyst preparation and characterization puts me in an excellent position to forge the **next generation of tools for design of catalysts**.

My research career started as an undergraduate at University of Málaga (Spain) under the supervision of **Prof. Pedro Maireles Torres** and **Prof. Jose Santamaría González**. I worked on catalysis for the production of biodiesel and I gained preliminary experience on synthesis by traditional methods, characterization and evaluation of the activity of base catalysts.

I thoroughly enjoyed my research experience and I moved to Madrid (Spain) to do my PhD in one of the most prestigious groups in catalysis in Spain, that one lead by **Prof. José Luis García Fierro**, under the direction of **Prof. Manuel López Granados** (Institute for Catalysis and Petrochemistry research, CSIC). My PhD was focused on **base and acid catalysis** for biodiesel production, and the characterization of the catalysts was always a strong point. It helped us to identify the active sites of one of the most active base catalysts for biodiesel production. Characterization also provided us with insight about the causes of deactivation of solid acid catalysts during the reaction. Deactivation was avoided by the development of a novel catalyst that joined the advantages of homogeneous and heterogeneous catalysis. During my PhD, I did two short stays in Brazil and The Netherlands, which helped me to have a more international vision of the science.

My love for the knowledge, the catalysis and the concept of biomass conversion brought me to the **University of Wisconsin-Madison** to conduct **postdoctoral research** under the supervision of **Prof. James Dumesic** and **Prof. Manos Mavrikakis**. Here, I am focused on the **controlled synthesis of bimetallic and bifunctional catalysts** for reactions of interest in biomass conversion and for the direct synthesis of H_2O_2 from H_2 and O_2. I am familiar with many techniques of preparation and characterization of catalysts. I have been trained here for the use of several electronic microscopes, and I really enjoy matching structure with activity. In this respect, I am working to design catalyst architectures for each of my projects at Wisconsin, and I would like to continue this type of work in the future.

My future research directions will combine my experience on characterization and preparation to develop active and selective catalysts for reactions of interest. I love to work on biomass conversion, but since many chemical reactions need catalysts, my future research has diverse applications.

Finally, I have a particular interest in teaching, especially in teaching catalysis and experimental skills. In addition, I would feel confident teaching other subjects to chemical engineering students. When I was an undergraduate student, I taught various subjects to younger students. I developed a teaching technique to keep students engaged and involved in the material. I have also supervised people in the lab and the experience has been very rewarding. Active and cooperative learning will be central to whichever courses I teach, and my students will find from me the encouragement that they need to achieve their goals.

393677 Engineering Approaches to Elucidate Mechanisms of Physiologically Relevant Cellular Behaviors in Cancer

Abstract:

Approximately 90% of cancer related deaths worldwide occur because of metastasis of the primary tumor to other organs. The tumor microenvironment plays a key role in metastatic disease progression; however the associated mechanisms are far from being unraveled. The goal of my research program is develop tools to systematically study these processes and dissect relevant mechanisms *in vitro* and *in vivo*. During my doctoral research at the Ohio State University, I developed brain mimetic biomaterial systems (e.g., hydrogels, electrospun nanofibers) that mimic several features of the *in vivo* tumor microenvironment for investigating migration of patient derived human glioblastoma multiforme (GBM) tumor cells. These three dimensional (3D) cell culture systems recapitulate *in vivo* like morphology, migration, and molecular signatures of GBMs as opposed to traditionally studied 2D cell culture systems. Such biomimetic cell culture models will allow for patient tailored bioassays where patient tumor biopsies can be examined and several drug combinations can tested for personalized medicine, thereby guiding patient care and treatment decisions with the potential to transform current cancer treatment. During my postdoctoral research at Northwestern University, I developed microporous poly(ε-caprolactone) (PCL) biomaterial scaffolds for early detection and treatment of breast cancer metastasis. I also discovered and demonstrated the efficacy a novel therapeutic combination of cyclin dependent kinase-2 (CDK2) inhibitors and Eribulin for the treatment of triple negative breast cancer. To investigate the mechanism of drug action, I utilized a live cellular array that provides high-throughput analysis of dynamic transcription factor activities. This poster will highlight the application of these tools *in vitro* and *in vivo,* which will provide a foundation to elucidate mechanisms of tumor progression and subsequently develop early detection and therapeutic strategies.

383806 Composite Conjugated Polymer/Fullerene Dispersions for Organic Photovoltaic Applications

Abstract:

Controlling the structural morphology of conjugated polymer/fullerene composites is an important aspect in improving the performance of polymer solar cells. The efficiency of exciton dissociation and carrier transport to the working electrodes are both a function of the size and distribution of polymer and fullerene domains within the active layer. In the current solution processing paradigm, these characteristics are intrinsically linked to the detailed process history of the film. This fact compromises the promise of polymer solar cells as an inexpensive renewable energy technology because performance gains made in the laboratory may not translate well to scaled-up manufacturing processes. Aqueous dispersions of conjugated polymer/fullerene nanoparticles have the potential to address this challenge. Similar to their thin-film counterparts, the performance of photovoltaic devices derived from conjugated polymer/fullerene composite nanoparticles (CNPs) is a function of the structural morphology within each nanoparticle. However, whereas the structural morphology of an active layer produced using solution phase deposition is a function of the detailed history of the coating process, the structure of CNPs is fixed during their formulation. Because of this fact, the structure can be identified in the dispersed phase and tied, using single particle characterization, to its photovoltaic performance before it is deposited into a device. Therefore, the optimization of devices derived from CNPs can be broken into two discrete problems: improving the **intrinsic** properties of a nanoparticle as a function of how it is produced and optimizing the **extrinsic** properties of devices through the deposition of optimized CNPs. Our work has addressed both of these challenges. First, using P3HT/PCBM nanoparticles, we show that internal phase segregation is a complex function of the way the particles are produced and their composition. We also show that the resulting structure can be identified and optimized for photovoltaic performance. Second, we outline the principles that affect the deposition of these CNPs onto device relevant electrodes, and provide insight into the production of high-quality active layers from CNP dispersions. The combination of these two separate approaches opens up the potential for novel roll-to-roll processes to produce high performance photovoltaic modules of arbitrary complexity with flexible form factors.

385735 Branched Wormlike Micelles Under Dynamic Flow Conditions Using Spatiotemporally-Resolved Small Angle Neutron Scattering

Abstract:

The nonlinear rheology and segmental alignment of a model series of WLMs is studied to determine the effects of branching on shear flow phenomena. Branching in the mixed cationic/anionic surfactant CTAT/SDBS is controlled via the addition of the hydrotropic salt sodium tosylate. The shear-induced segmental alignment of the micelles is measured for local gap positions and time under large amplitude oscillatory shear (LAOS) in the flow-gradient plane using novel flow-SANS instrumentation [1] as well as temporally in conjunction with rheology in the flow-vorticity plane [2]. Our ability to resolve the microstructure via scattering projections in multiple planes of flow provides new insights into the effects of flow on the WLM microstructure under time-dependent shear flow. Local segmental orientation and alignment in the flow-gradient plane is found to be a complex function of the branching level, radial position, and time during the oscillation. Spatiotemporal SANS demonstrates that LAOS can elicit simultaneous elastic and viscous material responses in localized regions across the gap, thereby corroborating recent theoretical predictions with experiment for the first time [3]. Furthermore, the maximum alignment in the flow-gradient plane under LAOS is found to be significantly higher than that observed in steady shear at the same instantaneous shear rate. This nontrivial hyperalignment under dynamic deformation is interpreted in terms of the dynamics of the microstructural rearrangement during LAOS. This research quantitatively links micellar microstructure and topology to the measured nonlinear shear rheology of WLM solutions, providing a more complete data set to enable interpretation of the rheology and rigorous testing of microstructure-based constitutive equations.

[1] A.K. Gurnon, et al., J. Visual Exp. 84 (2014).
[2] C.R. Lopez-Barron, et al., Phys. Rev. Lett. 108 (2012).
[3] J.J. Stickel et al., Journal of Rheology 57 (2013).

375885 Droplet Microfluidics in Physical and Biological Systems

Abstract:

The exploitation of microdroplets produced within microfluidic environments has recently emerged as a new and exciting technological platform for applications within the chemical and biological sciences. Interest in microfluidic systems has been stimulated by a range of fundamental features that accompany system miniaturization. Such features include the ability to process and handle small volumes of fluid, improved analytical performance and low unit cost. Moreover, microfluidic systems that generate and utilize a stream of sub-microlitre droplets dispersed within an immiscible continuous phase have the added advantage of allowing microscopic investigation of physical properties of complex fluids such as emulsions, suspensions and foams. I will describe here three different studies. All of which use the microdroplet platform to establish fundamental understanding of different physical systems as well as to develop devices for biomedical applications. First I will explore the transition from reversible to chaotic behavior in the oscillatory shear flow of water-in-oil emulsions. The emulsion is being flown through a microchannel and is forced to rearrange due to a central constriction. The surprising results show that while at the Stokes flow limit, the emulsion exhibit behaviors that vary from complete reversible to complete irreversible depending on the volume fraction, velocity and viscosity ratio. I will present here the first direct visualization of the phenomenon. This study is an important step in understanding the microscopic rearrangements of droplets and particles near jamming. The second study deals with the break-up of droplets in a concentrated emulsion during its flow as a 2D monolayer in a microchannel consisting of a narrow constriction. The probability of droplet break-up increases with flow rate and the rate of deformation of the drops. The probabilistic nature of the break-up process arises from the stochastic variations in the packing configuration of the drops as they enter the constriction. The results of this study shed new light on the stability of concentrated emulsions and the resulting applications in droplet microfluidics and two-phase flow in porous media. Finally, I will describe the development of a new method for rapid detection of tuberculosis using droplet-based microfluidics. Ultimately, this tool will be able to detect tubercle bacilli in a sensitive, rapid, specific and quantitative manner at low cost. This tool will be particularly important in resource-limited settings where TB is the most prevalent.

389654 Advanced Materials for Catalysts, Membranes and Sorbents Applications and Renewable Energy Technologies

Abstract:

My research focuses on the interface of chemistry, chemical engineering and materials science. Over the past six years, I have excelled in my research efforts related to the synthesis, characterization and application of organic/inorganic materials for catalysts, membranes and sorbents applications and renewable energy technologies. My PhD research experience at University Putra Malaysia involved development of a cost-effective method for synthesizing novel mixed-metal oxide catalysts being used in light alkane dehydrogenation and oxidation applications, in particular in n-butane to maleic anhydride process, whereby I adapted recent developments in reaction science to advance the development of novel catalysts. My PhD work resulted in 3 patents and 7 papers that have been published in high quality catalysis and chemical engineering journals.

During my previous research at Lulea University of Technology (LTU) in Sweden, I was involved in catalytic production of methanol, dimethyl ether and biofuel from black liquor synthesis gas. The work focused on catalyst synthesis (mainly mixed metal oxides and zeolites), characterization and evaluation into two different catalytic process; (1) conversion of black liquor synthesis gas to methanol (high-pressure hydrogenation reactor), and (2) methanol conversion to dimethyl ether and gasoline-range hydrocarbons (high-temperature dehydration reactor). As part of a multidisciplinary team of researchers and graduate students from LTU university and industry partners, for the first time in the world, we produced bio-methanol from black liquor synthesis gas.

Currently, I am a post-doctoral fellow at Georgia Tech, with Prof. Bill Koros working on membrane gas separations by tailoring membrane characteristics through investigation of new materials and structures to selectively control mass transport at the molecular level. In particular, the current research is focused on producing novel mixed-matrix hollow-fiber membranes and sorbents for post-combustion CO2 capture. This work addresses the economics of CO2 capture with easier scalability.

My research interests in the near future will focus on the fundamentals and applied aspects of catalysis and separation employed in science and engineering for sustainability. The proposed research will seek to design, synthesize and evaluate advanced catalytic materials to allow for revolutionary advances in separation and reaction science to address critical issues in pollution control, waste-to-energy and renewable energy and fuels.

18 Total publications, 13 first author, h-index=9

385802 Evaluating Issues in Environmental Adsorption: Mercury Oxidation and Shale Pore Characteristics

Abstract:

The research presented here will focus on recent work focused around the field of environmental adsorption. The topics are: 1) identification of adsorbed mercury species in applications related to coal-fired flue gases and 2) evaluation of analytic adsorption techniques for the determination of pore characteristics in shale gas. Both of these topics have implications in the field of clean energy: coal-fired power remains the primary source of electric energy in the United States, China and India and the primary anthropogenic source of emitted mercury, while the production of natural gas from non-traditional sources, such has black shale, has outpaced the information available about these sources and the long-term implications of the their development.

1.) Mercury exists in various forms in the flue gases of coal combustion, from its elemental state (Hg^0) at the boiler exit, to various oxidized forms (Hg^{2+}) as it interacts with fly ash particle surfaces as the flue gas cools. The exact species of the oxidized mercury is, at this point, an outstanding question. Thermodynamic calculations suggest that the favored form in coal combustion flue gas is $HgCl_2$, but HgO and $HgBr_2$ could exist under certain conditions, such as the introduction of Br into the combustion process or in cases when coal is burned and contains trace amounts of Br.

An electron ionization quadrupole mass spectrometer (EI-QMS) has been modified for the direct measurement of elemental and oxidized Hg at industrial relevant concentrations ($< 15\ \mu g\ m^{-3}$ or 2 ppbv) from a methane combustion flue gas. Ultimately, the extent of homogeneous and heterogeneous oxidation may be measured, with the Hg species directly identified through mass spectrometry. In addition, the oxidized species created upon exposure to surfaces present in coal combustion, such as fly ash and activated carbon, are identified through the use of packed-bed experiments. A variety of sorbents have been evaluated, with the MS used to track any desorbed species. This information is critical for evaluating the extent of reaction on a sorbent surface and the potential for desorption of various oxidized Hg species. By investigating the desorbed species as a function of temperature provides indication of the relative strength of adsorption of the various Hg species and subsequently their relative stabilities on the various sorbents tested. These types of investigations will also increase the fundamental understanding of the surface reaction of Hg on a variety of sorbents and provide further understanding that can lead to advances in Hg capture.

2.) The rapid ascent of shale sourced gas has created a need for analytical evaluation of shale, in order to determine permeability and gas-in-place. Researchers have focused on the adsorption properties of shale, in order to evaluate potential gas-in-place and understand how gas transports from micropores to macropores to fractures. The literature, to this point, has been unfocused, and pointed towards evaluating as many shale samples from the broadest perspective as possible. However, analytical adsorption approaches appropriate for investigating homogenous samples, such as activated carbon or zeolitic material, has limitations when applied to highly heterogeneous samples. Shale, which has high variability at all scales, from the formation level to the microscopic level, is a particularly difficult challenge.

This work focuses on creating a standard analytical method for the evaluation of gas shale, focused on low-pressure adsorption techniques. The outgas temperature will be evaluated to determine effects on adsorption isotherms, with emphasis on potential changes in pore volume. The samples will be evaluated with 3 probe gases (N_2 at 77 K, Ar at 87 K and CO_2 at 273 K). Isotherms will be evaluated with DFT techniques to determine pore size distribution, while using information from Fourier transform infrared spectroscopy (FTIR), scanning electron microscopy (SEM), elemental analysis and total and inorganic carbon contents to support the gas adsorption observations. The thorough analytical approach presented in this work will present a method that researchers can use to properly evaluate pore characteristics of shale gas, which can be used to support research in flow models and gas-in-place estimates.

389397 Engineered Nanomaterials for Energy Harvesting

Abstract:

Due to a continual rise in demand for energy worldwide, significant attention is being provided to new environmentally clean energy resources such as solar energy on one hand while large efforts are also concentrated on thermoelectric devices which convert thermal energy directly into electrical energy and hold promising contributions in waste heat recovery. For both approaches, strategies are underway to identify materials and device architectures that are inexpensive and scalable for widespread use yet efficient compared to existing technologies. Nanotechnology enables novel approaches to energy conversion that may provide both higher efficiencies and simpler manufacturing methods. I plan to develop a research program which will focus on tuning and improving existing properties and harnessing novel properties through nanostructuring, doping and interface engineering in these materials.

Electronic, optical, mechanical and thermal properties of metals and semiconductors (SCs) in nano-sized systems can be controlled appreciably by varying their size and give rise to potentially new phenomena not observed in their bulk counterparts. Since the properties of the material are directly related to its crystal structure, the ability to control the crystal phase could also lead to new behavior. Nanostructures offer a potential method to obtain additional phases that would not exist in the bulk. Nanostructures can also lead to shifts in phase-transition (PT) temperatures. This effect is not only interesting for fundamental reasons, but the ability to modify the PT temperature allows one to stabilize (or destabilize) desirable (or undesirable) phases, depending on the goals. Because the properties of the nanostructures change through the PT, this can have implications for potential applications.

The introduction of trace intentional impurities (or doping) is central to controlling the behavior of SC materials. Dopants possess the ability to strongly modify the optical, magnetic, and electronic properties of bulk SCs. Modern SC-based technology owes its existence, in large part, to the fact that these materials can be doped. It is the ability to control precisely the number of carriers available in the SC by doping, which has expedited the advance in SC-based electronic and optoelectronic technology. The advantage of doped SCs is that they provide the device engineer with a wide range of mobilities, so that materials are available with properties that meet specific requirements. Hence, it is natural to extend the versatility of nanostructures by adding dopants.

While interfaces are typically viewed as a dividing layer between two neighboring materials, the properties of the interface are in-fact, at times, radically different from the constituent materials. Nanomaterials, in particular, where nearly 20-30% of the total atoms are located at or very close to the surface, provide an ideal platform to study these effects. Rational engineering of the interfaces between two different components in a nano-composite, presents an opportunity for creating materials with novel properties that could not be achieved otherwise. Hybrid nanocrystal/conducting polymer materials have extensively been used in optoelectronic devices such as photovoltaics and light-emitting diodes, and recently for thermoelectric applications as well.

My PhD research focused mostly on zero-dimensional nanostructures (nanocrystals typically less than 10-nm in size). I developed a versatile synthesis that yields high quality silver chalcogenide nanocrystals (NCs) with size-dependent phase behavior, optical and transport properties. However, by changing the surface of the NCs, I was able to modify the phase behavior of these NCs which allowed us to completely tailor both the optical and electronic transport properties of the material. Additionally, I developed a facile approach to lightly dope SC NCs (CdSe and PbSe) with a controllable amount of electronically active impurities. The addition of even a single impurity per NC has a dramatic effect on their optical properties. Furthermore, studies of the electrical transport through the NC films show complex behavior in the Fermi level as a function of dopant concentration. The results demonstrate that dopant behavior in NCs is not as simple as one might expect. Thus, these experiments begin to reveal the properties of a novel class of NC materials in which both interface and impurities play a key role that may be important for future NC devices.

Multicomponent materials allow us to unite the benefits of both components into one hybrid material and design systems that do not necessarily conform to standard mixing theories for carrier transport. My postdoctoral research aims towards designing and characterizing novel solution processable organic/inorganic nanocomposites (inorganic component being one-dimensional materials or nanowires) and gaining a functional understanding and control of carrier transport at these soft/hard interfaces primarily for thermo-electric applications. We observe unusual thermo-electric transport in a high-performance organic/inorganic thermoelectric composite leading to an optimized power factor in the composites that exceed that of the individual components, suggesting non-effective medium behavior. We increase the conductivity in these hybrid systems further by doping the individual components separately. Finally, by changing the nature of the interface, we can completely switch the nature of charge carriers in these material systems which are particularly attractive for building real thermoelectric modules. My results underscore the importance of understanding and controlling interfacial phenomena in hybrid organic/inorganic systems, and provide a general route for enhancing carrier transport in hybrid materials and devices.

Further work will seek to enhance this interfacial phenomenon in layered two-dimensional materials (nano-sheets) where essentially the entire material is at the interface. In the process, we hope to gain a fundamental understanding of these hybrid systems that would provide a general and robust framework to guide materials design and allow us to tailor suitable material systems for energy applications. Future research plans will be discussed in addition to the above topics.

382256 Experimental Fluid Dynamics for Advanced Materials and Biological Systems

Abstract:

Experiments investigating fundamental transport phenomena are the foundation for my past and present research, and they are my primary interest and focus for future research. I have conducted experiments in fluid dynamics in different areas of science and technology including biological systems, advanced materials, and materials processing. During my doctoral work I investigated the properties of lignocellulosic biomass from a rheological perspective. The fluid and mechanical properties of biomass are still of great interest to those processing biomass to biofuels and other bioproducts, and much work remains to link the physiochemical structure of biomass to its rheological properties. My postdoctoral work has focused on transport phenomena in fluid-fluid interfaces like those found in foams, emulsions, and biological membranes. In this area there is enormous opportunity to advance knowledge in synthetic biology by characterizing the fluid and mechanical properties of synthetic cell membranes with microrheological techniques. Most recently I have investigated a novel processing technique for advanced materials called fractal structuring. The technique involves a confined, multi-component flow through a simple but uniquely designed mold of iterative baker's transformations that results in highly structured composites of co-continuous phases with potential applications in photovoltaics, fuel cell membranes, and gas separations. Fluid dynamics simulations and experiments are required here to fully exploit the technique and design the next generation of fractal structuring molds. In this presentation I will highlight important results from my past and present research and indicate where additional investigations of fundamental transport phenomena in each area hold promise for future discoveries.

380066 Engineering Solutions for Systems Biology: Microfluidics and Unbiased Statistical Tools to Uncover Genetic Relationships through Phenotyping

Abstract:

Systems biology focuses on studying living organisms as an ensemble of molecular, dynamic networks that are orchestrated to perform biological functions. The main questions are directed to understanding how all the players interact to accomplish certain goals within a cell or within an organism. This new approach to biology necessitates new experimental paradigms that enable perturbing biological networks and extracting information in a multidimensional scale. Just like systems biology aims at finding how biological ensembles work together (and what happens to them with disease), assays and experimental platforms should focus on extracting ensemble results rather than individual features. Experimental approaches need to be able to look deeper into the information provided by experiments and need to allow better control of the perturbations assayed.

Here I show how we have applied an integrated approach to perform biological studies enabled by microfluidics, automation, mathematical algorithms and machine learning with the nematode *C. elegans* (an excellent model for systems biology). I present how microfluidics can be coupled to unsupervised data extraction tools to obtain deep information and enable high-content studies of complex biological systems. Specifically, I discuss how integrated engineering platforms can be used to elucidate relevant questions in neuroscience, such as synaptic plasticity, learning and the effects of aging on neurological health. We have applied one such system to perform automated genetic screens and to identify differences in synaptic patterns, hidden to the naked eye. These new experimental platforms are promising in our challenging search for a systems-level understanding of biological processes.

389788 Applied Mechanics Studies of Complex Fluids for Pharmacokinetic Application

Abstract:

My research interest lies primarily in non-Newtonian fluid's behavior in response to external field in bulk and near wall. A considerable portion of earlier research work addressed problems related to modeling response of colloidal hybrids, effect of surface charge and contact mechanics, and capillary sintering during drying of colloidal dispersion. My current research involves development of next generation pharmacokinetic models for therapeutic targeted drug delivery.

A class of hybrid complex fluids is made of colloid hard-core with polymer brushes. For battery research they are used for ion transportation. For gas absorption purposes they act as porous support to screen molecules and hold active absorption agents. In drug delivery hybrid complex fluids (which we refer to as nanogels or deformable nanocarriers) are used as vehicles to carry important drugs and target them to specific tissues that are either inflamed or diseased. In this case the core is made of native proteins such as lysozyme and the brush is made out of biocompatible polymers such as dextran or hyaluronic acid. The dimensions of these nanostructures range from 10-100 nm, which puts them below the diffraction limit, making experimental investigations challenging.

Novel Materials for Targeted Drug Delivery: The deformable nanocarrier has a large molecular weight (10^2-10^5kDa) and it is difficult to predict its structural dynamics using molecular dynamics techniques. There is also a lack of mechanistic insight to predict selectivity of binding of the nanogels to inflamed or diseased tissue under physiologically relevant conditions in vasculature. In order to model the deformable gel-based NC, the structure of the deformable gel is coarse-grained based on theoretical models from polymer physics and simulated annealing protocol, which mimics the experimental method of synthesis of this novel material. The Brownian dynamics framework is incorporated to include hydrodynamic interactions, which led to the calculation of equilibrium and steady-shear rheological properties of the nanogel. Brownian dynamics with near-wall interactions along with rheological response in bulk gives important insight that is essential for drug design and predicts the targeting capabilities to specific tissues, which is important for clinical translation.

Novel Materials for Energy Applications: Nano organic hybrid materials are made of silica cores and polyethylene glycol (PEG) brushes. Most of the applications like flow in a semi-confined space is related to complex flows. Primary goal of my research work has been understanding the properties of these materials at Brownian and mesoscale level by developing theoretical model to characterize these complex materials. Based on these models the steady shear rheological properties and transient response[1] have been investigated. A constitutive model for complex flows exhibiting a variety of local linear flows is derived and applied to a complex stochastically varying flow of the fluid through fixed bed of fibers. If the ratio of size of the core to polymer brushes decreases, the structural anisotropy increases and the material shows sign of strain thinning. Such material acts as an additive, which reduces the internal stresses of the fluid in a conduit and the risk of fracturing of underground pipelines (which is a common problem in fracking process) reduces.

Prior research work was primarily on consolidation and cracking in drying colloidal dispersion. The work addressed fundamental understanding related to colloidal films formation broadly desiccation processes of water-soluble colloids. Water-soluble colloids are important ingredient in paint industry (for reduction of volatile organic components) and for continuous manufacturing of drugs in pharmaceutical processing. Drying of water-soluble colloids and its film formation were investigated through experiments and theory. If a dispersion of colloidal particle is allowed to dry, particles packing under a convective field (set by the evaporation) is in compression and the solvent from bulk, pass through porous bed of particles to reach the interface. Charge nature of the particles changes the conventional drying-consolidation profile[3] while the presence of flaws in the packed region is effected due to an interplay between strain energy, stored in the particle network, and interfacial energy.[2] Experiments were performed in rheometer (semi-confined geometry[4,6]) and in microchannels (confined geometry[5]). In semiconfined geometry the pressure required to open a crack from flaw was predicted while in confined geometry the additional effect, due to presence of boundaries and channels, were investigated and hence depicted an overall consolidation behavior.

My research experience is mainly in applied mechanics studies of complex fluids. I would like to further extend the methodologies developed for amphiphilic carriers such as rodlike micelles[7] to include wall-effects and resolve internal and inhomogeneous / anisotropic stresses-strains near wall. In biomedical applications, rodlike micelles can have long circulating times, which are essential in sustained delivery of drugs required for some medical conditions. Rodlike micelles can also be an important self-assembly for drag reduction in oil pipelines and hot water transportation lines. I plan to carry out experiments in 4-roll mill to understand the response of rodlike micelles in a variety of flows and extend the present understanding of soft glassy fluid elements to rodlike micelles, in general FENE dumbbells. I am also interested in packing and sintering of alumina film formation with or without a heat source. Alumina is widely used as refractory coatings, catalyst and more recently transparent optics. I would like to study alumina film formation through theory and experiments to predict packing and its effect on film properties.

I wish to acknowledge funding support for my work through Cornell University-KAUST center and NIH 1R01EB006818-05 & NIH U01-EB016027, made available to me through my postdoctoral supervisors.

References

1. Sarkar, A. and D. L. Koch, "Brownian dynamics simulations of a local-fluid-volume preserving model of solvent-free nanoparticle fluids," The Society of Rheology 84th Annual Meeting (2013).
2. Sarkar, A. and M. S. Tirumkudulu, "Asymptotic analysis of stresses near a crack tip in a two- dimensional colloidal packing saturated with liquid," Phys. Rev. E, 83, 051401 (2011).
3. Sarkar, A. and M. S. Tirumkudulu, "Consolidation of charged colloids during drying," Langmuir, 25(9), 4945–4953 (2009).
4. Sarkar, A. and M. S. Tirumkudulu, "Critical stress required to open a flaw in a two dimensional colloidal packing saturated with solvent," International Symposium on Recent and Emergent Advances in Chemical

Engineering (REACH).
5. Sarkar, A. and M. S. Tirumkudulu, “Delamination of drying nanoparticle suspensions,” Soft Matter, 7, 8816–8822 (2011).
6. Sarkar, A. and M. S. Tirumkudulu, “Ultimate strength of a colloidal packing,” Soft Matter, 8, 303–306 (2012).
7. Discher Dennis E. and Eisenberg Adi,"Polymer Vesicles", Science 297, 967 (2002).

382321 Bridging the Gap Between Chemical and Biological Catalysis to Produce Biorenewable Chemicals

Abstract:

Biomass is a renewable source of carbon that has the potential to replace fossil-based resources as the feedstock for producing fuels and chemicals. An attractive option for obtaining biorenewable chemicals from biomass leverages the high efficiency of traditional heterogeneous chemical catalysts to upgrade highly functionalized molecules produced via recent advances in metabolic and enzyme engineering. By such coupling of chemical and biological catalysis it is possible to access molecules not easily obtainable by either discipline alone.

My doctoral research has focused on the challenges and opportunities associated with bridging the gap between chemical and biological catalysis. One key aspect of such strategies is to elucidate reaction networks that stem from a particular biologically-derived intermediate. For example, cortalcerone, which can be derived enzymatically from glucose, can be dehydrated over a Brønsted acid catalyst, such as HCl, to yield furylglyoxal hydrate (FH). In turn, FH has appropriate functionality to be converted to furylglycolic acid (FA), a hydroxy-acid amenable for co-polymerization with lactic acid to yield biorenewable polymers reported to have polystyrene-like properties. FA can be produced from FH via an intramolecular Meerwein-Ponndorf-Verley-Oppenauer hydride shift using a Lewis acid catalyst, such as a Sn-containing beta zeolite. Furthermore, FA can be produced in one pot from cortalcerone using a bifunctional Al-containing Sn-beta zeolite possessing both Brønsted and Lewis sties. In another example, triacetic acid lactone (TAL), a highly functionalized molecule obtained via polyketide biosynthesis, can be upgraded to 11 potential end products and intermediates, including bifunctional ketones via decarboxylation of TAL or its hydrogenated analogue, lactones via dehydration of hydrogenated TAL, and unsaturated acids, including the common food preservative sorbic acid, via ring opening of these lactones.

Another key aspect of coupling chemical and biological catalysis is to elucidate the influence of biogenic impurities on heterogeneous catalysts. For example, amino acids can result in deactivation of the metal catalysts used for TAL hydrogenation. This effect can, however, be mitigated by appropriate catalyst design. Overcoating a traditional Pd catalyst with poly(vinyl alcohol) (PVA) results in the formation of microenvironments surrounding the metal nanoparticles that prevent adsorption of such species. For example, the Pd catalyst used in TAL hydrogenation loses 83% of its activity after 14 hours of exposure to methionine (a representative, sulfur-containing amino acid), but a PVA-overcoated catalyst is stable under the same conditions. Indeed, combining the use of a PVA overcoat to prevent adsorption of amino acids with the use of PdAu bimetallic nanoparticles to prevent deposition of carbonaceous residues from TAL decomposition, it is possible to decrease the deactivation rate constant by an order of magnitude when upgrading TAL recovered from spent cell culture medium. My future research will continue to focus on these types of challenges and opportunities that are associated with integrating chemical and biological catalysis, with the ultimate goal of efficiently producing high-value chemicals from biomass.

385612 Computational Design of Advanced Materials to Meet Health, Environmental and Energy Challenges

Abstract:

Society is facing health, environmental and energy challenges, including fast disease diagnosis, affordable supply of clean water, and sustainable energy supply and storage. Rational design of advanced materials is essential to meet these challenges. The fast development of computational technologies and simulation methods enables us to study properties inaccessible to experiments, provide molecular-level principles to guide experiments, and perform large-scale screening of molecular structures for desired properties.

My PhD and postdoctoral research illustrates how computational simulations can be applied in materials design. My PhD work at the University of Washington focused on exploring the differences between zwitterionic and non-ionic materials, and the structure-property relationships of zwitterionic materials. Using quantum mechanical calculations, atomistic simulations and enhanced sampling methods, I have investigated (a) the effects of zwitterionic and non-ionic materials on the structures of chymotrypsin inhibitor 2 and on the association of two hydrophobic sheets, and (b) the hydration, ionic interactions, and self-association of different zwitterionic molecules. My study explained the different effects of zwitterionic and non-ionic materials on bioactivities of proteins, and provided the principles for designing nonspecific protein-resistant zwitterionic materials. A material that I designed computationally was synthesized and tested in my PhD advisor's group.

My postdoctoral research at the North Carolina State University focuses on developing computational approaches to predict the toxicity of different types of nanoparticles. The small sizes of nanoparticles enhance their ability to enter the body, and concerns have arisen over their toxicity. The protein corona on nanoparticles serves as the signature of the nanoparticles in biological systems and determines the interactions between nanoparticles and biological systems. My research aims to develop coarse-grained models that can be used to predict protein-nanoparticle interactions, and to answer fundamental questions on nano-bio interactions.

In my future work, I plan to apply my expertise in computational simulations to develop advanced materials with the aim to address the health, environmental and energy challenges that we face today. I am interested in designing computationally (1) peptide surfaces that resist nonspecific protein adsorption but bind specific proteins for disease diagnosis, (2) zwitterionic porous materials that have high salt-rejection rates and nonfouling properties for desalination, and (3) polymeric electrolytes that have high ionic conductivity for lithium-ion batteries.

Omics-based, Automated Disease Modeling

Jason E. Shoemaker*, Yoshihiro Kawaoka

Department of Microbiology and Immunology
The Institute of Medical Science
University of Tokyo

A systematic understanding of the interactions within biological systems promises to guide the design of next generation disease diagnostics and therapies. In the past decade, systems biologists have successfully applied traditional process systems and controls engineering techniques to analyze the highly complex environments of biological systems; leading to major contributions in the fields of diabetes, drug target prioritization, and metabolite optimization. Yet, many challenges remain for creating medically significant models. The most prominent of challenges is the explosion in data that has occurred since the development of microarray and next generation sequencing techniques. The wealth of data and high degree of biological variance can quickly overwhelm traditional modeling techniques.

My work currently focuses on integrating several computational techniques from the fields of systems engineering, machine learning and bioinformatics for rapidly developing disease models. Specific challenges include:

- Signal deconvolution in *in vivo* whole genome and protein data ('omics' data) to determine the changes local lymphocyte demographics. We have deployed a web tool, CTen, related to this work
- System Inference Microarray Analysis (SIMA) in which we employ nonlinear, biologically inspired models to predict gene expression across experimental conditions
- Applying machine learning to the topological characteristics biological networks to identify safe drugs

Additionally, I am involved in several systems software projects, including Garuda which aims to provide a single platform to unify systems biology tools.

The foundation of my research program will be developing machine learning and artificial intelligence algorithms that can automatically process large scale, dynamic data into effective biological models. In this poster, the successful application of the three tools mentioned above is discussed in the context of influenza virus infection and interferon therapy. Then we review methods and motives to improve these tools and how they can be implemented in long term strategy for developing omics-based, automated disease modeling platforms.

388835 Electrodeposition and Activity of Electrocatalysts in Ionic Liquids

Abstract:

Applications of electrocatalysts can be found in sensors, electro-organic synthesis, waste-water treatment and energy conversion devices. The activity and selectivity of the electrocatalysts can be fine-tuned by applying potential or current. In addition, the electrochemical devices employing the electrocatalysts are not limited by Carnot efficiency. The requirement for triple-phase boundary for electrochemical reactions to occur in these devices necessitate that the electrocatalysts be deposited in a site-specific manner for maximum utilization. Electrodepostion of the catalysts meets this requirement. However, in traditional aqueous system, electrodeposition is limited by the narrow electrochemical window of water (1.2 V). In ionic liquids (ILs), the electrochemical window can be as high as 6 V and the metals having reduction potential negative than that of H_2 evolution could be reduced. Moreover, product selectivity and reaction efficiency can be enhanced in ILs for reactions such as CO_2 reduction. However, studies in these fields are limited. My graduate work in nitrogen-doped ordered mesoporous carbon (NOMC) supported Pt for oxygen reduction reaction (ORR) and my postdoctoral work in electrodeposition of fission platinoids (Pd, Rh, Ru) in ILs puts me in a favorable position to advance these fields.

My graduate work focused on creating strong metal-support interaction (SMSI) between carbon and Pt. Although various types of advanced carbon supports such as carbon black, carbon nanotubes, aerogels, carbon fibers and graphene have been investigated as the support, none have been shown to increase intrinsic ORR activity of Pt. To overcome this, I explored nitrogen-doped carbon supports. Nitrogen has been shown to change electronic and microstructural properties of carbon. However, there had been few works on Pt-supported nitrogen-doped carbon for ORR and most of it was limited to model systems such as highly oriented pyrolytic graphite (HOPG). In my work, NOMC was synthesized by templating on SBA-15 with pyrrole as the precursor. Pt nanoparticles were deposited on the NOMC by electroless reduction in ethylene glycol. I showed how pore-order, graphiticity and nitrogen content plays a role in ORR activity of the Pt. This has provided guidance for designing of high-performing nitrogen-doped carbon supports.

Currently, I am working on recovery of fission platinoids (Pd, Rh, Ru) through electrodeposition in ionic liquids. Current processes for recovery of fission products is based on extraction in highly flammable volatile organic solvents (VOCs) such as kerosene and dodecane. ILs provide safer alternative to the VOCs because of their negligible vapor pressure. ILs also have complexing ability, hence metals could be dissolved in ILs without introducing additional complexing agents. Also the metals solvated could be recovered directly through electrodeposition in ILs. I am studying electrochemical and electroreduction behavior of Pd in many ILs previously not studied. The ILs affect the electroreduction behavior by controlling speciation, electrolyte/electrode interface, and the mass-transfer which in turn affect the thermodynamics, kinetics and morphology of the deposits. This work will expose fundamental underpinnings of electrodeposition of Pd, Rh and Ru in ILs.

In my poster, I will summarize my graduate and postdoctoral work and talk about how I will exploit them to develop the field of electrodeposition of catalysts and their activity in ILs for applications in low temperature fuel cells and CO_2 utilization.

Scalable Nanofabrication of Functional Architectures through Hybrid Lithography

Jonathan P. Singer

Chemical and Environmental Engineering, Yale University, New Haven, CT, 06511

In order to advance patterning capabilities for future applications, innovative lithographic strategies that focus on scalable and cost-effective techniques must be developed. To address this challenge, a growing number of alternative patterning methods to conventional photolithography for the fabrication of complex structures have emerged, all of which possess certain advantages and disadvantages, generally in the form of tradeoffs between speed and control. My research has focused on developing hybrid lithographic nanofabrication techniques to overcome the limitations of individual techniques—specifically, the development of processes to enable the scalable fabrication of previously impossible architectures.

In my doctoral work in the lab of Edwin Thomas at MIT, I designed techniques that combine large area patterning methods (*i.e.* interference lithography and self-assembly) with 3D laser direct write and that can simultaneously exploit emerging developments in enhanced resolution. This research has resulted in hybrid techniques that have enabled (i) the single-shot fabrication of plasmonic resonators, (ii) the local fill-fraction control of complex periodic and quasiperiodic photonic and phononic architectures with deliberate defects, and (iii) the use of thermocapillary forces for the controlled dewetting of polymer thin films and alignment of block copolymer networks for applications in affordable pattern transfer. In my current role as postdoctoral associate in the lab of Chinedum Osuji at Yale, I have continued work on hybrid methods with an increasing focus on functional materials, such as fluorescent, plasmonic, and semiconducting nanoparticles and conjugated polymeric systems, with applications in, for example, organic-inorganic bulk heterojunction photovoltaics. These materials are difficult to deliberately pattern through conventional means, so my work pursues large-scale alternative methods including (i) magnetic field alignment of nanowires, (ii) high-aspect ratio nanoimprint with metallic glass molds, and (iii) durotaxis-directed electrospray of polymers and nanoparticles.

Moving forward, I will continue to couple experiments and simulations in the full cycle of process-to-device development to create new strategies for patterning that incorporate materials with functionalities in their molecular and supermolecular structures into the final complex architectures that are compatible with high-throughput fabrication processes.

Selected Manuscripts

1. Pelligra, C. I.; Singer, J. P.; Huang, S.; Mayo, A.; Mu, R. R.; Osuji, C. O.; " Scalable High-Fidelity Growth of Semiconductor Nanorod Arrays with Controlled Geometry for Photovoltaic Devices using Block Copolymers," *Manuscript in submission.*
2. Hu, H.; Singer, J. P.; Osuji, C. O.; "Morphological Development in Thin Films of a Lamellar PS-b-P4VP Block Copolymer Deposited by Electrospray," *Manuscript in submission.*
3. Shao, Z.; Liu, Z.; Singer, J. P.; Li, H.; Liu, Y.; Osuji, C. O.; Schroers, J.;" Enhanced Non-Polymorphic Crystallization Kinetics of a Bulk Metallic Glass through Shear," *Manuscript in preparation.*
4. Singer, J. P.; Shao, Z.; Gopinadhan, M.; Schroers, J.; Osuji, C. O.; "High-Resolution Nanoimprint of Functional Polymers with Bulk Metallic Glass Molds," *Manuscript in preparation.*
5. Singer, J. P.; Gotrik, K. W.; Lee, J.-H.; Kooi, S. E.; Ross, C. A.; Thomas, E. L.; "Alignment and Reordering of a Block Copolymer by Solvent-Enhanced Thermal Laser Direct Write," *Polymer*, **55**, 7, 1875-1882, 2014.
6. Singer, J. P.; Lin, P.-T.; Kooi, S. E.; Kimerling, L. A.; Jurgen, M.; Thomas, E. L.; "Direct-Write Dewetting of Polymer Thin Films by a Laser-Induced Thermal Gradient," *Advanced Materials*, **25**, 42, 6100-6105, 2013.
7. Lee, J.-H.; Veysset, D.; Singer, J. P.; Retsch, M.; Saini, G.; Pezeril, T.; Nelson, K. A.; Thomas, E. L.; "Dynamical Responses of Layered Nanostructure Composites at High Strain Rate Deformation," *Nature Communications*, **3**, 1164, 2012.
8. Singer, J. P.; Lee, J.-H.; Kooi, S. E.; Thomas, E. L.;" Rapid Fabrication of 3D Terahertz Split Ring Resonator Arrays by Novel Single-Shot Direct Write Focused Proximity Field Nanopatterning," *Optics Express*, **20**, 10, 11097-11108, 2012.
9. Singer, J. P.; Kooi, S. E.; Thomas, E. L.; "Focused Laser Spike (FLaSk) Annealing of Photoactivated Chemically Amplified Resists for Rapid Hierarchical Patterning," *Nanoscale*, **3**, 7, 2730-2738, 2011.

Design, optimization, monitoring and control of continuous pharmaceutical manufacturing plant for QbD and PAT based next generation of efficient manufacturing

Ravendra Singh
NSF Engineering Research Center for Structured Organic Particulate Systems (C-SOPS), Chemical and Biochemical Engineering, Rutgers University, Piscataway, NJ, USA

A novel manufacturing strategy based on continuous processing integrated with inline/online monitoring tools (PAT) coupled with efficient automatic advanced feedback control system is highly desired for efficient QbD based manufacturing of the next generation of pharmaceutical products with optimal consumption of time, space and resources. However, the particulate systems are still poorly understood and there is different level of complexities involved, therefore to achieve this paradigm shift, new techniques and tools need to be developed through extensive fundamental and applied research. This area of research and education are highly encouraged and supported by pharmaceutical companies, academic institutions, regulatory authority (FDA) and national science foundation (NSF). For example, recently national science foundation of USA has established a 100 million dollar research center for structured organic particulate systems (C-SOPS) at Rutgers University New Jersey. Furthermore, more than 40 companies are supporting this center economically and providing scientific inputs so that the research outcomes can be directly applied to the companies.

In this work, a scientific foundation for the optimal design, optimization and control of the continuous process involving structured organic particulate systems has been developed. The mathematical models for continuous manufacturing of active pharmaceutical ingredients and final product tablet, has been developed for virtual experimentation and applied for plant design, optimization and control. Advanced modelling techniques such as PBM, DEM and CFD have been applied for pharmaceutical process modeling. The developed continuous tablet manufacturing plant and process model has been adapted by several pharmaceutical companies. One of the challenges is to monitor and control the process efficiently in real time for QbD based manufacturing. A flexible monitoring and control system incorporating different PAT techniques and tools and control strategies for continuous pharmaceutical manufacturing process has been designed and implemented to the plant situated at Rutgers University. Different control strategies such as PID, PID coupled with dead time compensator, and MPC has been compared and evaluated practically for pharmaceutical process control in terms of set point tracking and disturbances rejection ability, robustness, and stability/instability. A new hybrid MPC-PID control strategy has been developed and implemented into the pharmaceutical tablet manufacturing plant and its performance has been found better than the other control strategies. Principal components analysis (PCA) based additional supervisory control layer has been also integrated with the hybrid feedback control system. For the first time, the closed-loop operation of the tablet manufacturing plant using NIR sensor has been demonstrated, which is a significant advancement in pharmaceutical manufacturing as endorsed by the Food and Drug Administration (FDA) and other regulatory authorities as part of the Quality by Design (QbD) paradigm. A systematic generic framework including the methods and tools through which the cost and profit can be estimated and the Dynamic Real Time Optimization (DRTO) can be integrated with hybrid control system has been also developed. The integrated DRTO provides the optimal operational set points for the plant control system in real time optimizing an objective function. DRTO integrated with hybrid control strategy ensures the maximum possible profit irrespective of the market demand fluctuations.

Prior to this, I worked in a large scale European pharmaceutical project (F^3 Factory project), which involves 25 partners from industry and Academia; where my responsibility was to manage and lead the DTU's parts of the F^3 Factory project and also to coordinate with the other partners associated with it. My doctoral research work, done at Department of Chemical and Biochemical Engineering, Technical University of Denmark, has been awarded for prestigious EFCE Excellence Award 2010, given in

Recognition of an Outstanding PhD Thesis, from European Federation of Chemical Engineering. I have exemplary academic records, previous teaching experiences, and an established trail of valued research accreditations'. I have published several peer reviewed scientific international Journal articles, popular reputed magazine articles, book chapters and presented in more than 47 international conferences.

The objective of this presentation is two-fold. First to give the overview of my research and teaching experience, and second to demonstrate the design, optimization, monitoring and control of continuous tablet manufacturing plant.

Some selected references

1. **Singh, R.,** Sahay, A, Muzzio, F., Ierapetritou, M., Ramachandran, R. (**2014**). Systematic framework for onsite design and implementation of the control system in continuous tablet manufacturing process. Computers & Chemical Engineering, DOI: 10.1016/j.compchemeng.2014.02.029.
2. **Singh, R.,** Barrasso, D., Chaudhury, A., Maitraye Sen, Ierapetritou, M., Ramachandran, R. (**2014**).Closed-Loop Feedback Control of a Continuous Pharmaceutical Tablet Manufacturing Process via Wet Granulation. Journal of Pharmaceutical Innovation, 9, 16-37.
3. **Singh, R.,** Ierapetritou, M., Ramachandran, R. (2013). System-wide hybrid model predictive control of a continuous pharmaceutical tablet manufacturing process via direct compaction. European Journal of Pharmaceutics and Biopharmaceutics, 85(3), Part B, 1164-1182.
4. **Singh, R.,** Ierapetritou, M., Ramachandran, R. (**2012**). An engineering study on the enhanced control and operation of continuous manufacturing of pharmaceutical tablets via roller compaction. International Journal of Pharmaceutics, 438 (1-2), 307-326.
5. **Singh, R.,** Godfrey, A., Gregertsen, B., Muller, F., Gernaey, K. V., Gani, R., Woodley, J. M. (2013). Systematic substrate adoption methodology (SAM) for future flexible, generic pharmaceutical production processes. Computers & Chemical Engineering, 58, 344 - 368.
6. **Singh, R.,** Gernaey, K. V., Gani, R. (2010). ICAS-PAT: A Software for Design, Analysis & Validation of PAT Systems. Computers & Chemical Engineering, Volume 34, Issue 7, 1108-1136.
7. **Singh, R.**, Gernaey, K. V., Gani, R. (2010). An ontological knowledge based system for selection of process monitoring and analysis tools. Computers & Chemical Engineering, Volume 34, Issue 7, 1137-1154.
8. **Singh, R.,** Gernaey, K. V., Gani, R. (2009). Model-based computer-aided framework for design of process monitoring and analysis systems. Computers & Chemical Engineering, Volume 33, Issue 1, 22-42. (Citations>45).
9. Sen, M., Barrasso, D., **Singh, R.**, Ramachandran, R. (**2014**). A Multi-Scale Hybrid CFD-DEM-PBM Description of a Fluid-Bed Granulation Process. Process Journal, 2(1), 89-111.
10. Sen, M., **Singh, R.,** Ramachandran, R. (**2014**). Simulation based design of an efficient control system for the continuous purification and processing of active pharmaceutical ingredients. Journal of Pharmaceutical Innovation, 9, 65-81.
11. Sen, M., **Singh, R.,** Ramachandran, R. (**2014**). A hybrid MPC-PID control system design for the continuous purification and processing of active pharmaceutical ingredients. PROCESS Journal, accepted.
12. Sen, M., Dubey, A., **Singh, R.,** Ramachandran, R. (**2013**). Mathematical Development and Comparison of a Hybrid PBM-DEM description of a Continuous Powder Mixing Process. Journal of Powder Technology, http://dx.doi.org/10.1155/2013/843784.
13. Sen, M., Chaudhury, A., **Singh, R.,** John, J., Ramachandran, R. (**2013**). Multi-scale flowsheet simulation of an integrated continuous purification–downstream pharmaceutical manufacturing process. International Journal of Pharmaceutics, 445 (1-2), 29-38.
14. Sen, M., Rogers, A., **Singh, R.,** Chaudhury, A., John, J., Ierapetritou, M., Ramachandran, R. (**2013**). Flowsheet optimization of an integrated continuous purification-processing pharmaceutical manufacturing operation. Chemical Engineering Science, 102, 56 – 66.
15. Sen, M., **Singh, R.,** Vanarase, A., John, J., Ramachandran, R. (**2012**). Multi-dimensional population balance modeling and experimental validation of continuous powder mixing processes. Chemical Engineering Science, Volume 18, 349-360.
16. Samad, N. A. F. A., **Singh, R.**, Sin, G., Gernaey, K. V., Gani, R. (**2011**). A generic multi-dimensional model-based system for batch cooling crystallization processes. Computers & Chemical Engineering, Volume 35, Issue 5, 828-843.
17. **Singh, R.,** Cameron, I, Gani, R. (2011). Process and Product Modelling: A case study approach. Book chapters 12.2-12.6, Publisher: Elsevier, pp 380-430.
18. **Singh, R.,** Ierapetritou, M., Ramachandran, R (**2013**). Hybrid advanced control of a flexible multipurpose continuous pharmaceutical tablet manufacturing process via direct compaction. Computer Aided Chemical Engineering, Volume 32, 757-762.
19. Muzzio, F., **Singh, R.,** Chaudhury, A., Rogers, A., Ramachandran, R. Ierapetritou, M. (**2013**). Model- predictive design, control and optimization of pharmaceutical process. PharmTech magazine Europe, Volume 37, Number 6, 40-41, 77.
20. **Singh, R.,** Sahay, A., Oka, S., Liu, X., Ramachandran, R., Marianthi Ierapetritou, Fernando Muzzio, F. (**2013**). Online monitoring, advanced control and operation of robust continuous pharmaceutical tablet manufacturing process. BioPharma magazine Asia, Volume 2, Issue 5, October 2013.
21. **Singh, R.**, Raquel Rozada-Sanchez, R., Dean, W., Perkins, J., Muller, F., Godfrey, A., Gernaey, K. V., Gani, R., Woodley, J. M. (**2012**). A generic process template for continuous pharmaceutical production. Computer Aided Chemical Engineering, Volume 31, 715-719.
22. **Singh, R.,** Boukouvala, F., Jayjock, E., Ramachandran, R. Ierapetritou, M., Muzzio, F. (**2012**). Flexible Multipurpose Continuous Processing. PharmPro Magazine, Pharmaceutical Processing, July 2012, Vol. 27, Issue 6, p22-25.
23. **Singh, R.,** Boukouvala, F., Jayjock, E., Ramachandran, R. Ierapetritou, M., Muzzio, F. (**2012**). Flexible Multipurpose Continuous Processing of Pharmaceutical Tablet Manufacturing Process. GMP news, European Compliance Academic (ECE), http://www.gmp-compliance.org/ecanl_503_0_news_3268_7248_n.html.
24. Samad, N. A. F. A., **Singh, R.**, Sin, G., Gernaey, K. V., Gani, R. (**2011**). Systematic Procedure for Generating Operational Policies to Achieve Target Crystal Size Distribution (CSD) in Batch Cooling Crystallization. IEEE, 1-6.
25. Gani, R., **Singh, R.**, PAT Slowly Yielding to Broader QbD (2011). Genetic Engineering & Biotechnology News, March 15, Vol. 31, No. 6, pp 40-41. http://genengnews.com
26. Samad, N. A. F. A., **Singh, R.**, Sin, G., Gernaey, K. V., Gani, R. (**2010**) . Control of process operation and monitoring of product qualities through generic model-based in batch cooling crystallization. Computer Aided Chemical Engineering, Volume 28, 613-618.

27. **Singh, R.**, Rozada-Sanchez, R., Wrate, T., Muller, F., Gernaey, K. V., Gani, R., Woodley, J. M. **(2010)**. A retrofit strategy to achieve "Fast, Flexible, Future (F^3)" pharmaceutical production processes. Computer Aided Chemical Engineering, Volume 29, 291-295.
28. Samad, N. A. F. A., **Singh, R.**, Sin, G., Gernaey, K. V., Gani, R. **(2010)**. Systematic Modeling of Generic Multi-dimensional Model-based System for Batch Cooling Crystallization Operations. Computer Aided Chemical Engineering, Volume 29, 86-90.
29. **Singh, R.,** Gernaey, K. V., Gani, R. **(2009)**. A software tool for design of process monitoring and analysis systems. Computer Aided Chemical Engineering, Volume 26, 321-326.
30. **Singh, R.,** Gernaey, K. V., Gani, R. **(2008)**. Off-line design of PAT systems for on-line applications. Computer Aided Chemical Engineering, Volume 25, 423-428.

386030 Materials and Systems Design for Healthcare and Energy Applications

Abstract:

My research is focused on developing mathematical, computational, and experimental methods for rapid materials screening and systems design for pharmaceutical and solar fuels applications. My research plan in these two different areas is described as follows.

Research Area 1: **Designer Crystals for Healthcare Applications**

Current Collaborators: Prof. Doraiswami Ramkrishna (Purdue U.) and Dr. Hsien-Hsin Tung (AbbVie Inc.)

Description: The crystal structure and morphology are two critical determinants of the physicochemical properties such as solubility, hardness, cleavage, optoelectronic properties, electrical and heat conductivities, and piezoelectric effect of crystalline materials. The polymorphs (crystal structure) are determined in the early stage of crystallization process whereas the morphologies depend on growth, dissolution, and milling conditions. Figure 1 shows a scheme that probes processes at different length scales using a combination of experimental and mathematical tools to control crystal structure, shape, and size by tuning various environmental conditions. The proposed research is fundamental to crystal growth in different areas (*e.g.* the semiconductor industry, catalytic applications, and the pharmaceutical industry).

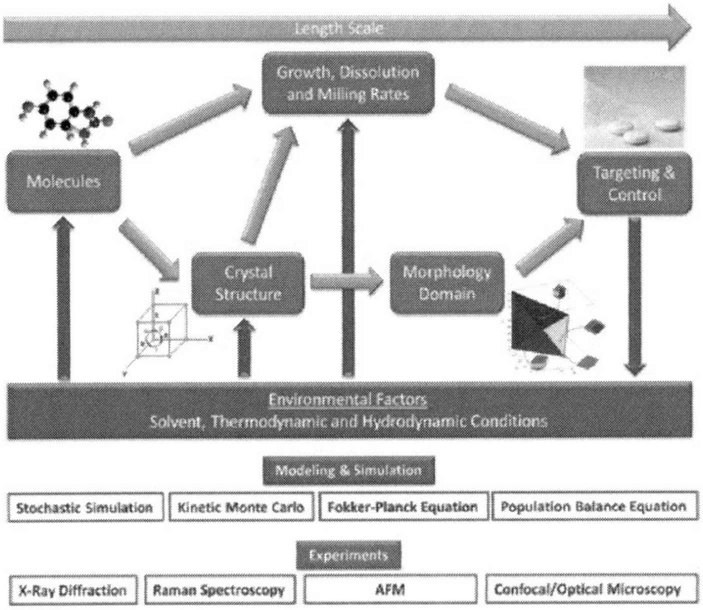

Figure 1: Research Plan for Identification of Structure-Property Relationships and Directed Synthesis of Crystalline Materials

Selected Publications:

Meenesh R. Singh, Jayanta Chakraborty, Nandkishor Nere, Hsien-Hsin Tung, Shailendra Bordawekar and Doraiswami Ramkrishna, "Image-Analysis-Based Method for Measurement of 3D Crystal Morphology and Polymorph Identification using Confocal Microscopy," Crystal Growth & Design, 12 (7), 3735-3748, 2012

Meenesh R. Singh, Parul Verma, Hsien-Hsin Tung, Shailendra Bordawekar and Doraiswami Ramkrishna, "Screening Crystal Morphologies from Crystal Structure," Crystal Growth & Design, 13 (4), 1390-1396, 2013.

Meenesh R. Singh and Doraiswami Ramkrishna, "A Comprehensive Approach to Predicting Crystal Morphology Distributions with Population Balances," Crystal Growth & Design, 13 (4), 1397 – 1411, 2013.

Meenesh R. Singh and Doraiswami Ramkrishna, "Dispersions in Crystal Nucleation and Growth Rates: Implications of Fluctuation in Supersaturation," Chemical Engineering Science, 107 (7), 102-113, 2014.

Doraiswami Ramkrishna and **Meenesh R. Singh**, "Population Balance Modeling. Current Status and Future Prospects," Annual Review of Chemical and Biomolecular Engineering, 5 (1), 2014

Research Area 2: **Artificial Photosynthesis for Solar Fuels**

Current Collaborators: Dr. Nathaniel Lynd (LBNL), Prof. Rachel A. Segalman (UC Berkeley), Prof. Alexis T. Bell (UC Berkeley) and Prof. Nathan S. Lewis (Caltech)

Description: Harnessing solar energy to produce liquid fuels is a promising route to support future energy demands. One possible route is to first produce hydrogen, which can be generated by photoelectrochemically splitting water. The solar fuel generators such as photoelectrochemical cells (PECs) involve a two-step conversion of sunlight to hydrogen using a photovoltaic (PV) cell directly (wireless) or indirectly (wired) connected to an electrochemical cell. My research in this area is focused on identifying materials and methods to develop an efficient, robust, cheap and durable PECs for water-splitting and carbon dioxide reduction. One of the design objectives here is to minimize the potential losses in the device using interface and systems engineering approaches.

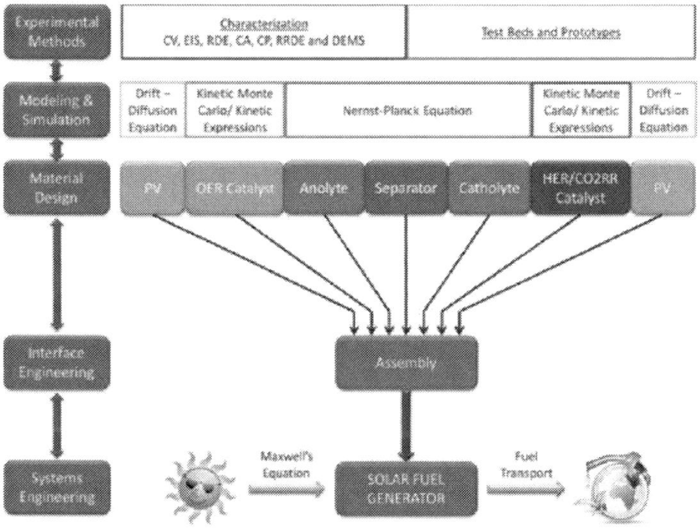

Figure 2: Research Plan for the Design of Cheap, Efficient, and Durable Solar Fuel Generators

Selected Publications:

1. **Meenesh R. Singh**, John C. Stevens, and Adam Z. Weber, "Design of Membrane-Encapsulated Wireless Photoelectrochemical Cells for Hydrogen Production", *Journal of The Electrochemical Society*, 161 (8), 2014

2. Jian Jin, Karl Walczak, **Meenesh R. Singh**, Chris Karp, Nathan S. Lewis, and Chengxiang Xiang, "Experimental and Modeling/Simulation Evaluation of the Efficiency and Operational Performance of an Integrated, Membrane-Free, neutral pH solar-Driven Water-Splitting System," *submitted*

3. Christopher M. Evans, **Meenesh R. Singh**, Nathaniel A. Lynd, and Rachel A. Segalman, "Improving the Gas Barrier Properties of Nafion via Thermal Annealing: Evidence for Dual-Mode Diffusion," *submitted to Macromolecules*.

380723 Engineering Microbial Production Platforms for Efficient Carbon Utilization

Abstract:

Microbial production platforms promise to be a sustainable alternative to today's petrochemical industry. By harnessing the breadth and flexibility of cellular metabolism to process available energy sources, 'living foundries' can be engineered that convert renewable sugar feedstocks directly into a number of enantiopure commodity and high-value specialty chemicals. Despite isolated successes, however, these platforms have been unable to scale to meet the gap created by increasing oil prices due to challenges inherent in driving carbon flux towards the desired product. Drawing on tools from synthetic biology, metabolic engineering, systems biology, and next generation sequencing and synthesis, my research has focused on the development of molecular strategies to increase the viability of these processes, and the elucidation of cellular responses to such intervention.

Here, I present my work towards the engineering of more efficient bacterial and eukaryotic systems. In my doctoral research (with Kristala Prather, Chemical Engineering, MIT), I developed transcriptional and post-transcriptional tools that dynamically divert carbon from central metabolism into a pathway of interest in *E. coli*. These tools pave the way for the development of new product classes within *E. coli* and allow for the optimization of existing pathways by reducing carbon waste. As a postdoc (with Michelle O'Malley, Chemical Engineering, UCSB), I am studying the effects of compartmentalized expression of metabolic pathways on productivity. This approach has tremendous potential to both locally concentrate metabolites and enzymes leading to greater productivity, and limit carbon loss to metabolic side reactions localized elsewhere. While these examples pursue more efficient carbon utilization, there still remains the challenge of sustainable, cost-effective feedstocks for these processes. Towards that end, I am also analyzing next generation sequencing data to identify novel biomass-degrading enzymes from anaerobic gut fungi and elucidate their hydrolytic strategies against diverse lignocellulosic substrates. Such research strives to secure sustainable feedstocks for biotechnology and presents a unique opportunity to examine novel design strategies in a non-model organism.

In the future, my research program will focus on the development of next generation microbial platforms with applications in sustainability, health and energy. My approach is to use the tools of data-driven science to formulate novel hypotheses and elucidate emergent systems-level properties of native systems, and engineer innovative solutions with synthetic biology which exploit robust and versatile features of microorganisms.

388992 Engineering Immunity: Design and Development of Customized Nanomaterials with Controlled Immunostimulatory Effects for Biomedical Applications

Abstract:

Engineered nanobiomaterials (ENMs) are rapidly emerging as a new class of materials that allow tuning of their physicochemical properties for biomedical and pharmaceutical applications, e.g., drug delivery, tissue engineering and vaccines. ENMs could interact with biological entities at the nano-bio interface, which could induce cascades of signaling pathways leading to the activation of immune responses including both innate and adaptive immunity. It has been identified that hydrophobicity, size, and surface charge are the possible parameters affecting the biocompatibility of ENMs. However, the detailed mechanisms by which ENMs could generate immunostimulatory effects at the nano-bio interface are still not fully understood. In light of this, my research will focus on immunoengineering, which will involve developing ENMs with tunable physicochemical properties, investigating the interaction of ENMs with immune systems, and further customizing their applications in vaccine development, tissue engineering, and cancer immunotherapy.

A Spotlight on My Research on Chemical Vapor Deposition

Aravind Suresh
Assistant Professor-in-Residence, Chemical and Biomolecular Engineering Department
University of Connecticut, Storrs, CT 06279
E-mail: aravind@engr.uconn.edu

Background: The research for my doctoral dissertation involved the development of high-temperature oxides with the dual functionality of catalytic activity for hydrogen generation and mixed conductivity for hydrogen purification. Most of my work was done on cobalt-doped barium cerate-zirconates [1, 2]. I also started the investigation on iron-doped barium zirconates, which was then continued by my advisor Prof. Benjamin Wilhite's group. We are now awaiting the publication of the results of the structural and electrochemical characterization of the material.

I have been an Assistant Professor-in-Residence at the University of Connecticut in charge of teaching the junior and senior laboratory courses in the department. During that time, I have also been working on different research projects in collaboration with various faculty members. For this poster, I am focusing the attention on some of my work in the field of Chemical Vapor Deposition (CVD) of polymers.

Spotlight: Initiated CVD (iCVD) and photo-initiated CVD (piCVD) are emerging techniques for creating conformal coatings of polymers on a variety of substrates, providing good control on functionality and minimizing the use of strong solvents [3, 4]. The challenge in iCVD involves decoupling the heated filament and the cold stage, whereas that in piCVD involves the identification of suitable photo-initiators.

During the course of my research on photo-initiated CVD (piCVD), I discovered the property of auto-polymerization in hexyl acrylate vapor when subjected to ultraviolet (UV) radiation. I studied the rate of deposition with respect to process parameters and reported the results through a communication [5]. I am now working on understanding the mechanism of the deposition process, which appears to be unique with respect to the literature on piCVD and also with respect to the polymerization and deposition of the same monomer using iCVD, in that the observed rate of deposition increases with time with an exponential trend and increases with film thickness with a linear trend. The behavior suggests that the deposited poly hexyl acrylate film itself acts as a source of free radicals driving the process. If that is indeed confirmed to be the case, hexyl acrylate vapor or the deposited poly hexyl acrylate film can be used as an initiator in piCVD for depositing polymers whose monomers may not themselves exhibit auto-polymerization under UV radiation.

I am working on the deposition of thermally stable, highly cross-linked polymers using iCVD in collaboration with Prof. Daniel Burkey at the Universtity of Connecticut and Dr. Salmaan Baxamusa at the Lawrence Livermore National Laboratory (LLNL). The focus of the project is to develop CVD polymers that are optically clear and resistant to attack by atmospheric oxygen on aging. CVD techniques that rely on plasmas to initiate the polymerization produced colored films that are susceptible to aging in the ambient atmosphere and iCVD is being explored as an alternative. The preliminary results thus far have been encouraging and testing will continue over the summer of 2014.

I am applying the technique of iCVD to understand the bonding characteristics at the interface between metals and polymers in collaboration with a group at the Institute of Materials Science at the University of Connecticut. They are working more on depositing metals on polymer films; I presented the option of depositing polymers on metal sheets, where the only restriction would be that the monomer of the desired polymer must be vaporizable under the working conditions of the instrument. The project, in its nascent stage, represents situations where CVD can be a valuable tool in a larger research project rather than the focus of the project itself.

Summary: I am looking for tenure-track positions in universities, with a focus on establishing a research direction encompassing two ends of the materials spectrum – thin-film polymer deposition and high-temperature oxides. I would also be interested in a position that combines an emphasis on teaching along with the opportunity of building a research group.

References:

[1] Suresh, Basu, Carter, Sammes, Wilhite, Journal of Materials Science (2010) **45**: 3215-3227

[2] Basu, Suresh, Carter, Wilhite, Journal of the European Ceramic Society (2011) **31**: 1421-1429

[3] Lau, Gleason, Macromolecules (2006) **39**: 3688-3694

[4] Baxamusa, Gleason, Advanced Functional Materials (2009) **19**: 3489-3496

[5] Suresh, Anastasio, Burkey, Chemical Vapor Deposition (2014) **20**: 5-7

390479 Fundamental Studies of Nanoscale Bio-Interactions

Abstract:

Nanoscale bio-recognition is an essential requirement in many biological processes (e.g. enzyme-subtract interaction, immune response, and signaling pathways). Although our knowledge of the molecular and structural basis of these interactions is limited, scientists have found a way to apply these interactions to several applications like affinity chromatography, bio-sensors, and drug delivery. My Ph.D. work focused on affinity membranes for bio-separations. It centered around the development (selection of ligands, modification of membranes) and testing (batch and dynamic adsorption experiments, modeling, estimation of binding capacity, dissociation and kinetic constants, effect of ionic strength and flux) of affinity membranes for the purification of different classes of biomolecules: monoclonal antibodies, lectins and MBP-fusion proteins. More recently, during my post-doctoral studies, I have continued to work on several projects with the common focus of protein-protein interactions, from separation to aggregation. The poster will give an overview of the research conducted in the following topics: (i) specificity of nuclear transport using QCM-D; (ii) insights into the lytic mechanism of antimicrobial peptide using supported lipid bilayer to mimic bacterial membranes; (iii) the use of AFM in force mode to measure the binding forces between different split-protein partners; (iv) biophysical characterization of amyloid aggregates, from dimers to fibrils; (v) charged membranes for the separation of proteins with similar molecular weight and charge. I have broadened my background as a chemical engineer, integrating it with a variety of bio-tools. I plan to continue my career maintaining this interdisciplinary approach in studying protein-protein recognition both as basic science (e.g. better understanding of biological mechanisms) and applied science (e.g. developing bio-sensors).

389776 Reaction Engineering: Bridging Fundamental Engineering Approaches with Biomass Pyrolysis and Transport in Zeolites

Abstract:

Biomass pyrolysis has been widely explored for its potential to generate a sustainable chemical source capable of producing synthetic fuels and chemicals. Lignocellulosic biomass specifically presents great promise with its carbon rich backbone and widespread prevalence. The high temperature thermal conversion of biomass to pyrolysis oil occurs on the order of milliseconds and converts the long chain biopolymers to a carbon-rich liquid crude. The kinetics of biomass pyrolysis is greatly complicated by significant heat and mass transport challenges. The complex fluid dynamics of the reactive liquid intermediate are only beginning to be understood, including the existence of vigorous bubbling and aerosol generation. Unlike traditional crude oil, the condensed bio-oil is highly oxygenated and acidic which often requires significant upgrading prior to end use. The reactive nature of the bio-oil complicates the upgrading process by presenting difficulty in heating and vaporization where the fluid properties can drastically effect the product distributions.

Additionally, the use of zeolites is widely utilized to catalytically upgrade fuels for hydrodeoxegenation, hydrogenation, and cracking. While new materials are being synthesized with transport lengthscales that are increasingly smaller, diffusional transport limitations are presenting in the small particles. The presence of these transport limitations remains a significant technical challenge. However, the use of several experimental and computational techniques is utilized to probe the mechanistic nature of these barriers, and a greater understanding is beginning to be revealed. In this work, original findings are overviewed covering biomass pyrolysis with aerosol generation mechanisms and rate-dominant diffusion limitations in hierarchical zeolites.

Materials for 4D biology: spatial and temporal control of the stem cell niche

Mark W. Tibbitt and Robert Langer, Massachusetts Institute of Technology

Interactions between mammalian cells and the surrounding extracellular matrix (ECM) are critical in the regulation of stem cell differentiation, cell migration, tissue morphogenesis, and disease progression. Cell decisions are often made by integrating these complex interactions, which are dynamic in 4D: three-dimensional space and time. Synthetic polymer-based hydrogels have emerged as an attractive cell culture platform to investigate the influence of these cell-matrix interactions on cell function in a systematic manner. Hydrogels are appealing mimics of the ECM on account of their high water content, tissue-like elasticity, and facile transport of nutrients, waste, and soluble factors. Furthermore, poly(ethylene glycol) (PEG) hydrogels can be formed under mild, cytocompatible conditions and are easily modified to present various biochemical and mechanical signals to cells during culture.

While cell-laden PEG hydrogel constructs have been employed for regenerative medicine and fundamental biological applications, there is still a lack of materials that allow the user to recapitulate the spatially heterogeneous and dynamic environment that cells interact with in vivo. My Ph.D. research has focused on the development, characterization, and application of photoresponsive PEG-based hydrogels that afford the user control of the mechanical and biochemical properties of the scaffold in both 3D space and time. This class of materials facilitates experiments that probe the relationship between spatially heterogeneous and dynamic signals from the ECM and cell function in a well-defined environment to study the fundamentals of the cell-material interactions, stem cell differentiation and plasticity, as well as the cellular response to local changes in soluble factor concentration.

Specifically, we have developed photoresponsive materials for the dynamic modulation of the ECM during 2D and 3D culture. These materials have been used to investigate stem cell response to dynamic changes in elasticity and the plasticity of stem cells during microenvironmental mechanotransduction. Importantly, we uncovered fundamental information about mechanical dosing and memory in mesenchymal stem cells that is regulated by YAP/TAZ signaling. Further, we have utilized photoresponsive hydrogels to develop photodegradable microspheres for the targeted delivery of soluble factors to study tissue morphogenesis and chemotaxis. Thin films of photodegradable hydrogel have been employed for the selective capture and release of cells for diagnostic application.

In my Postdoctoral research, we have extended these paradigms to develop smart and responsive polymeric systems for controlled delivery of therapeutics to intervene during disease progression. Specifically, we are focusing on the use of shear-thinning injectable hydrogels to deliver therapeutics for cardiac valve disease and cancer in a controlled manner.

390144 Utilizing Optimally Designed Peptide-Anchored Liposomes for Specific B Cell Activation and Vaccine Formulations

Abstract:

Lipid-based nanoparticles have been widely characterized for the delivery of small molecules, peptides and proteins. Liposomes containing surface-bound antigens are similarly being developed as vaccines and have been shown to induce strong immune responses when compared to soluble antigens by facilitating cross-presentation and enhancement of humoral responses. Yet so far no studies have focused on how specific liposome design considerations modulate immune responses at the single cell level to enhance CD4+ T cell help and antibody production through interaction and activation of specific B cells. Using an antigen-specific 3-83 Ig transgenic mouse model, we evaluated the role of liposome composition on the direct binding and internalization of liposomes with surface-displayed peptide antigens to antigen-specific B cells. Initially we focused on the role of surface PEGylation and antigen tethering. We hypothesized that while surface PEGylation has been shown to improve liposome stability in serum and enhance lymph node targeting in vivo, it may hinder binding and internalization through specific B cell receptors. Our studies showed that unilamellar PEGylated liposomes containing maleimide-functionalized PEG-lipids conjugated to cysteine-containing antigenic peptides (p31 and p0 peptides with high and no binding affinities, respectively) bound with varying intensities to B cells as a function of PEGylation (10-0% PEG): 5% PEG liposomes bound with the highest intensity to B cells, while 0% PEG liposomes containing peptide antigen anchored directly to the bilayer surface had significantly reduced binding as a result of steric hindrance. These conditions as well as a third condition (5% PEG with surface anchored peptide antigen which resulted in even lesser binding compared to 0% PEG liposomes) were evaluated for their ability to activate antigen-specific B cells in vitro. B cell activation was assessed at the single-cell level using flow cytometry to quantify the up-regulation of early B cell activation markers, such as CD69, CD86, CD40, and MHCII. As predicted, 5% PEG liposomes with tethered p31 resulted in the strongest antigen-specific B cell activation, while the weakest binding 5% PEG liposomes with surface anchored p31 resulted in the lowest levels of activation comparable to unstimulated cells. The incorporation of TLR4 agonist, monophosphoryl lipid A (MPLA), into the liposome bilayer significantly enhanced early activation (day 1) and helped sustain activation and B cell viability at later times (up to 4 days) in an antigen-specific manner. However, MPLA incorporation into liposomes interestingly provided no enhancement on B cell binding despite the presence of TLR4 on the surface of B cells, signifying MPLA-triggered activation is dependent on specific antigen-mediated uptake of liposomes by B cells. Finally, we evaluated the effectiveness of these 3 liposome conditions to induce antigen-specific humoral responses in naïve mice. Antigen-specific antibody titers in mice immunized with antigen-bearing liposomes containing MPLA were found to correlate directly to B cell binding profiles, with 5% PEG liposomes with tethered p31 resulting in titers ~log10-fold higher than 5% PEG liposomes with surface anchored p31. Current studies are focused on understanding the relationship between B cell binding and activation and the ability of these B cells to present helper peptides to CD4+ T cells, as well as how these liposomes can be utilized to probe and expand low-frequency antigen-specific B cells in spleens and lymph nodes post-immunization. We believe the results obtained in these studies will further the development and understanding of effective particle-based peptide vaccines.

New Materials for Chalcogenide Based Solar Cells

Tosun, B. Selin, University of Washington, Chemical Engineering

Global energy consumption is estimated to increase ~50% in the next 30 years (United States Energy Information Administration). Currently, the majority of the world's energy demand is met by burning fossil fuels, generating CO_2, a greenhouse gas that contributes to global warming. Covering only 0.1% of the earth's surface with 10% efficient solar cells can meet the energy demand of the entire world. However, in 2013, solar cells provided only 0.1% of the world's total energy supply. High cost of solar cells still limits their wide-spread use. Currently, the market is dominated by solar cells based on expensive, high purity, single-crystal silicon (c-Si) wafers, which are produced using high temperature processing. Solar cells that are based on thin film semiconductors, such as copper indium gallium diselenide (CIGS), are becoming less expensive than c-Si solar cells because their production consumes less energy and less material. With the excitement generated as a result of their high potential in terawatt-scale energy conversion, thin film solar cells have reached to high overall conversion efficiencies (~20.4%), nearly as high as c-Si solar cells (20-25%). Hence, the thin film solar cell market share is increasing very quickly, from 0.5% to 6% for CIGS during the last 6 years. Innovative products are emerging rapidly, such as Dow Solar's roof shingles, which are expected to increase widespread-use in the coming decade.

During my PhD at University of Minnesota, Chemical Engineering and Materials Science, under the supervision of Prof. Eray S. Aydil, I focused on balancing the industrial needs with long-term solutions, mainly on CIGS solar cells. Thin film CIGS solar cells are built layer-by-layer on substrates (glass or flexible steel foil) through a series of thin-film deposition processes. Typically, substrates are coated with ~1 μm thick molybdenum layer, then by ~2 μm thick light-absorbing CIGS film. Following, a ~50 nm thin cadmium sulfide (CdS) film, also called as the buffer layer, is deposited through chemical bath deposition (CBD) to form the p-n junction. The photogenerated charges are separated at the CIGS-CdS interface to create electric current. While Cd is toxic, the high-efficiency solar cells are obtained only with CBD-CdS. The solar cells are completed by depositing ~100 nm thick zinc oxide (ZnO) window layer followed by a ~200 nm thick conductive transparent tin-doped indium oxide (ITO) film. The CIGS solar cell is very sensitive to moisture; their efficiency degrades with time, particularly when the temperature is high. Sealing with glass increases the weight, the cost, and precludes the possibility of making flexible solar cells. My thesis concerned the layers above the CIGS absorber, particularly developing alternative materials for CdS and ZnO layers that (i) increases their long-term stability, and (ii) minimizes Cd usage. Towards this end, I worked in two parallel directions:

First, I developed an alternative material to ZnO that stops grain boundary diffusion of water to the CdS-CIGS interface: I hypothesized that facile diffusion of water through grain boundaries of polycrystalline ZnO film is the culprit in diminishing the damp-heat stability of solar cells. Accordingly, I worked on using semi-crystalline tin dioxide (SnO_2) films without grain boundaries in lieu of ZnO. The damp-heat stability test emulates the aging of the solar cells under high humidity (85% relative humidity) and high temperature (85 °C) conditions. I used magnetron sputtering to deposit SnO_2 films with various types of microstructure and studied the correlation between the deposition conditions and the film properties. My best results showed over 95% preservation of the peak efficiency even after 240 hours at 85 °C and 85% relative humidity. In comparison, under identical test conditions, the solar cells without the SnO_2 layer lost nearly 80% of their initial efficiency within 24 hours after commencing the test, where 100 hours is standardized as 20-year-lifetime for outdoor efficiency. A patent is filed by the University of Minnesota and Dow Solar Solutions; the results have been published in three journal articles and presented at two international conferences.

Second, I studied CBD of $Zn_{(1-x)}Cd_{(x)}S$ ($0 \leq x \leq 1$) to replace CdS. It is desirable to reduce, or even eliminate, the toxic Cd. CBD of transparent zinc sulfide (ZnS) has been the highest efficiency buffer layer among the alternatives, but is not as efficient as CBD-CdS. A ternary alloy of these two materials, $Zn_{(1-x)}Cd_{(x)}S$, can provide the benefits of both layers to achieve higher efficiencies. I demonstrated a CBD method of $Zn_{(1-x)}Cd_{(x)}S$ where the composition could be changed controllably. I studied the structure and composition variation along the film thickness using a suite of material characterization methods including x-ray diffraction, spectroscopic ellipsometry and depth profiling by Auger electron spectroscopy on these films. An additional distinguishing feature of my work was the addition of ethylenediaminetetraacetic acid (EDTA) in the solution. EDTA chelates metals, slows the deposition and improves the film's crystallinity. Using a combination of Auger depth profiling and spectroscopic ellipsometry, I discovered that a CdS rich layer forms at the film/substrate interface due to the faster reaction of Cd than Zn. In parallel, I also worked on the improvement of the traditional CBD approach by developing a continuous-flow CBD system (CF-CBD) to minimize toxic waste generation. Although the traditional CBD-CdS provides high efficiency solar cells, it is a batch process and wastes materials. Homogenous nucleation and growth of CdS in the liquid solution competes with the deposition reaction. I designed and built a novel CF-CBD reactor that allows the heating of the substrate with minimal heating of the liquid solution. This reduces homogeneous nucleation and growth of CdS particles and thus the precursor waste and the production cost while depositing films uniformly on large substrates, 10 cm in diameter, and also provides excellent control over the compositional depth profiles. These results are presented in three international conferences, and published in three peer review articles.

In the last year of my PhD, I started research on the synthesis of an earth-abundant material that can replace CIGS. While CIGS solar cells have high efficiency, consumption of indium in flat panel display market and the low abundance of this element limits the cost reductions in CIGS solar cells. CZTS is an earth-abundant equivalent to CIGS and has recently shown to yield over 12.6 % efficiency. A thin CZTS film can be deposited using nanoparticles capped with organic ligands dispersed in organic solvents, such as toluene. However, organic ligands are resulting in an undesired C incorporation. I eliminated organic solvents by a ligand exchange procedure and achieved colloidal dispersions of CZTS nanoparticles in water forming environmentally friendly ink. A patent has been filed by the University of Minnesota, published in *Chemical Communications*, and presented in an international conference.

Since my graduation, I have been pursuing my studies in thin films solar cell area as a Post-Doctoral Researcher at University of Washington, Chemical Engineering, under the supervision of Prof. Hugh W. Hillhouse. Here, I focus on 2 projects: (i) improving the optoelectronic properties of solution processed $CZT(S_xSe_{1-x})_4$, and (iii) alternative solution processing methods on Halide Perovskite (HP) based solar cells. Even though, the $CZT(S_xSe_{1-x})_4$ solar cells developed very rapidly, the current device efficiencies still do not come close to single-junction detailed balance (~31%), mostly due the low open-circuit voltages (V_{OC}). I studied improving the optoelectronic properties, and therefore in V_{OC}, at various processing conditions by reducing the potential fluctuations originating from non-stoichiometries in the film. This work is going to be presented in the Materials Engineering and Science Division, and a manuscript is under preparation. Secondly, the recent breakthroughs on solar conversion efficiencies (15%) with HP materials attracted a great attention. Currently, the poor microstructure of HP solar cells is believed to be the limiting factor for higher solar conversion efficiencies. In the light of that, I am focusing on various solution processing methods thought different in-situ and post deposition techniques. In addition, I am working with a graduate student to build and develop a custom-made Multicolor Photoluminescence instrument for identifying the optoelectronic properties of the materials as a function of absorber film depth.

My professional aim is to contribute my research skills toward the improvement of technology, in particular, solar cells as faculty member while training graduate and undergraduate students. During my PhD, I have instructed two core Chemical Engineering classes: Reaction Kinetics and Reactor Engineering, and Mass Transport and Separation Processes; and I have mentored six undergraduate students, three of which joined to PhD programs. As a Post-Doctoral associate, I have been mentoring two undergraduate students, and working with three graduate students on various projects. I believe that my accumulation on topics as Reaction Kinetics, Material Science & Chemistry, Material Characterization Techniques, Semiconductors Fundamentals, Transport Phenomena, Thermodynamics, and experience on industrial collaborations prepare me well for a faculty position in energy related Engineering programs. I consider solar cells as an economically viable energy source for the very near future. As a faculty, I am planning to conduct research on dilute magnetic semiconductors for solar cell applications, especially via solution processing with my background in both PhD and Post-Doctoral studies.

393732 Enabling the Spectroscopic Tools That We Need to Get the Hidden Information We Want

Abstract:

Catalysis lies at the heart of the chemical industry and could be considered as the society's gateway to the sustainable and petroleum-free future for the fuel and chemical production. Since the energy needs of the developed world are currently over-dependent on the utilization of finite mineral resources, enormous research efforts aim to the shift from petroleum supply to natural gas and renewables, such as biomass, for fuel and chemical production in an economic and efficient manner. In that sense, it is common knowledge that catalysts matter and can provide better ways to make new materials from various sources. Spectroscopy is the enabling tool for the knowledge based and hence time-effective catalyst design, which goes beyond the scope of an empirical observation. Monitoring the events taking place in such materials under realistic conditions is crucial for understanding the reaction mechanism of many important chemical processes and would allow the rational design of new or improve the existent catalysts.

Willing to investigate in depth the structural changes of supported metal oxide catalysts, my research as a graduate student relied on the utilization of vibrational spectroscopies under realistic reaction conditions. Operando Raman spectroscopy of MoO_3/TiO_2 catalysts for the Oxidative DeHydrogenation of ethane revealed that the terminal Mo=O functionalities present on the dispersed species could be responsible for the non-selective ethane to CO_x reactions underlining the importance of the anchoring Mo–O–*Support* bonds as the active site. Armed with my already acquired knowledge supported by my inherent eagerness to explore new scientific areas, my postdoctoral research lies on the critical issue of solvent selection in the field of biomass to fuels and chemicals applications. By coupling thorough spectroscopic measurements with ab-initio DFT calculations, we were able to provide an explanation of the enhanced 5-Hydroxylmethylfurfural (HMF) stability if polar aprotic solvents used as a co-solvents.

What I envision in my research plans is the rational utilization of spectroscopic techniques coupled with simultaneous measurements of catalytic reactivity (Operando spectroscopy). In this respect, my ultimate goal is to unravel novel structure-function relationships for heterogeneous and homogeneous catalytic systems. My approach lies first at the design of a multi-spectroscopic reactor that would be able to monitor under realistic reaction conditions the structure of the catalyst by means of Visible Raman, Surface Enhanced Raman and/or FTIR spectroscopies. However, one crucial drawback regarding Operando spectroscopy lies on the fact that cannot be directly applied for the study of homogeneous catalytic systems. For that reason, a new alternative will be proposed namely Operando Solvation that will allow to monitor changes in solvent structure that could be directly correlated with data of the catalytic reactivity for biomass to biofuels and chemical reactions.

Refernces

(1) Tsilomelekis, G.; Boghosian, S. *Phys Chem Chem Phys* **2012**, *14*, 2216.

(2) Tsilomelekis, G.; Boghosian, S. *Catal Sci Technol* **2013**, *3*, 1869.

(3) Tsilomelekis, G.; Josephson, T. R.; Nikolakis, V.; Caratzoulas, S. *Chemsuschem* **2014**, *7*, 117.

385917 Engaging Students in the Capstone Laboratory Using New Processes & Novel Pedagogy

Abstract:

One of the greatest problems facing engineering education in the 21st century is developing ways to engage modern students in the learning process, as these students seem less receptive to the traditional transmission model of lectures and experiments. Further challenges present themselves as engineering disciplines continue to evolve, both in terms of technological advances and scope, necessitating updates to engineering curricula at the university-level. During my PhD research with Prof. Jeffrey R. McCutcheon and Prof. Daniel D. Burkey, I was able to explore possible solutions to these challenges by changing the way the capstone laboratory course was taught at the University of Connecticut, both by building new experiments and by using learner-based pedagogies.

Updating Experimental Curriculum

The capstone laboratory class at the University of Connecticut was a standard unit operations laboratory with fluid mechanics, heat transfer, and pilot-scale separations experiments. In order to better prepare students for the broader applications of modern chemical engineering, several experiments were added that include chemical vapor deposition and bacterial fermentation kinetics in a bioreactor. While I assisted in the development of these experiments, I was principally responsible for the development of several membrane separations experiments, including a reverse osmosis system, a forward osmosis test system, and an osmotic heat engine capable of generating electricity from salinity gradients. All three of these systems serve dual roles as both educational platforms and as research tools. Additive manufacturing elements were also added to the course by the development of a small-scale reactor design experiment, which allow students to test continuous flow reactors they design and print with a 3D printer. For all new experiments, it was important that they all be based on some chemical engineering fundamental, to allow students to see how what they have learned can be used once they leave school.

Improving Engagement with Gamification

In order to improve student attitudes toward the laboratory course, optional game-based elements were introduced. Students were divided into teams that would all complete a common set of experiments. Students were encouraged to complete optional tasks designed to assist with laboratory report preparation. Completion of these tasks gives student teams non-grade points. At the end of the semester, the team that had collected the most points would receive a non-grade based reward. Students have overwhelmingly embraced this system, saying it made the class feel memorable and kept them interested in the course throughout the semester. Other faculty in the University of Connecticut Chemical Engineering Department have adopted some elements of this course into their own.

Teaching & Research Interests

I seek a position that primarily focuses on teaching. I am comfortable teaching a capstone laboratory course, as I have served as a teaching assistant for the course since Fall 2009 and have co-taught the course with the instructor of record since Fall 2011. I also have an interest in teaching courses at other levels, particularly fluid mechanics, unit operations/separations, process design & safety, or first-year engineering courses, as I have recently helped teach a pilot version of a project-based freshman design course. My previous experience also makes me highly qualified to teach specialty courses such as membrane separations, water quality engineering, or engineering ethics. My research interests include applying novel engineering pedagogies such as the ones I have developed as part of my doctoral work. I am very interested in finding collaborators and doing long-term studies with faculty members at other universities. My other research interests are related to membrane separations, with desalination and engineered osmosis process engineering as a major focus.

377937 Nonequilibrium Self-Assembly and Structures

Abstract:

My research focuses on nonequilibrium self-assemblies (NESAs) that require energy supply to maintain their order or, as cells, to "survive". Nonequilibrium assemblies are energy (external field) driven systems that reside in steady states or long-lived metastable structures, some of which can be highly dissipative. DNA transcription machinery, self-propelling micelles, liquid crystals under external fields, cytoskeletal fibers, and molecular motors are all examples of various aspects NESAs. In my research, these complex assemblies are studied using the tools of nonequilibrium thermodynamics. Until now, NESA in artificial systems has been achieved by modifying the interactions between the constituent objects/parts. In the proposed methodology, I will look at the problem from a fundamentally different perspective – namely, I will keep the interactions between the components intact but instead change the agitation mode to achieve different structures. For this purpose, I aimed to control the particle SA in nonequilibrium regime by applying local anisotropic vibrations (for the solvent or particles or both), which effectively translating into anisotropic temperature. According to the equipartition theorem at thermal equilibrium, kinetic energy of the particles is equally divided into all their degrees of freedom translating into isotropic temperature. This theorem does not hold outside of equilibrium and, in principle, one can envision situations in which different degrees of freedom have different "thermal" energies. In my research, I have first studied such hypothetical situations by means of modeling. These experiments validated my hypothesis that the ultimate shape of the assemblies can be controlled by the mode of anisotropic agitation. With this foundation, my current and proposed research will focus both on understanding the fundamental implications of having anisotropic temperature as well as implementing such NESA systems in practice. In terms of theory and modeling, I will implement multiscale simulations that bridge the gap between the molecular (or nano) scale of the non-equilibrium agitating particles moving anisotropically, and the larger particles that are being agitated. During the simulations, I will perform agglomeration tests by measuring the orientational ordering (to probe and quantify the structure of the growing assemblies). I will also calculate energies and entropies at various stages of NESA and compare the results with those of nonequilibrium coarse-grained field theories. Simultaneously, I will implement several system in laboratory practice – these systems will be based on larger assembly components being surrounded by smaller magnetic particles that can be anisotropically "jiggled" by time-varying external fields. These magnetic particles are the non-Brownian particles effectively creating an anisotropic temperature field experienced by the larger particles. The ultimate objective – and, indeed, great hope – of my work, is that I will be able to engineer self-assembling structures of arbitrary shapes and properties by designing appropriate agitation "schedules". Of course, magnetic agitation is but the first step in these directions, and I also envision other classes of systems based on acoustic or pressure waves, and also on rapidly varying external temperature gradients. This work will create a novel experimental test-bed for studying non-equilibrium self-assembly under well-defined conditions. It can also lead to practical development in designing efficient bottom-up micro- and nanomanufacturing schemes.

Molecular dynamics simulation has employed to test the above hypothesis. Several simulations with various settings were performed including change of the long-range interactions between the beads, changing the density of the system, relative diameters. The anisotropy in temperature would act as an external field that can form various patterns of self-assembled (SA) structures including hexagonal and rod shaped considering hydrodynamic and friction effects. It is found that different agitation modes play crucial form in obtaining different SA patterns. These patterns are found in DNA-coated particles, colloids, polymers, and supercooled fluids.

As a faculty candidate, I would apply my beyond equilibrium knowledge to the SA in magnetic fluids in the presence of a magnetic field as an external force and study different effects such as static and alternative field to develop a complet theoretical approach for dynamic self-assembly far from equilibrium. The results of this study would be a basis for more complex NESA systems such as photoswitchable nanoparticles or to control NESA systems that are coupled with chemical reaction networks.

374568 Constructing, Screening, and Evolving Therapeutic Proteins

Abstract:

Protein-based therapeutics are beginning to come of age, but the development of proteins tailored for specific diagnostic and therapeutic applications remains an ongoing challenge. Yeast surface display (YSD) and other display technologies provide powerful approaches to the high-throughput isolation and evaluation of proteins with characteristics suitable for therapeutic settings. However, these techniques require judicious library construction, reformatting of candidate binding proteins prior to most biological characterizations, and are normally limited to screening with the chemical diversity encoded in the 20 canonical amino acids. The work presented here describes the combination of YSD with synthetic antibody libraries and noncanonical amino acids (ncAAs). These efforts have yielded 1) an extremely efficient approach to isolation of binding proteins in conventional YSD format; 2) a switchable display/secretion technology for rapid isolation and immediate production and characterization of candidate binding proteins in soluble form; and 3) a platform for the construction and evaluation of bioconjugates on the surface of yeast.

The discovery of new antibodies remains an important aspect of therapeutic discovery. Yeast surface display (YSD) and other in vitro technologies have emerged as powerful alternatives to conventional immunization and hybridoma technologies, but these in vitro approaches require the construction of antibody libraries with amino acid diversities and loop length variations that enable the efficient discovery of high affinity antibodies. Synthetic antibody libraries described by Sidhu and coworkers have previously been shown to yield high affinity binding proteins when employed in a phage display format. We utilized one of these libraries to construct a billion-member synthetic antibody library in a YSD format in order to take advantage of the quantitative screening capabilities afforded by YSD. Initial screening with model antigens indicated that the Sidhu library yielded full-length binding proteins more efficiently than a YSD nonimmune antibody library previously constructed in the Wittrup laboratory. We have used the synthetic antibody library to isolate a suite of antibodies targeting cancer-associated fibroblasts (CAFs), an understudied component of many solid tumors. We screened the synthetic antibody library for proteins recognizing both the human and murine forms of fibroblast activation protein (FAP), a membrane-bound protease selectively overexpressed on the surface of CAFs. Our screening campaign yielded more than 50 unique antibodies recognizing FAP, with close to 30 exhibiting crossreactivity for both the human and murine form of the protein. Secondary screens for binding enabled efficient identification of candidates recognizing multiple epitopes of FAP, and further development has resulted in a suite of full-length murine antibodies. Pharmacokinetic experiments in mice have confirmed that these antibodies are suitable for in vivo work, and further evaluation of the antibodies as agents for the therapeutic targeting of CAFs is underway. The general antibody isolation and characterization approaches used in this work are accessible to many single-investigator laboratories and should be applicable to the development of antibodies against numerous therapeutic targets.

While the combination of conventional YSD and high-quality synthetic antibody libraries is quite powerful, conventional display technologies are limited by the requirement of converting displayed proteins into soluble form prior to characterization at the end of a screening campaign. Toward this end, we have developed a switchable display/secretion platform that enables the rapid production of soluble reagents immediately following screening efforts without the use of any cloning steps. In our implementation, we use a ncAA-mediated amber suppression event to control the display of an antibody-like scFv-Fc on the yeast surface. Flow cytometry results have confirmed that the presence of the ncAA in the induction media enables display of full-length scFv-Fcs on the yeast surface. This system has the necessary characteristics for screening and soluble reagent production: in model enrichment experiments, a single round of bead-based selections enables approximately 500-fold enrichment of clones binding the target antigen, while yields of secreted scFv-Fcs are consistently upwards of one milligram per liter. Encouraged by these initial results, we have constructed synthetic antibody libraries in the switchable format and screened them against epidermal growth factor receptor (EGFR), a validated cancer antigen. ScFv-Fcs secreted from enriched populations of the library specifically recognize mammalian cells expressing EGFR, confirming both the success of our screening and the recognition of EGFR in its native form. These results suggest that this switchable screening format will enable rapid characterization of additional biological effects of candidate binding proteins (e.g., receptor clustering or downregulation, cellular proliferation, cellular differentiation, or apoptosis) in the future. The implementation of the switchable display/secretion technology enables cloning-free approaches to the generation of specific binding reagents and functional evaluation of these proteins for cancer targeting and other applications.

Although established display technologies have yielded a startling array of engineered proteins with therapeutic properties, the limited chemical repertoire normally found within proteins can impede efforts to tailor proteins for use in specific therapeutic applications. For example, conventional display technologies are rarely suitable for the characterization or engineering of antibody-drug conjugates or other bioconjugates. To overcome this limitation, we have adapted the switchable system described above to enable the incorporation of ncAAs with reactive functional groups into proteins displayed on the yeast surface. We first used the display/secretion system to confirm that several ncAAs, including ncAAs containing ketone or azide groups, can suppress the amber codon in the switchable system in cells expressing an aminoacyl-tRNA synthetase of appropriate specificity; this provides indirect proof of the display of proteins containing reactive side chains on the surface of yeast. Flow cytometry experiments further indicate that cells displaying proteins containing these functional groups can be selectively modified using oxime ligations (ketones), copper-catalyzed azide-alkyne cycloadditions (azides), or strain-promoted cycloadditions (azides) as determined by detection of reaction products on the surface of yeast. Thus, this approach enables both efficient display of proteins containing ncAAs on the surface of yeast and facile, site-specific protein modification, giving this system all of the necessary components for constructing bioconjugates on the surface of yeast. We anticipate that this approach will offer unprecedented throughput for constructing, evaluating, and evolving antibody-drug conjugates and other small molecule-protein conjugates, opening up new possibilities for the development of therapeutics that simultaneously exploit the favorable properties of both proteins and small molecules.

Multiscale Modeling and Simulation of Biological and Polymeric Systems

Balaji Iyer Vaidyanathan Shantha
Swanson School of Engineering, University of Pittsburgh, Pittsburgh, PA-15261

Engineering novel devices with self-regulated intelligent functional response involves seamless integration of materials with multiple components and different properties. Knowledge of structure-function relationships at various length and time scales and harnessing of these relationships to establish specific design criteria is a key requirement to achieve the goal. In view of this requirement, there is an indispensable need for development of sophisticated theoretical and computational models, which could be utilized as valuable tools for designing intelligent multifunctional materials. Recent advances in nanoscale synthesis, functional self-assembly and molecular cell biology has further fueled this demand specifically in macromolecular and biological soft matter systems.

My research interests are in the field of structure-function relationships and multiscale modeling of polymeric and biological systems. This interest stemmed from an intriguing question on dynamics of ring polymers. Notably, the absence of free ends and the complex static structure of ring polymers in presence of topological constraints posed specific challenges different from that of their linear counterparts. I earned my doctorate degree in Chemical Engineering for the study aimed at developing scaling arguments and constitutive relations that connect structure and dynamic response of ring polymeric systems. The experimental verification of some of our theoretical predictions emerged from advances in use of fluorescently tagged DNA strands as model polymers. These advances opened up new exciting avenues for addressing problems in molecular biology from a polymer physics perspective. Specifically, the complex multiscale problem of packaging a meter long DNA within the cell nucleus only a few microns in diameter piqued my interest. I joined Prof. Gaurav Arya's group, Department of NanoEngineering, University of California, San Diego, to devise coarse-grained models for facilitating the study of eukaryotic genome organization within the interphase cell nucleus. Currently, I am seeking ways to combine polymer theory with biologically inspired material design in Prof. Anna Balazs' group, Department of Chemical and Petroleum Engineering, University of Pittsburgh. Computational design of synthetic nanoparticle networks having the desirable features observed in natural dermal armor of animals is the motivation behind this project.

My research goal is to conceive and develop theoretical models and computational tools to establish design criteria for multifunctional and intelligent self-regulated systems at the interface of materials and biology. In this pursuit, I envision the development of multiscale approaches that are inherently multifunctional and can be readily adapted for studying dynamic response. In particular, development of these approaches in synergy with experiments can be valuable to the growth of new technologies that enhance the quality of life. From an academic perspective, the development of multiscale approaches to modeling provides ample opportunities for students to gain a fundamental understanding of interaction between components of a system at different scales, while simultaneously equipping them with the firm knowledge base for developing future technology.

For more information please visit: http://iyerb.weebly.com/

388037 Bioengineering New Solutions for Pediatric Diseases: Platforms to Improve the Treatment of Brain Tumors and Juvenile Diabetes

Abstract:

Pediatric brain tumors and juvenile diabetes (Type I) represent two of the most devastating diseases afflicting children. While significant advances have been made to improve the survival rate of patients afflicted with both diseases these solutions are often at a cost to the overall quality of life for survivors. We foresee tremendous opportunity to utilize the latest advances in Biomedical and Chemical Engineering to develop new solutions that specifically address current challenges faced in the management of pediatric brain tumors and juvenile diabetes.

The first part of my poster will describe the work completed during my Ph.D. at the University of Washington, which was aimed towards the development of multifunctional nanoparticles to aid in the diagnosis and treatment of medulloblastoma (MB) tumors, the most common form of brain tumor afflicting children. Currently, children afflicted with MB require aggressive neurosurgery and the success for tumor resection depends on the accuracy of tumor margin delineation, of which current techniques demonstrate variable success. Biological barriers continue to limit access of therapeutics and contrast agents to tumor cells. To overcome these limitations, we are developing tumor-specific, optical/MR imaging nanoparticles specifically engineered to transverse biological barriers and aid in the diagnosis, staging, resection of brain tumors, and delivery of therapeutics to brain tumors.

The second part of my poster will focus my current postdoctoral research project in Professor Robert Langer's lab at MIT aimed towards the engineering of a bioartifical pancreas. Type 1 diabetes accounts for 10% of all diabetes cases, is typified by a complete deficiency of insulin production, and stems from an autoimmune response in affected individuals that leads to the T-cell mediated destruction of insulin producing β-cells localized in the pancreas. Exogenous cell therapy, in the form of islet transplantation has been intensely investigated as a method to restore glycemic control. However, due to host rejection of transplanted cells and the lack of suitable encapsulation materials or specific immunosuppressive therapies it's widespread clinical application has been limited. Our efforts have been two fold and focused on both understanding the mechanism by which current encapsulation materials become rejected and innovating novel superbiocompatible materials that can mediate foreign body reactions and fibrosis. Significantly, we have developed a prototype device that is able to treat diabetic rodents for over 6 month.

Novel Material Design via Optimized Multiscale/Multiphysical Models

Sesha Hari Vemuri, Myung S. Jhon, Lorenz T. Biegler

Department of Chemical Engineering, Carnegie Mellon University, Pittsburgh, PA 15213

The confluence of engineering and science in the form of nanotechnology has unleashed tremendous opportunities for novel materials and systems bringing about a paradigm shift to the human civilization in the 21st century. New materials and systems have helped us to achieve what was previously though impossible in a broad range of areas including energy production and storage, water purification, information system, and complex fluids systems. Obtaining a thorough understanding of how the atomistic building blocks of the materials influence the overall performance at the systems scale is the holy grail of nanotechnological research. Researchers face a multitude of challenges which include disparate time and length scale interactive phenomena that span across multiple physical realms. Although there have been significant enhancements in experimental and theoretical methodologies over the past few decades to investigate these multiscale and multiphysical processes, it can often be expensive in both time and resources.

In my research, I try to solve the challenge of providing a complete picture of the atomistic to system level integration by developing multiscale/multiphysical modeling framework combined with optimization tools to predict and identify novel materials in a given engineering system via atomistic architecture to provide desired properties [1]. The modeling methodology is provided via case studies in polymer electrolyte fuel cells (PEMFC), micro/nano hydrodyanamic systems, thin films [2], and NEMS. We utilize the mesoscale modeling methodologies originating from the kinetic theory approach called lattice Boltzmann method (LBM) [3] as our centerpiece and integrate the lower level (atomistic/molecular) and higher level (continuum/systems scale) models, thereby obtaining seamless information exchange without losing physical essence of the system. In PEMFC, we were able to identify design criteria such as the atomistic structure of possible membrane materials, minimum catalyst loading & catalyst micro-geometry, and hydrophobicity distribution in gas diffusion layer (GDL). We began by performing computationally intensive atomistic simulation of hydration and proton conduction phenomena in candidate polymer electrolyte membrane materials and combined the information to molecular dynamic simulation via reduced order models (ROM). The complex multiphase interactions between the gaseous diffusing reactants and the liquid water/vapor mixture in complex porous gas diffusion layer are simulated via LBM. Since the kinetic based theory assumes a pseudo-particle picture, these interparticle interactions can be easily realized. We were then able verify previous studies [4] on how liquid water blocks the path for the oxygen to reach the cathode catalyst layer via GDL, thereby introducing further mass transfer limitations to the PEMFC performance. Furthermore, we integrate the atomistic/molecular/LBM model with ROMs and our optimizing solver for providing design criteria of minimum catalyst (Pt) loading while maintaining the same reaction kinetics.

In addition, I will also present our benchmark case study of thin film polymeric systems we also developed hybridized LBM/MD model which can efficiently simulate complex flow in confined geometries via exchanging the velocity and forces at the mesoscale/molecular model interface.

[1] P.S. Chung, R. Smith, S.H. Vemuri, Y.I. Jhon, K. Tak, I. Moon, L.T. Biegler, and M.S. Jhon, *J. Appl. Phys.*, 111, 07B712 (2012)

[2] S.H. Vemuri, P.S. Chung, R. Smith, G.Y. Yeom, Y.I. Jhon, N.E. Lee, L.T. Biegler, and M.S. Jhon, IEEE Trans. Magn., 48, 4277 (2012)

[3] S.H. Vemuri, H.M. Kim, P.S. Chung, and M.S. Jhon, *J. Appl. Phys.*, 115, 17B749 (2014)

[4] L. Hao, P. Cheng, Journal of Power Sources, 195 3870 (2010)

Massachusetts Institute of Technology

Ab initio discovery of energy conversion pathways

Lee-Ping Wang
Ph.D. Advisor: Troy Van Voorhis, MIT. Postdoctoral Advisors: Todd J. Martínez, Vijay S. Pande, Stanford U.
Department of Chemistry, Stanford University. Email: leeping@stanford.edu

STANFORD UNIVERSITY

Discovering new reaction pathways with an *ab initio* nanoreactor

Introduction: Recent methodological and computational advances harken the advent of theory and computation playing a principal role in discovery. I introduced the *ab initio* nanoreactor—a highly accelerated first-principles molecular dynamics simulation of chemical reactions that discovers new molecules and mechanisms without preordained reaction coordinates or elementary steps.

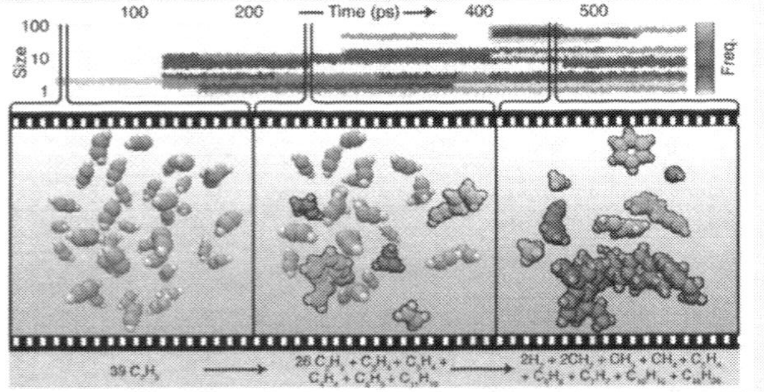

Figure 1: Timeline of a nanoreactor simulation movie. The simulation begins with a collection of acetylene molecules (C, teal; H, white). New molecules are automatically highlighted with molecule-specific colors to indicate the observed reactivity.

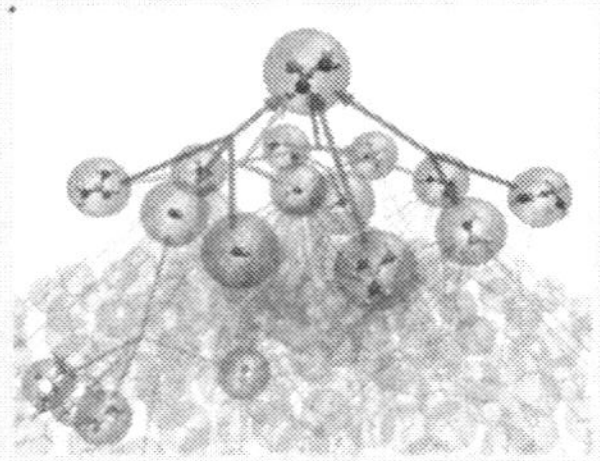

Figure 2: The nanoreactor discovered over 600 compounds, connected by over 1000 pathways. We represent this vast reaction network as a pyramid with a molecule of interest at the top tier (red sphere; $CO(NH_2)_2$). Colored arrows are reaction pathways that link the top molecule with reactants and products on the second tier (blue spheres).

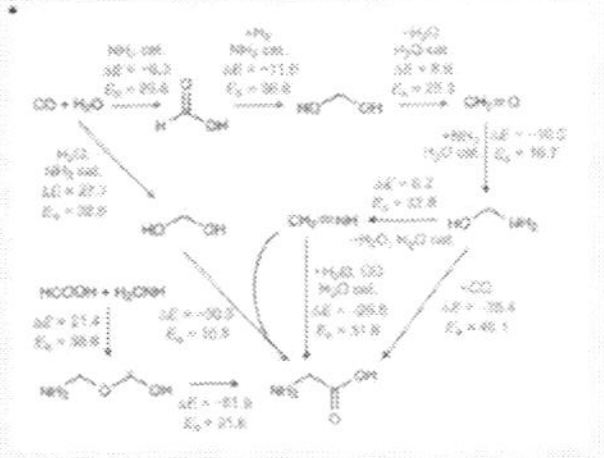

Figure 3: The nanoreactor reveals new pathways for amino acid synthesis from primitive compounds proposed to exist on the early Earth, providing new insight into the classic Urey-Miller experiment. Here the amino acid glycine (bottom) is formed via four different pathways from the starting materials: CO, H_2O, CH_4, NH_3 and H_2.

Outlook: I will use the pathways discovered by the nanoreactor to parameterize kinetic models and formulate design principles for designing improved catalysts for energy conversion. Furthermore, I anticipate that the nanoreactor will contribute to our future understanding of complex reactivity in natural systems by providing novel hypotheses for reaction mechanisms in arenas as diverse as combustion chemistry, atmospheric chemistry and astrochemistry.

Publication: Wang L-P, Titov A, McGibbon R, Liu F, Pande VS and Martínez TJ. "Discovering chemistry with an *ab initio* nanoreactor", *Nature Chemistry* **2014**, doi:10.1038/nchem.2099 (also News of the Week in Nov. 10 issue of *C&E News*.)

Renewable energy conversion mechanisms

Introduction: I have a long-standing interest in renewable energy and the efficient conversion of solar energy into chemical fuels. During my Ph.D. research, I found that ruthenium and cobalt based catalysts for the oxygen evolution reaction (OER) operate via different mechanisms, implying that fundamentally different models are needed to describe the OER kinetics.

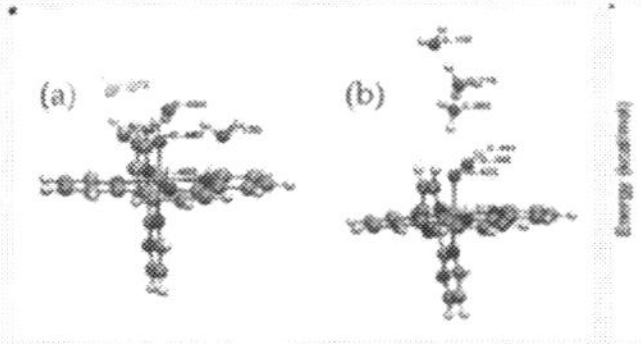

Figure 4: The ruthenium water splitting catalyst operates via an acid-base mechanism where water attacks Ru^{V}.

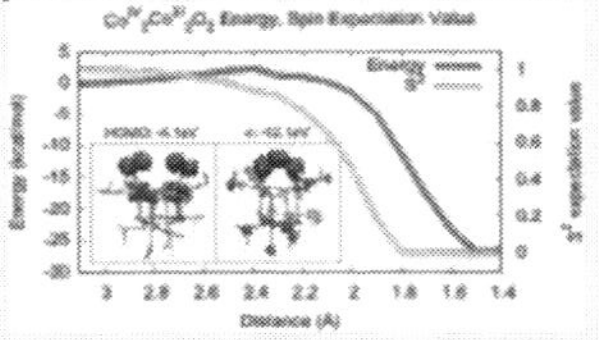

Figure 5: The cobalt oxide water splitting catalyst operates via a direct-coupling mechanism involving two Co^{IV} groups.

Outlook: I am interested in understanding the thermodynamics and kinetics of the most efficient energy conversion catalysts, which are often heterogeneous and nanostructured with many participating pathways. I will approach these challenging problems by applying methods that I developed during my postdoc (left, right, below).

Publications (6 total): Wang L-P, Wu Q and Van Voorhis T. "Acid-base mechanism for ruthenium water oxidation catalysts." *Inorg. Chem.* **2010**, *49*, 4543.

Wang L-P and Van Voorhis T. "Direct-coupling O_2 bond forming a pathway in cobalt oxide water oxidation catalysts." *J. Phys. Chem. Lett.* **2011**, 2, 2200.

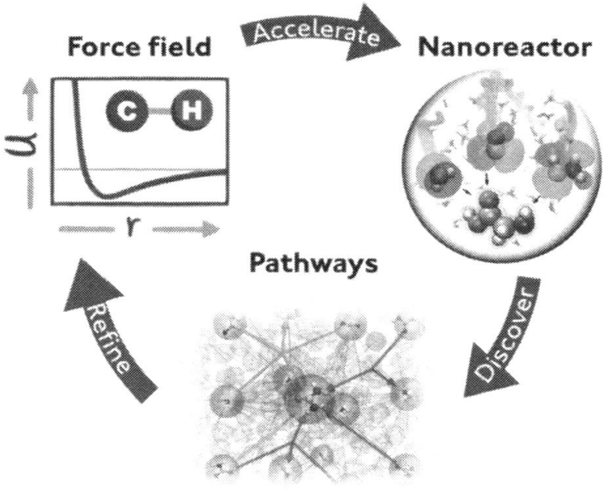

Predict kinetics
Experimental verification

I will take a discovery-based simulation approach toward understanding the complex reaction networks of energy conversion catalysts.

Accelerating simulations via systematic parameterization of accurate force fields

Introduction: *Ab initio* simulations are computationally costly because they treat the electrons explicitly. In protein dynamics simulations and other areas, the electrons are treated implicitly using force fields. During my Ph.D. and postdoctoral research, I introduced the ForceBalance method for systematically deriving force field parameters from combinations of experimental and *ab initio* data. ForceBalance has been applied to improve the accuracy of water, protein, and lipid bilayer models (*in prep.*)

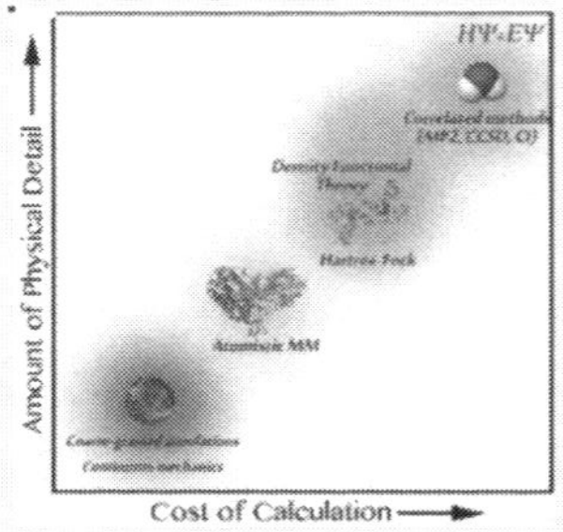

Figure 6: Molecular mechanics force fields can access larger system sizes and time scales than ab initio methods.

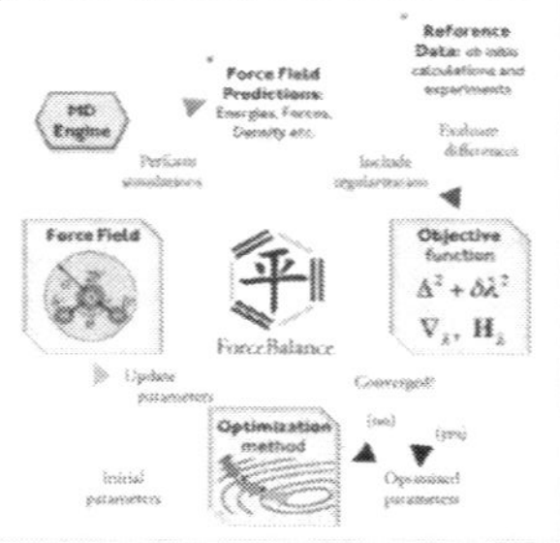

Figure 7: The ForceBalance optimization procedure for systematically building approximate model potentials.

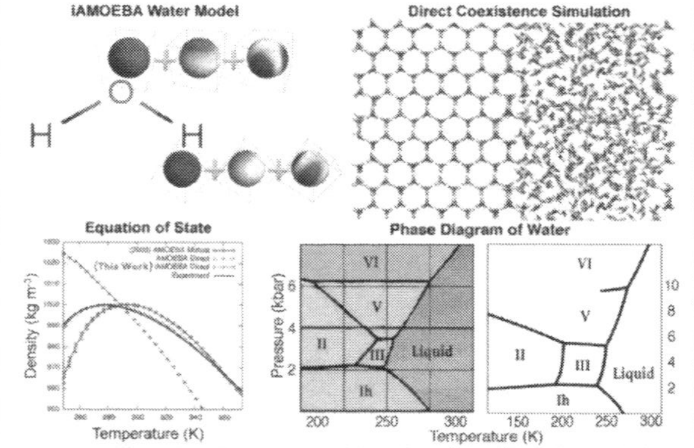

Figure 8: ForceBalance was used to parameterize the iAMOEBA polarizable water model (upper right), which accurately reproduces the equation of state (lower left) and predicts the liquid-ice phase diagram (lower right).

Outlook: I will apply ForceBalance to develop reactive force fields, starting with established models such as ReaxFF and moving towards new functional forms. These force fields will accelerate simulations for discovering new reaction pathways.

Publications (8 total, 3 submitted / in preparation): Wang L-P, Martinez TJ and Pande VS. "Building force fields: An automatic, systematic and reproducible approach." *J. Phys. Chem. Lett.* **2014**, *5*, 1885.

Wang L-P, Head-Gordon T, Ponder J, Ren P, Chodera J, Eastman P, Martinez TJ and Pande VS. "Systematic improvement of a classical molecular model of water." *J. Phys. Chem. B* **2013**, *117*, 9956.

390528 Dynamics of Entangled Rod-Coil Block Copolymers

Abstract:

Polymer science is exploring advanced materials which combine functional domains such as proteins and semiconducting polymers with traditional flexible polymers onto the same molecule. While thermodynamic assemblies of different geometries introduce many interesting new phenomena such as entropic packing and liquid crystalline interactions, dynamic effects are also important to understand for optimal design of material mechanics, processing kinetics, and because of the new physics that directly arises from the motion of multiple domains of dissimilar geometries.

We have recently proposed a theory for the molecular mechanisms of motion in entangled rod-coil block copolymers as a model for this wider class of functional polymeric materials. The large geometrical differences between rigid rods and Gaussian coils cause significant nonlinearities in dynamical behavior as these two motifs are combined on the same molecule. In particular, our theory hypothesizes that the motion of rod-coils is slowed relative to rod and coil homopolymers because of a mismatch between the curvature of the rod and coil entanglement tubes. Dual relaxation mechanisms are predicted in the small and large rod limits, where either the rod or the coil block is expected to determine the motion of the overall molecule.

In this poster we describe the mechanisms of rod-coil block copolymers in detail, while providing significant evidence from both simulation and experiment in support of our theory. We first consider the tracer diffusion of symmetric coil-rod-coil triblocks using molecular dynamics simulation and forced Rayleigh scattering experiments. These results are then generalized to consider rod-coil diblock copolymers, whose asymmetry requires additional analysis. Self-diffusion studies then provide an important step in translating the new physics into practical applications. Finally, additional simulations using coarse-grained models provide a more direct understanding of the underlying mechanisms.

383115 Computer Simulations of Nanoparticle Delivery: From Nanoparticle Design to Cell Membranes

Abstract:

The delivery of therapeutics to disease sites via nanocarriers represents one of the active research areas in the applications of nanotechnology to medicine. It is a rather complicated process involving several different steps ranging from the synthesis of nanocarriers with therapeutics, to the circulation of nanocarries in the body, the interactions of nanocarriers with cell surfaces, the cell membrane penetration and internalization of nanocarriers, and the release of therapeutics from the nanocarriers, etc. Each of these steps, and thus the effectiveness of the overall therapeutics delivery, are affected by different factors, such as the design of nanocarriers, the therapeutics-nanocarriers interactions, the specific composition of the disease cell surfaces and so on. Using computer simulation and modeling, we have investigated the problem from the perspectives of nanoparticles design and cell membranes: 1) the design of the functionalized spherical nanoparticles affects their specific interactions with cell surfaces of interest. Such specific interactions between the functionalized nanoparticles and cell surfaces are achieved using the ligand-receptor interactions. Different nanoparticle design factors, including the characteristics of polymer tethers, nanoparticle size, the ligand density and architectures, the ligand-receptor interaction strength, and so on, are explored in order to generalize their influences on the interaction affinity between nanoparticles and cell surfaces. The 'dilemma' of the interaction affinity and interaction specificity are illustrated and some strategies to overcome the 'dilemma' and achieve both high affinity and specificity of nanoparticles-cell surface interaction are discussed; 2) the architecture of lipid tails influences the formation and the morphologies of water pores inside the lipid membranes. The resultant formation of either hydrophobic or hydrophilic water pores inside the lipid membrane possibly dictates the transport of the materials across of lipid membranes. The knowledge gained from these studies enables the fundamental understanding of the delivery of therapeutics via nanocarries and hopefully serves as guidance for the future experimental design of nanocarriers to obtain optimal therapeutic effects.

387311 Electrochromic Nanocomposites with Endured Energy Storage Properties

Abstract:

Sustainable energies from transient and renewable energy reservoirs such as wind and solar are in urgent demand due to the ever increasing energy crisis arising from the limited reserves of fossil fuels. Given the intermittent nature of these renewable energy resources, reliable energy storage systems are critically needed to store and supply energy in a stable way.[1-5] Meanwhile, electrochromic materials for energy saving have attracted equivalent attention from both the academia and industry, for example, controllable light-reflective or light-transmissive devices for optical information and storage, glare-reduction systems for offices, smart windows for use in cars and buildings.[6-8] In this context, devices with integrated functions of energy storage and electrochromism are highly desirable from the point of view of sustainable development.[9]

Recently, electrically conductive polymers (CPs) have received intensive attention owing to their high pseudocapacitances arising from the rich redox reactions and the corresponding color switching, as well as their facile preparation, environmental stability, high electrical conductivity, and low cost, which make them promising candidates for both energy storage and electrochromic applications. However, structural degradation as a result of the swelling/shrinkage during the intercalation/deintercalation of counterions in the redox reactions remains a challenge for pure CPs thin films.

Targeting to achieve stable conductive polymer nanocomposite thin films with desirable energy storage and electrochromic properties, polyaniline nanocomposite thin films incorporated with different nanofillers, i.e., tungsten oxide and graphite oxides have been prepared by in-situ electropolymerization of the monomers onto the nanofiller coated ITO glass.[10, 11]These nanocomposite thin films have demonstrated much more enhanced stability due to the interactions between the polymer matrix and the nanofillers, which help buffer the stress developed in the polymer matrix. Meanwhile, the nanocomposites exhibit multi-color electrochromism at different potentials, high color contrast, fast switching speed, and high energy efficiency while possessing significant energy storage properties in terms of high energy densities and power densities. The study opens a door for fabricating conductive polymer nanocomposite thin films with integrated energy storage and electrochromism functions and improved stability for future sustainable energy applications via facile electropoloymerization techniques

References

1. J. Zhu, M. Chen, H. Wei, N. Yerra, N. Haldolaarachchige, Z. Luo, D. P. Young, T. C. Ho and Z. Guo, *Nano Energy*, 2014, **6**, 180–192.

2. H. Wei, H. Gu, J. Guo, S. Wei, J. Liu and Z. Guo, *The Journal of Physical Chemistry C*, 2013, **117**, 13000-13010.

3. H. Wei, H. Gu, J. Guo, S. Wei and Z. Guo, *ECS Journal of Solid State Science and Technology*, 2013, **2**, M3008-M3014.

4. H. Wei, H. Gu, J. Guo, S. Wei and Z. Guo, *Journal of The Electrochemical Society*, 2013, **160**, G3038-G3045.

5. J. Zhu, M. Chen, H. Qu, X. Zhang, H. Wei, Z. Luo, H. A. Colorado, S. Wei and Z. Guo, *Polymer*, 2012, **53**, 5953-5964.

6. R. J. Mortimer, *Chem. Soc. Rev.*, 1997, **26**, 147-156.

7. H. Wei, X. Yan, Y. Li, H. Gu, S. Wu, K. Ding, S. Wei and Z. Guo, *The Journal of Physical Chemistry C*, 2012, **116**, 16286-16293.

8. H. Wei, X. Yan, Y. Li, S. Wu, A. Wang, S. Wei and Z. Guo, *The Journal of Physical Chemistry C*, 2012, **116**, 4500-4510.

9. H. Wei, X. Yan, Q. Wang, S. Wu, Y. Mao, Z. Luo, H. Chen, L. Sun, S. Wei and Z. Guo, *Energy and Environment Focus*, 2013, **2**, 112-120.

10. H. Wei, J. Zhu, S. Wu, S. Wei and Z. Guo, *Polymer*, 2013, **54**, 1820-1831.

11. H. Wei, X. Yan, S. Wu, Z. Luo, S. Wei and Z. Guo, *The Journal of Physical Chemistry C*, 2012, **116**, 25052-25064.

SELECTED PEER-REVIEWED PUBLICATIONS (10 out of 33)

1) J. Zhu, M. Chen, **H. Wei**, N. Year, N. Haldolaarachchige, Z. Luo, D. P. Young, T. C. Ho, S. Wei* and Z. Guo*; Magnetocapacitance in Magnetic Microtubular Carbon Nanocomposites under External Magnetic Field, Nano Energy, 6,180–192 (2014).

2) **H. Wei**, D. Ding, X. Yan, J. Guo, L. Shao, H. Chen, L. Sun, H. A. Coloado, S. Wei* and Z. Guo*; Tungsten Trioxide/Zinc Tungstate Bilayers: Electrochromic Behaviors, Energy Storage and Electron Transfer, Electrochimica Acta, 132, 58-66 (2014).

3) **H. Wei**, D. Ding, S. Wei and Z. Guo; Anticorrosive Conductive Polyurethane Multiwalled Carbon Nanotubes NanoComposites; *Journal of Materials Chemistry A*, 1, 10805-10813 (2013).

4) **H. Wei**, H. Gu, J. Guo, S. Wei and Z. Guo; Multiwalled Carbon Nanotubes with Tuned Surface Functionalities for Electrochemical Energy Storage; *ECS Journal of Solid State Science and Technology*, 2(10), M3008-M3014 (2013).

5) **H. Wei**, H. Gu, J. Guo, S. Wei, J. Liu and Z. Guo; Silica Doped Nano-polyaniline with Endured Electrochemical Energy Storage and the Magnetic Field Effects; *Journal of Physical Chemistry C*, 117(25), 13000-13010 (2013).

6) **H. Wei**, H. Gu, J. Guo, S. Wei and Z. Guo; Electropolymerized Polyaniline Nanocomposites from Multi-walled Carbon Nanotubes with Tuned Surface Functionalities for Electrochemical Energy Storage; *Journal of the Electrochemical Society*, 160(7), G3038-G3045 (2013).

7) **H. Wei**, J. Zhu, S. Wu, S. Wei and Z. Guo; Electrochromic Polyaniline/Graphite Oxide Nanocomposites with Endured Electrochemical Energy Storage; *Polymer*, 54(7), 1820-1831 (2013).

8) **H. Wei**, X. Yan, S. Wu, Z. Luo, S. Wei and Z. Guo; Electropolymerized Polyaniline Stabilized Tungsten Oxide Nanocomposite Films: Electrochromic Behavior and Electrochemical Energy Storage; *Journal of Physical Chemistry C*, 116(470), 25052-25064 (2012).

9) **H. Wei**, X. Yan, Y. Li, H. Gu, S. Wu, K. Ding, S. Wei, and Z. Guo; Electrochromic Poly(DNTD)/WO3 Nanocomposite Films via Electropolymerization; *Journal of Physical Chemistry C*, 116(30), 16286-16293 (2012).

10) **H. Wei**, X. Yan, Y. Li, S. Wu, A. Wang, S. Wei and Z. Guo; Hybrid Electrochromic Fluorescent Poly(DNTD)/CdSe@ZnS Composite Films; *Journal of Physical Chemistry C*, 116 (7), 4500-4510 (2012).

AWARDS/HONORS:

3rd Place Award in Chinese-American Chemical Society (CACS) – SW Fall Event,

2012.

APPLICATION AND MECHANISM UNDERSTANDING OF NANO-STRUCTURED CATALYSTS IN BIOFUEL PRODUCTION

Cun Wen, University of South Carolina, Columbia, SC

Department of Energy has set the goal of biofuel production to 36 billion gallons by 2022, while the present production rated at 1.1 billion gallons at 2011. The capacity needs more than 3200% increase in total or 40% increase in annual for the next 10 years. One of the limiting factors in biofuel production is the low throughput of the batch reaction mode. Though it has been long known that heterogeneous processes can accelerate the biofuel production, the actual progress is hindered by poor understanding of reaction mechanisms and requires sophisticated material design and synthesis. The reaction mechanism research is intricate with numerous potential reaction pathways, similar to the possible routes to overcome Himalayas. Our strategy to tackle this issue is to combine the mechanism research with kinetics studies, which serves as loyal contour-mapper and tells us the most feasible way. This strategy has been proven to be effective.[1-12] Herein, I bring in nano-structured heterogeneous catalysts with our strength in advanced nanomaterial synthesis and fundamental reaction-mechanism studies to expedite the biofuel production.

To upgrade biomass to biofuels, the aromatic rings and the C-O bonds need to be broken down and C-C/C-H bonds need to be build up, on which stage heterogeneous catalysts can show off. Firstly, 2-methyfuran acid catalysis reaction is chosen as a model reaction to exemplify our idea of heterogenerizing biofuel production. The 2-methylfuran is a mode molecular representing biomass feed stocks, which usually need acid catalysts for deoxygenating because of oxygen as heteroatom and need chain growth for generating biofuel because of low carbon numbers (<C6). Usually concentrated sulfuric acid is used as acid catalyst to serve the purpose, but it is unfavorable for large scale industrial production. Our results show that the zeolite catalyst can replace the concentrated sulfuric acid with even better production selectivity in terms of cetane number. Furthermore, the cetane number of the product can be increased by decrease the acid strength of the zeolites. To understand the relationship between acid strength and product cetane number, the reaction mechanism on the zeolite catalyst is probed by *in situ* techniques, such as FTIR. The reaction mechanism study reveals that high acid strength increases the adsorption energy, and thus prolongs the resident time of the intermediates on the zeolite. Because longer resident time facilitates the C-C chain growth, the branched products with poor cetane number are favored on zeolites with high acid strength.

Secondly, the products from 2-methfuran acid catalysis reaction need further hydrogenation to upgrade to final biodiesel, and Co is reported to be active for the hydrogenation reaction, whose activity can be further promoted by Cu doping. Thus, bimetallic core-shell nanoparticle Cu@Co with alloy of CoCu on the shell (for high activity) and Cu as the core (for low catalyst cost) would be optimum for this reaction. Traditionally, the core-shell bimetallic nanoparticles are synthesized in two steps. In the first step, the core is formed with one metal, and then in the second step the core is coated with second metal as the shell. Thus, the two-step method not only is poor in economic, but also yields mono-metal instead of alloy on the surface, while alloy is needed for high activity in this hydrogenation reaction. Herein, I demonstrated a one-pot methodology for mono-dispersed

Cu@Co Core-shell nanoparticle synthesis with alloy of CoCu on the shell as catalyst for hydrogenation reaction. Furthermore, I can not only synthesis the nanoparticle in one-step, but also keep a tight control over the size distribution (relative standard deviation 7.9%).

Thirdly, biomass upgrading to biofuel is not limited to the acid and hydrogenation reaction, and other process can also serve the purpose, such as partial oxidation. The partial oxidation has been proved to be an effective way to produce biofuels from biomass aligned with Fischer-Tropsch synthesis. However, the long-standing problem in partial oxidation is the hot spot formation in the catalyst bed, which can lead to the catalyst deactivation or even explosion. It has been known that the hot spot is formed by the side reaction that is the total oxidation taking place beside the partial oxidation. The key factors guiding the reaction path through total or partial oxidation is unknown, which hinders the catalyst development. To address this issue, I looked into the mechanism of partial oxidation production of syngas ($CO+H_2$). Our research has revealed that the lattice oxygen reducibility of the catalyst is the key factors governing the reaction mechanism. With high lattice oxygen reducibility, total oxidation rather than the partial oxidation is favorite. By tuning the lattice oxygen reducibility of Er_2O_3 with Mn doping, I successfully shift the reaction mechanism from total oxidation to partial oxidation, and eliminated the hot-spot formation.

Read Me More:

1. Wen C, Dunbar D, Zhang X, Lauterbach J, Hattrick-Simpers J. Self-healing catalysts: Co3O4 nanorods for Fischer-Tropsch synthesis. *Chem Commun.* 2014;50:4575-4578.
2. Wen C, Barrow E, Hattrick-Simpers J, Lauterbach J. One-step production of long-chain hydrocarbons from waste-biomass-derived chemicals using bi-functional heterogeneous catalysts. *Phys Chem Chem Phys.* 2014;16(7):3047 - 3054.
3. Wen C, Liu Y, Guo Y, Wang YQ, Lu GZ. Strategy to eliminate catalyst hot-spots in the partial oxidation of methane: enhancing its activity for direct hydrogen production by reducing the reactivity of lattice oxygen. *Chem Commun.* 2010;46(6):880-882.
4. Wen C, Liu Y, Guo Y, Wang Y, Lu G. Synthesis of the rare earth compound nanosheets induced by lamellar liquid crystal. *Solid State Sciences.* 2009;11(11):1985-1991.
5. Wen C, Liu Y, Tao F. Integration of surface science, nanoscience, and catalysis. *Pure Appl Chem.* Jan 2011;83(1):243-252.
6. Liu Y, Wen C, Guo Y, Lu G, Wang Y. Modulated CO Oxidation Activity of M-Doped Ceria (M = Cu, Ti, Zr, and Tb): Role of the Pauling Electronegativity of M. *J Phys Chem C.* 2010;114(21):9889-9897.
7. Liu Y, Wen C, Guo Y, Lu G, Wang Y. Effects of surface area and oxygen vacancies on ceria in CO oxidation: Differences and relationships. *J Mol Catal a-Chem.* Feb 2010;316(1-2):59-64.
8. Liu Y, Wen C, Guo Y, et al. Mechanism of CO Disproportionation on Reduced Ceria. *ChemCatChem.* 2010;2(3):336-341.
9. **Meadia Highlights:**Jacoby M. Custom Catalyst Resists Oxidation. *C&EN News.* 2014;92(13):25.
10. **Meadia Highlights:**Pichon A. Catalysis: Lower reactivity means more hydrogen. 2010; http://www.nature.com/nchina/2010/100120/full/nchina.2010.7.html.
11. **Meadia Highlights:**Saxton C. Enhancing catalytic activity. 2010; http://www.rsc.org/Publishing/ChemTech/Volume/2010/02/enhancing_catalytic.asp.
12. **Meadia Highlights:**Chinese New Year. *Chem Commun.* 2010.

391971 Surface and Interfacial Properties in Ceramic and Inorganic Materials

Abstract:

Nano-scale materials have recently attracted great interest due to their ability to advance currently available materials to engineer next-generation technologies and devices. This can lead to lower cost of production, reduced carbon foot print, higher energy efficiency, and improved performance in a variety of applications. Since nanoparticles and nanocomposites have higher surface areas than their traditional micro-scale counterparts, it is becoming increasingly important to understand how surfaces and interfaces interact with their surrounding environments to contribute to the advanced properties and performances. Using a variety of characterization tools, my research focuses on identifying the molecular-scale surface and interfacial structures to explain the mechanisms that control the observed macroscopic properties.

My current and past research includes investigation of interfaces that exist in polymer-derived, nanocomposite ceramics and of surfaces of *di*calcium silicate nanoparticles. I completed my doctoral degree at the University of California, Davis under the advisement of Alexandra Navrotsky and Sabyasachi Sen, where I developed structural models for silicon-based polymer-derived ceramics that were synthesized from pyrolysis of a variety of preceramic polymers. The nano-scale and micro-scale structures are highly tunable and have been proposed for a wide variety of high temperature applications. It was found that the interfaces between the nano-scale domains in these ceramics contribute to high chemical and thermal stability that make these materials a desirable candidate for next generation high temperature ceramics. Furthermore, it was determined that the morphology contains percolation networks that may also contribute to interesting transport and mechanical properties. I am currently a University of California President's Postdoctoral Fellow at the University of California, Santa Barbara, working under the advisement of Bradley Chmelka. My postdoctoral research focuses on understanding the hydration mechanisms in *di*calcium silicates, a minor component in anhydrous cement powders. Sol-gel derived β-*di*calcium silicate nanoparticles were recently synthesized to increase the hydration rate, with respect to conventionally prepared, micron sized β-*di*calcium silicates. Increasing the overall hydration rate of *di*calcium silicate is considered to be an attractive option for replacing a portion of *tri*calcium silicate (the major component in anhydrous cement powders) to achieve similar mechanical strength properties with lower total calcium contents. Correspondingly, this could reduce the carbon foot print of the cement manufacturing process due to the release of carbon dioxide during the refinement of calcium. The molecular-scale surface structures of anhydrous and hydrated β-*di*calcium silicates synthesized by the sol-gel and conventional methods are currently under investigation to identify the surfaces chemistry and interactions that cause the increased hydration rate.

My future research interests include molecular-scale investigations on nanostructured ceramic and inorganic materials with applications in fuel cell electrodes and membranes, battery electrodes, supercapacitors, and high temperature ceramics. My experimental approach to investigate these systems primarily involves the use of high-resolution, solid state nuclear magnetic resonance spectroscopy, which enables identification of molecular-scale structures in inorganic and ceramic materials. Additionally, intermolecular interactions that occur at surfaces and interfaces can be elucidated to provide knowledge of the fundamental chemistry that is involved in the enhanced properties of next generation engineering materials.

393338 Multi-Level Engineering Approaches for Manipulating Plant Metabolism in Culture

Abstract:

Plants are a source of a vast diversity of natural products that have commercial applications as flavors, fragrances, pesticides and pharmaceuticals. These products are often the result of sophisticated secondary metabolic pathways that are specialized to plant cells. Due to the complex structures of these compounds, chemical synthesis is often infeasible, necessitating the use of biological production platforms. In this project, methods were developed and employed to characterize and engineer plant cell systems at both the environmental and molecular levels to decipher mechanisms for promoting plant secondary metabolism. Studies utilized suspension culture of the non-model plant system, *Taxus*. These cultures are utilized for commercial production of the anti-cancer agent paclitaxel (Taxol^{TM}, Bristol-Myers Squibb), which is used in the treatment of breast, lung and ovarian cancers, as well as AIDS-related Kaposi's sarcoma.

Environmental manipulation: When plant cells divide, the cell walls do not fully separate, leading to the formation of aggregates, which can range in size from two to thousands of cells. These aggregates result in the formation of subpopulations within a culture. This heterogeneity is believed to be a major source of the inherent variability associated with plant production systems. To aid in the understanding of culture aggregation, a method was developed to reduce aggregation through the application of mechanical shear. A population balance model was established to determine the amount of shear required to reach a target aggregate size distribution. By applying shear over multiple generations of cell growth, a more heterogeneous culture with a constant mean aggregate size was developed with no affect on cell health.

Molecular-level manipulation: Although plant primary metabolism is well characterized and highly conserved across plant species, plant secondary metabolism is often species specific and, as a result, poorly defined. Additionally, the interface between primary and secondary metabolism is poorly understood, limiting metabolic engineering efforts. High-throughput methods were developed to evaluate the flux distribution through specific pathways involved in *Taxus* metabolism. It was found that cultures that produce paclitaxel also divert carbon flux towards the production of phenolics and flavonoids. A long-term study revealed high levels of variability in culture aggregation and secondary metabolite accumulation. These studies are amongst the first to study global secondary metabolism in a non-model plant species and provide valuable insight into the design of effective metabolic engineering strategies to promote production of a particular class of secondary products.

Due to the high level of interaction between cells in these complex cellular systems, successful engineering efforts must look beyond the level of the metabolic pathway. This project is the first of its kind to characterize and manipulate *Taxus* cultures on a multi-scale level to allow for the effective engineering of cultures for increased paclitaxel production. Moving forward, manipulation of culture aggregation through mechanical shearing and cellular metabolism through metabolic engineering could lead to more homogeneous cultures with reduced variability and increased paclitaxel accumulation.

Polyhedral Oligomeric Silsesquioxane (POSS)-Based Hybrid Nanostructured Thermoplastic Polyurethanes (TPUs)
---- Synthesis, Processing and Biomedical Applications

Polyhedral Oligomeric Silsesquioxane (**POSS**) is a class of well-defined, highly symmetric molecules that contain an inner inorganic framework made up of silicone and oxygen $(SiO_{1.5})_x$ and externally covered by eight organic substitutes. The hybrid organic-inorganic framework renders POSS thermally and chemically robust nature, and the eight vertex R-groups feature the versatility in chemistry. In a subtle way, nano-scale POSS moieties were introduced into conventional thermoplastic polyurethanes(TPUs) to form a new generation of hybrid TPUs nano-composites. By tailoring the biocompatible soft segments (e.g. poly(ethylene glycol)(PEG), poly(caprolactone)(PCL), poly(d,l-lactide)[P(d,l)-LA], etc), we have developed a series of novel hybrid POSS-TPUs with a wide variety of biomedical applications, including mechanical robust hydrogels, nano-structural antimicrobial hydrogel webs with long-term duration, enzymatic biostable polyester-based hybrid TPUs for long-term biomedical implantation, hybrid TPUs with high-drug (i.e. Paclitaxel) loading and tailored drug release for biodegradable stent coating, among others.

The cornea is the transparent outermost layer of the eye. Trauma, bacterial and viral infections, and heritable conditions lead to loss of corneal function and visual impairment in over ten million individuals world-wide. Recapitulating the microstructure of native human corneal stromal tissue is believed to be a key feature in successfully engineering corneal tissue. Employing a strategy of surface contact guidance and growth factor supplementation, we demonstrated that on aligned fibrous substrates made from biodegradable poly(ester urethane) urea (PEUU), human corneal stromal stem cells (hCSSCs) could be induced to secrete and organize a type-I collagen-based extracellular matrix (ECM) abundant in characteristic human corneal stromal ECM components. Spatial self-organization of the collagen-based ECM by hCSSCs featured stratified multilayered collagen-fibril lamellae with orthogonal orientation, and uniform fibril size and inter-fibril spacing in a pattern mimicking human corneal stromal tissue. The approach of combining substrate cues with growth factor augmentation offers a new means to engineer well-organized, collagen-based constructs with appropriate nanoscale structure for corneal repair and regeneration.

Heterogeneous Catalysis for Renewable Energy and Green Chemistry

Xiaofang Yang, Chemistry Department, Brookhaven National Laboratory

Exhaustion of fossil-based fuels and the concern of climate change due to carbon dioxide emission have generated growing interests in searching for renewable energy sources and clean energy technologies. The future success of these technologies largely relies on advancing catalyst development for many catalytic processes such as biomass conversion, dry reforming, oxygen reduction and alcohol oxidation in fuel cells, hydrogen production from biomass or water splitting reaction, etc. To meet the challenges in catalyst development, my research is primarily focused on the study of surface science, electrochemistry and heterogeneous catalysis with the aim to discover high performance catalysts in energy-related catalysis and green chemistry.

My doctoral research extensively explored Pt monolayer (ML) and novel non-Pt model electrocatalysts for proton exchange membrane (PEM) fuel cells and electrolyzers. This work sought insight and discovery of new materials by establishing clear structure-activity relationships in electrocatalysis. My study combined the synthesis and characterization of single crystal surfaces in ultrahigh vacuum (UHV) and evaluating their catalytic performance in electrochemical conditions. Single crystal surfaces are characterized by well-ordered surface structures, which serve as ideal models to study fundamental behavior or properties for the related functional materials. Supplemented with theoretical simulations, this approach created an excellent platform to study the relationship between catalytic activity and surface geometric structure and electronic properties of electrocatalysts.

My postdoctoral research focuses on developing high performance catalytic systems for CO_2 conversion and alkene epoxidation and addresses the key scientific issues through theoretical and experimental studies on both model surfaces and powder catalysts. Successful understanding of the principles to control the selectivity and activity allows me to design better catalysts for propylene epoxidation and converting CO_2/CH_4 to syngas. Epoxidation of propylene is an important industrial reaction. However, a lack of selectivity to propylene oxide (PO) is the major challenge in the direct epoxidation of propylene by O_2. My studies on mixed oxides demonstrate the feasibility of designing novel materials as active and selective catalysts which could significantly impact the production of PO in the future. Recently, rising CO_2 concentration in atmosphere has been associated with adverse global climate changes. An alternative way to alleviate this problem is utilizing CO_2 as inexpensive and readily available feedstock to produce other chemicals or fuels. Thus, my research efforts are to identify active catalysts for converting CO_2/CH_4 to syngas. My preliminary findings of the highly active carbide surfaces could potentially lead to the design of high performance catalysts for CO_2 conversion.

My future research will continue in the area of catalysis for novel fuels, green chemistry and fuel cell catalysis, in the context of research and development of alternative energy technologies. Specifically, by engineering model catalysts of alloys, metal oxides and nanostructures, my research will address the key factors such as geometric structures and electronic properties that impact catalytic properties of these materials. The goals of my studies are to provide an improved in-depth understanding of the origin of the enhanced catalytic activity and to tailor the catalyst performance for specific applications.

Ph.D. Advisor: Prof. Bruce E. Koel

Postdoc Advisors: Prof. Jingguang G. Chen and Dr. Jose Rodriguez

PUBLICATIONS

20. "Molybdenum carbide as alternatives to precious metal catalysts for selective conversion of CO_2 by hydrogen, M.P. Porosoff , **X. Yang** and J.G Chen, *Angew. Chem. Internat. Ed.*, **2014**, in press.
19. "Theoretical and experimental studies of ethanol decomposition and electrooxidation over Pt-modified tungsten carbide" T. G. Kelly, A. L, Stottlemyer, **X. Yang**, J. G. Chen, *J. Electrochem. Soc.* **2014**, *161, E3165-E3170*
18. "Nanostructures of Au sub-monolayer: synthesis and its application for reduction of molecular oxygen", **X. Yang** and B.E. Koel, *Nat. Mater.*, **2014**, to be submitted.
17. "Stabilization of catalytically active Cu^{+} surface sites on titanium-copper mixed oxide films", A. Baber, **X. Yang**, H. Kim, K. Mudiyanselag, A. Al-Mahboob, J. Sadowski, J. Rodriguez, P. Liu, J. Chen, and D. Stacchiola, *Angew. Chem. Internat. Ed.*, **2014**, in press.
16. "Geometric requirements for catalytic sites on Pt-Sn alloy surfaces", J Gao, H.Zhao, **X. Yang**, B.E. Koel and S.G. Podkolzin, *Angew. Chem. Internat. Ed.*, **2014**, in press.
15. "Opportunities and challenges in utilizing metal carbides in renewable energy applications", **X. Yang** and J.G Chen, *ACS Catal.* **2013**, to be submitted.
14. "Structural origin of anisotropic transport in electrically conducting dichloroacetic acid-treated polymers", J. M. Mativetsky, J. Tarver, **X. Yang**, B. E. Koel, Y. Loo, *Organic Electronics* , **2014**, 15, 631–638.
13. "A New Class of Electrocatalysts for Oxidation of Ethylene Glycol by Supporting Monolayer Pt on Engel-Brewer Alloy Substrates", **X. Yang**, M. Li, B. Koel and J.G. Chen, *Angew. Chem. Internat. Ed.*, **2014**, Submitted
12. "Controlling acetylene adsorption and reaction on Pt-Sn catalytic surfaces", J. Gao, **X. Yang**, S.G. Podkolzin and B.E. Koel, *ACS Catal.*, **2013**, 3, 1149-1153.
11. "WO_3-α-Fe_2O_3 composite photoelectrodes with low onset potential for solar water oxidation" , P, Zhao, C. Kronawitter, **X. Yang,** J. Fu and B.E Koel, *Phys. Chem. Chem. Phys,* **2014**, 16, 1327-32.
10. "Highly stable Pt–Au@Ru/C catalyst nanoparticles for methanol electro-oxidation", Q. He, B. Shyam, M.Nishijima, **X. Yang**, B. Koel, F. Ernst, D. Ramaker, and S. Mukerjee, , *J. Phys. Chem.* **2013**, 117 (3), 1457-1467.
9. " Electrochemical and spectroscopic study of novel Cu and Fe-based catalysts for oxygen reduction in alkaline media", Q.He, **X. Yang**, X. Ren, B. Koel, and S. Mukerjee, *J. Power Sources,* **2012**, 213, 169-179.
8. "Activation of tungsten carbide catalysts by use of an oxygen plasma pretreatment", **X. Yang**, Y.C. Kimmel, J. Fu, B.E. Koel, and J.G. Chen, *ACS Catalysis* **2012**, *2*, 765–769.
7. "Nano-faceted C/Re(11-21): fabrication, structure and template for synthesizing nanostructured model Pt electrocatalyst for hydrogen evolution reaction", **X. Yang**, H. Wang, W. Chen, B.E. Koel and R.A. Bartynski, *ACS Nano.*, **2012,** *6*, 1404-1409.
6. "A novel CuFe based cathodic catalyst for oxygen reduction reaction in high pH environment", Q.He, **X. Yang**, X. Ren, B. Koel, S. Mukerjee, and R. Kostecki, *J. Power Sources,* **2011**, 196, 7404.
5. "Role of surface Fe in enhanced activity for the oxygen reduction reaction on a $Pd_3Fe(111)$ single crystal alloy", **X. Yang**, J. Hu, J. Fu, R. Wu, and B. E. Koel, *Angew. Chem. Internat. Ed.,* **2011**, 50, 10182-10185.
4. "Formation of Pd monomers and dimers on single crystal $Pd_3Fe(111)$ surface", **X.Yang**, J. Hu, R. Wu, B. Koel, *J. Phys. Chem. Lett.,* **2010**, 1(16), 2493–2497.
3. "Influence of phosphate anion adsorption on the kinetics of oxygen electroreduction on low index Pt(hkl) single crystals", Q. He, **X. Yang**, W. Chen, S. Mukerjee, B. Koel, and S. *Chen, Phys. Chem. Chem. Phys.*, **2010**, 12(39),12544-55.
2. "Surface structure of $Pd_3Fe(111)$ and effects of oxygen adsorption", **X. Yang**, L. A. Welch, J. Fu and B. E. Koel, in *Catalytic Materials for Energy, Green Processes and Nanotechnology*, edited by C-Y. Mou, J. Liu, H.H. Kung, S. Dai (*Mater. Res. Soc. Symp. Proc.* Vol. 1217, Warrendale, PA, 2010), 1217-Y08-43.
1. "Improving electrocatalysts for O_2 reduction by fine-tuning the Pt–support interaction: Pt monolayer on the surfaces of a $Pd_3Fe(111)$ single-crystal alloy," W.P. Zhou, **X. Yang**, M.B. Vukmirovic, B.E. Koel, J. Jiao, G. Peng, M. Mavrikakis, and R.R. Adzic, *J. Am. Chem. Soc.*, **2009**, 131, 12755–12762.

389126 Engineering Advanced Membrane Materials for Gas Separation and Biofuels Production

Abstract:

My research work focuses on design and development of novel high performance hollow fiber membranes and sorbents with complex morphologies out of advanced polymers and nanostructured materials, and their applications in membrane-based gas separation, and biofuels purification. During my PhD at University of Chinese Academy of Sciences with Prof. Yinhua Wan, I introduced a new process for improving the affinity between nano-sized particles and polymer matrix through chemical bonds and hence suppressed the formation of microvoids at polymer–inorganic interface, which was used to develop novel organic permselective pervaporation membranes for biofuels purification from biomass fermentation broth. This work has been successfully resulted in a new research program around $2.0M at Chinese Academy of Sciences funded by Chinese Government as well as incubated several collaborations with companies to carry out the pilot scale application for bio-ethanol and bio-butanol production via fermentation/pervaporation coupled process based on cellulose and Cassava renewable feed stock, respectively.

Currently, I am a research engineer and lab manager at Georgia Tech, where I am expanding my expertise in membrane-based gas separation by engineering the economical membrane materials. I also provide mentorship and instruction to the undergraduates, PhD students and post-doctoral fellows in a group of over 20 researchers. This position has allowed me to expend my expertise from flat-sheet membranes to high performance hollow fiber membranes, and also from membrane-based separations to combined hollow fiber sorbents adsorption and membrane-based coupled separations. By using advanced cross-linkable polymers, functional polymers, and several commercially available polymers, I have created high performance hollow fiber membranes for aggressive sour gas separations, CO_2 removal from flue gas, and biofuels purification.

My future research will address both fundamental issues and applied research related to membrane-based gas separations, organic permselective pervaporation membrane technology for biofuels purification, and recovery and/or removal of volatile organic compounds (VOCs) from industrial wastewaters. Using the knowledge of the industrially available material properties developed in the fundamental studies, I will strongly focus on conducting cutting-edge research in engineering the advanced polymer structures and economical processes to produce high performance membranes with the emphasis on global CO_2 reduction, energy conversion, and clean technology which will be met industrial applications.

385622 Production of Furan-Based Products from Lignocellulosic Biomass

Abstract:

Lignocellulosic biomass has great potential for producing alternative fuels and chemicals. In particular, furan-based products such as HMF and furfural are attractive intermediates of biofuels and chemicals. Various conversion technologies for furan-based products have been investigated; however, there are still some challenges including low productivity, formation of by-products, complexity of reaction, high cost of catalysts, and difficulty of product separation. In order to overcome the aforementioned barriers, two novel processes for furan-based products were developed.

The first approach is a biphasic process for production of 5-bromomethyl furfural (BMF). BMF is a furan derivative with high reactivity and easy to be converted to other chemicals, and it is also an intermediate of furan-based biofuels. The reported in this study was a single step biphasic system including an aqueous phase of molten salt hydrate with mineral acid as catalyst and an organic phase of extracting solvent. The BMF was produced in the aqueous phase and immediately extracted into the organic phase to prevent the further decomposition and condensation. The production of BMF from different substrates including monosaccharides, polysaccharides, and real biomass was investigated. Under the optimum conditions, BMF yield from cellulose reached 90.3 %. The effects of reaction factors including acids loading, concentration of halide salt, and volume ratio of organic to aqueous phase were investigated. Different molten salt hydrates solutions (LiBr, LiCl and $CaBr_2$) were also compared to convert cellulose into halidemethyl furfurals (BMF/CMF). The application of biphasic system improved the products selectivity and prevented the formation of byproducts (humins).

The second approach for production of furan-based products from biomass is a one-pot process (HDA) integrating hydrolysis of polysaccharides to monosaccharides, dehydration of the monosaccharides to furans, and aldol condensation of the furans with acetone to produce furan acetone adducts as hydrocarbon precursors. The conversion was completed in a LiBr/acetone system without pretreatment or preprocessing of the biomass. Acetone was not only used as a solvent but also a reactant involved in aldol condensation reactions. Halide salt (LiBr) and hydrochloric acid (HCl) were used as catalysts in the process. According to the experimental results, halide salts play crucial roles in all three reactions above. A series of experiments were performed to optimize reaction factors for efficient conversion of corn stover. A product yield of 78% was obtained under the optimum conditions. In the meantime, lignin was significantly depolymerized and separated from other components during the process as a potential valuable co-product (HDA lignin). Most of LiBr and HCl (99%) were recyclable.

385611 Predictive Theoretical Modeling of Complex Fluids: From Advanced Materials to Engineering Nanomedicine

Abstract:

Theoretical modeling provides insights into fundamental understanding of many complex systems ranging from organic-inorganic hybrid materials to biological fluids. A suitably chosen coarse-grained theory that captures the essential physics of the system can be predictive and guide the materials design for a variety of applications before conducting exhaustive experimental efforts or computational tests of the full complex problem. Here I demonstrate the application of classical theoretical modeling in two nanoparticle systems followed by my proposed research plans.

Nanoparticle-organic hybrid materials (NOHMs) are a new type of complex fluid consisting of 10 nm diameter spherical inorganic core particles surface-functionalized with oligomeric organic molecules with no other solvent. The absence of a solvent, the small size of the nanocores and oligomers, and the incompressibility of the tethered oligomeric fluid make the oligomer-mediated interactions non-pairwise-additive. In my PhD research, I developed a classical density-functional approach for model hard spheres with tethered bead-spring oligomers that allows a direct description of the system free energy as a functional of the probability densities of cores and oligomers without assuming pairwise additivity. Based on the coarse-grained model and my theories, I predicted a striking result of solvent-free NOHMs that the structure factor must go to zero at zero wave number, which can be used to characterize the polydispersity in experimental systems. I also discovered that the carbon capture capacity, disorder-order phase transition, and transport properties of these materials are governed by the oligomer-configurational entropy as the tethered hairs have to uniformly fill the interstitial space. These predictions have been verified by relevant molecular dynamics simulations and experiments.

The use of ligand-functionalized nanocarriers that bind to specific receptor proteins is gaining translational prominence in targeted vascular drug delivery. Owing to the wide spectrum of time and length scales involved in the hemodynamics and nanocarrier-endothelium binding landscape, a unified theoretical framework bridging the microscale many-body hydrodynamic interactions in the vessel tube and the nanoscale nanocarrier adhesion on the vessel wall would provide useful insight into the tailored design of functionalized nanocarriers for specific targeted biophysical environments. As a postdoctoral researcher, I developed a theoretical formalism for studying nanocarrier-cell adhesive dynamics in the presence of red blood cells (RBCs) using a combined framework of generalized Langevin equations and dynamical density-functional theory. Analyzing the RBC-driven margination and the position/velocity autocorrelation functions of the nanocarriers for different vessel diameters, blood flow rates, and binding free energies would allow the quantification of nanocarrier binding affinity and the prediction of biodistribution. The predicted correlation functions for nanocarrier motion have been validated by finite-element direct numerical simulations of Navier-Stokes fluid around a moving nanoparticle at different separations from the vessel wall.

My proposed research plans are aimed at addressing some of the key challenges in engineering nanomedicine and advanced materials design using nanoparticle-based systems. By combining classical theories and computational techniques, my research group will explore how the active mass transport of the magnetic nanoparticle is affected by physiologically relevant interactions in blood flows, how the drug-loaded nanoparticle may overcome biophysical barriers in the extracellular matrix to reach the target site, and how the functionalization of the particle can be optimally designed to achieve self-assembling bioactive nanostructures. Concurrently, I will develop and apply new theoretical and simulation approaches to advance fundamental studies of systematic materials processing, complex networks characterization, and ordered directed self-assemblies for energy applications.

385492 Microfluidic Approaches Towards Metabolic and Genetic Engineering

Abstract:

Microbes provide a promising platform for broad and sustainable chemical production because of their potential to produce a wide range of molecules with high specificity from renewable feedstocks. Although metabolic engineering has existed for several decades, leaps in progress have been driven in part by tangential technological advancements in DNA synthesis, sequencing and automation. These advancements have streamlined editing genomes towards desirable phenotypes and have spurred the development of seminal technologies like trackable multiplexed recombineering (TRMR), multiplexed automated genome engineering (MAGE) and phage-assisted continuous evolution (PACE). Even entire genomes are now possible to synthesize *de novo*. The ever-improving simplicity of manipulating organisms and the demonstrable effectiveness of automation technologies towards this end, suggest that that the future of metabolic engineering and synthetic biology will soon require a drastically different focus, shifting from the basics manipulating and editing genomes to how to effectively engineer and optimize organisms from interdisciplinary approaches. My research effort will utilize an interdisciplinary approach from a fundamental chemical engineering perspective, combining microfluidic and automation technology development to address the needs of metabolic engineering and synthetic biology. This approach will enable more efficient and effective methods to optimize organisms thereby drastically improving the scope, throughput and basic understanding of genetic and metabolic engineering. My planned research efforts are as follows:

Proposed Research Projects

Parallelization and miniaturization of chemical processes as a broad screening approach: Correlating the phenotype of an engineered cell to its genotype by screening limits throughput in the design-build-test phase of genetic engineering. Identifying extracellular products produced, like biofuels, can involve chemical processes in series (production, separation, detection) which reduces the throughput of genotypes that can be tested and thereby limits the effectiveness of engineering organisms. The research proposed here is to develop a broad approach toward parallelizing the production, separation and detection of extracellular products using multiple emulsions in a microfluidic system. This idea can be used to screen for a variety of chemicals and capable of throughputs up to 10^3-fold greater than conventional approaches. By coupling separations and colorimetric reactions in a multiple emulsion format, a method could be developed to perform processes in parallel enabling high-throughput analysis single-genotype analysis. With high-throughput screening, genome-wide search technologies and methodologies will be applied to understand the metabolic engineering of strains whose investigation is currently limited by screening technologies.

Miniaturized, long-write DNA synthesis methodology and development: Many genetic engineering tools and technologies use massive quantities of diverse synthetic oligonucleotides and as such have relied on advancements in DNA synthesis. But, the high and somewhat stable costs of gene synthesis, caused by harsh chemistry and inefficient synthesis, have limited the further applicability of DNA synthesis. The research proposed here is developing a concurrent approach of enzyme and chemistry development with microfluidic system engineering to develop a novel and superior approach to DNA synthesis with respect to the current industry standard. Specifically an enzyme-based DNA synthesis chemistry will be developed in an electrophoresis-driven microfluidic platform allowing for millisecond transport and reaction times, precise temperature control and low reagent use. By coupling molecular biology and protein engineering along microfluidic system development, I envision a bench top system could be developed to improve de novo oligonucleotide and gene synthesis.

Automation of functional pathway discovery: Integrating heterologous genes in non-native hosts can allow for the production of a wide-range of chemicals by either adding a new metabolic pathway or bypassing the host cell's native regulation scheme. However, this is largely a trial and error process and heterologous gene expression typically can result in incompatible enzyme products. To overcome this, I propose an automated system of chemostats, coupled with novel microfluidic transformation technologies in series to continuously create combinations of different pathways of heterologous genes in *E.coli*. This system will be use to rapidly construct entire pathways which can be screened for chemical production. By integrating an automation approach to recursive postivite/negative selection I will develop an open-sourced continuous flow directed evolution system capable of constructing pathways.

Previous Research

I studied under Professor Mark Burns in the Ph.D. program at the University of Michigan to develop technologies enabling active control of flow, particles and chemicals in microfluidic systems. There, I integrated chemical engineering concepts governing microfluidic design (transport and fluid dynamics), system engineering and microfabrication. My Ph.D. resulted in three first author research publications, three contributing author publications and a book chapter contribution. I was awarded the competitive Microfluidics Training Program Fellowship, an NSF graduate honorable mention, and doctoral qualifying exam distinction having the top research and grades for my Ph.D. qualifying exam. After completing my Ph.D. I worked at Becton Dickinson on their BD Max platform - an open-source automated molecular diagnostic robotic system which utilized microfluidic technology developed from the Burns lab.

Currently, I am a research associate in Professor Ryan Gill's research group at the University of Colorado studying methods to relieve throughput bottlenecks in the design-build-test cycle which is core to synthetic biology and metabolic engineering. My most notable work has been developing a method to track populations of combinatorial mutants by utilizing core chemical engineering concepts along with next-generation sequencing technologies. This work has improved the throughput of analyzing combinatorial engineered populations by over 10^4-fold. Another of my publications uncovered implications of a fluid phenomena in microchannels - an effect that arose while developing microfluidic systems for genetic engineering. I have also been involved in authoring several successful grants including the DARPA living foundries grant and DoE biosystems design grant.

Teaching and Professional Activities

Teaching is an important focus of mine and as a professor I will aim to impart students with the knowledge, skills and attitudes to ensure successful transition from education to application. Aside from positively contributing to the undergraduate and graduate educational structure through the ABET requirements, I would also like to apply concepts and ideas published in educational literature like, *Chemical Engineering Education*. I have experience teaching undergraduate discussion classes in Material and Energy Balances and Transport Phenomena. Through these experiences, and other instructional experiences outside of the classroom setting, I've worked closely with a variety of learning styles giving me a good perspective on how best to instruct and interact with students. In addition, I was able to work with many B.S. and M.S. engineers at Becton Dickinson, and was able to observe the skills that I saw promoted success from an engineering perspective. I would like to use my opportunity as a professor to support rich undergraduate extracurricular work. I have had extremely positive experiences in the past with undergraduates and REU students and hope to continue giving opportunities for independent and directed hands-on learning for students. I would also like to support undergraduate students by making them aware of the opportunities available to them for industry, graduate school and various internships (SURF, REU, etc). Finally, I was trained on the Microfluidics in Biomedical Sciences Training Program at Michigan and saw it as a fantastic experience. I would like to support programs like this and initiate the development of collaborative training programs.

386935 Unique Physicochemical Phenomena in Polymer Nanocomposites

Abstract:

Unique physicochemical phenomena in polymer nanocomposites (PNCs), including giant magnetoresistance (GMR), negative permittivity, interfacial polarization and reduced viscosity attracted more attention owing to their significant effects on both the scientific community and the practical applications. However, how to achieve the properties by tailoring the structure of the PNCs is still unknown. Understanding the structure-property relationship is the key solution to the challenge, which strongly relies on mechanism study of these phenomena. Semi-conductive nanocomposites with the aforementioned unique properties have been developed and are applied to explore the working mechanism. In composition optimization and structure design, materials with different magnetic property, electrical conductivity and morphologies will be integrated to achieve the targeting performances and reveal the structure-property-performance relationship.

Targeting the GMR and negative permittivity, a surface initiated polymerization method has been developed to fabricate the polyaniline (PANI) nanocomposites with ferrite and perovskites nanoparticles, the polymerization process takes place on the surface of the nanoparticles and the microstructure study reveals a distinctive structure with the insulated nanoparticles coated by the conductive PANI. Depending on the localization or delocalization function of the nanoparticle plays on PANI, PNCs show different GMR value. PNCs with ferrite nanoparticles show negative GMR and positive real permittivity, however, the PNCs with perovslites nanoparticles have positive GMR and negative real permittivity. The positive and negative GMR were analyzed theoretically from the wave functional shrinkage model and forward interference model, respectively. And the negative permittivity was attributed to the delocalization of charges and the metallic state in PANI.

Targeting the reduced viscosity and interfacial polarization, PNCs with different kinds of conductive nanofillers including polyaniline nanofibers and nanospheres, polypyrrole nanofibers and nanospheres, carbon coated iron (Fe@C) nanospheres, carbon nanotubes, graphene nanosheets and graphene nanosheets decorated with core-shell nanoparticles have been fabricated. The embedded nanoparticles in polymer matrix were found to favor the disentanglement of the polymer chain. And the reduced viscosity of PNCs was attributed to the attraction between the nanofiller and polymer matrix caused by physical or chemical mechanism. The interfacial polarization were found to depending on the specific surface area of the nanofiller, an enlarged interfacial area between the nanofiller and polymer matrix allows more charge carriers blocked at the interfaces and leads to enhanced interfacial polarization.

SELECTED PEER-REVIEWED PUBLICATIONS (7 out of 28):

1. **X. Zhang**, V. Bitaraf, H. A. Colorado, S. Wei and Z. Guo, Vinyl Ester Resin: Rheological Behaviors, Curing Kinetics, Thermo-mechanical and Tensile Properties, ***AIChE Journal***, 60(1), 266-274 (2014).
2. **X. Zhang**, O. Alloul, Q. He, J. Zhu, M. J. Verde,Y. Li, S. Wei and Z. Guo, Strengthened Magnetic Epoxy Nanocomposites with Protruding Nanoparticles on the Graphene NanoSheets, ***Polymer***, 54(14), 3594–3604 (2013).
3. **X. Zhang**, O. Alloul, J. Zhu, Q. He, Z. Luo, H. A. Colorado, N. Haldolaarachchige, D. P. Young, T. D. Shen, S. Wei and Z. Guo, Iron Core Carbon Shell Nanoparticles Reinforced Electrically Conductive Magnetic Epoxy Resin Nanocomposites with Reduced Flammability, ***RSC Advacnes***, 3, 9453-9464 (2013).
4. **X. Zhang**, Q. He, H. Gu, S. Wei and Z. Guo, Polyaniline Stabilized Barium Titanate Nanoparticles Reinforced Epoxy Nanocomposites with High Dielectric Permittivity and Reduced Flammability, ***Journal of Materials Chemistry C***, 1, 2886-2899 (2013).
5. **X. Zhang**, Q. He, H. Gu, H. A. Colorado, S. Wei and Z. Guo, Flame Retardant Electrical Conductive Nano-polymers based on Bisphenol F Epoxy Resin Reinforced with Nano Polyanilines, ***ACS Applied Materials & Interfaces***, 5(3) 898-910 (2013).
6. **X. Zhang**, S. Wei, N. Haldolaarachchige, H. A. Colorado, Z. Luo, D. P. Young and Z. Guo, Magnetoresistive Conductive Polyaniline - Barium Titanate Nanocomposites with Negative Permittivity, ***Journal of Physical Chemistry C***, 116(29), 15731–15740 (2012).
7. **X. Zhang**, J. Zhu, N. Haldolaarachchige, J. Ryu, D. P. Young, S. Wei, and Z. Guo, Synthetic Process Engineered Polyaniline Nanostructures with Tunable Morphology and Physical Properties, ***Polymer***, 53(10), 2109–2120 (2012).

AWARDS/HONORS:

1. 3nd Place Research Excellence Award in the *2013 Chinese-American Chemical Society Southwest Chapter (CACS-SW) Annual Poster Competition,* Houston, TX, USA, October 2013.
2. Honorable Mention Award in student poster competition at International Polyolefins Conference 2013, Houston, TX, USA, February 2013.
3. Scholarship Award, SPE (Society of Plastics Engineers) Thermoplastic Materials & Foams Division, at International Polyolefins Conference 2013, Houston, TX, USA, February 2013.
4. 2nd Place Research Excellence Award in the *2011 Chinese-American Chemical Society Southwest Chapter (CACS-SW) Annual Poster Competition,* Houston, TX, USA, October 2011.

383850 Rational Engineering of Microbes for Metabolite Overproduction

Abstract:

It has been a century since industrial scale microbial fermentation processes were established, which produced simple natural metabolites, such as ethanol, acetone, glycerol and butanol. During the past two decades, advances in DNA technology have enabled transfer of heterologous pathways to microbial host cells and manipulation of a specific step of an endogenous pathway, leading to great improvements in diversity and titer of the metabolites that microbes can make. For example, gram-per-liter titer of plant therapeutic molecules (e.g. paclitaxel and artemisinin precursors) has been achieved in *Escherichia coli* or *Saccharomyces cerevisiae*. However, these amazing successes still heavily relied on trial and error experiments, which are very time-consuming; this is because the interactions between heterologous pathways and cellular metabolism and those among parts of a foreign pathway were not well understood. The faculty candidate is interested in understanding details of these interactions for developing fast and rational approaches to improve the metabolite overproducing strains, and inventing novel practical means of preventing negative interactions among pathway segments - such as spatially separating incompatible pathway modules into different cells or compartments of a cell. During the poster session, the candidate will present our projects that embody these research ideas, and discuss his research plan with search committees.

Computational Systems Biology of Metabolism in Single- and Multi-Species Microbial Systems

Ali R. Zomorrodi

Graduate Program in Bioinformatics and Department of Biomedical Engineering, Boston University, Boston, MA

Metabolism is defined as the set of all biochemical reactions in a living organism required to maintain life. The efficient engineering of metabolism in natural and synthetic microbial systems for biotechnological applications requires a systems-level understanding of the metabolic interactions. This calls for the development of systems-level analysis tools and modeling techniques capable of providing a quantitative understanding of the metabolism as well as quantifying the system-wide impact of engineering manipulations.

Here, I review my recent work on the development of efficient computational systems biology tools to shed light onto the less-understood aspects of metabolism in single- and multi-species microbial systems, and to identify promising network rewiring strategies for biotechnological applications. I first introduce a workflow that combines the ensemble modeling of metabolic networks with customized optimization-based algorithms for creating large-scale kinetic models of metabolism. This is exemplified by constructing a large-scale kinetic model of *E. coli* core metabolism satisfying multiple sets of mutant flux data. I next introduce k-OptForce, a computational strain design protocol integrating the available kinetic information with the genome-scale stoichiometric-based models of metabolism to identify system-wide metabolic interventions leading to the targeted overproduction of a biochemical of interest. Results and lessons learned by the application of k-OptForce to overproduce a number of biochemicals and biofuels (e.g., fatty acids, triacetic acid lactone, L-serine etc.) in *E. coli* and *S. cerevisiae* will be shared.

The focus is then shifted from single-species systems to microbial ecosystems, where multiple microorganisms are involved and communicate through the exchange of biochemical cues. Here, I introduce a multi-level and multi-objective optimization framework, called OptCom, for the metabolic modeling and analysis of microbial communities using genome-scale metabolic models. Next, I introduce an extension of this procedure, called d-OptCom, which enables capturing the impact of dynamic changes in the environment on the function and structure of microbial communities. This is followed by a discussion of ongoing efforts on the game-theoretic analysis of microbial communities using genome-scale metabolic models in dynamic spatially structured environments. Overall, the wide array of computational and systems modeling tools introduced in this poster highlight their utility in model-driven biological discovery and in complementing metabolic engineering efforts for biotechnological applications.

393772 Plasma Catalysis for Materials and Energy Challenges

Abstract:

Low-pressure plasmas can be used for gas phase activation via electron-molecular reactions producing ions, radicals, metastables and photons that can be beneficial for catalysis of several technologically important reactions. First of all, when used along with heterogeneous catalysts, the plasma activation could lower the activation energy further. Such synergistic effect is termed here as plasma catalysis Plasma catalysis is emerging as one of the most promising alternatives to address energy and environmental challenges.

However, it is not clear whether one needs the same type of catalyst materials even for plasma environments or not. For example, the low-melting metals in the presence of plasmas can catalyze the growth of one-dimensional materials. In the absence of plasma activation, the molten metals do not exhibit any catalytic activity towards selective dissolution even with application of high temperatures. The reasons for such synergic effect with plasma activation and metal clusters suggest that such metals could be interesting for plasma catalysis research to address environmental/energy challenges in addition to materials growth. In this research program, using results from our prior research, new catalysts will be designed for studying carbon dioxide reduction using plasma catalysis approach. CO_2 is a cheap, non-toxic and non-flammable, renewable feedstock compared to oil or coal for producing valuable chemicals and transportation fuels. However, the existing methods for CO_2 reduction are not economically viable. The production of chemicals from CO_2 could have a significant positive impact on the global carbon balance.

Hydrogen electrocatalysis: from mechanistic study to advanced catalyst development

Wenchao Sheng

ws2440@columbia.edu
Department of Chemical Engineering, Columbia University, New York, NY 10027

My research interests are centered on addressing challenges in renewable energy research, and mainly focused on electrochemical energy storage and conversion applications such as electrolyzers, fuel cells, flow batteries, and electrochemical synthesis of high-value chemicals. I aim to bridge the fundamental understanding in electro-catalysis and surface science constructed on model electrode systems to the design and development of high-performance nano-scale materials and structures for clean energy applications.

One of the intriguing topics in renewable energy research is the utilization and generation of hydrogen energy. Hydrogen oxidation and reduction reactions (HOR/HER) are two of the key components in low-temperature fuel cells and electrolyzers. The pH values of the electrolytes have a great impact on the reaction kinetics such that the HOR/HER on the state-of-the-art Pt catalyst in alkaline media decreases by two orders of magnitude than that in acids. The slower HOR kinetics on Pt is projected to impose an additional overpotential (~130 mV at 1.5 A/cm^2_{MEA} with a Pt loading of 0.05 mg_{Pt}/cm^2_{MEA}) in alkaline or alkaline membrane fuel cells (AFCs/AMFCs), thus the use of Pt catalyst at the anode in AFCs/AMFCs requires a higher catalyst loading.[1, 2] With the development of many non-precious metal catalysts for the oxygen reduction reaction in both acid and alkaline electrolytes, showing comparable activity to Pt, the demand for highly active and cost-effective HOR catalyst in alkaline medium is arising and becomes one of the crucial challenges to make AFCs/AMFCs more practical.

Rational design of highly efficient HOR/HER catalysts requires identification of the reaction descriptor for the HOR/HER in alkaline electrolyte. It is shown that the HOR/HER exchange current density in alkaline media on a wide range of monometallic surfaces follows a volcano type of relationship with the calculated hydrogen binding energy (HBE) of the metals with Pt having the optimum HBE value.[3] This discovery indicates that HBE values can serve as the designing principle for developing efficient electrocatalysts for the HOR/HER in base. Motivated by this finding and the fact that Ni exhibits certain HOR/HER activity in base, we calculated the HBE of NiMo bi-metallic and CoNiMo multi-metallic surfaces and found that they have similar HBE values as Pt. We subsequently synthesized NiMo and CoNiMo alloys by electrodeposition and performed HOR/HER tests in 0.1 M KOH at different temperatures. The results show that the HOR/HER on CoNiMo outperforms electroplated Ni by 20 times.[4] At sufficient catalyst loading, this promising material will alleviate the catalyst cost issue raised by higher Pt loading in alkaline electrolytes, and will facilitate the development of the AFCs/AMFCs technology.

References:

1. K. C. Neyerlin, W. B. Gu, J. Jorne and H. A. Gasteiger, *J Electrochem Soc*, 2007, 154, B631-B635.
2. W. C. Sheng, H. A. Gasteiger and Y. Shao-Horn, *J Electrochem Soc*, 2010, 157, B1529-B1536.
3. W. C. Sheng, M. Myint, J. G. G. Chen and Y. S. Yan, *Energ Environ Sci*, 2013, 6, 1509-1512.
4. W. C. Sheng, A. P. Bivens, M. Myint, Z. B. Zhuang, R. V. Forest, Q. R. Fang, J. G. Chen and Y. S. Yan, *Energ Environ Sci*, 2014, 7, 1719-1724.

393802 Polymer Supported Catalysts for Heterogeneous Catalysis

Abstract:

Research Interests:

Selective and efficient conversion of oil or gas feedstock into highly valued bulk and fine chemicals is one of the major focal points in heterogeneous catalysis. Traditional inorganic catalyst supports suffer from mass transfer limitations, coking and low interaction with organic molecules. In particular, catalytic reactions in fine chemical synthesis require a suitable environment of ligands and solvent, which provide the required geometry and stabilize the transition state of the reaction. Polymers create a 'solvent'-like organic reaction environment, onto which ligands and metal complexes can be grafted, or metal nanoparticles can be loaded to provide the catalytic functionality. Polymers also swell when placed in a solvent enabling new mass transfer pathways. However, most of polymers do not offer high thermal stability and resilience to mechanical stress, which consequently leads to a limited use in industrial reactions. My core competence is to design and synthesize thermally and mechanically stable polymer supports, to introduce functional groups into these polymer supports, to test them in model reactions and to apply them to practically relevant industrial processes.

PI Experience:

In November 2012, I joined Prof. Ferdi Schüth's group as a postdoc. I discussed the idea of ultrasonic spray pyrolysis as a method to synthesize catalysts with him and he agreed to give me the financial support but I had to find a lab facility to conduct the research. With his help, I was offered laboratory space from Prof. Christof Schulz and Dr. Hartmut Wiggers at the University of Duisburg-Essen. I applied and obtained an Alexander von Humboldt fellowship to add to the financial support given by Ferdi Schüth. After proposing the idea, three PhD students were very eager to join the lab and work with me on this interesting work. Currently, mesoporous sulfonated carbon spheres have been synthesized via one step ultrasonic spray pyrolysis using cheap precursors and will be tested in dehydration reactions.

Research Experience:

My PhD thesis focused on controlled synthesis of colloidal nanoparticles, heterostructures and their applications in catalysis. For instance, plasmonic-catalytic nanostructures have been synthesized to utilize solar energy in heterogeneous catalysis. Entering my postdoc, I broadened my field into the catalyst-support interaction. I have developed a polyphenylene support as an excellent platform for heterogeneously catalyzed reactions, which are normally carried out under homongeneous conditions. High thermal stability of the polyphenylene support further enables gas phase continuous partial oxidation of benzyl alcohol to benzaldehyde. I have also built an *in situ* electron paramagnetic resonance setup to study the redox properties of $CuO\text{-}CeO_2$. Using this set-up combined with online mass spectroscopy, the three step redox mechanism could be directly observed for the first time.

Teaching Experience:

I have two years' experience as a teaching assistant in Chemical Principles and three years' experience as a teaching assistant in Astronomy. I taught the exercise courses and tutorials, off-class activities, and also main lectures. I have also tutored 4 PhD students. Currently, I supervise an exchange student, Jacob Oskar Abildström, from The Technical University of Denmark in our group.

Future Research:

My core competence as a faculty member in chemical engineering department is to design polymer supports for heterogeneous catalysis. In the short term, my research comprises of polymer synthesis, studing their chemical and thermal stability, mass transfer, functionalization, and activity in model reactions. Cross coupling polycondensation, cyclotrimerization, radical or acid/base catalyzed polymerizations are the major routes to obtain conjugated and non-conjugated polymers. Metal nanoparticles are impregnated via swelling-impregnation method developed by myself. In addition, the spray drying technique offers a continuous way to synthesize polymer catalysts. Chemical stability of polymers can be studied by treatment with redox gas or liquids at reaction conditions. Their relations with polymer structure will be investigated by varying the ratio between aromatic and aliphatic units. Selective hydrogenation of alkynes with different molecular sizes will be used to investigate the mass transfer limitation. The polymer supported catalysts will be tested in several model reactions both with loaded metals, ligands or a combination of both. The focus for the application of the reactive polymers will be cross-coupling reaction, selective oxidation and hydrogenation. If selected to obtain tenure, I will aim for more challenging tasks with polymers, such as inorganic-organic hybrid catalysts, gas phase heck reactions, and benzene to phenol conversions.

Selected Publications:

(14 total, 7 first author, 427 total citations, 1 patent – ResearcherID F-6893-2014)

1. **Feng Wang**, Jerrik Mielby, Felix Herrmann Richter, Guanghui Wang, Gonzalo Prieto, Takeshi Kasama, Claudia Weidenthaler, Hans-Josef Bongard, Søren Kegnæs, Alois Fürstner, Ferdi Schüth "A Polyphenylene Support for Pd Catalysts with Exceptional Catalytic Activity" *Angew. Chem. Int. Ed.* **2014**, DOI: 10.1002/anie.201404912.
2. **Feng Wang**, Si Cheng, Zhihong Bao, Jianfang Wang "Anisotropic overgrowth of metal heterostructures induced by site-selective silica coating" *Angew. Chem. Int. Ed.* **2013**, *52*, 10344-10348.
3. **Feng Wang**, Chuanhao Li, Huanjun Chen, Ruibin Jiang, Ling-Dong Sun, Quan Li, Jianfang Wang, Jimmy C. Yu, Chun-Hua Yan "Plasmonic harvesting of light energy for Suzuki coupling reactions" *J. Am. Chem. Soc.* **2013**, *135*, 5588-5601.
4. **Feng Wang**, Ling-Dong Sun, Jun Gu, Wei Feng, Yi Yang, Jianfang Wang, Chun-Hua Yan "Selective Heteroepitaxial Nanocrystal Growth of Rare Earth Fluorides on Sodium Chloride: Synthesis and Density Functional Calculations" *Angew. Chem. Int. Ed.* **2012**, *51*, 8796-8799.
5. **Feng Wang**, Chuanhao Li, Ling-Dong Sun, Chun-Hu Xu, Jianfang Wang, Jimmy C. Yu, Chun-Hua Yan "Porous Single-Crystalline Palladium Nanoparticles with High Catalytic Activities" *Angew. Chem. Int. Ed.* **2012**, *51*, 4872-4876.

6. **Feng Wang**, Chuanhao Li, Ling-Dong Sun, Haoshuai Wu, Tian Ming, Jianfang Wang, Jimmy C. Yu, Chun-Hua Yan "Heteroepitaxial Growth of High-Index-Faceted Palladium Nanoshells and Their Catalytic Performance" *J. Am. Chem. Soc.* **2011**, *133*, 1106-1111.
7. **Feng Wang**, Ling-Dong Sun, Wei Feng, Huanjun Chen, Man Hau Yeung, Jianfang Wang, Chun-Hua Yan "Heteroepitaxial Growth of Core–Shell and Core–Multishell Nanocrystals Composed of Palladium and Gold" *Small* **2010**, *6*, 2566-2575.

393894 Electrokinetic Instability in Microchannel-Nanochannel Devices: A Comparison of Slip-Velocity and Full-Formulation

Abstract:

formulation

In systems containing ion-selective membranes or nanochannels, concentration polarization (CP) under DC voltage beyond the classical Levich limit leads to the breakdown of local electroneutrality over micron or larger scales at the salt-depleted interface. This manifests itself in the growth of an extended space charge (ESC) region, which becomes unstable above a critical voltage drop. The instability results in the formation of a fast-flowing vortex system with complex, often chaotic, dynamics. In unconfined systems, i.e. large electrolytic cells, this contributes strongly to the overlimiting conductance (OLC) of the system. However, both the role of the instability in OLC as well as its origin and onset become more complicated in highly confined systems such as microchannel devices. The problem of instability under geometric confinement has been studied both analytically and numerically using two different approaches. The first approach considers the ESC to be negligibly thin. Thus it is replaced by a slip boundary condition for the hydrodynamics and boundary values for the concentration and potential obtained from asymptotic analysis. This yields directly a marginal stability curve and critical parameters for the onset of instability. The second approach employs a direct numerical calculation of the stability problem resulting in a growth-rate surface from which marginal stability and critical parameters may be extracted. While both show suppression of the instability with increasing confinement, there are some interesting discrepancies between the features which merits further consideration.

392943 DNA-Controlled Partition of Carbon Nanotubes in Polymer Aqueous Two-Phase Systems

Abstract:

Sorting single-wall carbon nanotubes (SWCNTs) of different chiralities is both scientifically interesting and technologically important. Specific short single-stranded DNA sequences have been identified to show recognition abilities for specific chiralities of SWCNTs from their synthetic mixture. In addition, polymer aqueous two-phase (ATP) extraction has been demonstrated recently as a very effective way to achieve nanotube sorting. In this work, we report a systematic study of sequence-dependent partition of DNA-wrapped SWCNTs in several polymer ATP systems. This unique approach of combining DNA sequence and polymer ATP system enabled the purification of many single-chirality SWCNT species at high purity. In addition, the DNA-controlled partition of SWCNTs in ATP system suggests the sensitive dependence of hydration energy on the exact spatial distribution of hydrophilic groups which are determined by the DNA sequence and the SWCNT chirality (n,m).

393908 Design of Biomimetic Functional Nanomaterials to Study Integral Membrane Proteins

Abstract:

Integral membrane proteins (IMPs) reside within the cellular membrane and play a key role in modulating a wide range of physiological processes, from neuronal signaling to immune and hormonal response. Their malfunctioning has been implicated in various diseases, including cancer and cardiovascular diseases, which makes them an important pharmaceutical target. The native lipidic environment is imperative for the stability and activity of IMPs, which makes *in vitro* studies of these proteins very challenging. The overarching goal of my research has been to develop novel biomimetic interfaces for reconstituting IMPs to facilitate their structural and functional studies. For example, during my postdoctoral research at the NIST Center for Neutron Research, I fabricated a functionalized nanotubular membrane for reconstituting IMPs, and used this platform in combination with neutron scattering techniques to determine the IMP's structure. Additionally, during my graduate research, I developed a microarray-based platform to selectively immobilize IMPs for high-throughput target screening. Currently, my research at the University of Delaware is focused on studying the role of surfactant micelles in the stabilization and crystallization of IMPs. Surfactant micelles are widely used as a bilayer mimic for reconstituting IMPs in protein-detergent complexes (PDCs). We are employing nuclear magnetic resonance and small-angle neutron and x-ray scattering (SANS and SAXS) techniques to study the stability of PDCs over molecular and microscopic length scales, respectively, which we expect will enable the rational design of PDCs and IMP crystallization trials. Using my interdisciplinary research background in biomaterials, nanotechnology, neutron scattering techniques, and IMPs, I intend to develop an independent research program on *biomimetic functional nanomaterials* for: biophysical characterization of IMPs using neutron scattering; biosensors to monitor IMPs; and high-throughput drug screening platforms.

394054 A Systems Level Approach Towards Rational Strain Development for Microbial Bioprocesses

Abstract:

Microbial based biochemical and/or biofuel production presents an attractive and sustainable potential source of industrial chemicals and transportation fuels have been the research focus in the recent past. The attractiveness of this approach is greatly enhanced by the use of renewable feedstock such as lignocellulosic materials, waste product and byproduct streams such as crude glycerol derived from biodiesel production and gaseous effluents (CO_2 and CO) from industrial plants. Generation of value added products from waste streams not only offers an efficient and effective means to recycle, reuse and/or dispose the waste stream but also improves the overall economics of a sustainable biorefinery. However, there are challenges associated with the efficient utilization of the materials such as lignocellulosic compounds and other waste product, thereby lowering the efficiency of the bioconversion process. Moreover, the production of biofuels and biochemicals from microbial fermentation has been severely limited due to the accumulation of products, such as butanol, which are toxic to the cells producing them, resulting in a stress response associated with growth and survival inhibition, and other cellular physiological events such as sporulation observed in the genus of *Clostridia*, thereby limiting the productivity of the desired end product.

The metabolic potential and regulation of the microbial metabolism can be comprehensively interpreted using the tools of systems biology and the information obtained from such study can be used for rational design of the metabolic pathways and regulation. By linking the changes in genome (structural genome) for strains exhibiting desired characteristics from directed evolution and/or chemical mutagenesis to the (functional genome) transcriptome, proteome and metabolome of the cells thereby, enabling targeted genome engineering towards the development of highly productive and/or tolerant strain.

Meet the Faculty Candidate Poster Session

AIChE 2014 Atlanta, GA

Blake N. Johnson, Department of Mechanical and Aerospace Engineering, Princeton Institute for the Science and Technology of Materials, Princeton University, Princeton, NJ, USA

The development of a biofabrication method which offers the ability to precisely assemble biologically functional multi-component devices and materials could allow for both revolutionary fundamental biological discoveries using advanced *in vitro* models for tissue regeneration and could ultimately lead to new tissue engineering-based applications for treatment of disease and complex injury.

Along these lines, my postdoctoral research has been dedicated the development of novel three-dimensional (3D) bioprinting approaches for creating anatomically conformal nerve guidance channels capable of regenerating complex peripheral nerve injury. Additionally, the approach was adapted to develop complementary nervous system-on-chip models which remain highly faithful to the composition and structure of peripheral nerve tissue. A light scanning-based model acquisition technique known as 3D structured-light scanning was used to generate high resolution computer models of the sciatic nerve tibial branch which served as the template for the final 3D printed nerve guide. The bioprinting approach also facilitated incorporation of supporting physical and biochemical cues which included longitudinal guidance cues and gradients of neurotrophic factors. The selective regeneration of the sensory and motor nerve pathways bifurcating from mixed nerve was also examined by incorporating internal hydrogel matrix which contained guiding biochemical cues in the different pathways. In order to elucidate the critical factors affecting regeneration of the bifurcating mixed nerve injury, we have developed complementary nervous system-on-chip platforms for the studying the influence of surface topology and biochemical cue spatial distribution on various important factors critical to successful peripheral nerve regeneration, including the support and guidance of axonal growth and the migration of Schwann cells.

My PhD research was focused on the developing ultra-sensitive biosensor technology for real-time monitoring of protein-protein, DNA-DNA, and DNA-RNA interactions in complex backgrounds, such as circulating fluids and environmental settings. Specifically, the research involved the modeling of sensor physics, as well as the fabrication and characterization of piezoelectric-based hybrid mechanical and electrochemical label-free biosensors. Significant biosensing achievements included the label- and preparation-free detection of a class of important circulating biomarkers, known as microRNA, directly in human serum and the detection of toxin-producing cyanobacteria, also known as blue-green algae, in environmental source waters. In addition to biosensing applications, my research also developed means of using sensor-based approaches for modulating the surface binding energy of immobilized molecules and cells using electromechanical and electrochemical energy.

As a new faculty member, I would continue to develop novel lab-on-chip devices for neuroregenerative therapy in the peripheral nervous system (PNS) as well as attempt to expand towards solving problems targeting fundamental biological questions related to central nervous system regeneration and degradation. Using the nervous system-on-chip platforms, I would also plan to examine new approaches for sensor-based stimulation of neuroregenerative response, as such devices provide a controlled setting for examining cellular responses to mechanical, electrical, and drug stimuli.

394061 Packings and Assemblies for Continuous Families of Polyhedra

Abstract:

Nanoparticles and colloids of various polyhedral shapes are synthesized and used as building blocks for self-assembly. It has been shown in simulations that a plethora of complex crystals (including quasicrystals) can form entropically solely due to the anisotropic shape of the particles. At the limit of maximum density, the densest packings, as have been studied in math for centuries, becomes the relevant quantity. However, a general understanding of how changes in shape affect packings and assemblies remains largely unexplored. Here, we study continuous families of polyhedra to demonstrate richness and complexity in behavior as a function of shape. We investigate connections between assemblies and densest packings, discuss the possibility of predicting one from the other and outline general guidelines for experiments.

394727 Co-culture based modular engineering for aromatic and aromatic-derived compound production in Escherichia coli

Abstract:

Traditional metabolic engineering studies focus on engineering target metabolic pathways in one single cell. Despite having gained many successes, this approach suffers from several disadvantages which limit its wide application in the future metabolic engineering research. Here, we developed a novel bacterium *E. coli-E. coli* co-culture approach to completely modularize and optimize the biosynthetic pathway in two separate cells. In particular, a membrane-bound transporter was utilized to facilitate the intermediate mass transfer between the two modules/cells so that the two cells worked as an integrated whole. The co-culture system was shown to produce value-added aromatic and aromatic-derived compounds with high titer and yield. Moreover, the co-culture system can be engineered to efficiently utilize sugar mixtures, which is challenging for the single cell strategy. The established method in this study offers a novel perspective for conducting metabolic engineering studies in the future.

396659 Power and Remediation with Photovoltaics

Abstract:

My research interests are focused on the fundamental understanding and practical application of photovoltaics. The prospective work will focus on the synthesis of new materials and better understanding of the underlying properties that determine the effectiveness of different materials for photovoltaic and photoelectrochemical processes.. I believe that the projects described below are feasible and will provide new insights in their respective fields.

Remediation of carbon dioxide is a problem without a clear practical solution. Sequestration is a temporary answer and may not be providing even the minimal results originally predicted. Electrochemical reduction of carbon dioxide is possible but requires such a significant energy input that it is not practical at this point. Using "free" energy from sunlight to reduce carbon dioxide with a photoelectrochemical system is an appealing solution. P-type semiconductors are able to directly reduce molecules in solution. Accordingly, although the direct reduction of CO_2 requires approximately 1.9V of input, this could be reduced with co-catalysts and the use of strategic electrolytes (acetonitrile, ionic liquids).

Oxynitride photocatalysts are known in the literature. While their synthesis is highly energy and time intensive, they show promise in both photovoltaic and water splitting applications. Synthesizing oxynitride photocatalysts at lower temperatures and shorter times would enhance the ability to seek out new oxynitride photocatalysts. The use of a local exothermic reaction will meet these goals. The use of a localized exothermic reaction will generate the heat needed to synthesize the oxynitride materials. This procedure may also allow for high-throughput synthesis techniques that will speed the search for new and better materials even more.

Hybrid organic-inorganic perovskite materials have recently garnered a great deal of interest due to their low cost and high efficiency in photovoltaic applications. All of the focus, however, has been on Pb-halide based materials. Although cheap, using lead in the material could be problematic. This work would focus on finding other perovskite-type photovoltaic materials by substituting the metal and halide. Experiments would include full physical and spectroscopic (time-resolved ultrafast) characterization of the material as well as application testing for photovoltaic, water splitting, and remediation purposes.

These projects will allow for collaboration in a number of fields, including materials synthesis techniques, energy storage methods (batteries, flow cells, etc.), and reaction kinetic measurements (steady-state and transient).

396658 First Principles-based Multiparadigm, Multiscale Strategies for Simulating Complex Materials Processes

Abstract:

My research focuses on, (1) development of advanced computational algorithms for atomistic modeling, and (2) their applications to current problems in chemistry and chemical engineering. The research lies at the interface of physics, chemistry, chemical engineering, and numerical and algorithms analysis. The aim of my research is to study processes that are of current interest from both small- and large-scale perspectives. The research is motivated by diverse applications including energy technologies, nanotechnology, biomaterials, catalysts, and organic materials where experimental approaches are inaccessible and/or insufficient. In principle, any material is a collection of particles including nuclei, electrons, and phonons. To engineer alternative materials for improved or new functionalities, it is crucial to understand chemistry and energetics of the materials, which govern their mechanical, electronic and thermal properties. I use *ab-initio* quantum mechanics (QM) and density functional theories (DFT) to achieve the goals of my research. Such methods fail, however, when more realistic systems must be studied. For instance, micro-second studies of systems involving large number of particles and high temperature are far beyond the scope of QM and DFT methods. Much of my research is focused on bridging first-principle simulations, normally feasible for small systems, and large-scale simulations necessary for studying systems of practical interest. I have extensive experience in development of reactive force fields, such as ReaxFF as well as invoking reactive force fields to study real-life problems.

I will present an example of such problems by discussing a process-based molecular model of nano-porous silicon carbide (SiC) membrane. A broad class of important materials, such as carbon molecular sieves, SiC, and silicon nitride, are fabricated by temperature-controlled pyrolysis of preceramic polymers. In particular, the fabrication of SiC membranes by pyrolysis of a polymer precursor that contains Si is quite attractive for separation of hydrogen from other gases. It has been quite difficult to extract atomistic-scale information about such SiC membranes since they are amorphous. I will discuss development of a reactive force filed and its application for the thermal decomposition of hydridopolycarbosilane (HPCS) to form SiC nanoporous membranes. I will also highlight my future research plans which I believe are of substantial scientific merit and national interest which fit well with a range of research programs launched by different funding agencies.

Directed Self-Assembly of Sub-10 nm Particles: Role of Driving Forces and Template Geometry in Packing and Ordering

Shafigh Mehraeen[1,2], Mohamed Asbahi[3], Wang Fuke[3], Joel K. W. Yang[2,3], Jianshu Cao[1], Mei Chee Tan[2]

[1]Department of Chemistry, Massachusetts Institute of Technology, Cambridge, Massachusetts 02139, United States
[2]Pillar of Engineering Product Development, Singapore University of Technology and Design, Singapore 138682
[3]Institute of Materials Research and Engineering, Agency for Science, Technology and Research (A*STAR), Singapore 117602

Abstract

By comparing the magnitude of forces, a directed self-assembly mechanism has been suggested[1] previously in which immersion capillary is the only driving force responsible for packing and ordering of nanoparticles, which occur only after the meniscus recedes. However, this mechanism is insufficient to explain vacancies formed by directed self-assembly at low particle concentrations. Utilizing Monte Carlo and Brownian dynamics simulations, we developed a theoretical model, and proposed another mechanism. In our proposed mechanism, the competing driving forces controlling the packing and ordering of sub-10 nm particles are (1) the repulsive component of the pair potential, and (2) the attractive capillary forces, both of which apply at the contact line. The repulsive force arises from the high particle concentration, and the attractive force is caused by the surface tension at the contact line. Our theoretical model also indicates that the major part of packing and ordering of nanoparticles occurs before the meniscus recedes. Furthermore, utilizing our model, we are able to predict the various self-assembly configurations of particles as their size increases. These results lay out the interplay between driving forces during directed self-assembly, offering a better template design now that we know the importance and the dominating driving forces in each regime of particle size.

References

1. Liddle, J. A.; Cui, Y.; Alivisatos, P. *J Vac Sci Technol B* **2004,** 22, (6), 3409-3414.

396344 Microfluidics for Polymer Self-Assembly in Confined Quasi-2D Geometries and Enzyme Design towards Plastic Recycling and Biofuel Production

Abstract:

My postdoctoral work with Prof. David Weitz (Harvard University, 2010-present) involves high throughput microfluidics for performing chemical reactions and biochemical assays in picoliter drops. Taking advantage of material behaviors at liquid-liquid interfaces, enables me to engineer complex materials of hierarchical orders with novel functionalities. For example, I created triggered-release systems using stimuli responsive polymers and enzyme-based functional materials in collaboration with Advanced Energy Consortium and BASF companies. Additionally, in collaboration with New England BioLabs, I genetically engineered proteins and also performed single molecule assays using DNA templates and an *in vitro* transcription-translation system. Moreover, part of my duties as a postdoc included mentoring and supervising the research of undergraduates, graduates and postdocs on a large number of projects.

Thus, I believe that my training and expertise position me advantageously for becoming a faculty member.

As a new faculty member, my first thrust is to take advantage of confinement to engineer novel structures that otherwise are unattainable in bulk. It has been demonstrated that confinement is a powerful tool to break symmetry on the scale of single molecules, allowing polymers to form new phases that are not possible using bulk self-assembly. However, the phase behavior of polymer blends under confinement has not been characterized experimentally in detail due to the lack of available techniques. We recently developed a novel microfluidic method for fabricating monodisperse double emulsions with ultra-thin polymer shells, less than 100 nm in thickness. Due to the quasi-two dimensional physical confinement this method offers exquisite control over the phase behavior of the polymer blend enabling non-equilibrium assembly to create membranes with controlled porosity and permeability. In addition, I will synthesize new polymers and investigate how changing their molecular structure affects self-assembly to obtain quantitative structure-function correlations rules. Applications range from encapsulation and controlled release to membranes for separation and purification.

My second thrust addresses the huge environmental challenges arising from plastic waste and energy production. The current approach for dealing with plastic waste is incineration, which releases large amount of CO_2. In contrast, enzymatic degradation is a possible solution for efficient transformation into high-value fine chemicals without significant greenhouse gas emissions. However, naturally occurring enzymes are inefficient at processing plastic substrates. To increase their efficiency for plastic degradation controlled mutations can be utilized; this approach necessitates generating an enormous library of enzyme mutants. However, developing reliable high-throughput screening platforms to screen for functional variants in libraries with large size is very challenging.

I will use drop-based microfluidics to develop a high-throughput screening platform to overcome these difficulties by using new assays based on *in vitro* transcription and translation (IVTT) system. I will adapt this approach to engineer enzymes and enzyme-polymer conjugates through mutagenesis and study their interaction with natural and non-natural substrates. The results of this study will be applied to the design of enzymes for new applications: such as the biochemical recycling of plastics and production of renewable energy through degradation of cellulose.

Bio-inspired Engineering: Formulation Principles and Design Strategies for First-in-class Therapeutics

Rahul K. Keswani, PhD
Research Fellow, University of Michigan, Ann Arbor

Recent advances in pharmaceutical engineering have entailed the development of multifunctional materials that produce tailored pharmacological responses. Concurrently, scientists have taken advantage of naturally occurring materials to design components for incorporation with synthetic agents to produce super-efficient, targeted and intelligent therapeutics that provide sustained therapy. In this poster, I will provide proof-of-concept and design strategies for *two novel, efficient and biologically relevant therapeutics*.

1. Typically, drugs have unique sites of action and mechanisms that are dependent on formulation and administration routes. The associated chemical interaction of drugs with the human body leads to either pharmacologically relevant metabolites or alternative biochemical forms of the compound. One such drug is the riminophenazine dye, clofazimine that is known to have biocrystallized forms. These have been clinically reported but the mechanism of formation has never been investigated. I will describe the *first-in-class biocrystal-based theranostic agents* that mimic the stabilized chemical drug formulation of clofazimine and that pharmacologically, is both a potent anti-inflammatory medication and a fluorescence and photo-acoustic signaling-based diagnostic imaging agent. Most importantly, the design principles are easily translatable for other pharmaceuticals with varied chemical properties such as basicity, lipophilicity and solubility.
2. Most gene therapeutics rely on using either non-viral vectors for packaging genetic cargo or employ modified viruses for their natural ability to insert genetic cargo into cells. I will provide proof-of-concept and design principles of combining enhanced functional capabilities of naturally sourced biomaterials with the remarkable efficiency of viruses into *first-in-class superlative hybrid gene therapeutics.*

Selected References:

1. **RK Keswani**, K Su & DW Pack, 2014, "Efficient In Vitro Gene Delivery by Hybrid Biopolymer/Virus Nanobiovectors", Journal of Controlled Release, 192, 40-46.
2. **RK Keswani**, IM Pozdol & DW Pack, 2013, "Design of Hybrid Lipid/Retroviral-Like Particle Gene Delivery Vectors", Molecular Pharmaceutics, 10(5), 1725-1735.
3. **RK Keswani**, J Baik, C Hitzman, L Yeomans, N Rodriguez-Hornedo, KA Stringer, GR Rosania, 2014, "A concentrative chloride transport pathway parallels massive accumulation of a weakly basic drug in mouse macrophages", Journal of the American Chemical Society, submitted.

396061 Engineering synthetic microbial cell factories for fuels and green chemicals

Abstract:

Global energy demand, rising petroleum prices and environmental concerns have stimulated increasing efforts to develop more sustainable and cost-effective approaches to produce carbon-neutral fuels and green chemicals. Exploitation of the diverse microbial pathways based on metabolic engineering and synthetic biology frameworks provides a promising solution to synthesis of fuel and pharmaceutical molecules that are previously derived from the petroleum-based resources. As an emerging discipline, synthetic biology is becoming increasingly important to design, construct and optimize metabolic pathways leading to desired overproduction phenotype. We will highlight some of our most recent efforts in applying synthetic biology tools such as combinatorial pathway engineering strategies to fine tune the expression level and balance various sections of a metabolic pathway. Some of the synthetic biology techniques that will be highlighted include the construction of the *ePathBrick* vectors that support the modular assembly of multi-gene metabolic pathways and development of synthetic malonyl-CoA switches that allow dynamic tuning of metabolic flux in *E. coli*. Efficient genome engineering tools such as CRISPR-cas9 directed multiplex combinatorial pathway engineering will also be covered. Applications of such strategies have resulted in the robust production of important compounds that are of particular interest to the pharmaceutical and biofuel industry.

Swelling in polymeric membranes for natural gas applications

Leslie Schulte, Chemical Engineering, Kansas State University
leslieschul@gmail.com

Natural gas production in the US has increased dramatically with the expansion of fracking. These operations require separations to remove CO_2, H_2S and higher hydrocarbons from CH_4 before natural gas can be placed in the pipeline and delivered to the consumer. Polymeric membranes are an attractive technology for this separation because they are modular in nature, have a small footprint and do not require a phase change for separation. Unfortunately, the different components in the raw natural gas mixture swell polymers. This swelling, or plasticization, limits the separating ability of the membrane. Many researchers have worked to develop an understanding of why swelling occurs and there are several solutions including thermal and chemical treatments. These treatments, while suppressing swelling, also decrease the productivity of the membrane. *My research goals are to understand how plasticizing molecules affect membrane materials. I'll use this understanding to develop novel techniques that prevent the decline in selectivity when membranes are exposed to plasticizing environments.*

My PhD project focuses on fabricating and characterizing flat sheet membranes. The primary goal of this research was to design a more chemically and thermally stable polymeric membrane with gas separation properties. This was accomplished by blending commercially available polymers. My work is part of an NSF-funded IGERT program. This program brings together an interdisciplinary group of engineers, agronomists, and sociologists working on the conversion of biomass to fuel.

Thermochemical Conversion of Biomass to Renewable Fuels

Paige A. Case

University of Maine Department of Chemical and Biological Engineering

Doctoral Advisors: William J. DeSisto and M. Clayton Wheeler

I am passionate about research and enjoy working in groups with a diverse set of people. My current research focuses on development of simple, robust processes to deoxygenate and stabilize bio-oils to produce drop-in transportation fuels. I plan to continue research in the field of sustainable fuels and chemicals.

As part of a multidisciplinary, multi-institute research group working on thermochemical conversion strategies for production of fuels and chemicals from woody biomass, I developed methods and techniques that have been adopted by other researchers. Over the course of my research career, I have published four peer-reviewed journal articles and have given several papers and posters at conferences. Each of my poster presentations has been awarded special recognition.

I have extensive experience in non-traditional education and outreach. I assisted with grant writing and was responsible for coordinating UMaine's curriculum for the Renewable Energy minor. I have enjoyed mentoring several undergraduate students, and my work with an REU student last year resulted in a peer-reviewed publication about value-added products from tannins. I have been active in K-12 education outreach as well, and have developed activities for several organizations and represented our department at recruiting events. Perhaps most notably, I have spent three weeks each summer developing design projects and leading teams of high school juniors in laboratory experiments for UMaine's Consider Engineering program.

Publications

"Formate Assisted Pyrolysis of Pine Sawdust for in-situ Oxygen Removal and Stabilization of Bio-oil"
Case, Paige A., Wheeler, M. Clayton, DeSisto, William J. Bioresource Technology. In Press.

"Effect of Residence Time and Hot Gas Filtration on the Physical and Chemical Properties of Pyrolysis Oil"
Case, Paige A., Wheeler, M. Clayton, DeSisto, William J. Energy & Fuels. 28 (6): 3964–3969 (2014)

"Pyrolysis of Pre-treated Tannins Obtained from Radiata Pine Bark"
Case, Paige A., Bizama, Constanza, Segura, Christina, Wheeler, M. Clayton, Berg, Alex, DeSisto, William J. Journal of Analytical and Applied Pyrolysis. 107: 250-255 (2014)

"Liquid hydrocarbon fuels from cellulosic feedstocks via thermal deoxygenation of levulinic acid and formic acid salt mixtures"
Case, Paige A., van Heiningen, Adriaan R.P., Wheeler, M. Clayton. Green Chemistry. 14: 85-89 (2012)

Awards
Third place, TCS2014 Frontier Young Scientist Poster Award
"*Proposed mechanisms for pyrolysis of calcium pretreated pine*"
Poster presentation at TCS2014, Denver, CO (September 2014)

Second place, TCBiomass2013 Student Poster Challenge
"Formate-Assisted Pyrolysis of Biomass"
Poster presentation at TCBiomass2013, Chicago, IL (September 2013)

395739 Functional Nano-Surfaces for Transformative Materials in Biological & Energy Applications

Abstract:

My research interests are in surface science, thin films, and nanostructures. Surface science has a very broad impact on industrial, environmental, electronic, biotechnological, biomedical, and military applications. The focus is to modify and design synthetic functional nano-surfaces in two different areas: (1) biological, and (2) energy-related applications. The first area of research includes designing biologically active surfaces for tissue engineering, regenerative medicine, drug/nanoparticle delivery, and antimicrobial surfaces. Our aim is to design biocompatible surfaces in which nanoparticles and drugs are embedded in various biodegradable polymers for combinational therapy. Our main goal is to create controlled staggered-release delivery vehicles to embed and deliver various biologically active constituents. The second area of research is to modify surfaces to create smart materials for energy applications. We will study the structures in complex fluids in terms of aggregation, precipitation, and adsorption to modify surfaces against corrosion and fouling in complex fluid transportation. We will employ microfluidics to study these complex systems.

First-Principals Modeling of Methanol Fuel Cells: Kinetics and Catalyst Design

Glen R. Jenness

The decreasing supplies of petroleum resources coupled with an increase in energy demand has created a need for an alternative power supply. Proton-exchange membrane fuel cells (PEMFCs) are an attractive prospect, due to the high energy density and low emissions. Previous PEMFC technology relied on the oxidation of H_2, but due to the difficulties of storing H_2, this route is not economically favorable. By replacing H_2 with an alternative renewable proton source we can alleviate the cost of storing and transporting H_2, as well as making the PEMFC technology more environmentally friendly. Methanol is an attractive proton source due to low storage costs, good thermal stability, high energy density, and is readily produced from biomass resources. While methanol fuel cells offer a high energy density, the efficiency remains low, limiting the use of methanol fuel cells. As a consequence, there has been an increased effort in recent years to find an efficient catalyst for the oxidation of methanol. Previous catalysts employed a platinum nanoparticle with an oxide support, such as γ-Al_2O_3. However, recent experiments indicate that a platinum nanoparticle supported on a carbon-based support, such as carbon nanotubes (CNT), have led to an increase in the efficiency of methanol fuel cells[1].

Several challenges still remain to be resolved in order to increase the efficiency of methanol fuel cells. Firstly, the CNT itself can carry an intrinsic reactivity, with the reactivity being determined by the metallic or semi-conducting nature of the CNT, arising from the chirality. However, the effect of the chirality on the reactivity of the CNT support is still debated[2]. I have shown that the choice of nanoparticle support can result in a charge state being induced on the nanoparticle, which in turn affects the bonding/anti-bonding interactions between the reactants/products and the catalyst, with the resulting charge state being related to the band gap of the support material[3, 4]. As the band gap of the CNT is driven by the chirality, the Pt–CNT interaction will also be driven by the chirality. Additionally, the efficiency of CNT supported catalysts can be increased via the doping of the CNT with nitrogen[5], which in turn changes the electronic properties of the CNT (i.e. band gap), as well as introduces a Lewis acid center into the support. As the mechanism for the oxidation of methanol on the Pt/CNT catalyst is still in question, the effect of introducing a Lewis acid center is currently unknown. I have also shown that the strength of the Lewis acid center can be related to the atomistically resolved conduction band properties of the Lewis acid center, with the reactivity being related to the strength of the Lewis acidity[6]. Finally, it should be mentioned that one factor preventing the wide spread use of methanol fuel cells is the cost of the Pt metal, and it would desirable to use a base metal such as nickel or iron, without sacrificing efficiency.

Through the use of computational tools (i.e. state-of-the-art *ab-initio* methods such as density functional theory), I seek to resolve these issues. In particular, a) how the metallic vs. semi-conducting nature of the carbon nanotube affects the methanol oxidation process; b) how the effect of doping the CNT affects the electronic properties of the catalytic system; c) how the introduction of a Lewis acid center into the support affects the mechanism and kinetics of the methanol oxidation process. The investigation of these effects from an atomistic level will allow us to understand the origin of these interactions, which enable a rationale and systematic improvement of the efficiency of methanol fuel cells.

References

[1] Sharma, S.; Pollet, B. G. Support Materials for PEMFC and DMFC Electrocatalysts—A Review. *J. Power Sources* **2012**, *208*, 96–119.

[2] Denis, P. a.; Iribarne, F. On the Applicability of Cluster Models to Study the Chemical Reactivity of Carbon Nanotubes. *J. Comput. Chem.* **2011**, *32*, 2397–403.

[3] Jenness, G. R.; Schmidt, J. R. Unraveling the Role of Metal–Support Interactions in Heterogeneous Catalysis: Oxygenate Selectivity in Fischer–Tropsch Synthesis. *ACS Catal.* **2013**, 2881–2890.

[4] Hermes, E. D.; Jenness, G. R.; Schmidt, J. Decoupling the Electronic, Geometric and Interfacial Contributions to Support Effects in Heterogeneous Catalysis. *Mol. Simul.* **2014**, 1–11.

[5] Maiyalagan, T.; Viswanathan, B.; Varadaraju, U. Nitrogen Containing Carbon Nanotubes as Supports for Pt—Alternate Anodes for Fuel Cell Applications. *Electrochem. commun.* **2005**, *7*, 905–912.

[6] Jenness, G. R.; Christiansen, M. A.; Caratzoulas, S.; Vlachos, D. G.; Gorte, R. J. Site-Dependent Lewis Acidity of γ-Al_2O_3 and Its Impact on Ethanol Dehydration and Etherification. *J. Phys. Chem. C* **2014**, *118*, 12899–12907.

395320 Mechanism of non-disruptive membrane fusion by amphiphilic, monolayer-protected gold nanoparticles

Abstract:

The cellular membrane is the biological interface responsible for controlling the transport of material into the cell. Thus, it is a critical barrier that must be overcome to successfully deliver desired molecules to the cell interior. Recently, gold nanoparticles (NPs) protected by a binary mixture of purely hydrophobic and anionic end-functionalized alkanethiol ligands were observed to spontaneously penetrate cell membranes via a non-endocytic, non-disruptive mechanism. This direct penetration process is highly promising for applications in drug delivery, bioimaging, and biosensing as NPs can reach the cytosol without being trapped in endosomal compartments or inducing cytosolic leakage. However, no mechanism has yet emerged to explain the experimental observations, limiting the ability to rationally engineer NPs for optimal internalization.

Inspired by these findings, I present here research conducted at MIT with Prof. Alfredo Alexander-Katz in close collaboration with Prof. Darrell Irvine at MIT and Prof. Francesco Stellacci at EPFL. I used a combination of coarse-grained and atomistic simulation methodologies to show that these amphiphilic NPs can spontaneously insert into and fuse with lipid bilayers to obtain configurations resembling transmembrane proteins. The full insertion pathway is remarkably similar to the mechanism of vesicle-vesicle fusion due to the physicochemical similarity between these synthetic and biological systems, allowing us to take cues from biology to understand NP behavior. Predictions from my simulations were confirmed experimentally leading to the development of design guidelines to optimize fusion efficacy. Furthermore, experimental results demonstrated that the same NPs predicted to fuse with bilayers are internalized by the non-endocytic process described above, establishing a link between favorable bilayer interactions and intracellular localization. These findings yield significant physical insight into behavior at this important nano-bio interface and illustrate the power of such simulations in guiding NP design for biomedical applications.

In my future research, I will utilize detailed simulations across multiple time and length scales to generalize this strategy to the study of various membrane-active biological and synthetic systems in close collaboration with experimentalists. In particular, I am interested in understanding mechanisms of post-translational membrane protein insertion, capturing the role of lipid interfaces in membrane binding, and using novel optimization algorithms to tune NP-membrane interactions. My goal is thus to both study fundamental questions in membrane biophysics and develop novel engineered nanomaterials for biomedical applications.

395023 Characterizing the Structure and Function of a Protein Fusion Complex of Photosystem I and Hydrogenase Enzyme

Abstract:

Proteins are large, complex molecules, and their interactions within even the simplest of microorganisms are a complex and intriguing process, requiring a mechanistic understanding at the single molecule level that often defies the capabilities of current science and technology. Gaining insight into proteins and their interactions is crucial to our understanding of the biological world, and could provide technological advancement in alternative energy, drug discovery, and a wealth of other areas. Membrane proteins in particular comprise ~30% of open reading frames, yet represent only a small fraction of known protein structures due to a lack of understanding of the effects of protein-environment interactions on their structure and stability. Here, we are interested in the study of structural effects on the stability and activity of redox proteins. In particular, we are focused on generating protein fusions between the integral membrane protein photosystem I (PSI) of *Synechocystis* sp. PCC 6803 and the membrane-bound hydrogenase (MBH) of *Ralstonia eutropha* H16, a unique *in vitro* complex capable of harnessing solar energy for the production of molecular hydrogen (H_2). We present our study of the detergent-solubilized solution structure of PSI using small-angle neutron scattering (SANS) experiments and all-atom molecular dynamics (AAMD) simulations, and discuss our results in applying a similar approach to studying PSI-MBH protein fusions. By utilizing a combination of computational modeling and experimental characterization of these fusion complexes, we hope to provide a novel method for probing the structure-function relationship that governs redox proteins.

394917 Self-assembly, self-organization, and the design of new materials

Abstract:

My research focuses on understanding and controlling the self-assembly and self-organization of nanoparticle-based materials, which find applications ranging from photonics and plasmonics to industrial formulations such as coatings. A central aim of my work is to *develop quantitative design strategies for controlling dynamic pathways to self-assembly*. At the intersection of physics and engineering, this research utilizes synthetic chemistry, advanced imaging techniques, micromanipulation, and statistical physics. I will discuss the development of a suite of non-equilibrium pathways to self-assembly that are implemented using DNA reaction networks, including self-limiting growth of finite-sized structures, assembly of novel periodic lattices due to specific next-nearest-neighbor interactions, and design of smart materials that can reconfigure themselves in response to an external cue. These pathways can be understood quantitatively using core chemical-engineering concepts, including chemical kinetics and reaction engineering, transport, and thermodynamics.

Engineering human pluripotent stem cell-derived neurovascular systems to model neurological disease and explore therapeutic strategies

Ethan Lippmann, Ph.D.

My research focuses on engineering human pluripotent stem cell (hPSC) fates for applications in developmental and disease biology, drug screening, and regenerative medicine. During my graduate studies, I developed a process for differentiating human pluripotent stem cells (hPSCs) to endothelial cells possessing a blood-brain barrier (BBB) phenotype. In conjunction with renewable sources of pericytes, astrocytes, and neurons, we created a fully-human, scalable, BBB co-culture model that possesses physiological barrier properties and is being evaluated for its ability to predict brain uptake of small molecule drugs. For my postdoctoral work, I have been developing methods to improve the diversity and scope of neural differentiation from hPSCs. We devised a novel platform for generating pure neural stem cells (NSCs) from hPSCs in a fully defined and scalable manner and then used this platform to uncover biological cues that deterministically regulate posterior neural patterning, thereby generating uniform populations of NSCs possessing fixed positional identities corresponding to specific hindbrain and spinal cord domains. We are currently incorporating these defined NSCs onto custom surfaces that facilitate spatiotemporal control over cell behavior in an effort to improve the derivation of neuronal subtypes such as motor neurons. As an independent investigator, I will integrate the technology developed during my graduate and postdoctoral studies to engineer hPSC-derived, disease-specific *in vitro* neurovascular models that will serve as tools for dynamically assessing drug efficacy, identifying novel druggable targets, and studying progressive neurodegeneration in a dish.

Education
Ph.D., Chemical Engineering, 2012, University of Wisconsin, Madison, WI
B.S., Chemical Engineering, 2006, University of Illinois, Urbana, IL

Selected publications
1. **Lippmann ES**, Al-Ahmad A, Azarin SM, Palecek SP, and Shusta EV. A retinoic acid-enhanced, multicellular human blood-brain barrier model derived from stem cell sources. *Scientific Reports* 4:4160 (2014).
2. **Lippmann ES**, Estevez-Silva MC, and Ashton RS. Defined human pluripotent stem cell culture enables highly efficient neuroepithelium derivation independent of small molecule inhibitors. *Stem Cells* 32:1032-1042 (2014).
3. Sha J, **Lippmann ES**, McNulty J, Ma Y, and Ashton RS. Sequential nucleophilic substitutions permit orthogonal click functionalization of multicomponent PEG brushes. *Biomacromolecules* 14:3294-3303 (2013).
4. **Lippmann ES***, Azarin SM*, Kay JE, Nessler RA, Wilson HK, Al-Ahmad A, Palecek SP, and Shusta EV. Derivation of blood-brain barrier endothelial cells from human pluripotent stem cells. *Nature Biotechnology* 30:783-791 (2012). (*Equal contributions)
5. **Lippmann ES**, Weidenfeller C, Svendsen CN, and Shusta EV. Blood-brain barrier modeling with co-cultured neural progenitor cell-derived astrocytes and neurons. *Journal of Neurochemistry* 119:507-520 (2011).

394808 Rational design of multi-input controlled protein switches towards the development of target specific therapeutic enzymes and biosensors

Abstract:

Protein switches are engineered proteins that are composed of an input domain that recognizes and responds to an input signal and an output domain whose function is regulated by the state of the input domain. Engineered protein switches have a number of exemplary properties for sensing and therapeutic applications including a large dynamic range, high specificity for the activating ligand, and a modular architecture that will facilitate fine-tuning of the desired properties. A modular design of protein switches allows the possibility of developing versatile protein switches, in which different input domains can be coupled to the output domain. Thus, protein switches that can be regulated through exogenous or endogenous inputs have a broad range of biotechnological and biomedical applications such as the next generation therapeutic enzymes and biosensor systems. For enhanced target specific controls and improved specificity of the engineered protein switch, it is desirable to establish a means to regulate protein activity with multiple input controls; yet, engineering of such a complex regulatory system is challenging. Here we describe the design of switchable enzymes that requires multi-input signals such as an effector molecule, temperature, pH, redox agent, or electrochemical signal for activation. First, we inserted an enzyme domain into an effector-binding domain such that both domains remained functionally intact. Second, we induced the fusion to behave as a switch through the introduction of a linker designed to respond to different environmental conditions including temperature, pH, or redox potential. We confirmed the multi-input switching behavior *in vitro* and *in vivo*. We also demonstrated that some of these engineered switches could function as protein logic gates. Structural and thermodynamic studies supported the hypothesis that switching resulted from manipulation of allosteric control mechanism via change in conformational equilibria or dynamics of the fusion domains. These results embody a general strategy for the rational design of complex protein switches with multi-level controls.

Engineering polymer materials for the energy-water nexus: new chemistries, in-situ diagnostics, and directed self-assembly

Christopher G. Arges, Ph.D

Institute for Molecular Engineering
University of Chicago
carges@uchicago.edu; http://ime.uchicago.edu/nealey_lab/people/christopher_arges/

Polymer electrolyte membranes (PEMs) are found at the heart of membrane electrode assemblies (MEAs) and they are central to the performance of electrochemical energy conversion and storage devices in polymer electrolyte fuel cells, water electrolyzers, flow batteries, supercapacitors, and reverse electrodialysis. They are also essential to electrodialysis processes used for water desalination and purifying industrial effluent streams. Two of the most stringent requirements for PEMs are exceptional ionic conductivity, which is paramount to device efficiency, and longevity. This poster will cover research strategies for addressing PEM chemical stability and ionic conductivity through i.) the synthesis of new PEM chemistries, ii.) elucidating material performance and degradation in electrochemical systems using *in-situ* diagnostics, and iii.) the adoption of block copolymer directed self-assembly (DSA) for creating ordered, periodic architectures with high degree of functionality. These strategies will be illustrated through three different case studies. The first two parts of the poster stem from my Ph.D. work with Professor Vijay Ramani at the Illinois Institute of Technology, while the latter part highlights my current postdoc work with Professor Nealey at the University of Chicago.

Part 1 of the poster describes the use of a micro, *in-situ* bifurcated optical fluorescence probe for monitoring the reactive oxygen species (ROS) generation in an operating PEM fuel cell. Chemical degradation of the proton exchange membrane observed by fluorescence response correlated with the macroscopic fluoride ion flux rate collected from the water emanating from the fuel cell. The latter technique substantiated the accuracy of the *in-situ* probe for real-time monitoring of PEM degradation. This part of the poster will also describe mitigation of reactive oxygen species degradation through the incorporation of mixed-metal oxide nanoparticles that scavenge the free radicals in the PEM.

In part 2, the chemical and mechanical stability of anion exchange polymer electrolyte membranes (AEMs) for alkaline fuel cells and water electrolyzers and redox flow batteries will be highlighted. Understanding AEM alkaline stability will be demonstrated through a best practices approach that leverages multidimensional NMR spectroscopy, membrane mechanical analysis, examination of AEM electrochemical cell stability using a variety of electroanalytical techniques, and post-mortem NMR analysis of MEAs. Additionally, chemical structure-property relationships for mitigating chemical degradation of AEMs in alkaline electrochemical systems will be given.

The final part of the poster will showcase research for improving PEM conductivity by orienting periodic, ion-conducting domains in diblock copolymer electrolytes perpendicular to electrode surfaces using DSA. Other extensions of this powerful technique, such as creating periodic patterned arrays of electrocatalyst nanoparticles and fabrication of miniaturized electrochemical devices (e.g., micro-supercapacitors) will be disseminated.

My future academic career seeks to compliment my background in electrochemical engineering and polymer chemistry from my Ph.D training with polymer physics and lithography/nanolithography learned in my postdoc work. From my point of view, amalgamating these different skill sets will lead to new classes of functional and resilient materials for the energy-water nexus and efficient micro-electrochemical systems for portable consumer products. The independent, academic program I am aim to build will cut across a multitude of different disciplines spanning the fields of chemical engineering, materials science, physics, chemistry, and mechanical engineering. My vision is to construct a team with students and research professionals with diverse backgrounds in addition to fostering collaborative alliances with researchers worldwide.

394799 Modeling and Simulation of Corona Phase Molecular Recognition Sensors

Abstract:

Corona Phase Molecular Recognition (CoPhMoRe) has been recently introduced as a means of generating synthetic molecular recognition sites onto nanoparticle surfaces. A synthetic heteropolymer is adsorbed and confined to the surface of a nanoparticle, forming a corona phase capable of highly selective molecular recognition due to the conformational imposition of the particle surface on the polymer. In this work, we develop a computationally predictive model for analytes adsorbing onto a polymer corona phase around a single walled carbon nanotube surface using a 2D equation of state that takes into consideration the analye-polymer, analyte-nanoparticle, and polymer-nanoparticle interactions using parameters determined independently from molecular simulation. The curvature dependence of the interaction energies is determined to be a weak contribution to the overall interaction energy. Overall, the resulting model is able to correctly predict 83% of a 374 analyte-polymer library generating experimental fluorescence responses within 20% error of the experimental values. This modeling framework is an important step to the design of such polymers to target a specific analyte.

Universality of Block Copolymer Melts from Metadynamics and Multi-GPU Simulations

I will discuss how we used massively parallel molecular dynamics simulations on multiple GPUs and the well-tempered metadynamics method to locate the ODT of diblock copolymers. Simulation results for five different coarse-grained models of symmetric diblock copolymers demonstrate both a universal equation of state and parameter $(\chi N)_{\mathrm{ODT}}$ at the order-disorder transition, for invariant degrees of polymerization as low as $\overline{N} \sim 200$. In this regime, Fredrickson-Helfand theory must break down. I will also present HOOMD-blue v1.0, a high performance particle simulation framework, which runs in parallel on over 3,000 GPUs.

Nanotherapeutics for Neurological Disorders

Elizabeth Nance, PhD
Anesthesiology and Critical Care Medicine, Center for Nanomedicine
Johns Hopkins University, Baltimore MD

Neurological disorders account for 13% of the global burden of disease. Due to the complex nature of delivering drugs to the central nervous system (CNS), therapeutics for neurological disorders take 35% longer to develop than drugs for other diseases. The design and manipulation of nanoparticle platforms for drug delivery is an increasingly prominent area of interest for medical applications; however, while many drug delivery platforms have shown efficacious pre-clinical results, this has not translated to a significant number of FDA approved, clinically applied nanoparticle systems. This is especially true for neurodevelopmental disorders, such as autism and cerebral palsy, which are life-long chronic disabilities that begin in early childhood. They have no effective cure, and are often underserved by novel drug delivery technologies, which primarily focus on adults. Therefore, there is great opportunity to bring nanotherapeutic approaches to neurodevelopmental disorders, with results that can be translated to adult brain disorders. *My research goals focus on understanding nanoparticle platform interaction, both as biophysical probes and imaging biomarkers, within disease physiology and pathology, with a focus on neurodevelopmental diseases. This knowledge will then be used to better design and implement therapeutic nanoparticle platforms in clinically relevant models of both pediatric and adult CNS diseases.* My laboratory will provide training for engineering students, applied neuroscience students, and clinical fellows. My long-term goal is to transition my laboratory into a Translational Nanomedicine Center, focused on understanding neurological diseases as they relate to therapeutic applications, and engineering nanomedicine platforms that lead to symptom free survival in patients diagnosed with neurological disease. This multidisciplinary work will involve a team of experts in engineering, nanomedicine, chemistry, materials science, pediatrics, neurology, and neuroscience.

I am uniquely suited to establish a laboratory that integrates training in engineering nanoparticle platforms with an understanding of nanoparticle formulation, biophysics, and physiology and progression of neurological disorders. During my doctoral work with Dr. Justin Hanes, I developed nanoparticles for use as biophysical probes in the CNS, a now patented technology, to better understand drug delivery limitations in both normal and diseased brain. Throughout my graduate studies, I led collaborations with faculty and students in the departments of neurosurgery, oncology, pathology, and radiology. My postdoctoral training with Dr. Sujatha Kannan and Dr. Michael Johnston has built my foundation in neuroscience and pediatric neurology, leading to experience in understanding neuroanatomy, neurophysiology and neural development, as well as development of clinically relevant animal models of pediatric CNS diseases. Based on my combined doctoral and postdoctoral work, I was recently awarded the Burroughs Wellcome Fund Career Award at the Scientific Interfaces, an award that provides $500k over 5 years for transitioning from postdoctoral level training to a tenure track faculty position.

Education

Ph.D., 2012, Johns Hopkins University, Chemical and Biomolecular Engineering
B.S., 2006, North Carolina State University, Chemical and Biomolecular Engineering

Selected Publications

1. **Nance E**, Zhang C, Shih T-Y, Xu Q, Schuster B, Hanes J. Brain penetrating nanoparticles improve paclitaxel efficacy in malignant gliomas following local administration. *ACS Nano*, Under review.
2. **Nance E***, Timbie K*, Miller WG, Song J, Louttit C, Klibanov AL, Shih T-Y, Swaminathan G, Tamargo RJ, Woodworth G, Hanes J, Price RJ. Noninvasive delivery of stealth, brain-penetrating nanoparticles across the blood-brain barrier using MRI-guided focused ultrasound. *J Control Rel.*, 189C:123-132 (2014).
3. Kannan R, **Nance E**, Kannan S, Tomalia D. Emerging concepts in dendrimer nanomedicine: translational aspects. J Internal Medicine, doi: 10.1111/joim.12280. (2014).
4. Balakrishnan B, **Nance E**, Johnston M, Kannan S. Nanotechnology in cerebral palsy: current developments. *Int J Nanomedicine* 8:4183-4195 (2013).
5. **Nance E***, Woodworth G*, Sailor K, Shih T-Y, Swaminathan G, Xiang D, Eberhart C, Hanes J. A dense poly(ethylene glycol) coating improves penetration of large polymeric nanoparticles within brain tissue. *Sci. Transl. Med.* 4, 149ra119 (2012). **Selected Cover Image

394783 Micro- and nano-engineered advanced materials for renewable energy, thermal, and electronic applications

Abstract:

Research Interest

My multidisciplinary research interests are in thin films and nanoscale processing for advanced materials – from materials to devices to circuits, blurring the boundaries between chemistry, electronics, physics, materials science and mechanics. With the state-of-the-art micro/nano fabrication techniques and surface engineering, I will explore both the fundamental and theoretical nature of new materials and devices. As an engineer and scientist, I view the area of micro- and nano-engineered advanced materials and next generation electronics as a diverse discipline full of formidable challenges that will engage the research community for decades to come.

Previous and Current Research

Synthesis of graphene films and organic compounds for developing electronic devices: Taking on the challenge to develop inexpensive materials with enhanced functionalities for electronics and in general energy applications, I have focused on low-cost direct synthesis of wafer-scale graphene on a variety of substrates for touch screens, displays, solar cells, and next generation transistors. Graphene, a one-atom-thick sheet of sp^2-bonded carbon atoms in a honeycomb structure, has attracted significant interest due to its outstanding mechanical, thermal, and electrical properties. I have synthesized graphene films on a variety of substrates including silicon, quartz, and sapphire, through a single-step thermal annealing process of a Cu/Ni bilayer pre-deposited on the substrate in a low pressure environment through chemical vapor deposition (CVD) technique. My method eliminated the need for solution process transfer of graphene to the target substrate, which usually induces defects in its structure. In addition, doping of graphene and organic thin films is important, as it gives rise to p-type and n-type materials, and it adjusts the work function of the film that is necessary to control charge injection and collection in devices such as solar cells, displays, and printed and flexible electronic devices. I developed a low temperature method to fabricate p- and n-type graphene and graphene p-n junctions by modifying the interface between graphene and its dielectric substrate with covalently bonded self-assembled monolayers (SAMs). In a multidisciplinary collaboration work, I showed that surface modifiers based on polymers containing simple aliphatic amine groups and redox-active solution processable metal-organics substantially adjust the work function of conductors, including metals, transparent conductive metal oxides, conducting polymers, and graphene. These methods can pave the way to simplified manufacturing of low-cost and large-area organic electronic technologies.

CVD polymers-surface modification: Conventional polymer deposition (for e.g. ink-jet printing, spin-coating or dip coating) uses solvents that may have adverse effects during the deposition process—such as degradation of the underlying layers or alteration of mechanical/electrical properties—and that cannot be applied for coating of surfaces with high aspect ratio patterns. Chemical vapor deposition (CVD) as a polymer thin film deposition technique offers a versatile platform for fabrication of a wide range of polymer thin films preserving all the functionalities.[5] The solvent-free, conformal vapor phase deposition technique enables the integration of polymer thin films or nanostructures into micro- and nanodevices for improved performance. In my work at MIT, I synthesized thin films of polymers via a substrate-independent and all-dry-initiated chemical vapor deposition (iCVD) technique. Surfaces coated with fluoropolymers exhibited high advancing water contact angle (WCA) and very low WCA hysteresis. These coatings reduce the strength of ice and clathrate hydrate adhesion to the substrates up to five times compared to bare substrates and are very desirable for flow assurance strategies aimed at reducing the occurrence of blockages in oil and gas pipelines. In addition, anti-reflective coatings are significantly enhancing the contrast of today's high resolution displays, but their usage in touch applications suffers from the issue of fingerprint visibility on the surface of such low-reflective substrates. In general, the appearance of fingerprints and oil residues are of major concern for numerous industries involved with maintaining display clarity for applications where handling occurs. To this end, I am working on developing anti-fingerprint and anti-reflective coating using bio-degradable CVD polymers that exhibit high water/oil contact angle. In order to use such coatings in a touch application it must withstand hundreds of thousands of finger touches and rubs. Additional advantages of the CVD technique for polymer deposition is that the mechanical properties of the synthesized polymers against wear and erosion can be improved via in-situ grafting, multilayer composite polymers, and post-synthesis treatment of polymers. Films with mechanical properties comparable to glass were obtained using this technique.

Proposed Research

Development of functional polymers:

Low-cost and scalable CVD polymers for high efficiency heat recovery: Thermoelectric (TE) materials have recently attracted significant interest for converting thermal energy to electric energy in the automobile industry, for recapturing energy currently lost from the hot effluent of power plant smokestacks, and for harvesting the heat currently generated but not used by photovoltaic cells. By understanding how heat and energy flow through materials, energy conversion mechanisms and processes can be integrated into functional devices. The performance of the thermoelectric material is evaluated by the dimensionless figure of merit (ZT) that is a function of electrical conductivity, thermal conductivity, absolute temperature, and Seebeck coefficient—the ratio of the induced voltage over the temperature gradient across the thermoelectric device. A general principle for engineering good TE materials is to separate electrical from thermal conductivity—in particular, to make them dependent on structures at different length scales. Today's thermoelectric materials of choice for low temperature energy conversion are bismuth antimony telluride alloys. The constituting atomic elements have a low natural abundance and are therefore expensive. Additionally, low natural abundance goes hand in hand with some level of toxicity for the environment. Therefore, in today's scenario it is unthinkable to enable a widespread use of thermoelectric installations for waste heat recovery. In that context, widely available, although less efficient, organic thermoelectric materials have become the subjects of interest to the scientific community. Polymers as TE materials have been developed via electro spinning, inkjet printing, and electrochemical deposition. However, these methods are not conformal, use solvents, and require additional treatments to dope the polymer before integrating it into thermoelectric generators. I plan to develop polymer TE materials through oxidative CVD where the conformal, all-dry, substrate-independent, and low-cost characteristics of conducting polymers can be leveraged to make scalable thermoelectric generators. The extent of doping and electrical conductivity of such polymers can be tuned throughout the synthesis process by changing growth conditions. In addition, a state-of-the-art mechanism for grafting such polymers to surfaces provides the robust adhesion required for industrial application. Understanding thermal and electrical properties of such polymers is inherently interdisciplinary, involving backgrounds in synthetic chemistry, polymer physics, condensed matter physics, and materials characterization. Through better understanding of transport in amorphous and disordered materials, it is possible to tailor the thermal and electrical conductivity in synthesized polymers and integrate them into low-cost, scalable, and environmentally friendly organic thermoelectric generators.

Mechanical properties of CVD polymers: CVD polymers fully retain the rich chemistry of organic monomers enabling technology for surface modification of membranes, microfluidic structures, and biomedical devices. These CVD polymeric layers possess mechanical flexibility and conformal coverage attributes and can be integrated into device fabrication schemes for optoelectronics and sensors. While promises have been demonstrated in terms of photovoltaics, organic displays, and sensors, the reliability of these devices ultimately depends on their ability to retain their functionality during operation. In terms of organic electronics, this will present a challenge when considering their need for enhanced mechanical properties as well as the potential for fracture and loss of adhesion at critical interfaces. In this work, I will continue to build on my expertise in studying mechanical properties of ultrathin polymers, synthesized using CVD technique. I will investigate the use of functionalization layers containing hydroxyl, vinyl, and fluorine groups to graft these polymers to the desired substrate to enhance their adhesion. I will also seek to improve the mechanical properties, for example elastic modulus, hardness, and toughness through synthesis of stacked ultrathin multilayer structures, cross-linking of various monomers, and post-synthesis treatment of polymers. The onset cracking strain will be related to film thickness, film composition as determined by XPS/FTIR, mechanical properties will be examined through nanoindentation/nanoscratching, and film defects determined from SEM and optical microscopy techniques. XPS and spectroscopic FTIR will be critical in understanding the early stages of growth of these films and linking this to their formation and mechanical properties. I will also make multilayer stretchable and conductive CVD polymer films and integrate them in flexible and organic electronic/photonic devices and examine their reliability under enhanced flexing and stretching conditions.

394345 Experimental Evolution of Heterologous Pathways in Microbes

Abstract:

Our ability to engineer microbial metabolic pathways is improving rapidly, enabling the efficient and sustainable biosyntheses of both bulk and fine chemicals. Tools from synthetic biology have supported these advances and allowed the use of diverse natural and engineered enzymes. However, novel enzymes frequently require significant optimization to function efficiently in a new production host. More precisely, optimization is a complex co-adaptation process, as the host must accommodate its new pathway and the pathway adjust to its new host. Many metabolic engineers seek to use rational and semi-rational strategies to predictably design highly-productive pathways in heterologous hosts. Unfortunately, these approaches are limited by our inability to account for the complexities involved in genome-scale engineering of microbial metabolism. When knowledge is limiting, evolutionary approaches can be powerful tools, both for forward optimization and as learning opportunities to investigate potential design strategies.

In my graduate research, I used RNA switches to perform directed enzyme evolution in live cells and systems biology to optimize an engineered microbe. My postdoctoral research has applied strategies from metabolic engineering to study natural microbial evolution, looking at how microbes evolve after horizontal gene transfer.

My independent research laboratory will investigate strategies for optimizing the interactions between a host and a novel enzyme or pathway using three related approaches. In my first project, I will extend my postdoctoral research to construct and optimize synthetic mobile genetic elements for bioremediation, effectively performing metabolic engineering in indigenous microbes. Next, I will study adaptation after horizontal gene transfer in a pathogen, looking at how previously-benign bacteria can evolve to become virulent. My final proposal describes a novel selection strategy using synthetic competitive ecosystems in microfluidic droplets. These ecosystems will be used to optimize complex pathways, such as those required to produce antibiotics, while studying the evolution of cooperation and competition in microbial communities. In combination, these projects will offer a novel perspective on important questions in health and ecology while also producing forward design strategies for synthetic biology and metabolic engineering.

388584 Rationally Designed and Deterministically Engineered Electrodes for High-Performance Energy Storage Applications

Abstract:

Over the past two decades, the widespread miniaturization of portable electronic devices has led to the need for similarly scaled energy storage systems. However, the energy storage technology has not improved at the same rapid pace as the portable electronics technology, resulting in limited operation duration of the device on a single charge. The vast majority of the energy storage systems utilized in the portable electronic devices consist of batteries. Considering the currently available battery chemistries, a drastic improvement in the battery performance in terms of capacity is not expected in the near future. Unless a completely new battery chemistry with an extremely high energy density is developed, the only way to achieve longer operation times is through increasing the mass of the active material present within the battery, which would be incompatible with the current shrinking trend of the portable electronics. However, improvements can be achieved in the rate of the energy transfer, i.e. power density, which these batteries can deliver. A battery that can be charged very rapidly would still be quite desirable even if it possesses a relatively low energy density.

The batteries with rapid charge and discharge capabilities can only be realized via simultaneous minimization of certain inherent resistances encountered during the transport of the ionic and electronic species. There are both intrinsic and extrinsic factors affecting the transportation process of ions and electrons. Intrinsic factors, such as diffusivity and conductivity of the battery components, depend on the choice of materials to be utilized within the system. Extrinsic factors, including the surface area, diffusion and conduction path lengths, on the other hand, primarily rely on the size, dimension, and geometry of the system components.

The focus of our research has been on the realization of rationally designed and deterministically engineered three-dimensional electrodes with precisely controlled dimensions that address the aforementioned critical extrinsic factors determining the ultimate performance of the battery. MEMS technologies and electrochemical techniques have been utilized for the fabrication as well as characterization of these scalable, well-ordered, and high-surface-area electrodes with rapid charge and discharge capabilities. Utilization of micro- and nanofabrication techniques also enables the realization of electrodes for high-performance power sources for a wide variety of applications, ranging from autonomous microsystems to macroscale portable electronics. The resultant electrodes demonstrate an improved performance in not only batteries, but also other energy storage systems, such as capacitors and supercapacitors.

394233 Design of Genetically Encodable Protein Switches and Potent Metallobiocatalysts

Abstract:

Precise molecular recognition between proteins and small molecules is critical to cellular function. A rigorous test of our understanding of the physiochemical principles that govern molecular interaction phenomena is the creation of de novo protein-ligand interfaces with programmable properties. I recently developed a generalizable computational/experimental hybrid method for designing proteins that bind small molecules and used it to create selective binders for the steroid digoxigenin. This work and other exciting advances provide a platform for creating novel ligand binding proteins, but fundamental gaps in our understanding of structure-function relationships in proteins need to be addressed in order to robustly and routinely design molecular biodevices that sense the presence of the bound small molecule and respond by performing a specified function. The goal of my interdisciplinary research group will be to uncover the design principles necessary to develop next-generation systems-level protein technologies in which molecular recognition events are coupled with downstream biological or chemical functionalities. Design advancements will be motivated by and provide insight into the biochemical/biophysical principles that promote allosteric changes in proteins and/or enable the coordinated integration of multiple complex chemical events. In particular, we will focus on two major avenues: designing novel molecular switches for regulating metabolic pathways, and developing new metallobiocatalysts for challenging small molecule activation reactions related to green chemistry and energy. Through our design efforts, we hope not only to create novel functional biomolecules but also to uncover design principles that shed light on the underlying structure-function relationships and regulatory mechanisms in natural systems.

394100 Understanding Molecular Mechanisms of Rare Events: Applications in Nanofabrication in Semiconductor Industry, Novel Drug Delivery Systems and Fuel Cells

Abstract:

At molecular level, rare events are events which occur at timescales of a few nanoseconds to several days. These events underlie many observed physical phenomena, such as phase transitions, conformational changes in proteins and polymers, chemical and biological reactions and self-assembly. Rare events often have a rate limiting step, which lies at the maxima of the free energy along the reaction coordinate pathways. Since traditional molecular simulation methods are ineffective in sampling such events, special techniques have been developed to sample them, such as forward flux sampling, variants of string method and transition path sampling. Deciphering molecular mechanisms behind rare events is fundamental to our understanding of physical systems of interest and useful for many technological innovations. In the last four decades, semiconductor industry has seen a relentless scaling down of transistor dimensions in accordance with the Moore's law. The state of the art semiconductor technology is operating at critical dimensions of tens of nanometers. The traditional scaling of transistor dimensions, often called Dennard scaling, is no longer applicable at present length scales, and hence focus has moved to innovations in novel materials and new fabrication methodologies. One approach is to use a high K dielectric material in thin film transistors. Hafnium oxide based inorganic dielectric materials have many problems, such as poor interface properties with silicon causing development of interface traps, which affect reliability of devices. Molecular modeling methods can be utilized to design optimal polymeric substances which are capable of forming a good interface layer with silicon. High-K polymeric materials, such as poly(vinylidenefluride-co-trifluoroethylene) and poly vinyl phenol with varying molecular weight, grafting density and co-polymeric mixtures would be materials of interest for this study. Long relaxation times of these polymers as well as their conformational changes under electric field can be studied as rare events in molecular simulations. Another innovation which has improved performance of semiconductor devices is the introduction of strain in silicon to increase mobility of charge carriers. One way to induce strain is by first depositing thin films and then either curing them or by providing a free surface to allow them to expand or contract. Molecular modeling of structural modifications in thin films via rare event sampling techniques can provide useful insights on the role of chemistry of the film and its interaction with the underlying silicon crystal, which will help in better design of this process. Hydrogels are cross-linked network of hydrophilic polymers. Hydrogels can undergo gel-sol transition upon changes in environmental conditions, such as pH, temperature or light. Using environmental sensitive hydrogels as site specific drug delivery systems is a promising idea. The important parameters of investigation would be the sensitivity of sol-gel transition to stimuli, and the kinetics and thermodynamics of the transition. These investigations can most effectively be undertaken by simulating the sol-gel transition under varying environmental conditions using molecular simulations. Transport properties of membranes are functions of the structure of their polymer network. For fuel cell applications, it is desirable to have high proton conductivity with reduced water cross-over. Molecular modeling of proton transport as a function of the structure of the membranes would help develop more selective membranes with high permeability for protons for fuel cell applications. In my post-doctoral work at Princeton University, I applied forward flux sampling in molecular dynamics simulation of an all-atom water model to study kinetics of evaporation and condensation of water in hydrophobic confinement at ambient conditions, thereby simulating time-scales in order of few hours. Using transition path sampling, I determined transition state ensemble associated with liquid to vapor transition of water in hydrophobic confinement. In my doctoral work at Columbia University, I studied conformational changes in proteins upon adsorption to hydrophobic surfaces. In order to eliminate kinetic traps while sampling different conformational states of proteins, I did Monte Carlo simulation at different temperatures and surface hydrophobicity to study different regions of the phase space. The entire phase space is generated by using methods such as simulated annealing, weighted histograms and parallel tempering. In order to study kinetics of a system which can transition between two states, the phase space can be first sampled by using equilibrium sampling techniques such as parallel tempering, simulated annealing and weighted histograms. Once different minima in the phase space have been identified, rare event sampling can be used to study the kinetics of transition from one minimum to the other.

397143 The Flame Synthesis of Carbonaceous and Heterogeneous Nanomaterials, and Using Soot for the Fischer- Tropsch Synthesis High Value Chemicals

Abstract:

We have a limited understanding of highly desired industrial chemical processes that naturally occur in flames – one reason that soot production remains a global issue. Still, empirical knowledge has driven the development of flame synthesis techniques for widely used products like carbon black, TiO_2, and even specialized carbon nanomaterials. Although the flame synthesis of metal oxide nanoparticles has seen success because this process follows a decent on a Gibbs free energy landscape, carbon particle formation is dominated by chemical kinetics and remains largely unknown to this day. Thus, further understanding it will allow for greater control over carbon nanoparticle formation under flame conditions, which can scale better than other synthesis techniques like the sol-gel method. One of my research plans is the theory-aided design of flame synthesis environments to harvest self-assembled nanoparticles and quantum dots. Within flames, the self-assembly process of carbon particles is a direct result of fuel-rich combustion, the characteristics of which can be tuned by optimizing reagent mixtures and reactor conditions for a desired property. These characteristics include, but are not limited to: size, band gap, surface functional groups, and heterogeneity. Furthermore, reagent fuels may be doped with varying concentrations of metals and/or transition metals to create a new generation of catalysts, sensors, and components of targeted therapeutics, smart materials, and even quantum computers.

Another project of interest relies on the AFM, TEM and FTIR evidence that hydrocarbon chains polymerize on nascent soot surfaces. My earlier computational work strongly suggests that persistent free radicals and combustion byproducts play a role in this process. I plan to exploit this process to produce hydrocarbon compounds that may be used as chemical feedstock or even liquid fuels. A longer term goal is to realize a solar-thermal based energy conversion technology that operates much like reactors employing the Fischer-Tropsch process.

B.A. Chemistry, Virginia Tech

B.S. Engineering Science and Mechanics, Virginia Tech

Ph.D. Mechanical Engineering, University of Souther California (Hai Wang)

Princeton Roving Postdoctoral Fellowship:

High Temperature Gasdynamics Lab, Stanford (Ronald Hanson)

Chemical Engineering Department, MIT (William Green)

397162 Metabolic reprogramming of hypoxic cancer cells

Abstract:

Cancer cells survive under nutrient-limited conditions to surpass normal cell growth. For example, oxygen-limited conditions significantly influence cancer cell metabolism; hypoxia aggravates the Warburg effect by elevating the conversion from glucose to lactate significantly in cancer cells because one oncogene, hypoxia-inducible factor (Hif) increases the expression of metabolic genes in glycolysis. Consequently, it inhibits pyruvate dehydrogenase via activating pyruvate dehydrogenase kinase to decrease oxidative phosphorylation in the TCA cycle. These are key events in tumorigenesis and proliferating cells. Thus, the study of cancer metabolism under hypoxia provides valuable insights in understanding cancer progression and potentially discovering novel therapeutic targets. However, the simple measurement of extracellular metabolites or gene expression levels can provide only limited information on cancer metabolism because most *in vivo* metabolic fluxes consist of reversible reactions and metabolic pathways are also connected to each other. Here, we quantified the metabolic fluxes covering glycolysis, the pentose phosphate pathway and the TCA cycle using ^{13}C-metabolic flux analysis with human cancer cell lines. Furthermore, we validated the observed metabolic behaviors using shRNA knock-down and over-expression cell lines for specific genes. In this course of study, we identified the significant role of hypoxia in regulating cancer cell growth through molecular regulation and metabolic reprogramming by the Hif oncogene.

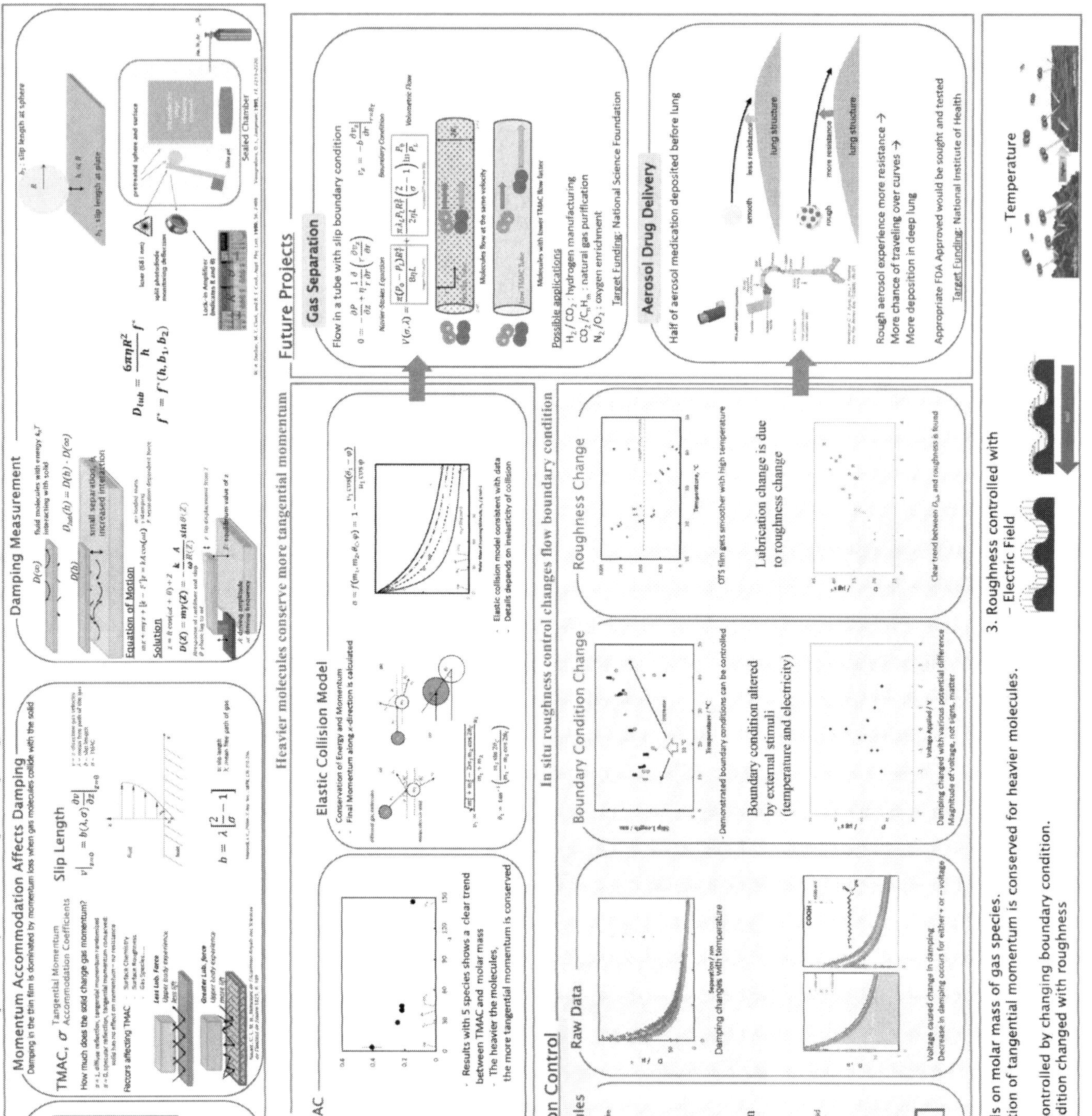
Measurement and Control of Gas Flow Boundary Conditions
Dongjin Seo and William Ducker
Department of Chemical Engineering, Virginia Tech, Blacksburg, Virginia 24061, USA
VirginiaTech
Invent the Future
Rarified Gas Flow
slip flow regime
Momentum Accommodation Affects Damping
Damping in the thin film is dominated by momentum loss when gas molecules collide with the solid
TMAC, σ Tangential Momentum Accommodation Coefficients
How much does the solid change gas momentum?
Factors affecting TMAC
Slip Length
Damping Measurement
Equation of Motion
Solution
Sealed Chamber
Gas Species and TMAC
Effect of Molar Mass to TMAC
Results with 5 species shows a clear trend between TMAC and molar mass
The heavier the molecules, the more tangential momentum is conserved
Elastic Collision Model
Conservation of Energy and Momentum
Final Momentum along x-direction is calculated
Elastic collision model consistent with data
Details depends on inelasticity of collision
Heavier molecules conserve more tangential momentum
Future Projects
Gas Separation
Flow in a tube with slip boundary condition
Navier-Stokes Equation
Boundary Condition
Volumetric Flow
Molecules flow at the same velocity
Molecules with lower TMAC flow faster
Possible applications
H2 / CO2 : hydrogen manufacturing
CO2 /CnHm : natural gas purification
N2 /O2 : oxygen enrichment
Target Funding: National Science Foundation
In situ Boundary Condition Control
Stimuli Responsive Molecules
Does a change in conformation of surface molecules alter the gas flow?
Electric Field
Raw Data
Damping changes with temperature
Voltage caused change in damping
Boundary Condition Change
Demonstrated boundary conditions can be controlled
Boundary condition altered by external stimuli (temperature and electricity)
Roughness Change
OTS film gets smoother with high temperature
Lubrication change is due to roughness change
In situ roughness control changes flow boundary condition
Aerosol Drug Delivery
Half of aerosol medication deposited before lung
smooth
less resistance
lung structure
rough
more resistance
Rough aerosol experience more resistance →
More chance of traveling over curves →
More deposition in deep lung
Appropriate FDA Approved would be sought and tested
Target Funding: National Institute of Health
Conclusion
1. TMAC depends on molar mass of gas species.
A greater fraction of tangential momentum is conserved for heavier molecules.
2. Flow can be controlled by changing boundary condition.
Boundary condition changed with roughness
3. Roughness controlled with
- Electric Field
- Temperature

397401 Advanced catalyst design strategies: Integrating machine learning and probability theory with quantum mechanics and descriptor-based analysis

Abstract:

Chemical processes driven by heterogeneous catalysts are central to a broad range of existing and emerging renewable energy technologies. The ability to engineer novel catalysts which are more active, selective, or inexpensive for a given chemical transformation has the potential to enable or radically improve processes for sustainable energy conversion. A particularly exciting development in the search for novel catalysts is the ability to design catalysts *in silico* using computational methods. A number of promising strategies for computer-aided catalyst design have been developed, including descriptor-based optimizations, genetic algorithms, and high-throughput computational screening. The previous success of *in silico* catalyst design demonstrates the potential power of catalyst prediction through fundamental understanding.

The problem of catalyst design can be considered in terms of two essential challenges: establishing a connection between atomic-scale models and observed reaction rates, and systematically predicting novel catalysts and processes based on these models. *My work will utilize machine learning and probability theory to more efficiently utilize existing theoretical and experimental results to provide fundamental understanding and robust predictions.*

Understanding the connection between atomic-scale models and observed rates is particularly difficult due to the ubiquitous nature of uncertainty in heterogeneous catalysis. The potential importance of transient species, rare events, and sparse active sites makes experimental investigations difficult; moreover, the complex electronic structure of surfaces and vast possibilities of active site geometries presents a challenge for even the most advanced computational methods. Using probability theory as an extended logic, *my research will seek to account for uncertainty in order to develop quantitative connections between atomic-scale models and catalytic activity.*

The development of predictive methods for catalyst discovery relies on access to large amounts of data and innovative approaches to data analysis. The current state of computational catalysis research places the field on the brink of a data-rich era, where research is limited by data analysis rather than data generation. The impact of available information can be maximized by the use of improved methods for data analysis and catalyst optimization. Descriptor-based analyses provide one route towards insight-based catalyst discovery. *My research will build on descriptor methods by using more advanced statistical methods and generalizing the design space beyond surface properties.*

397838 Catalytic Conversion of Transportation Fuel and Cellulosic Biomass Derivatives over Zeolite Catalysts

Abstract:

The current energy policies around the world have encouraged the development of fuel conversion processes and renewable energy sources. With this point of view, my Ph.D. research area can be classified into three regions; transportation fuel conversion, cellulosic biomass conversion into transportation fuel and value-added chemicals, and study of zeolite catalyst materials. For the first subject, various fuels, such as gasoline, diesel, and jet fuel including logistic fuel were investigated for production of hydrogen and light hydrocarbons in attempt to improve fuel efficiency and flexibility in remote area. For example, catalytic cracking of JP-8, logistic fuel in NATO (North Atlantic Treaty Organization) and the U.S. military, was conducted for the purpose of overcoming logical and functional limitation of existing devices such as diesel generator or unmanned aerial vehicle (UAV) in battle field by using liquefied petroleum gas (LPG) as power source of solid oxide fuel cell from JP-8 itself.

As aforementioned, the conversion of sustainable biomass for the production of energy and high value chemicals has been proposed as an enabling technology and received a great deal of interest for decades. Particularly, cellulosic biomass has been recently drawn attention as attractive feedstock due to abundance, low cost, and non-competing with food chain. Hence utilization of cellulosic biomass feedstock was allocated as another major research project. A variety of bio-feedstock, such as furfural, levulinic acid, gamma valerolactone, and pentanediol, was investigated over various zeolite catalysts, and finally, we were able to demonstrate a novel method for the production of LPG-like fuel, benzene, toluene, and xylenes from cellulosic biomass feedstock.

As catalytic fuel process and biomass conversion studies have been carried out using zeolite materials, synthesis of zeolite and zeolite based catalysts are essential part of my doctorial work. Several attempts, for instance varying synthesis temperature, time, organic template, etc., to modify crystal size, pore structure, and acid property of zeolite materials have been conducted. Furthermore, microwave-assisted heating method was compared with conventional hydrothermal heating method for the endeavor to development an efficient method for zeolite material synthesis.

A Route Towards Sustainability Through Multifunctional Hybrid Materials and Interfaces

B. Reeja Jayan, Karen K. Gleason[1], and Arumugam Manthiram[2]
(1) Chemical Engineering, Massachusetts Institute of Technology, Cambridge, MA
(2) Materials Science and Engineering, The University of Texas at Austin, Austin, TX

Areas of Expertise:

Solution and vapor phase synthesis of thin film materials (ceramic, polymer)
Materials discovery and advanced manufacturing
Hybrid (hard-soft matter) interfaces
Polymer photovoltaics
Lithium ion batteries

Abstract for Future Research:

The hallmarks of sustainable development include easy access to clean energy, clean air, potable water, adequate nutrition, and quality medical care. In order to solve the broad challenges we face in achieving a sustainable future, new **"hybrid" material systems** must be envisioned. These materials consist **of soft-hard matter interfaces** between organic and inorganic constituents at varying length scales (molecular to macro scale). Combining these materials incorporates the desirable properties of their organic (e.g. polymers, biomolecules) and inorganic (e.g. metals, ceramics) constituents; further, successfully modifying their compositions and interfaces can give rise to completely new functionalities that did not exist in the parent materials. Examples include organosilicon networks that can selectively conduct ions and block electrons in batteries,[1] and metal-organic frameworks that provide high surface area both to enhance light absorption in photovoltaic devices and to increase the attached fraction of binding biomolecules (proteins) in biosensors.

To enable applications in sustainability, I propose a research program with the following three goals: (1) **design** new multifunctional hybrid materials incorporating organic, inorganic and biological constituents at the molecular or nanometer scale; (2) **process** these materials into nanoscale thin films that form precisely engineered, covalently grafted interfaces with diverse materials like inorganic surfaces, polymers, and biomolecules (proteins); and (3) **investigate** how the morphology and chemistry of these materials and their interfaces influence mass and charge transport. The resulting knowledge will significantly advance technology development in energy generation, energy storage, and biosensing.

Postdoctoral Projects:

1. *Chemical Vapor Deposition of Conformal Nanoscale Polymeric Electrolytes for Three Dimensional Battery Architectures*

2. *Conducting Passivation of Silicon Solar Cells by initiated Chemical Vapor Deposition (iCVD) Polymerization*

Under supervision of **Professor Karen Gleason**, MIT

PhD Dissertation: *Development of Efficient, Stable Organic-inorganic Hybrid Solar Cells*

Under supervision of **Professor Arumugam Manthiram**, (UT-Austin)

Summary of Research Experience:

My **doctoral research** involved chemical and morphological engineering of the interfaces between thin films of organic (polymers) and inorganic (ceramic) materials. I probed the fundamentals of charge transfer phenomena at these interfaces using a combination of microscopic, spectroscopic, and device characterization techniques. My research led to improved hybrid polymer solar cells and lithium ion batteries. I also developed diverse thin film synthesis techniques for inorganic materials. In particular, my solution based process to crystallize ceramic thin films at low temperature using microwave radiation is extremely attractive for several functional material systems with applications in electronics, energy, and medicine.

My **postdoctoral research** addresses the key technological limitation to the development of three-dimensional (3D) battery architectures; the availability of a conformal thin film electrolyte. 3D batteries represent a novel approach to miniaturizing on-board electrochemical energy storage for many applications including sensing, actuation, communications, and medical implants. Using Initiated Chemical Vapor Deposition (iCVD) polymerization, I synthesized conformal nanoscale cyclic siloxane polymer films, which can be made to conduct ions at levels comparable to that of the most explored polymeric electrolyte, poly(ethylene oxide). This demonstration of a low-cost, thin film approach for creating highly stable siloxane-based compositions introduces a new direction to the field of polymer electrolytes, which has been quiescent for several years.

My research work has resulted in 15 publications in leading journals and 4 related patent applications have been filed.

Teaching, Leadership, and Outreach:

Over the years, I have demonstrated strong commitment to teaching, mentoring, and service to the scientific community. At UT-Austin and MIT, I assisted in teaching courses on Materials Engineering, Thermodynamics, Solid State Electronics, and Solar Conversion Devices. Additionally, I serve as reviewer for several scientific journals and as grant proposal review panelist for the *National Science Foundation (NSF).* I am also an Associate Technical Editor of *MRS Bulletin*, a leading publication in materials research. My leadership abilities are demonstrated by my roles as a founding member of the *MIT Think Tank* and as advisory committee member for the *MIT Innovation Initiative*.

Awards and Honors:

- Materials Science and Engineering Nominee for the Michael Granoff Outstanding Graduate Student Award, UT-Austin (2012)
- Graduate Poster Award, Nano Night, UT-Austin (2012)
- Cockrell School of Engineering Student Leadership Award, UT-Austin (2011)
- Travel Grant Award, Graduate Nano Portfolio Program, UT-Austin (2011)
- Doctoral Fellowship, American Association of University Women (AAUW) (2010)
- Mike Hogg Endowed Doctoral Fellowship, UT-Austin (2010)

- Professional Development Award, UT-Austin (2008, 2011)
- H.H. The Maharaja of Cochin Endowment Prize, University of Kerala, India (2002)

Selected Publications: ** denotes shared first-authorship*

N. Chen*, **B. Reeja-Jayan***, J. Lau, P. Moni, A. Liu, B. Dunn, and K. K. Gleason, "Conformal Cyclic Siloxane Electrolyte for Three-Dimensional Battery Architectures", *Nature Materials (under review).*

B. Reeja-Jayan, T. Adachi, R. J. Ono, D. A. Vanden Bout, C. W. Bielawski, and A. Manthiram, "Effect of Interfacial Dipoles on Charge Traps in Organic-Inorganic Hybrid Solar Cells", (2013), *Journal of Materials Chemistry A,* **1**, 3258-3262.

B. Reeja-Jayan*, K. L. Harrison*, K. Yang, Chih-Liang Wang, A. Yilmaz, and A. Manthiram, "Microwave-assisted Low Temperature Thin Film Growth in Solution*", (2012), Scientific Reports,* **2,** 1003.

B. Reeja-Jayan and A. Manthiram, "Understanding the Improved Stability of Hybrid Polymer Solar Cells Fabricated with Copper Electrodes", (2011), *ACS Applied Materials & Interfaces,* **3**, 1492-1501.

Jun Liu, **B. Reeja-Jayan**, and A. Manthiram, "Conductive Surface Modification with Aluminum of High Capacity Layered Li[Li0.2Mn0.54Ni0.13Co0.13]O2 Cathodes", (2010), *Journal of Physical Chemistry C*, **114**, 9528-9533.

Reviews:

B. Reeja-Jayan, P. Kovacik, R. Yang, H. Sojoudi , A. Ugur, D. H. Kim, C. D. Petruczok, X. Wang, A. Liu, and K. K. Gleason, (2014), "A Route towards Sustainability through Engineered Polymeric Interfaces", *Advanced Materials Interfaces*, **1**, 1400117.

398428 Instructing cell bioactivity using programmable DNA-peptide hybrid materials

Abstract:

DNA is one of the most promising molecules for the construction of nanomaterials, with programmable control over both nanostructure and dynamics. Despite this potential, DNA has not been used to a great extent in supramolecular biomaterials for regenerative medicine. My research uses novel DNA-peptide hybrids to create scaffolds for instructing cell behavior in ways not possible with current materials. I will present work on two main projects. The first uses a DNA nanotube as a scaffold for neural stem cell (NSC) attachment and directed differentiation. Modifying one of the strands that comprises the nanotube with a cell adhesion peptide results in a multivalent display of the biological signal on the nanotube surface. These nanotubes are potent scaffolds for NSC attachment compared with unmodified nanotubes, and selectively promote their differentiation to neurons. DNA allows for rational control over the structure obtained, and if nanotube formation is prevented, the NSCs no longer differentiation into neurons. My second project uses DNA as a handle to endow a biologically inert polymeric surface with bioactivity. DNA handles on an alginate coating allow for immobilization of peptides that promote fibroblast adhesion and spreading. By dynamically removing the peptide signal (using several DNA-specific triggers), it is possible to programmably reverse this bioactivity. DNA can also serve as a scaffold to present multiple signals with nanoscale control, promoting a distance-dependent synergy between two bioactive peptides or a peptide and a growth factor. These two projects highlight the power of DNA in both controlling nanostructure and dynamic properties of a cell scaffold, and are the foundation for a new area of DNA-biomolecule scaffolds for regenerative medicine.

399195 Micro and nanostructured interfaces for characterization of disease states in biological systems

Abstract:

The adhesion and aggregation of cells at solid-liquid interfaces plays a key role in biological function in many systems and is often mediated by both surface and environmental cues. For example, the formation of platelet aggregates onto the vascular sub-endothelium during *in vivo* clot formation is mediated by both the surface area available for adhesion and by shear stress, while the development of bacterial communities into biofilms is influenced by physicochemical surface properties and by interspecies interactions. Using innovative approaches in micro/nanofabrication and surface chemistry, we seek to engineer synthetic interfaces that recreate relevant *in vivo* situations in order to better understand how cell-surface and cell-cell interactions lead to disease states. These interfaces can then be used in clinical applications aimed at improving the diagnosis, monitoring, and treatment of disease. Specific applications of our engineered surfaces will include (1) measurement of platelet function for dosing anti-platelet agents, (2) measurement of bacterial interactions that occur during biofilm formation in the cystic fibrosis airway, and (3) high-avidity isolation of bacterial pathogens for rapid detection of bloodstream infections.

399283 Investigating the Interphase Regions Between Inorganic Nanomaterials and Thermosetting Polymer Backbones

Abstract:

The interfaces between a polymer and a metallic or ceramic nanoparticle in a polymer nanocomposite are generally not well understood. Polymer nanocomposites are typically approached on a case-by-case basis using chemistry based functionalization techniques on the particle surface to promote coupling to a tailored bulk polymer matrix phase. A great deal of difficulty lies in practically characterizing the resulting interphase region between the nanomaterial and bulk matrix, which has been demonstrated to play a significant role in the ultimate nanocomposite transient and/or dynamic mechanical response. During nanocomposite synthesis/fabrication, the interactions between the metallic/ceramic and polymeric phases are governed by a number of factors including mass transport from localized concentration gradients and/or differences in surface energy, and the kinetics of the reaction in a thermosetting type system. Practically, issues such as particulate flocculation and agglomeration, along with nanomaterial distribution within the composite, substantially influences bulk structure property relationships. The goal of this research is multifold and includes 1) development of novel characterization approaches to successfully probe the morphological evolution and final state of interphase regions in polymer nanocomposites and 2) fundamentally understanding the correlation between transport, kinetics, and thermodynamics that impacts the formation of the interphase region between polymer and nanoparticle. Ultimately, the goal is to develop predictive control of nanocomposite response as a function of nanomaterial integration approach for broad ranges of materials, particularly with respect to bulk composite dynamic or recoverable transient mechanical response.

399285 Interdisciplinary Engineering Fundamentals: Undergraduate Teaching and Learning

Abstract:

Interfacing nanotechnology and biology, my graduate research aims at understanding the physical and chemical mechanisms associated with the blood-borne spread of cancer which results in over 90% of cancer-related fatalities. Specifically, these studies further elucidate the roles of receptor-ligand interactions during breast cancer dissemination and investigate the efficacy of alternative engineered in-vitro platforms for studying the adhesion, migration and invasion of human cancers. The inherent breadth of this research in addition to my College Teaching and Learning Certificate Program at the University of Kentucky have inspired me to pursue a tenure-track faculty position in engineering education subsequent to my PhD defense in January 2015. Principally, I am passionate about undergraduate teaching and learning. My professional goals are to foster the personal and professional growth of future engineers by providing a solid foundation of introductory chemical engineering and mathematics as well as enhance communication and critical thinking skills. My specific teaching interests lie in early engineering-based coursework on process principles and balances, computational tools and fluid dynamics. In addition to these core courses, I am also interested in teaching elective courses on cell biology for engineers, cross-disciplinary journal clubs and micro/nanofabrication techniques. The biochemically-focused courses and undergraduate research I completed while earning a B.S. in Chemical Engineering at Rose-Hulman Institute of Technology as well as my participation on both the NSF Integrative Graduate Educational Research Traineeship (IGERT) and the NIH Cancer Nanotechnology Training Center (CNTC) grants at UK have shaped my firm views on the need for continued collaborations amongst STEM field majors and clinicians to ensure translational success from bench top experimentation to bedside patient care. In my experiences mentoring undergraduate researchers, I have seen that these opportunities and training can provide hands-on applications to supplement and enrich course materials as well as probe interest in designing innovative solutions to real-world clinically-relevant issues. Accordingly, it is also my desire to lead and participate in outreach and service efforts geared towards inspiring students, especially young women, to obtain degrees in STEM fields with a special emphasis on the importance of interdisciplinary collaborations.

399286 Multiscale Systems Analyses Advancing Drug Discovery Efforts and Therapeutic Development

Abstract:

Systems biology and biomedical control research has guided the rigorous assessment of regulation at the molecular, cellular, and tissue levels. Systems biology approaches — with its combination of computational, experimental and observational inquiry — has extended from model systems to human physiology and now to areas of pharmacology and human health. The Young Lab will combine expertise in systems biology, protein and cellular engineering, membrane biology, and synthetic biology to (a) identify and differentiate cell stress response pathways following local perturbations and (b) advance a mechanistic-understanding of spatiotemporal signaling that influences cell/tissue function. Our research pursuits are motivated by current challenges in drug discovery, therapeutic development, biopharmaceutical production, and limitations of human disease models.

The Young Lab will focus initially on three research objectives:

(1) Characterize diverse, global consequences of system disturbances leading to cell stress in a model eukaryote,

(2) Exploit these host-response pathways to facilitate biopharmaceutical production in yeast, *S. cerevisiae*, and

(3) Investigate the regulation of system perturbations resulting in the altered behavior of multicellular human tissue as a consequence of inflammation, including therapeutic intervention to facilitate drug discovery efforts in disease progression.

Due to the intrinsic complexity of biological systems, integration of experimental and computational approaches is crucial to transform the way we study cell physiology and behavior. To achieve a global understanding of how cellular processes and molecular interactions perform under external and internal perturbations, molecular measurements from many hierarchical cellular levels require integration within appropriate computational frameworks as recently described [1,2]. In the Young Lab, we will foster a research environment that is committed to excellence, encourages innovation and diversity, and advances scientific pursuits in drug discovery and development through multidisciplinary efforts.

[1] C. L. Young and A. S. Robinson, *Protein Folding and Secretion: Mechanistic Insights Advancing Recombinant Protein Production in S. cerevisiae*, **Current Opinion in Biotechnology**, 2014 Jul 14;30C:168-177.

[2] M. R. Ebrahimkhani, C. L. Young, D. A. Lauffenburger, L. G. Griffith, J. T. Borenstein *Approaches to In Vitro Tissue Regeneration with Application for Human Disease Modeling and Drug Development*. **Drug Discovery Today**, 2014 Jun;19(6):754-762.

Infections and diseases of the respiratory system contribute to morbidity and mortality in both the developing and the developed world. The therapeutic advantages of inhalable therapeutic treatment of respiratory diseases include the application of the therapeutic to the site of infection which enables a high localized concentration of therapeutic with low systemic exposure. However, the successful development of inhalable therapeutics requires the ability to successfully formulate the therapeutic into an inhalable form, the successful deposition of the material into specific respiratory regions, and controlled interactions of the material with extracellular and cellular environments. The solution to these challenges requires the expertise and interaction of multiple disciplines including: material science, engineering, molecular biology, immunology, pharmaceutics, and physiology.

My graduate and post-doctoral experiences in chemical engineering and pharmaceutics have uniquely prepared me to tackle these challenges. For drug delivery vehicle design, I have synthesized biodegradable polymers and used these to encapsulate antibiotics and proteins into particles on both the micron and nano scales. I have also used oligosaccharides from the surface of non-typeable *Haemophilus influenzae* and polyethylene glycol to modify the surfaces of polystyrene nanospheres to provide cell targeting and mucosal avoidance, respectively. The biodistribution, extracellular and cellular interactions of particles were evaluated utilizing *in vivo* murine models and *in vitro* models of the upper respiratory tract, mucosal secretions, and alveolar macrophages.

To supplement the *in vitro* experience, translational studies into particle aerosolization and aggregation of dry powder formulations have provided an understanding of deposition and distribution of inhalable therapeutic formulations. Combining this knowledge with my experience with human respiratory physiological measurements provides me with the tools required to address the challenge of designing inhalable therapeutic delivery systems for treatments of respiratory diseases.

374087 Dynamic Modeling of Colloidal Assembly

Abstract:

Understanding concentrated colloidal dynamics in the presence of different pairwise interactions and external fields provides a basis to predict the temporal evolution of colloidal microstructures in diverse phenomena including suspension rheology and colloidal assembly. However, a microscopic theory of concentrated colloidal dynamics does not yet exist that rigorously includes both statistical mechanical (configuration dependent free energy changes) and fluid mechanical (configuration dependent multi-body hydrodynamic interactions) contributions. This work shows the implementation of a novel approach to model colloidal assembly by means of analyzing the temporal evolution of microstructures, that can be observed in experiments and simulation, with a small number of order parameters that describe the state of the system. This methodology is used to analyze the structure evolution of isotropic and anisotropic colloidal building blocks. Knowledge of these dynamic models enable the fundamental understanding of assembly systems, as well as the optimization, design and control of assembly systems to produce low-defect colloidal crystals.

387166 Next Step in Computational Materials Design: Surface Properties of Metal Oxides

Abstract:

In the last decade, metal-oxide (MOs) based materials have found many important applications in batteries, as water-splitting catalysts or as cathodes of solid oxide fuel cells and other uses. Computationally, the bulk properties (i.e. thermodynamic stability, optical response) of these materials have already been qualitatively well captured. However, in real world situations, the processes of interest are often localized to a single site on a surface, reduced to nanostructures, or the bulk is self is altered under the contact with environment (i.e. Pourbaix Diagrams of metal-oxides). While for simple metals some of these challenges have been already addressed, the understanding of metal oxides in reduced dimensions is still in its infancy and offers great promise for future research. Especially important topics are a) types of possible surface terminations, b) relative energetic stability as function of environment, c) modification of properties due to localization and d) role of different active sites such as terraces vs. edge-sites for the process of interest.

My current post-doctoral research with Prof. Jens Nørskov (Stanford U.) and Prof. Alex Bell (UC Berkeley) already resulted in some progress in above topics. Very recently, in a series of papers, I was able to establish importance of types of surface structural motifs for water splitting activity of earth-abundant transition metal-oxides (TMOs: Fe, Co, Ni) [1,2]. As a results, I have identified highly active Fe sites in very promising mixed (Ni,Fe)OOH catalysts [2]. Currently, I continue to work on more general aspects of surface structure of TMOs. In my PhD work, I have also significantly contributed to progress of highly accurate quantum Monte Carlo methods [3,4], which serve as larger framework for widely used density functional theory and quantum chemistry.

My future research interests focus on significantly enhancing the understanding of metal-oxides in reduced dimensions. My knowledge and experience of three classes of state-of-the-art methods: periodical DFT methods, cluster-based quantum chemistry approaches and quantum Monte Carlo methods enables me to obtain high accuracy and cross-validated results, which are necessary to tackle the challenges in materials discovery. My objective is to establish 1) multi-purpose computational electronic structure laboratory including development of more accurate but scalable methods 2) classify and characterize the types of important structural motifs occurring in reduced dimensions such as in surfaces and 3) use these findings to discover new materials with enhanced surface related properties.

Selected Publications (22 total, 10 first author, h-index=10):

[1] Bajdich, M.; García-Mota, M.; Vojvodic, A.; Nørskov, J. K.; Bell, A. T., J. Am. Chem. Soc. 2013, 135, 13521–13530.

[2] Friebel, D.; Louie, M.; Bajdich, M; Sanwald, K.E.; Cheng, Mu-Jeng; Cai, Y.; Sokaras,D; Alonso-Mori, R; Weng, Tsu-Chien; Davis, R; Wise, A. M.; Bargar, J.; Bell, A. T.; Lercher, J. A.; Nørskov, J. K.; Nilsson, A.: Identification of highly active Fe sites in (Ni,Fe)OOH for electrocatalytic water splitting (submitted to Nat. Chem.)

[3] Bajdich, M.; Tiago, M. L.; Hood, R. Q.; Kent, P. R. C.; Reboredo, F. A., *Phys. Rev. Lett.* 2010, *104*, 193001.

[4] Bajdich, M.; Mitas, L. *Acta Phys. Slovaca* 2009, *59*, 81–168.

371890 Controlling Interactions at Nanoscale: A Versatile Tool for Assembling Multifunctional Future Materials

Abstract:

The interactions between nanoscale objects play an essential role in various industrial and biomedical processes. These interactions determine the material's physical properties such as tensile strength, flexibility, electrical and thermal conductivity of the resulting macroscopic material. My research focuses on the fundamental understanding of the interactions at the nanoscale in order to develop new multifunctional hybrid materials. During my doctoral studies at Technische Universität, Berlin in the group of Prof. Gerhard Findenegg, I investigated the effects of modulating the interaction between soft and hard nanomaterials. The influence of surface energy and curvature on the self-assembly of nonionic surfactant at silica nanoparticles[i] and in the nanopores of SBA-15 silica material[ii],[iii] were studied. In addition we also investigated the interaction of biologically relevant entities such as globular proteins with silica nanoparticles[iv]. We revealed how the adsorbed protein induces a pH-dependent completely reversible bridging aggregation of silica particles[v]. Earlier in 2014, I visited the group of Prof. Katsumi Kaneko at Shinshu University, Japan as a JSPS postdoctoral fellow for short period of time. My research in Japan was focused on the development of single walled carbon nanotube (SWCNT) and silica hybrid materials which can further be used as precursors for energy efficient electrodes. In this work we used the approach of dispersing the tubes in liquid and its subsequent drying to form highly porous hybrid aerogel[vi].

Since 2012, I am working in the group of Prof. Orlin Velev at North Carolina State University and Triangle Materials Research and Engineering Center (ΔMRSEC) deals with the programmed assembly of isotropic and anisotropic particles into chains as actuators for soft micro-robotic applications. We demonstrated that heteroaggregating particle pairs can be assembled into rigid permanent chains using dielectrophoresis (DEP) as a structure directing tool[vii]. We proposed a combinatorial based approach for predicting the chain length distribution and established the rules for permanent chain assembly[viii]. Recently, we developed a novel method for assembling superparamagnetic nanoparticles covered by lipid shell into magnetically responsive ultraflexible chains[ix]. Initial application of external magnetic field aligns the particles into linear chains, after which the particles are bound by the soft attractive potential induced by the surface condensed lipid. We demonstrated that the surface wetting lipid governs the formation of nanocapillary bridges and hence enables the ultraflexible chain assembly. We also established that the chains can be assembled in the form of self-repairing networks and a number of other hierarchal structures. In the future, this research field will be extended into the design of new functional materials with unusual optical, electrical and heat transport properties.

In future I plan to continue my research in the field of nanoscience with the aim to fabricate new materials with unusual physical and chemical properties. Based on my previous research experience on self- and directed assembly, I plan to establish my independent research team which would focus on both fundamental and applied aspects nanoscience and nanotechnology. In particular I am very interested in extending my research for assembling carbon based energy harvesting materials. In addition, I would also like to continue my research efforts towards developing a better understanding between nonentities and biological matter.

---

[i] B. Bharti; J. Meissner; U. Gasser; G. H. Findenegg *Soft Matter*, **2012**, *8*, 6573

[ii] B. Bharti; M. Xue; J. Meissner; V. Cristiglio; G. H. Findenegg *J. Am. Chem. Soc.*, **2012**, *134*, 14756

[iii] B. Bharti, T. Shin; B. Deme; G. H. Findenegg *in-preparation*

[iv] B. Bharti; J. Meissner; S. H. L. Klapp; G. H. Findenegg *Soft Matter*, **2014**, *10*, 718

[v] B. Bharti; J. Meissner; G. H. Findenegg *Langmuir*, **2011**, *27*, 9823

[vi] B. Bharti; R. Kukobat; D. Minami; K. Kaneko *in-preparation*

[vii] B. Bharti; G. H. Findenegg; O. D. Velev *Sci. Rep.* (*Nature*), **2012**, *2*,1004

[viii] B. Bharti; G. H. Findenegg; O. D. Velev *accepted*, **2014**

[ix] B. Bharti; A.-L. Fameau; M. Rubinstein; O. D. Velev *submitted*

378500 Design of Injectable Hydrogels for Regenerative Medicine

Abstract:

Cell transplantation via direct cell injection at the target site is a minimally invasive strategy for treating various injuries and degenerative diseases. However, cell viability is typically only 5% in these procedures, and the therapeutic success critically hinges on the survival and subsequent maintenance of the transplanted cells. To protect these cells, we designed an injectable hydrogel that undergoes two different physical crosslinking mechanisms. The first crosslinking step occurs *ex vivo* through peptide-based molecular recognition to encapsulate cells within a physical hydrogel, while the second crosslinking step occurs *in situ* through a thermal phase transition to form a reinforcing network. At room temperature, the physical hydrogel is shear-thinning and self-healing to facilitate gentle cell encapsulation and transplantation by syringe injection. At body temperature, the hydrogel forms a dual-network resulting in a 10-fold increase in shear modulus and significantly reduced erosion rates and prolonged retention time. Human adipose-derived stromal cells (hASCs) injected through a 28-gauge syringe needle were significantly protected from disruptive mechanical forces when encapsulated within the hydrogel compared to medium alone. *In vivo* subcutaneous injection of hASCs in a murine model demonstrated significantly improved cell retention when delivered in a double-network hydrogel compared to a single-network gel or medium alone. These cell-delivery materials are being further investigated for the ability to support functional tissue regeneration at sites of ischemia. In one strategy, the hydrogel mechanical properties are being tuned to enhance the secretion of pro-angiogenic growth factors by hASCs. In a second strategy, the hASCs are being co-cultured with induced pluripotent stem cell-derived endothelial cells (hiPSC-ECs) to promote tissue regeneration in a murine hindlimb ischemia model of peripheral arterial disease.

390757 Multi-Objective Modeling, Simulation, and Optimization for Economically and Environmentally Conscious Decision Makings

Abstract:

Chemical facilities, where large amounts of chemicals and fuels are processed, manufactured, and housed, are at high risk to originate air emission events, including intensive flaring and toxic gas release caused by various uncertainties such as equipment failure, false operation, natural disaster, or terrorist attack. Chemical plant concentrated regions may suffer localized and transient air pollution events that violate national ambient air quality standards (NAAQS). Flaring emissions, especially intensive start-up flaring emissions from chemical plants, have potentially significant impacts on local air quality.Furthermore, geographic allocation of chemical plants significantly affects industrial business sustainability as well as regional environmental sustainability. According to site selection rules, the air quality impact to surrounding communities of a newly constructed chemical plant must be taken into account. Therefore, it is very important to apply modeling, simulation, and optimization to provide economically and environmentally conscious decision makings and valuable quantitative engineering supports for multiple stake holders, including government environmental agency, regional chemical plants, and local communities.

In this paper, the general introduction will be given for our previously developed multi-systematic mathematical simulation and optimization models based on the methodology of multi-objective optimization. The first model will conduct the multi-plant start-up emission evaluation and help to provide control strategy. For any air quality violation is predicted to an AQCR, a multi-objective scheduling problem will be generated and solved to optimize the start-up sequence and start-up beginning time for all chemical plants. The scheduling model minimizes the overall air quality impacts to all of the AQCRs as well as minimize the total start-up time mismatch of all plants, subject to the principles of atmospheric pollutant dispersion. The second model aims to detect the possible emission sources (chemical plants) and identify the abnormal emission profile (emission source location, starting time, time duration, dynamic emission rate and total emission amount) from an accidental emission source responsible for an observed emission event based on an available air-quality-monitoring network so as to support diagnostic and prognostic decisions in a timely and effective manner.It provides valuable information for investigations of accidents and root-cause analysis for emission events; meanwhile, it helps evaluate the regional air-quality impact caused by such emission events as well. The third model evaluates the potential air-quality impacts from candidate sites of a new chemical plant in order to optimize the final site determination by minimizing its air-quality impacts. This study can not only determine the potential air-quality impact for allocation of multiple chemical plants with respect to regional statistical meteorological conditions, but also identify an optimal site for each new chemical plant with the minimal environment impact to surrounding communities. Case studies are employed to demonstrate the efficacy of the developed methodologies.

Engineered Natural Biomaterials for Understanding the Interplay Between Cells and Their Environment

Steven R. Caliari[1]
[1]Department of Bioengineering, University of Pennsylvania

Dynamic new biomaterial tools are necessary to address modern medicine challenges ranging from the repair and replacement of tissues and organs to the treatment of diseases such as fibrosis and cancer. In particular, the ability to engineer biomaterial milieus that re-create the heterogeneous microenvironments of physiological and pathological conditions is critically important. My PhD work in Prof. Brendan Harley's lab at the University of Illinois focused on the design of spatially-patterned collagen biomaterials to engineer orthopedic interfaces and direct cell behavior using approaches with broad applicability to a variety of regenerative medicine challenges. During my postdoctoral work in Prof. Jason Burdick's lab at the University of Pennsylvania I have developed a suite of mechanically dynamic hyaluronic acid hydrogel substrates useful for a range of fundamental mechanobiology studies, focusing in particular on understanding hepatic stellate cell behavior during liver fibrosis. As a principal investigator I will leverage my experience in chemical engineering, biomaterial fabrication and characterization, and mechanobiology to design regenerative templates for tissues including skin and skeletal muscle. On a more fundamental level, I will use these materials to investigate how cells interrogate their environment in disease states such as liver fibrosis and cancer.

Selected Publications (out of 15 first author published/submitted/in preparation):

7) **S.R. Caliari**, M. Perepelyuk, R.G. Wells, J.A. Burdick, 'Mechanically dynamic hydrogels for probing hepatic stellate cell behavior in fibrosis,' In preparation, 2014.

6) **S.R. Caliari**, B.A.C. Harley, 'Structural and biochemical modification of a collagen scaffold to selectively enhance MSC tenogenic, chondrogenic, and osteogenic differentiation,' *Adv Healthc Mater*, In press, 2014.

5) **S.R. Caliari**, L.C. Mozdzen, O. Armitage, M.L. Oyen, B.A.C. Harley, 'Periodically-perforated core-shell collagen biomaterials balance cell infiltration, bioactivity, and mechanical properties,' *J Biomed Mater Res A*, 102(4):917-27, 2014.

4) **S.R. Caliari**, B.A.C. Harley, 'Composite growth factor supplementation strategies to enhance tenocyte bioactivity in aligned collagen-GAG scaffolds,' *Tissue Eng Part A*, 19(9-10):1100-12, 2013.

3) **S.R. Caliari**, M.A. Ramirez, B.A.C. Harley, 'The development of collagen-GAG scaffold-membrane composites for tendon tissue engineering,' *Biomaterials*, 32(34):8990-8, 2011.

2) **S.R. Caliari**, B.A.C. Harley, 'The effect of anisotropic collagen-GAG scaffolds and growth factor supplementation on tendon cell recruitment, alignment, and metabolic activity,' *Biomaterials*, 32(23):5530-40, 2011.

1) T.A. Martin*, **S.R. Caliari***, P.D. Williford, B.A. Harley, R.C. Bailey, 'The generation of biomolecular patterns in highly porous collagen-GAG scaffolds using direct photolithography,' *Biomaterials*, 32(16):3949-57, 2011. *co-first authors

367093 Molecular Simulation As a Tool for Materials Synthesis, Catalysis, and Biocatalysis

Abstract:

The poster displays an overview of my PhD and postdoctoral computational research. Unifying themes include computational chemistry, molecular simulation, rare events, transition states, and structure-function predictions. Currently, at the Pacific Northwest National Laboratory my postdoctoral research is focused on materials and catalysis. In one project the mechanism of metal organic framework (MOF) synthesis is being investigated with computational chemistry focused on establishing detailed reaction schemes from metal ions and organic linkers to MOF building blocks. In a second project the kinetics and dynamics of water on metal oxide surfaces is studied. Using density functional theory (DFT) and *ab initio* molecular dynamics, energy barriers for individual adsorption, reaction, diffusion, and desorption events were obtained, and used in kinetic models, explaining experimental observables. My dissertation work at Iowa State University, as part of the Center for Biorenewable Chemicals, focused on enzymes. A bioinformatics approach was taken to find target enzymes with desired activities among the many ones existing in nature. To that end, a database that classified enzymes by structure was constructed. With experimental collaboration, a statistical approach identified enzymes with novel activity. Molecular dynamics and mixed quantum mechanics/molecular mechanics (QM/MM) simulations were done to understand in detail the mechanism of enzymatic catalysis; revealing transition state conformation, the roles of active site residues, and energy landscape.

380570 Multiscale Simulations of Polymer Dynamics, Polymerization Kinetics, and Polymer Blend Morphology

Abstract:

The availability of fast and large mulitnode computer clusters have made computer simulations of polymers routine. Development of parallel algorithms and availability of accelerators such as graphical processing units (GPUs) further enhanced our capability in performing these simulations. Despite these advances, simulating polymeric systems still remains a challenge because of the multiscale nature of polymers that spans multiple length and time scales. To illustrate I am presenting a vignette of my latest works where I used different coarse-graining techniques and atomistic simulations to simulate: (1) the dynamics of unentangled polymer melts confined in a highly adsorbing cylindrical walls, (2) the polymerization kinetics and chain characterization of bottle-brush macromolecule synthesized through the macromonomer approach, and (3) the morphology and characterization of organic photovoltaic blends with emphasis on the structure of the donor/acceptor interface. All simulation results were qualitatively, and in some cases quantitatively, in agreement with neutron scattering experiments.

For example in the simulations of polymer melts confined in pores, we used a polymer bead-spring model to simulate anodize aluminum oxide (AAO) infiltrated with polydimethyl-siloxane (PDMS) and found excellent agreement in the dynamic and static properties of the chains between neutron spin echo (NSE) and small angle neutron scattering (SANS) experiments, respectively. In the second example, we used a bead-spring model of a macromonomer and simulated ring opening metathesis polymerization (ROMP) by sequentially adding macromoners to form bottle-brush macromolecules. The structure of the bottle-brush was compared to SANS experiments. In the third example, a hindered-rotation model of a polymer was used to model poly(3-hexylthiophene) (P3HT). The simulation results of the P3HT/PCBM blend, in the presence of a substrate, shows qualitative agreement with neutron reflectivity (NR) experiments where a vertical composition profile indicating PCBM rich regions in the vicinity of the substrate and air interfaces. Furthermore, the orientation of the thiophene rings in the donor/acceptor interface was also investigated using atomistic molecular dynamics simulations.

My objective is to perform simulations that would tackle difficult areas in polymer physics such as charged polymers, liquid crystalline polymers, polymers at interfaces and polymer crystallization. In these future works, I will actively search for areas of collaboration with the goal providing insights that will provide a better understanding of these polymeric systems. The above examples illustrate my strategy to meet the challenge.

Graphene Electrode-Based ZnO Nanowire Hybrid Solar Cells

Sehoon Chang, Silvija Gradečak

Department of Materials Science and Engineering, Massachusetts Institute of Technology, Cambridge, Massachusetts, 02139;

Abstract

Growth of semiconducting nanostructures on graphene would open up opportunities for the development of flexible optoelectronic devices, but challenges remain in preserving the structural and electrical properties of graphene during this process. We demonstrate growth of highly uniform and well-aligned ZnO nanowire arrays on graphene by modifying the graphene surface with conductive polymer interlayers. Based on this structure, we demonstrate graphene cathode-based hybrid solar cells using two different photoactive materials – the colloidal quantum dots (CQDs), lead sulfide (PbS) QDs and the conjugated polymer, P3HT – with AM 1.5G power conversion efficiencies of 4.2% and 0.5%, respectively, approaching the performance of ITO-based devices with similar architectures. Our method preserves beneficial properties of graphene and demonstrates that it can serve as a viable replacement for rigid ITO in various flexible solar cell device configurations.

The continuous growth of global energy consumption has led to increasing demand in the development of next generation devices for energy storage and conversion. Proton exchange membrane fuel cells have attracted attention due to their potential as clean, eco-friendly power sources. The ability of a fuel cell to deliver power depends crucially on the rate of proton transport through a hydrated polymer electrolyte membrane (PEM) that is placed between the electrodes. Many people have studied proton transport properties in PEMs, but few have successfully linked properties with the structure and morphology of PEMs, especially in the hydrated states. This is largely due to the technical challenges associated with probing morphology of PEMs in the hydrated states.

My current research as a postdoctoral fellow at Lawrence Berkeley National Laboratory is focused on the fundamental investigation of morphology-conductivity relationships in model PEMs based on ion-containing block copolymers, comprising ionic and non-ionic blocks. Specifically, I study the water uptake and proton conductivity of a model block copolymer, polystyrenesulfonate-*block*-polyethylene-*block*-polystyrenesulfonate (S-SES), as a function of water activity in both humid air and liquid water. I observed a discontinuous increase in both water uptake and proton conductivity of S-SES membranes when nearly saturated humid air is replaced with liquid water. Using humidity-controlled synchrotron small angle X-ray scattering and cryogenic scanning transmission electron microscopy, I discovered the morphological underpinnings of these observations. The discontinuous increase in conductivity coincides with the emergence of heterogeneity in the hydrated channels: a water-rich layer is sandwiched between two polymer-rich brushes. While the possibility of obtaining heterogeneous hydrated channels in polymer electrolyte membranes has been discussed extensively, this is, to our knowledge, the first time that direct evidence for the formation of water-rich subdomains is reported.

This poster depicts the crucial findings of my postdoctoral work in establishing structure-property relationships in PEMs, along with an outline of water management in PEMs by introducing nano-scale pores into the proton conducting phase.

389629 Design of Thermodynamically Consistent Coarse-Grained Models in Soft Matter

Abstract:

The world's population growth is enhancing the need for efficient materials, however, the design, synthesis and processing of novel materials with optimal properties for applications is an important challenge, due to the manipulation of matter at ever decreasing length scales. Computer simulation methods are essential for improving the design of novel materials, because they can provide quick and cost-effective promising directions for materials synthesis, by indicating improvements in process design or identifying reasons for shortcomings. To date, my research has been in the broad area of soft matter and with particular emphasis on methods to control or enhance self-organization on those systems. Through my work I had tackled a series of challenging problems including, but not limited to developing models to successfully reproduce morphologies with long-range order of thin films of diblock copolymers under shear [1-4] (which show promising use in synthesizing nanowires, surface modification, and pattern transfer) and provide a fundamental understanding of the structural as well as the dynamical properties of solvent-free grafted nanoparticles [5-7] (which show promising use in carbon dioxide (CO_2) capture, energy capture and storage, as well as water purification). Most of these problems are multi-scaled ranging from the atomic scale to the macroscale. My current work, focus on the thermodynamics of the vapor-liquid equilibria of pure and mixed solvents of multi-functional amines, which is of great relevance for CO_2 capture processes. By coupling together a predictive equation of state (EoS) for the thermodynamics and phase behavior together with robust coarse-grained models for use in explicit molecular simulation greatly enhance predictive solvent design for CO_2 capture processes. The advantage of such approach is that a molecular-based equation of state can be used to obtain effective coarse-grained models that reproduce the macroscopic experimental thermophysical properties over a wide range of conditions, while conventional molecular simulation with these coarse-grained models can be performed to obtain properties (such as structure or dynamics) that are not directly accessible with the EoS. The realization of these exciting models and methodologies require a new level of understanding of the structure and properties of complex systems, as well as the capability to engineer and optimize their performance. More importantly they are not restricted to one field of research thus making them very versible. For specific molecular structures (e.g., block copolymers) there are well established methods and understanding of their behavior. We still lack a fundamental understanding of self-assembly as the building blocks deviate from the ideal molecular structures that theories are built on. Nevertheless, a more complex molecular structure would offer new possible venues for creating building blocks with selective functionality. One such example is hollow particles, e.g., nano-scale vesicles, protein cages. Such systems can be excellent candidates as nano-capsules for enzyme encapsulation and controlled release of drugs, which would have immediate relevance to many applications in physical and biological sciences. My interdisciplinary research program will therefore be dedicated to the science and engineering soft matter systems with optimal properties.

Selected Publications

[1] A. Chremos, K. Margaritis, A. Z. Panagiotopoulos, Soft Matter 6, 3588-3595 (2010).
[2] A. Chremos, P. M. Chaikin, R. A. Register, and A. Z. Panagiotopoulos, Macromolecules 45, 4406-4415 (2012).
[3] A. Chremos, P. M. Chaikin, R. A. Register, and A. Z. Panagiotopoulos, Soft Matter 8, 3803-3811, (2012).
[4] A. Chremos, A. Nikoubashman, and A. Z. Panagiotopoulos, J. Chem. Phys., 140, 054909, (2014).
[5] A. Chremos and A. Z. Panagiotopoulos, Phys. Rev. Lett. 107, 105503 (2011).
[6] A. Chremos, H. Y. Yu, D. L. Koch, and A. Z. Panagiotopoulos, J. Chem. Phys. 135, 114901 (2011).
[7] A. Chremos, D. L. Koch, and A. Z. Panagiotopoulos, J. Chem. Phys. 136, 044902 (2012).

Charge Carrier Dynamics in Thin Film Solid-State Solar Cells

Tailoring Solution-Processed Semiconductors for Energy Applications

Jeffrey A. Christians

Department of Chemical and Biomolecular Engineering and Radiation Laboratory, **University of Notre Dame**, Notre Dame, IN

The Clean Energy Demand

Global energy demand (currently ~13TW) is expected to more than double by 2050. To prevent dangerous anthropogenic climate interference, renewable and carbon neutral energy will become increasingly important in fulfilling this energy demand.

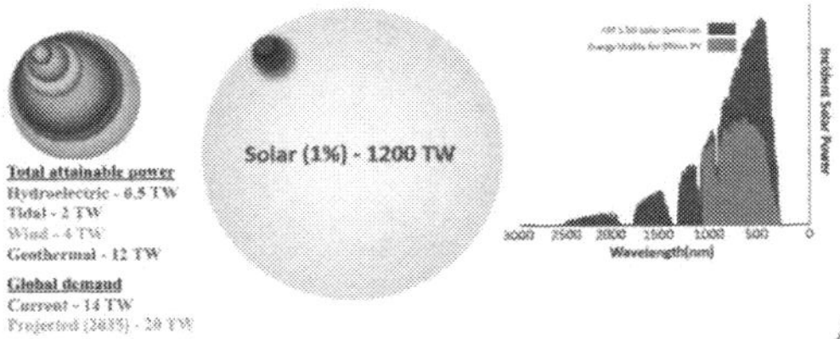

Research Overview

My doctoral research in Prashant V. Kamat's lab at Notre Dame has involved understanding the properties of semiconductor thin films for photovoltaics

(1) Constructing high efficiency, solution-processed photovoltaic devices

(2) Investigating charge transfer across semiconductor heterojunctions

(3) Elucidating semiconductor material properties

(4) Developing new methods for semiconductor thin film fabrication

Next Generation Photovoltaics

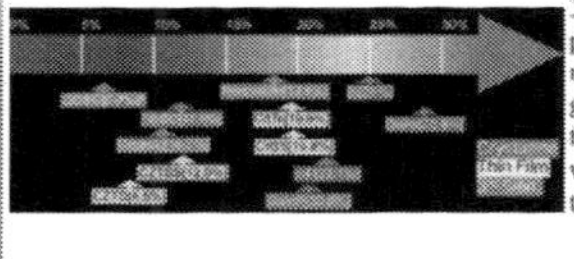

3rd gen. devices can be solution-processed, making them potentially much less expensive than 1st & 2nd gen. PV. Optimizing the charge transfer process in these solar cells will allow further efficiency gains for these materials.

Charge Transfer Processes in QDSCs

1. Electron transfer from quantum dot to metal oxide
2. Diffusion of electrons through metal oxide film
3. Regeneration of QD by electrolyte (hole transfer)
4. Electron-hole recombination at counter electrode

5–8. Negative electron-hole recombination processes

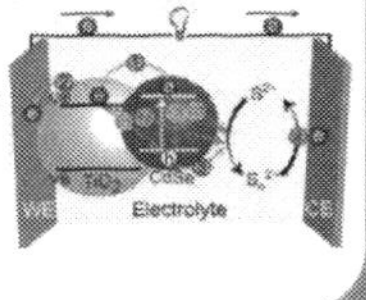

Extremely Thin Absorber Photovoltaics

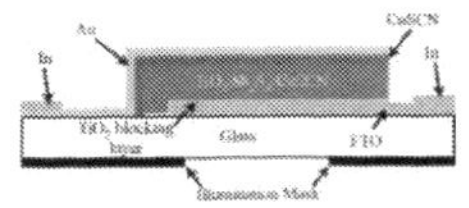

Sb_2S_3 ETA Solar Cells

The near-optimal band gap (1.7 eV), and low-cost processing make Sb_2S_3 a promising material for photovoltaics (below).

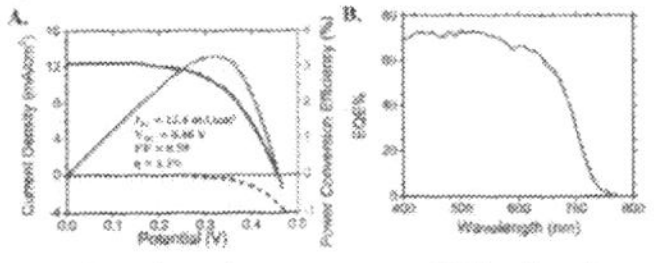

Transient absorption spectroscopy (TAS) offers the ability to study charge carrier dynamics on the femtosecond timescale (10^{-15} s). This can provide information of the rate of charge transfer and recombination reactions, providing valuable information for the further optimization of Sb_2S_3 solar cells.

Hole Transfer from Sb_2S_3 to CuSCN

Hole transfer rates were determined by tracking the sulfide radical induced absorption signal at 550 nm in the transient absorption spectra of Sb_2S_3 with and without the hole acceptor, CuSCN (below, A&B).

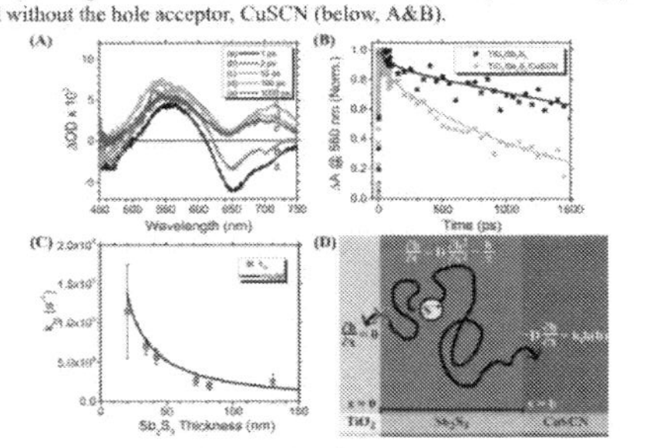

Modeling Interfacial Hole Transfer

By developing a model based on Fick's Laws of diffusion (above, C&D), we resolved the contributions of hole diffusion in Sb_2S_3 and hole transfer across the Sb_2S_3/CuSCN interface. This provides key information about the system and offers a guide for future studies to further improve PV performance.

Organometal Halide Perovskite Photovoltaics

Perovskite Solar Cells

Methylammonium lead iodide perovskites have swiftly advance beyond all other third generation materials, approaching CdTe and CIGS in terms of power conversion efficiency (right), while utilizing earth abundant materials and solution-processing fabrication methods.

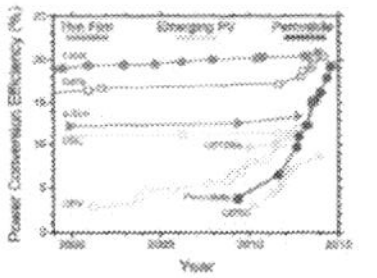

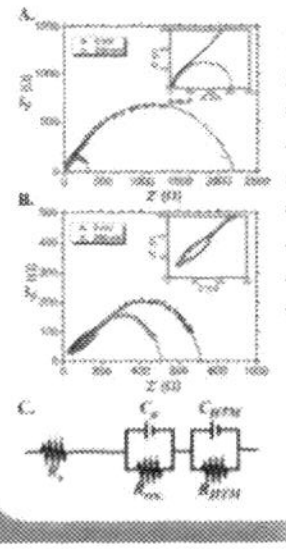

Inorganic Hole Conductors in Perovskite Solar Cells

CuI functions as an inexpensive hole conductor for organo-lead iodide perovskite solar cells, a promising alternative to organic hole conductors such as spiro-OMeTAD. It offers improved conductivity, as demonstrated by impedance spectroscopy (left), and efficiency of 6% compared to 7.9% for spiro-OMeTAD control devices (below).

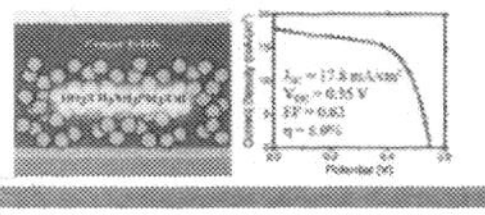

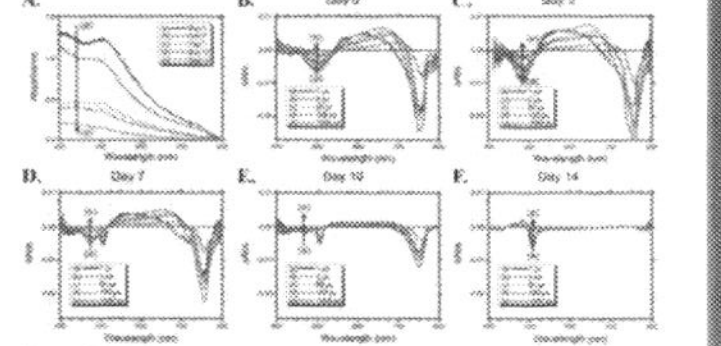

Humidity-Induced Excited State Transformation

TAS of $CH_3NH_3PbI_3$ exposed to humidity (above) elucidates how humidity exposure degrades the performance of perovskite solar cells (right). This work characterizes the nature of this degradation and reveals that changes in the optical and excited state properties occur much more slowly than the loss of PV performance.

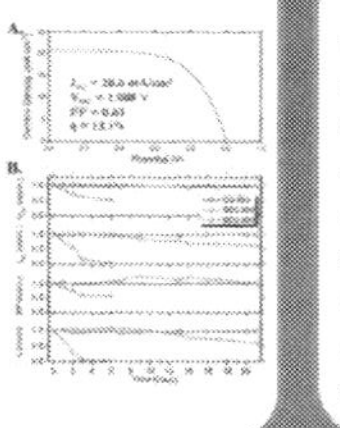

Proposed Future Projects

Solution-processed semiconductor devices have the potential to dramatically simplify the manufacturing of a wide variety of optoelectronic devices. Specifically, I am interested in developing solution-based techniques for the construction of photovoltaics and LEDs.

Solution-Processed Metal Chalcogenides

Recent advances have demonstrated how V_2VI_3 semiconductors can be deposited from solution.[1] This approach offers a facile method for creating $Sb_2Se_xS_{3-x}$ ETA solar cells with the ability to modulate the band gap between 1.7 eV and 1.3 eV.

Lead-Free Hybrid Perovskite PV

Current Sn-based hybrid perovskite show promise for photovoltaics but lack stability. Employing a layered hybrid perovskite structure (right) allows for increased stability and tunable material properties.[2]

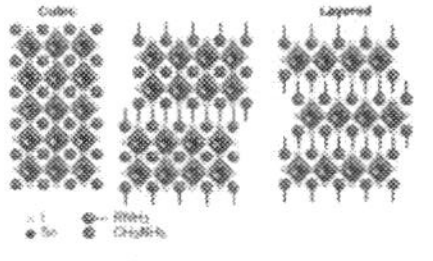

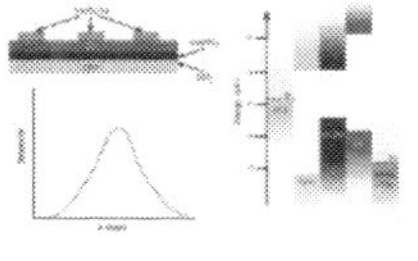

Hybrid Perovskite LEDs

The low contribution from trap-state and Auger recombination makes hybrid perovskites a promising material for LEDs.[3] Employing layered perovskites with high exciton binding energy should increase radiative recombination, leading to even better LED performance.

Acknowledgements

This work would not have been possible without Dr. Prashant V. Kamat, my doctoral advisor, the other members of the Kamat lab and my collaborators, and the members and staff of the Radiation Laboratory. This work was supported by the Office of Basic Energy Sciences of the U.S. Department of Energy, the Jana & Patrick Eilers Energy Research Fellowship, and the Bayer Predoctoral Research Fellowship.

Publications

1. Manser, J. S.; Christians, J. A.; Kamat, P. V. Current State of the Excited State Dynamics of Methylammonium Lead Iodide Perovskite. **2014** In Preparation (Review Article).
2. Christians, J. A.; Manser, J. S.; Kamat, P. V. Best Practices in Perovskite Solar Cell Efficiency Measurements. Avoiding the Error of Making Bad Cells to Look Good. **2014** In Preparation – To be Submitted 12/2014 (View Point).
3. Christians, J. A.; Miranda Herrera, P. A.; Kamat, P. V. Transformation of the Excited State and Photovoltaic Efficiency of $CH_3NH_3PbI_3$ Perovskite upon Controlled Exposure to Humidified Air. *J. Am. Chem. Soc.* **2014** Submitted.
4. Kamat, P. V.; Christians, J. A.; Radich, J. G. Quantum Dot and Nanowire Solar Cells: Hole Transfer as a Limiting Factor. *Langmuir* **2014**, 30 (20), 5729-5738 (Invited Feature Article).
5. Kim, J.-P.; Christians, J. A.; Choi, H.; Krishnamurthy, S.; Kamat, P. V. CdSeS Nanowires. Compositionally Controlled Band Gap and Charge Dynamics. *J. Phys. Chem. Lett.* **2014**, 5, 1103-1109.
6. Christians, J. A.; Leighton Jr., D. T.; Kamat, P. V. Rate Limiting Interfacial Hole Transfer in Sb₂S₃ Solid-State Solar Cells. *Energy Environ. Sci.* **2014**, 7, 1148-1158.
7. Christians, J. A.; Fung, R. C. M.; Kamat, P. V. An Inorganic Hole Conductor for Organo-Lead Halide Perovskite Solar Cells. Improved Hole Conductivity with Copper Iodide. *J. Am. Chem. Soc.* **2014**, 136, 758-764.
8. Christians, J. A.; Kamat, P. V. Trap and Transfer. Two-Step Hole Injection Across the Sb_2S_3/CuSCN Interface in Solid-State Solar Cells. *ACS Nano* **2013**, 7, 7967-7974.
9. Opperwall, S. R.; Divakaran, A.; Porter, E. G.; Christians, J. A.; Denhartigh, A. J.; Benson, D. E. Wide Dynamic Range Sensing with Single Quantum Dot Biosensors. *ACS Nano* **2012**, 6, 8078-86.

Molecular Engineering for Regenerative Medicine

Eun Ji Chung

Institute for Molecular Engineering, University of Chicago, Chicago, IL 60637, USA

When designing bioactive materials to direct desired outcomes for regenerative medicine, the appropriate structural and functional properties at all relevant hierarchical levels—including tissue, cellular, and molecular scales—must be incorporated. Furthermore, just as important is the intentional design of the biomaterial to be practical for the clinical setting, complementing available surgical techniques and emerging healthcare technology, while remaining safe and cost-effective.

To understand the biological and engineering criteria for multi-dimensional biomaterials design, my research-to-date has utilized biodegradable polymers and self-assembling assemblies for tissue engineering and theranostic applications. My graduate research at Northwestern University under the direction of Prof. Guillermo Ameer in the Dept. of Biomedical Engineering encompassed developing citric acid-based nanocomposites for orthopaedic tissues. Specifically, I investigated materials parameters that are optimal for stem cell differentiation and *in vivo* bone and ligament regeneration, and developed a novel method termed "low pressure foaming" to fabricate porous scaffolds. Upon receiving the IBNAM-Baxter Early Career Award, I continued my research at Northwestern as a postdoctoral fellow in Prof. Ramille Shah's lab in the Dept. of Materials Science and Engineering, where I fabricated *in situ*-forming, self-assembling structures derived from natural extracellular matrix proteins. Specifically, I focused on novel, collagen-hyaluronic acid membranes to act as bioactive coatings for drug delivery applications that can also complement minimally-invasive procedures. Moreover, I elucidated that the mechanism of self-assembly involved electrostatic interactions and osmotic pressure imbalances. To further diversity my experience in self-assembly and to integrate the rational design of nanoparticles for theranostic applications, I joined Dr. Matthew Tirrell's group at the Institute for Molecular Engineering at the University of Chicago as a recipient of the American Heart Association postdoctoral fellowship. My current research aims to develop peptide amphiphile micelles to target, diagnose, and treat a variety of diseases including atherosclerosis and cancer.

Building on my previous work and harnessing my biomolecular engineering background, I plan to establish an independent research program that aims to apply novel biomaterials strategies that addresses the limitations of medical solutions and laboratory models to enhance tissue regeneration and generate optimal clinical outcomes. Four areas of interest that I plan to pursue include: 1) Molecular assemblies for diagnostic and therapeutic applications, 2) Multifunctional, biomimetic polymers for biomedical applications, 3) Novel sources of stem cells for regenerative medicine, and 4) Design of *in vitro* models of disease.

Selected Publications (14 total):

1. **Chung, E.***, Cheng, Y.*, Morshed, R., Nord, K., Han, Y., Wegscheid, M., Wainwright, D.A., Lesniak, M.S., Tirrell, M.V. Peptide amphiphile micelles for targeting glioblastomas. Biomaterials 35(4): 1249-56 (2014).
2. **Chung, E.**, Jakus, A.E., and Shah, R.N. In situ forming collagen-hyaluronic acid membrane structures: Mechanism of self-assembly and applications in regenerative medicine. Acta Biomaterialia 9(2): 5153-5161 (2013).
3. **Chung, E.**, Sugimoto, M., Koh, J., Ameer, G.A. Low pressure foaming: A novel method for the fabrication of porous scaffolds for tissue engineering. Cover of Tissue Engineering Part C 18(2): 113-121 (2012).
4. Serrano, M.C.*, **Chung, E.***, Ameer, G.A. Advances and applications of biodegradable elastomers in regenerative medicine. Advanced Functional Materials 20(2): 192-208 (2010).
5. Wang, J., Singh, C., Liu, L., Irwin, R., Chen, S., **Chung, E.**, Thompson, R., Brinton, R. Allopregnanolone reverses neurogenic and cognitive deficits in mouse model of Alzheimer's disease. PNAS 107(14): 6498-6503 (2010).
**These authors contributed equally.*

For more information, please visit: http://tirrell.ime.uchicago.edu/People/Eunji_Chung.html

Holistic Multi-scale Approaches for Nanotechnology Convergence Systems

Pil Seung Chung
Department of Chemical Engineering, Carnegie Mellon University, Pittsburgh, PA 15213, USA

During the past decades, along with remarkable scientific and engineering achievements in many areas, the world has faced immense challenges. Increasing need for higher efficiency and reduction in system size motivates the investigation of novel nanoscale systems and materials. Various enhancements have been reported by tuning atomistic properties, chemical structures, and functionalities of advanced materials leading to eye-opening progress. The tremendous improvements in nanotechnology provide the necessary understanding of nanoscale physics to control and design the system properties, while at the same time focusing on the critical issue of integration with macroscopic system scales resulting in converging technology and phenomena at different time and length scales.

My research interest focuses on this challenging problem by introducing a novel holistic multi-scale integration in nanotechnology by hybridizing atomistic / molecular and mesoscale / continuum models, and developing unique methodologies for hierarchical integration of the complete system at different scales, and expanding to include multi-phenomena / physics (Fig.1). Information transfer from one scale to neighboring scales allows one to eventually predict macroscopic behavior or device scale functions from first principles in top-down / bottom-up / middle-out approaches. These approaches along with system optimization based on a clear understanding of nanoscale phenomena will lead to a novel multidisciplinary analysis paradigm.

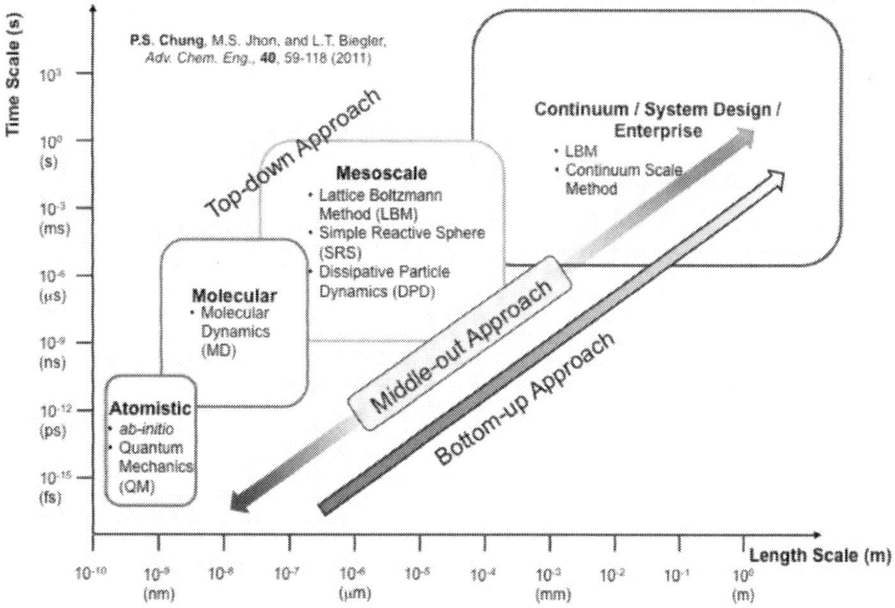

Fig.1. Hierarchy of multi-scale modelings and the integration methodology

In this poster session, I present case studies, which have been examined in the area of information technology and energy technology & sustainability for multi-scale modeling via a hierarchy of scale specific models coupled with optimization for the process system integration. Nano-tribology in the data storage system was chosen as a benchmark example utilizing atomistic / molecular simulations for the nanoscale film of functionalized oligomers / polymers with hybridization of novel graphene technologies and extended to mesoscale / continuum theories for the whole system integration [1,2]. In energy & sustainability area examples, we analyzed polymer electrolyte membrane fuel cell including membrane electrode assembly (polymer electrolyte membrane & catalyst layers) via atomistic / molecular simulations and transport phenomena in the gas diffusion layer via lattice Boltzmann method [3]. In addition, I will introduce and discuss the details of my future research topics and plan in this poster session.

Reference:
[1] P.S. Chung, S.H. Vemuri, S. Park, and M.S. Jhon, "Molecular rheological analysis on binary blends of perfluoropolyether lubricants," J. Appl. Phys., 115, 17B738 (2014).
[2] P.S. Chung, D.S. So, L.T. Biegler, and M.S. Jhon, "Nanotechnology convergence and modeling paradigm of sustainable energy system using polymer electrolyte membrane fuel cell as a benchmark example," J. Nanopart. Res., 14 (8), 853 (2012).
[3] P.S. Chung, R. Smith, S.H. Vemuri, Y.I. Jhon, K. Tak, I. Moon, L.T. Biegler, and M.S. Jhon "Multiscale/ multi-physical modeling in head/disk interface of magnetic data storage," J. Appl. Phys., 111 (7), 07B712 (2012).

386018 A Materials Genome Approach to Metal-Organic Frameworks: From Databases to Advanced Search Algorithms

Abstract:

My current research activity is at the intersection between molecular modeling and data science. In this poster, I will describe my postdoctoral work on metal-organic frameworks (MOFs) in Prof. Randall Snurr's group at Northwestern University and how these activities align with the emerging research area of data science and high-throughput molecular modeling.

In collaboration with researchers from Georgia Tech and Lawrence Berkeley National Laboratory, I have created a database of thousands of computation-ready, experimentally-derived metal-organic frameworks. Construction of the database required removal of impurities and solvents using graph-labeling and cluster-detection algorithms and text-mining. The atomic coordinates of the thousands of structures will soon be made available to the public through the Nanoporous Materials Genome Center. The database is expected to grow as new structures become available in the literature. Following the compilation of thousands of MOFs, we have computationally screened these materials using atomistic grand canonical Monte Carlo simulations to find the top performing MOFs for application in methane storage and hexane isomer separation as an initial demonstration of the utility of the database.

An emerging area in computational chemistry is the high-throughput computational screening of large numbers of structures to identify promising structures and to extract useful structure-property relationship for a given application. MOFs have been a particularly good target for such an approach because of the large number of synthesized structures in the literature. The early high-throughput computational screening of MOFs has been carried out in a brute-force manner, where a number of performance evaluations (e.g., adsorption isotherm at a given pressure) are carried out for every structure in the database. What enables such a high-throughput computational screening approach are several key simplifying assumptions, such as rigid frameworks and the use of generic force field parameters and classical atomistic simulation methods. The new frontiers of high-throughput molecular modeling will involve the removal of such simplifying assumptions and usage of high-level computational chemistry methods (e.g., DFT and QM/MM). These advanced approaches will require large amounts of computational resources, making it impossible to apply them for thousands of structures.

One way to avoid expensive performance evaluations for thousands of structures is to employ advanced search algorithms for metal-organic frameworks. I have developed an evolutionary search algorithm and applied it to a database of hypothetical MOFs to effectively find top MOFs in terms of importance performance criteria such as methane delivery capacity and volumetric surface area. The algorithm that has been developed here will not only be useful in terms of locating the best existing structures in a database but could help to find a new, as-yet synthesized MOFs for gas storage and separation application in the future. The approach could also be extended to optimize search for promising structures in other class of materials, such as metallic glasses that finds a widespread use in electronics and biomedical applications.

My future research direction will focus on the emerging area of high-throughput computation of metal-organic frameworks and other class of materials, with an emphasis on development of a platform for data science approaches and application of molecular modeling to search for best performing materials that could be used in energy and pharmaceutical applications (e.g., gas storage and drug delivery).

385098 Linear and Non-Linear Programming Techniques for System Identification and Green Engineering Applications

Abstract:

My research focuses on linear programming (LP) and non-linear programming (NLP) techniques for system identification and green engineering applications. To this end, I have explored interval analysis-based branch-and-bound techniques for parameter identification, assessment/synthesis of sustainable systems, graphical/analytical techniques for the global solution of the total annualized cost problem for a series of compressors, and the introduction of the novel concept of the attainable region for process networks. Throughout the course of my work, I have experienced first-hand the "curse of dimensionality" prevalent in non-linear optimization algorithms. Thus, I have applied analytical techniques, dimensionality reduction, convexification strategies, and functional analysis to identify global optima and to enhance the efficiency of their computation.

My global optimization expertise has prepared me well to pursue new research avenues in several areas, particularly optimal system identification, initial project feasibility assessment via attainable region concepts, and sustainability assessment and decision making. My strong mathematical and computational background will also allow me to contribute to collaborations that can benefit from this kind of expertise. Furthermore, I will disseminate all developed concepts and lessons learned into course-friendly materials to provide students opportunities to apply basic chemical engineering and optimization concepts to real-world green engineering problems.

385573 Engineered Nanostructures for Imaging, Sensing, Environmental Applications, and Their Implications to Human Health

Abstract:

The primary focus of my research will be on developing nanoscale materials for biomedical applications while also examining the implications through their interactions with biological systems. Light responsive nanomaterials offer a unique advantage in diagnostics and image-guided surgery. Therefore, one of my primary goals is to develop near infra-red light emitting nanomaterials for non-invasive sensative diagnostics tools. In my current postdoctoral research in Professor Ian Kennedy's laboratory at University of California, Davis, I worked on developing an X-ray based optical tomography system and associated contrast agent for deep tissue imaging. I also developed a novel system for early detection of prostate cancer using lanthanides via optical and magnetic resonance imaging (MRI) when I was doing one year of postdoctoral research jointly in the University of Victoria, Canada and the British Columbia Cancer Agency under supervision of Professor Frank van Veggel.

My other major research goals are to predictively determine deposition, translocation of the environmental particulate matters and inhaled drugs in the respiratory track, and to develop efficient materials & techniques for hazardous waste treatment. Very recently, I have developed a magnetic field aided gas phase synthesis technique that resulted in large scale production of extremely efficient magnetic nanochains for arsenic removal from water.

Selected publication (Out of 15):

Sudheendra, L.;* **Das, G. K.**;* Li, C.; Stark, D.; Cena, J.; Cherry, S. R.; Kennedy, I. M. X-ray Luminescence from Europium-doped Lanthanide Fluoride Particles. Chemistry of Materials**, 2014,** *26*, 1881–1888. **(*Co-First Authors)**

Das, G. K.; Johnson, N. J. J.; Cramen, J.; Blasiak, B.; Latta, P.; Tomanek, B.; van Veggel F. C. J. M. $NaDyF_4$ Nanoparticles as T_2 Contrast Agents for Ultrahigh Field Magnetic Resonance Imaging. The Journal of Physical Chemistry Letters, **2012**, 3, 524–529.

Das, G. K.; Zhang Y.; D'Silva, L.; Loo, J. S. C; Padmanabhan, P.; Heng, B. C.; Loo, J. S. C; Selvan S. T.; Bhakoo, K. K. and Tan T. T. Y. Single-Phase Dy_2O_3:Tb^{3+}Nanocrystals as Dual-Modal Contrast Agent for High Field Magnetic Resonance and Optical Imaging. Chemistry of Materials, **2011**, 23, 2439–2446.

Das, G. K.; Heng, B. C.; Ng, S.-C.; White, T.; Loo, J. S. C; D'Silva, L.; Padmanabhan, P.; Bhakoo, K. K.; Selvan S. T. and Tan T. T. Y. Gadolinium Oxide Ultranarrow Nanorods as Multimodal Contrast Agents for Optical and Magnetic Resonance Imaging. Langmuir, **2010**, 26, 8959 – 8965.

Polymer Membranes for Energy Storage and Delivery

Eric M. Davis

Materials Science and Engineering Division, National Institute of Standards and Technology, Gaithersburg, MD 20899

Developing a fundamental understanding of ion and water transport in ion exchange membranes (IEMs) is vitally important for the design and development of next-generation membrane materials for energy storage and delivery applications. One critical aspect is to understand how film thickness affects transport in these materials, since these materials are often times confined to nanometer size scales in commercial applications. Furthermore, obtaining a deeper understanding of how the addition of nanoparticles in these membranes affects water diffusion and polymer chain dynamics is highly desirable. Therefore, a key part of my future research as a new faculty would be developing and employing *in situ* characterization techniques to elucidate fundamental transport mechanisms in these IEMs in an effort to understand the relationship between material structure and device performance. With this, structural and molecular-level techniques, including small-angle neutron scattering (SANS) and time-resolved infrared spectroscopy, will play a central role in developing an understanding of how changes in material properties affect the mechanics and water/ion transport of IEMs, which ultimately affect fuel cell and battery performance.

My PhD work was primarily focused on the use of infrared spectroscopic techniques to accurately capture and measure water transport in glassy polymers. Additionally, I utilized nonequilibrium thermodynamic modeling to predict water sorption in these nonequilibrium, glassy polymers. My work on transport phenomena, infrared spectroscopy, and nonequilibrium thermodynamics will facilitate my future research endeavors, where I propose to employ these powerful tools to measure the transport of ions and water in IEMs and glassy polymers. During my post-doc, I have gained expertise in use of neutron scattering techniques, where these techniques are currently being utilized to investigate the structure of Nafion nanocomposite membranes. Understanding the location and size of these particles within the membrane is critical in understanding the mechanism by which these membranes exhibit improved performance. Additionally, my post-doc work has provided me with experience in the use of polarization modulation infrared adsorption spectroscopy (PM-IRRAS) and poroelastic relaxation indentation (PRI) to investigate water sorption and transport in polymer thin films. I believe that my PhD work on water transport in glassy polymers, as well as my post-doc work on structure and water/ion transport in IEMs, has provided me with a solid foundation in an area of research that is plagued with unanswered questions, and one with which I believe will be receiving an increasing amount of funding over the next decade.

R&D Management from Beginning to End

Jack Hipple
Principal
Innovation-TRIZ
Tampa, FL

Eldon Larsen
Professor of Engineering
Marshall University
South Charleston, WV

Extended Abstract

This workshop focuses upon managing R&D projects from beginning to end. Beginning refers to mental picture of an idea whereas end refers to solving the customers need and in the process generating profit for the organization. Using a case study, the speakers will delineate challenges involved at every stage of R&D project from a mental image of an idea to a profit making product. The talk also compares old paradigm of innovation management with the new proposed paradigm. The probability of success can be significantly improved using the new paradigm.

Successful Innovation Management

We have seen many idea management flow sheets, gate management processes, and prioritization recommendations starting with an idea that already exists, but no discussion of where the idea came from or what was necessary to generate and frame it. Moreover attention is not paid to shaping and propagation of the idea. This has been especially true over the past decade when processes such as Six Sigma came into vogue whose main thrust was to eliminate deviation. New concepts and innovation are, by their very nature, deviant! In order to generate fresh and innovative ideas we must create a new thought pattern that may challenge the existing status quo in any number of ways. These can even include, at an extreme, replacing one's current products and businesses with totally new approaches and models.

The key elements of successful commercial innovation include:

1. Continuous flow of unmet needs, both from current customers and related to core competencies of the organization ("outside in" as well as "inside out"
2. Using clear mental models of problems and potential solutions
3. Perseverance for idea propagation
4. Clear definition of the business opportunity

5. An appropriate product development and management team
6. Efficient process scale up and continued input from the market
7. Successful project management to insure implementation on schedule and within budget

We will look at each one and discuss the important issues.

1. Continuous Flow of Unmet Needs and Exploration of Uses for Core Competencies

An organization needs to have a process for how it decides what to work on. This may sound simplistic, when this is left to intuition or accident; it is unlikely that the "best" projects will be chosen.

Steps to make sure that an innovation effort is capturing the best opportunities:

- Talk to groups that are not your current customers but that might be able to use the functional capability that you have as a core competence. Customer competitors are another valuable source. The customers of your customers are another excellent source. This insures that you have awareness several levels above your product and customers and alerts you to potential long term needs and trends that may not be obvious to your direct customer or may not be readily disclosed by them.

In these discussions we want to make sure that we have a clear definition of what the "ideal result" that our customers are desiring. We can also look at parallel universes of technology and markets to explore opportunities.

2. Using Clear Mental Models of Problems and Solutions

It is one thing to "invent" a new product, but another to make sure that we understand how a potential customer is going to use the product as well as how the customer's customer is going to use their product. The days are gone when we can just make something, put it in a drum or tank car, and wait for the invoice to be paid. We must be intimately familiar with the product's use, its limitations (including regulatory issues if appropriate). We must understand the details of the physical use of the product and how it affects a customer's products and business.

It is also important to make sure that we do not have blinders on when thinking about use of our product. It is rare that there is only one way to accomplish a function. We must be constantly vigilant about competitive activities, not just from direct competitors, but from parallel approaches to meeting a customer's needs.

3. The People Side

Ideas, innovation, and projects never happen in a vacuum. They are accomplished with teams of people, all of which have different personalities and styles. Recognizing and using these pro-actively is critical to a project's success. How these are measure and used will be discussed.

4. The Business Opportunity

A new product is not a business unless we have a way of sustaining the product's sales, expanding its utility, and developing and enforcing an intellectual property portfolio. As we commercialize a new product, we must be constantly aware of, and be willing to address with resources and commitment, issues such as change in product form, packaging and shipping issues, product purity, and shelf life/stability. These issues are most adequately addressed by constant and continuous communication with the customer.

5. Product Development Teams

We frequently choose product team members based on "who's available". While this may be expedient, it may not serve the best interests of the new business development. The obvious component of team selection is function—marketing R&D, manufacturing, legal, sales, etc. but we frequently ignore the human side of team functioning. If team members spend too much time dealing with personal issues relating to other team members, that energy is not applied to successful new product launch. Tools that can assist in this include measurements such as the Myers Briggs family of instruments, the Hermann Brain Dominance instrument (HBDI), and the Kirton KAI assessment. Knowledge of these personality profiles will not change natural behavior, but can make team members aware of others' preferred styles and open discussion can eliminate hidden barriers to effective team functioning.

6. Efficient process scale up and continued input from the market

Often times, companies strive for launching the best product. It is important to realize that continued input from the market is critical in order to have that 'best product' in the market. It is necessary to efficiently scale up and launch a product sooner than later. It should be perfected using market feedback. Getting market feedback is the only way to close the R&D loop and serve your customers to the fullest.

7. Successful Project Implementation and Management

New product development does not end when the first semi-commercial sample has been OK'd by the customer. A new type of team is now required to oversee the details process design, plant construction (including all the sub issues of permitting, EPA regulations, etc.). Quick and efficient plant start up with appropriate contingency planning is also a part of this effort. Contingency planning, not only related to unforeseen plant construction and start up difficulties, but also to taking care of the customer's needs in these situations must be part of the plan.

Three keys to successful project management lead to successful project completion. These keys involve

- The importance of people
- The importance of defining and setting good objectives
- The importance of good planning

We will apply these three keys to successful project management. This is all about application, not just theory.

394219 Reverse Electrodialysis: Sustainable Energy from Hydraulic Fracturing Water Recycle

Abstract:

Due to the growing limitations in dependence on fossil fuels, attention has been turned towards searching for alternative means for harvesting energy. One emerging source is energy produced from natural salinity-gradients between freshwater and seawater. Two different processes used to accomplish this are pressure-retarded osmosis and reverse electrodialysis. Significant amounts of research are currently under way regarding pressure-retarded osmosis, however very few studies have been conducted on reverse electrodialysis due to heavy resistance in the diluate chamber of a reverse electrodialysis cell. The focus of this research is to reduce this resistance by using wafers and ion-exchange beads in order to maximize net power gain. Surrogate brackish/saltwater feed streams were used to measure overall power density.

By reducing resistance to levels previously deemed impossible, we are exploring the fundamental limits of separation technology. This research proposes an innovative resolution to high resistance levels through the use of ion exchange wafers to promote ionic movement. Here we show that such wafers have a great impact on the reverse electrodialysis process, and succeeded in achieving a sizeable gain in power density.

Traditional applications include harnessing energy from areas where freshwater bodies meet saltwater, as in an estuary. The future of reverse electrodialysis, however, lies in fracking. Produced water from a fracking well site in Oklahoma was used for power density calculations under realistic conditions. Preliminary results show that our system is capable of producing power densities of up to $10W/m^2$, which are five times higher than the current state-of-the-art. Environmental concerns will be all but eliminated if fracking can evolve to be a sort of cyclic process, where brackish fracking water is converted to energy to, in turn, power the overall operation. This poster presentation will model the power increase of electrode ionization and comment on future potential for improvement, especially in the realm of hydraulic fracturing applications.

394215 Modeling of the Aerobic Cometabolic Transformation of Chlorinated Ethenes By the Mycobacterium Elw-1

Abstract:

ELW1 is a recently isolated aerobic bacterium, in the mycobacterium family, that uses isobutene as a growth substrate. ELW1 was also discovered to have the ability to cometabolically transform substances with chemical structures similar to isobutene, such as chlorinated ethenes, into chloride ion and water. Chlorinated ethenes are notorious carcinogens, and groundwater pollutants, whose remediation has been a worldwide problem for over 40 years. Bioremediation of chlorinated ethenes by the augmentation of contaminated soils with microorganisms such as ELW1 is a proposed method of reducing contaminant concentrations in groundwater. To determine the viability of using ELW1 for groundwater remediation, an initial kinetic analysis of transformation rates must be performed. The focus of this research was to determine the rates of contaminant transformation when this organism was exposed to high concentrations of chlorinated ethenes such as trichloroethylene (TCE), *cis*-dichloroethene (c-DCE), and vinyl chloride (VC).

The ELW1 culture was grown and sustained in both batch culture and steady-state chemostat environments with isobutene as a growth substrate on a minimal growth media. Cells that were used for experiments were concentrated by centrifugation, washed with excess media and placed in a new media solution containing high concentrations of the compound to be tested. Initial rates of transformation were determined using gas chromatography to analyze 100μL samples of headspace gas over a period of one to three days to determine the rate of compound loss. The rates at which ELW1 can utilize isobutene were compared with rates of chlorinated ethene transformation, which overall quantifies the practicality of using this organism bioremediation purposes in the field.

Kinetic parameters of these cometabolic processes were determined using an analytical equation based on a modified Monod equation in order to account for decrease in cell activity over time. Results of this investigation indicate that the ELW-1 culture would be effective for the bioremediation of c-DCE and VC, but would be less effective for treated TCE.

394214 The Enzymatic Hydrolysis of Alfalfa Stalks for Use As a Biofuel Resource

Abstract:

Most biofuel resources, especially plant sources, have two key issues: farming space and slow growth. After overexpressing the sucrose phosphate synthase gene (SPS) in Medicago sativa (alfalfa) plants, large differences in phenotype and growth rate were seen in the plant. While the leaves of the alfalfa plants can be used for animal forage, the stems of the alfalfa can be fractionated into sugars that can then be fermented into bioethanols. This technique could solve the growth problem faced by bioethanol producers.

The purpose of the project was to fractionate and analyze the extractability of fermentable sugars from a control and several transgenic species that show overexpression in the SPS gene. Using a strong acid pre-treatment and enzymatic hydrolysis to break down the cellulose into glucose, the alfalfa stalks were fractionated then analyzed using ultraviolet spectroscopy with anthrone as an indicator for glucose content. Two of the three tested transgenic species were shown to have more glucose content per gram than the control species (2-4% more glucose/gram). A further development of the overexpression technique can lead to faster growth, increased concentration of fermentable carbohydrates, and larger crop yields.

UNIVERSITY OF SOUTH CAROLINA
College of Engineering and Computing

SAGE Power through Engineering

Synthesis and Characterization of FeAlCr via High Throughput Techniques

Randy Fang
Jonathan Kenneth Bunn and Dr. Jason Hattrick-Simpers
Department of Chemical Engineering
University of South Carolina
AICHE National Conference 2014
11/17/2014

UNIVERSITY OF SOUTH CAROLINA
Honors College

1

Fukushima Daiichi Nuclear Disaster

http://digginginthe clay.files.wordpress.com/2011/03/5525887859_e1934af238_o_d1.jpg

https://coto2.files.wordpress.com/2011/03/2-fuel-pellet-assembly.jpg

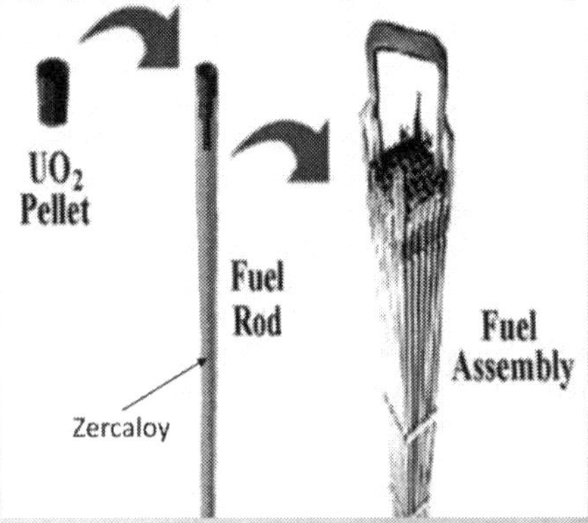

2

FeAlCr Alloy

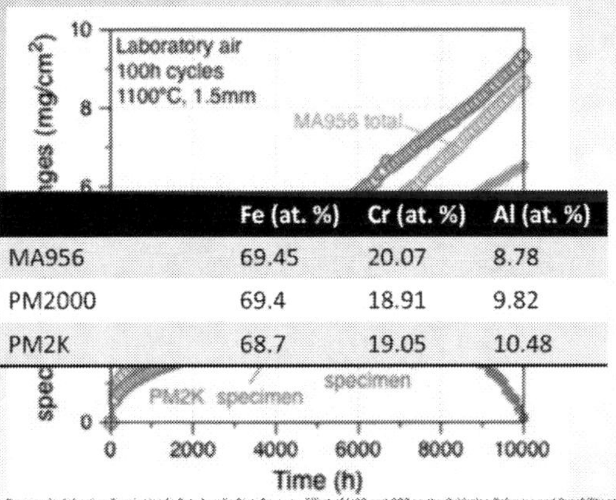

	Fe (at. %)	Cr (at. %)	Al (at. %)
MA956	69.45	20.07	8.78
PM2000	69.4	18.91	9.82
PM2K	68.7	19.05	10.48

Dryepondt, Sebastien Rouaix-Vande Put, Aurelie Pint, Bruce a., *Effect of H2O and CO2 on the Oxidation Behavior and Durability at High Temperature of ODS-FeCrAl*, Oxidation of Metals, Vol 79,2003

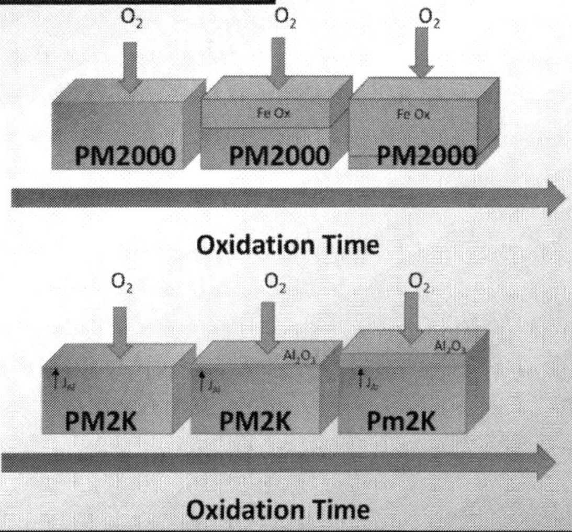

3

Introduction to High Throughput Techniques

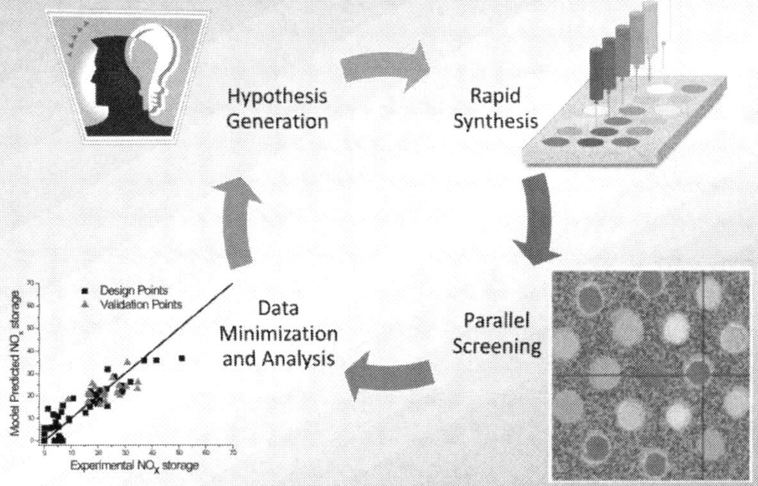

4

Research Objective

- **Implementing established high-throughput methods for FeAlCr high temperature oxidation resistance screening**
- Non-destructive characterization
- Gain comprehensive understanding of:
  - Time dynamics of oxide formation at high temperature
  - Role of metallurgical phase in promoting growth and continued growth of passivating oxide
- Identification of promising compositions for passivating oxide growth

5

High Throughput Approach to Explore the Ternary Phase Diagram

Qu Materials Science and Engineering A Vol 562 (2013)

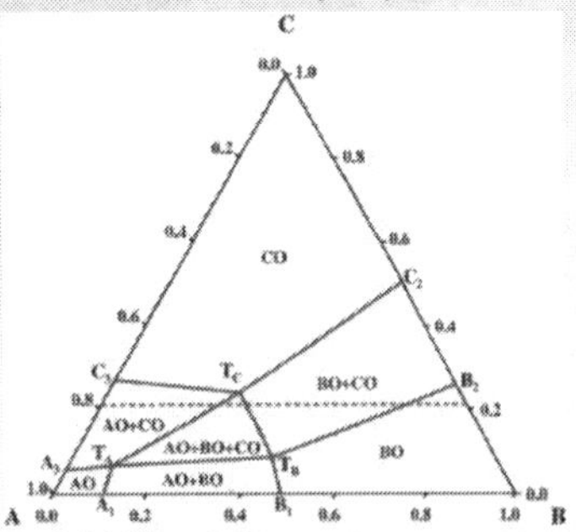

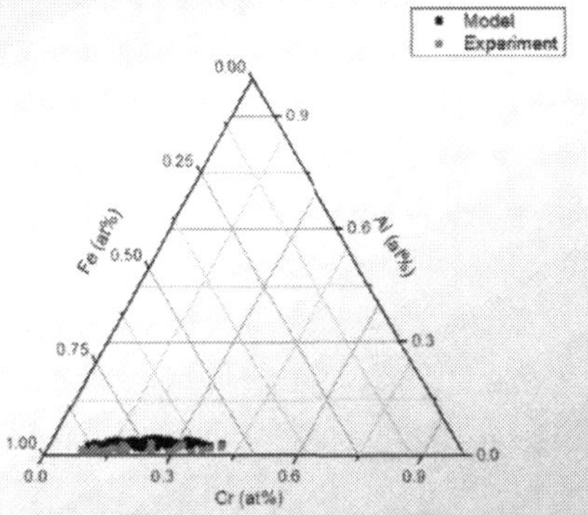

J. K. Bunn, C. J. Metting, and J. Hattrick-Simpers, "A Semi-Empirical Model for Tilted-Gun Planar Magnetron Sputtering Accounting for Chimney Shadowing", JOM, invited paper, Accepted January 2014 Special Issue

6

Sputtering Geometry and Sample Locations

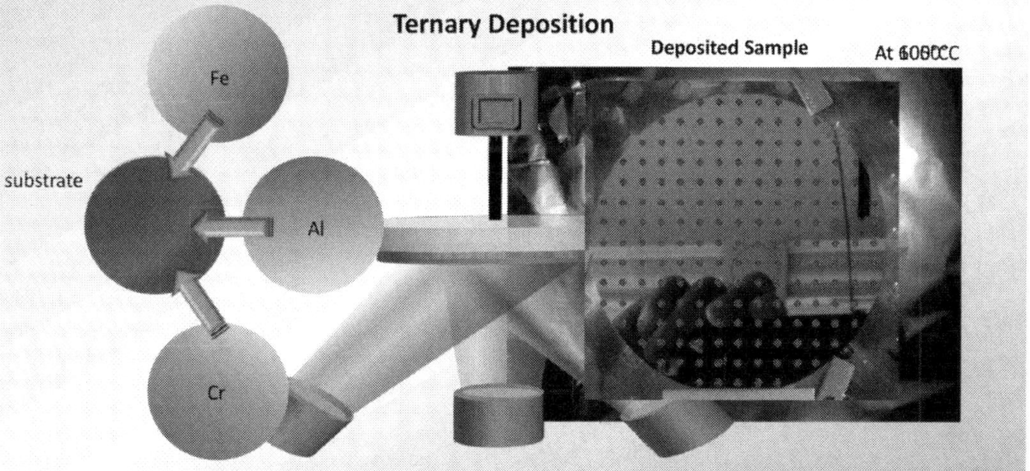

7

Magnetron Sputtering

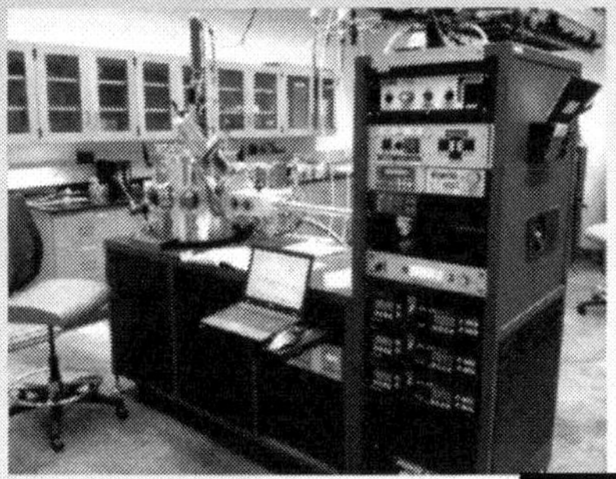

- 5 Guns (only used 3)
- DC/RF power sources
- Ultra high vacuum
- Deposition temperature at room temperature

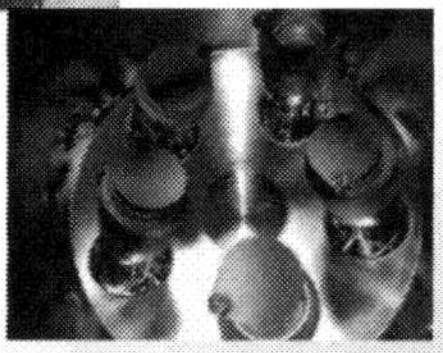

8

Experimental Design for High Temperature

Sample fabrication
-12 FeAlCr samples prepared on Al_2O_3 substrate and Ti with adhesion layer
-1 micron nominal thickness

Heat treatment
- Tube furnace treated
- Treated at 1050°C
- Air quenched

After deposition

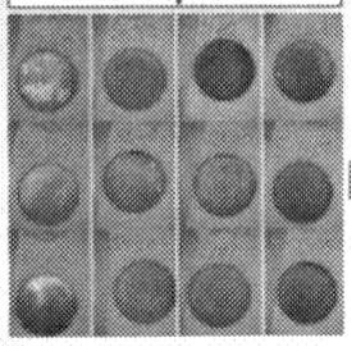

Cycled 7 times for cumulative heat treatments at 1050°C for:

-5 min	-10 min
-20 min	-40 min
-60 min	-120 min
-240 min	-360 min

After heat treatments

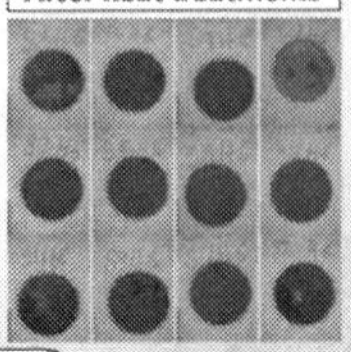

Characterization
-SEM/EDS
-XRD

Characterization Techniques
- **Raman**: Fe_2O_3, Fe_3O_4, and Al_2O_3
- **XRD**: FeCrAl, Surface Oxides, and the substrate
- **Optical Microscopy**

9

Oxide Formation at 5 minutes, 1050°C

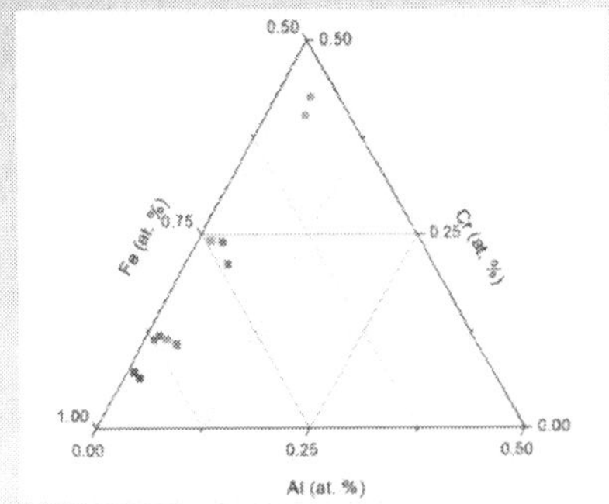

Blue: mostly Fe_2O_3, some Fe_3O_4
Red: mostly Fe_3O_4, some Fe_2O_3, some α-Al_2O_3
Green: very little Fe_2O_3, some α-Al_2O_3, some Fe_3O_4

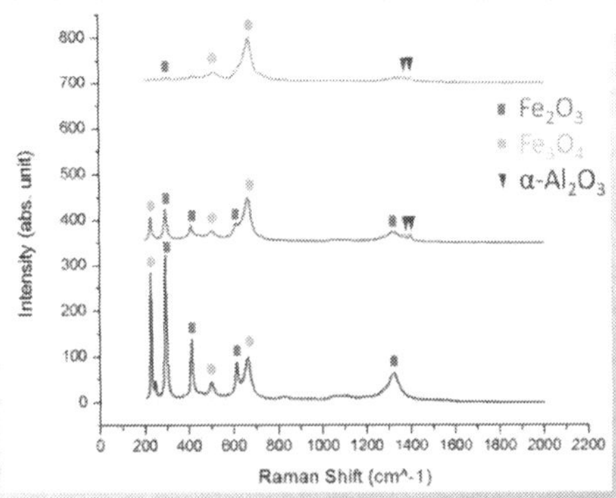

- No α-Al_2O_3 growth was observed when the composition is Fe heavy

10

Oxide Formation at 360 minutes, 1050˚C

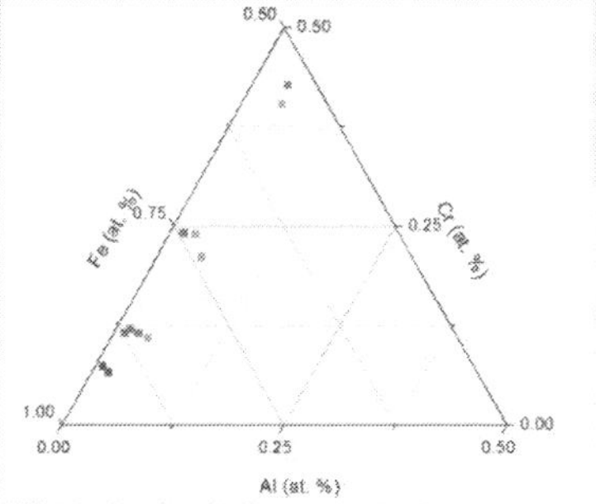

Blue: some Fe_2O_3 and Fe_3O_4, very little α-Al_2O_3
Red: some α-Al_2O_3, some Fe_3O_4
Green: majority α-Al_2O_3, some Fe_3O_4

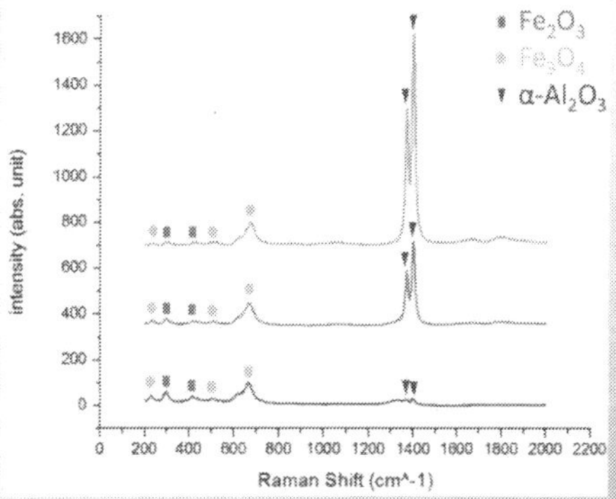

- Large amount of α-Al_2O_3 growth was observed at the Cr heavy region, indicating that Cr promotes the growth of the alumina oxides

Experimental Design at Low Temperature

Sample fabrication
-FeAlCr samples prepared on Si substrate with Ti diffusion barrier
-1 micron nominal thickness

Cycled 6 times for cumulative heat treatments 1 hour each (with the exception of 600°C):

-room temperature
-200°C
-300°C
-400°C
-500°C
-600°C (for 2hr)

Characterization
-*in-situ* GIXRD

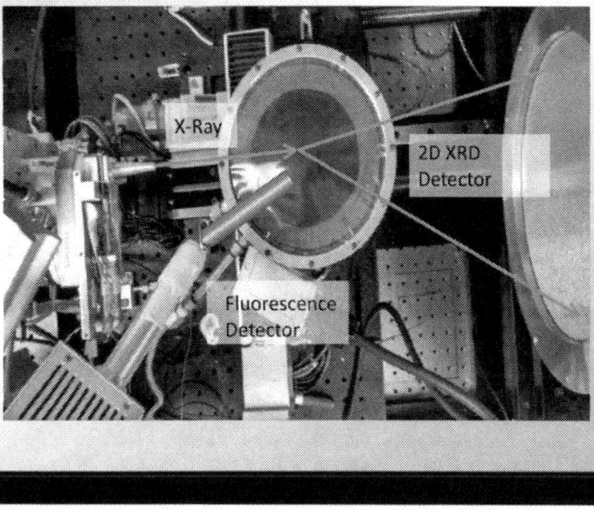

Oxide Growth Study through *in-situ* XRD

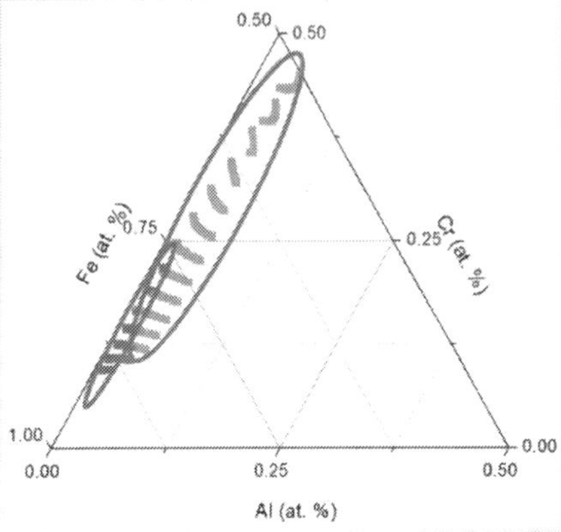

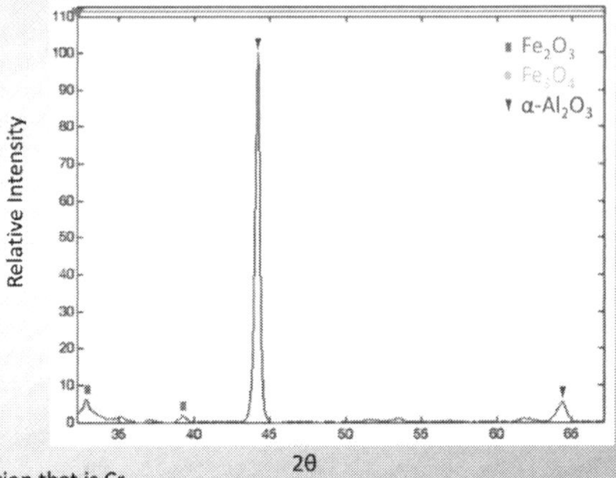

- Iron oxides are suppressed when the composition that is Cr rich (small amount are observed)

Comparison of Trends Between Low and High Temp Oxidation

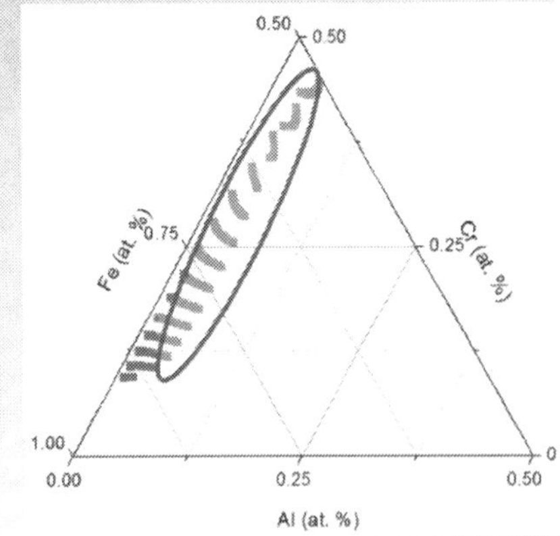

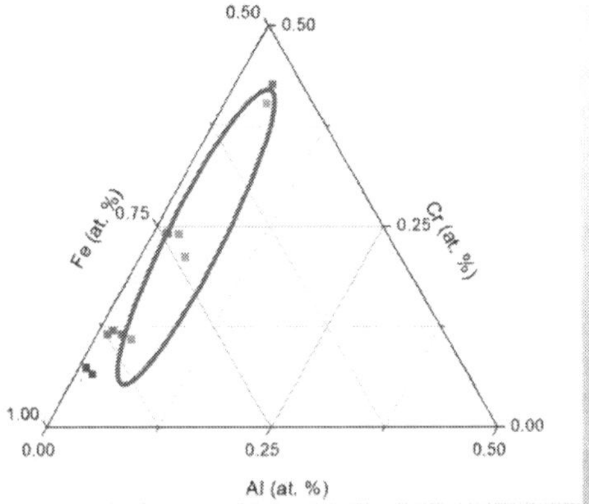

- XRD and Raman matched up where the more Al and Cr rich region are a lot easier to form α-Al_2O_3

Summary

- Probed the temporal effect of temperature on the metallurgic and oxide phase transitions
- A method for the **initial high-throughput and non-destructive screening** of FeCrAl thin film was developed
- **Cr promotes** α-Al_2O_3 growth on the surface of the samples to form a protective layer preventing further development of other oxides
  - This indicates that Cr rich FeCrAl are more favored to operate under high temperature conditions
- Oxidation studies indicate that severe oxide growth was not observed until 400°C
- The methodology can be applied to more complicated systems for high temperature oxidation resistant materials.
- Detailed studies of **oxidation and phase transitions of single compositions at different temperatures** will refine the results

15

Acknowledgement

- **Dr. Jason Hattrick-Simpers**
- **Jonathan Kenneth Bunn**
- **SmartState Center for Strategic Approaches to the Generation of Electricity**
- **USC Honors College (Science Undergraduate Research Fellowship Program)**
- **University of South Carolina**
  - **College of Engineering and Computing Science**

This project was funded in part by the South Carolina Honors College

16

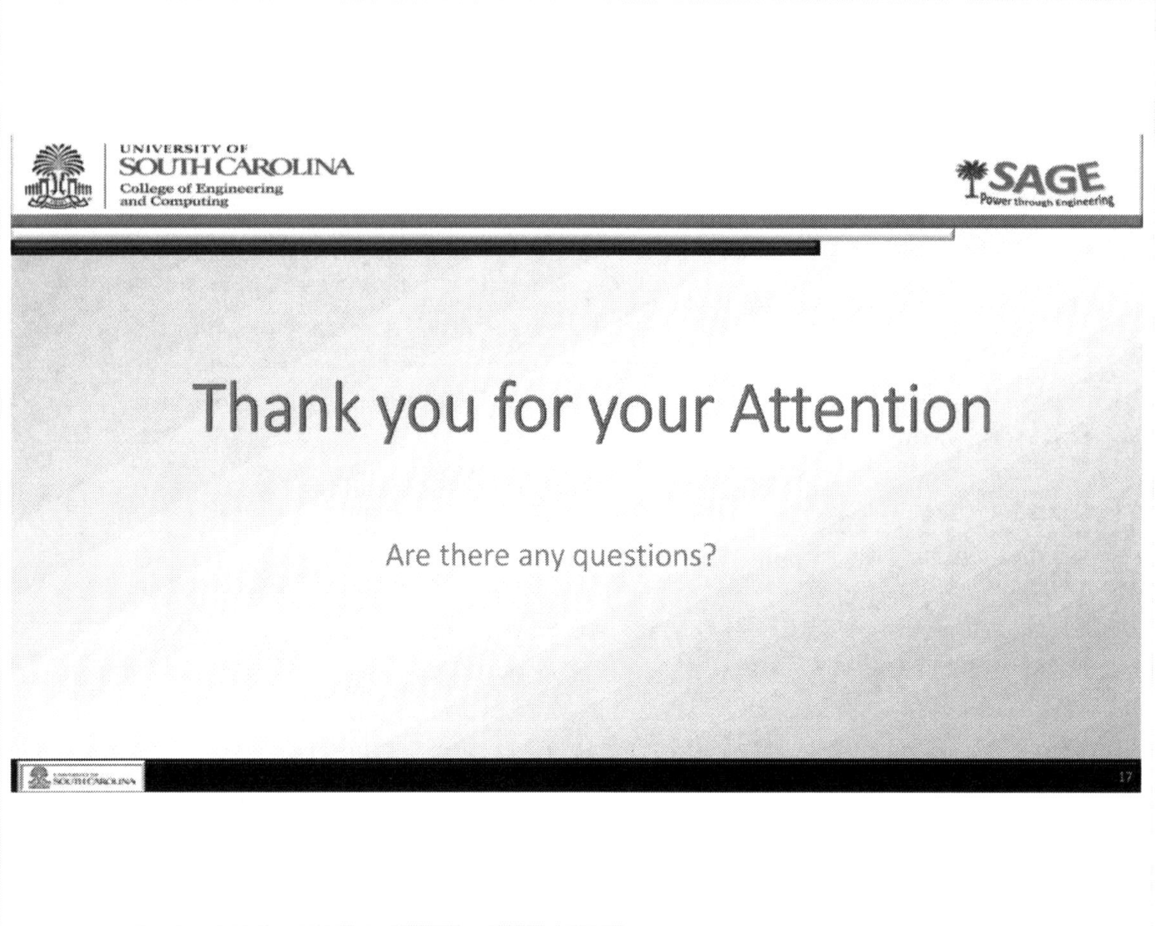
UNIVERSITY OF
SOUTH CAROLINA
College of Engineering
and Computing
SAGE
Power through Engineering
Thank you for your Attention
Are there any questions?
17

Crossing Borders and Communities for Healthy Water

Joshua Gomez, Meng Zhou, Dr. Shuguang Deng

Chemical and Materials Engineering Department, New Mexico State University, Las Cruces, New Mexico, 88003

In Partnership with:

NM Water Resources Research Institute

NMSU Chemical Engineering Department

The New Mexico Alliance for Minority Participation

New Mexico Environment Department

Border Partners (Non-profit)

EPA 2020

National Science Foundation

New Mexico State University

Puerto Palomas, Chihuahua, MX

- Located just south of Columbus, NM.
- Population of 4,688.
- Arsenic contamination (0.030-0.1 mg/L).
  - Horrible Carcinogen.

- Fluoride contamination (5-40 mg/L).
  - Bone Fluorosis and neurological damage.
- Concentration depends entirely on well water.
  - Naturally occurring concentrations of arsenic and fluoride exceed the permissible limits of Mexico and the U.S.

New Mexico State University

New Mexico Transboundary Aquifers

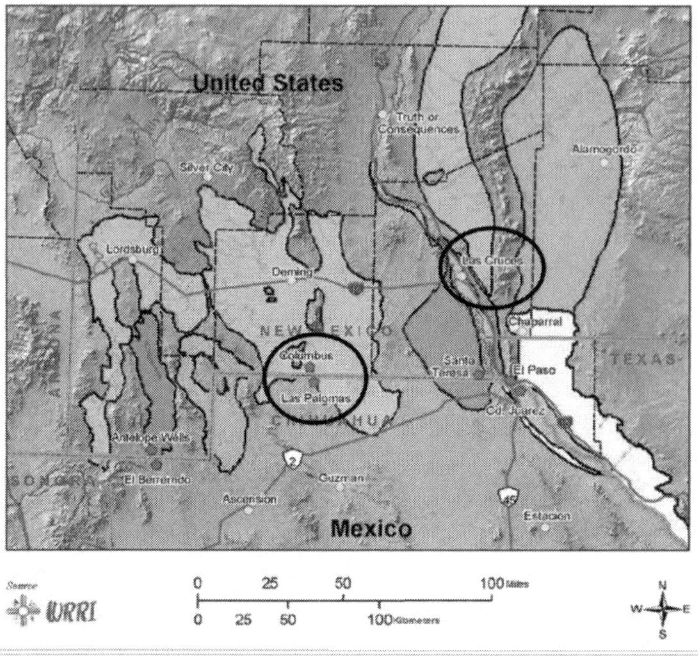

Proposed Solution

- Homemade water filter
  - Low cost <$15.00
  - Easy to assemble
  - Removes arsenic
  - Removes fluoride
  - Utilizing activated carbon (AC)
  - Utilizing activated alumina(AA)
    - Iron coated
  - Improves the taste

Filter Designs

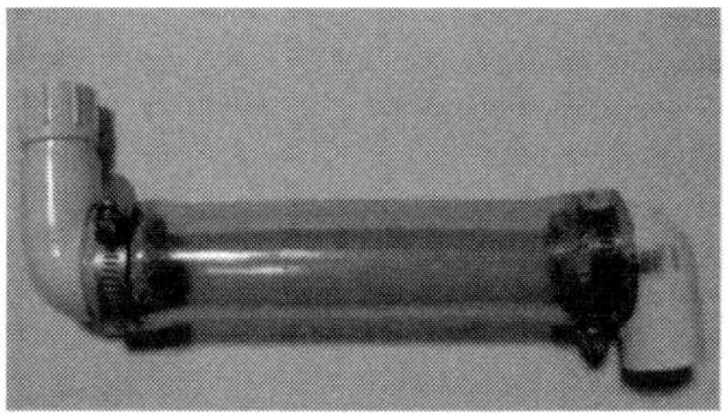

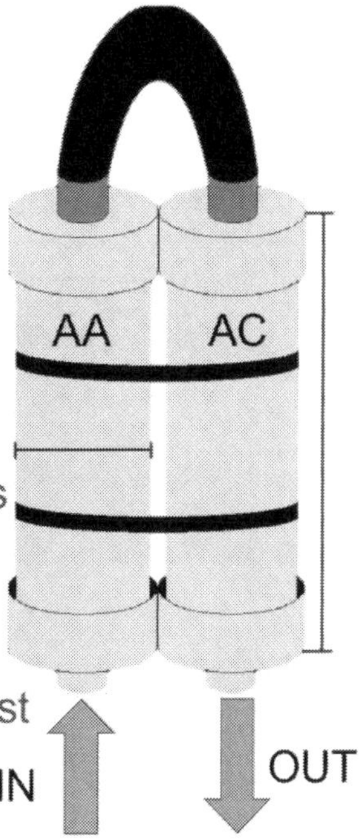

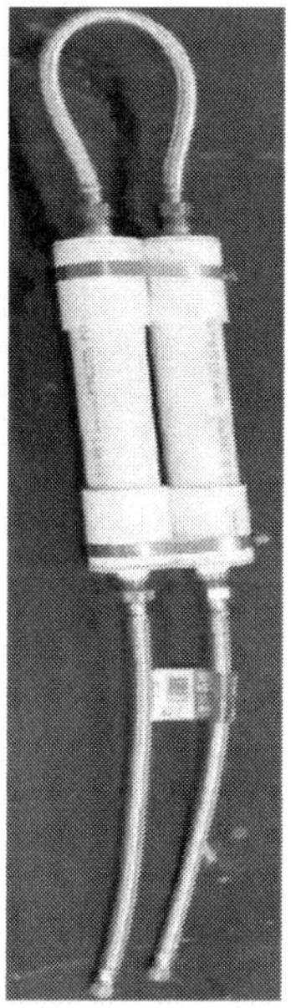

- 1 filter of the final design costs $5.00 to make.
- The Activated Alumina costs $1/kg.
- The Activated carbon used cost $35/lb.

New Mexico State University

Our Approach

NMSU demonstrated to Palomas science teachers how to assemble a simple water filter.

Science teachers Juan Carlos Huerta and Joel Carreon pose with the filter they assembled for installation at the Palomas primary school.

New Mexico State University

What's Working

Hall Environmental Laboratory, Albuquerque

CONTAMINANT	RESULTS	STANDARDS					
		U.S.	Exceeded	%	Mexico	Exceeded	%
Fluoride							
Unfiltered	4.90 mg/L	4.00 mg/L	Y	22.5	1.50 mg/L	Y	226.7
Filtered	0.240 mg/L	4.00 mg/L	N	-	1.50 mg/L	N	-
Arsenic							
Unfiltered	0.030 mg/L	0.01 mg/L	Y	200.0	0.02 mg/L	Y	20.0
Filtered	ND	0.01 mg/L	N	-	0.02 mg/L	N	-

ND = Not Detected

New Mexico State University

Optimization of Algae Derived Activated Carbon (AC)

- Low cost alternative to commercial activated carbon.
- More effective adsorbent than commercial carbon.
- Chlorella algae is used as a carbon source.
- Potassium Hydroxide and Iron (II) Chloride
  - Activating agent.
- Carbonization times of 500° C, 600° C, and 700° C.
- Shaker experiments to determine most effective adsorbent for Arsenic(III).

New Mexico State University

Carbon Experimental data

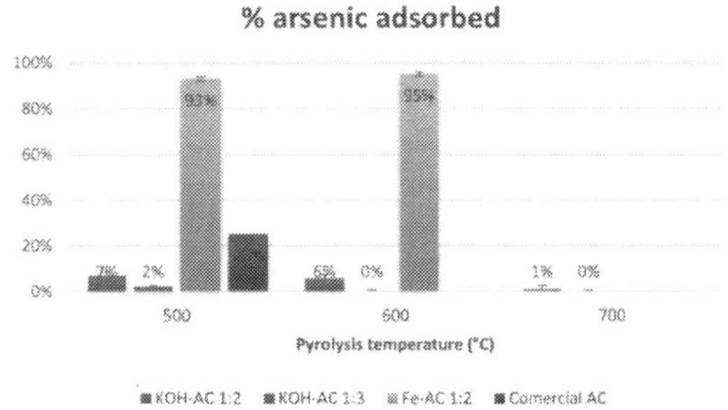

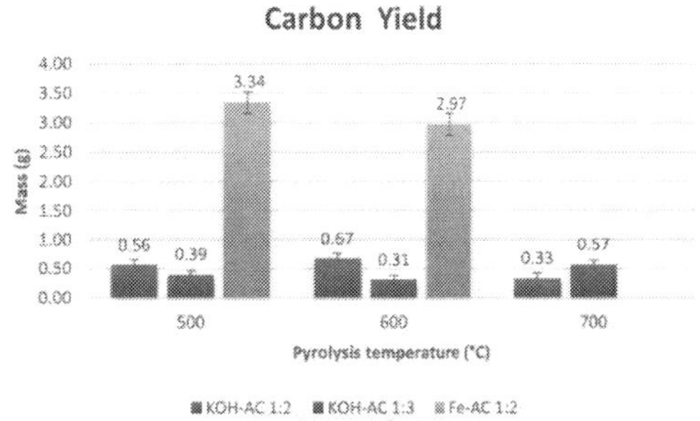

adsorption capacity (mg/g)			
	500°C	600°C	700°C
KOH-AC 1:2	0.082	0.067	0.010
KOH-AC 1:3	0.024	0.000	0.000
Fe-AC 1:2	1.143	1.168	-
Comercial AC	0.275	-	-

$$capacity = \frac{(InitialConcentration - FinalConcentration)Volume}{Mass\ of\ adsorbent}$$

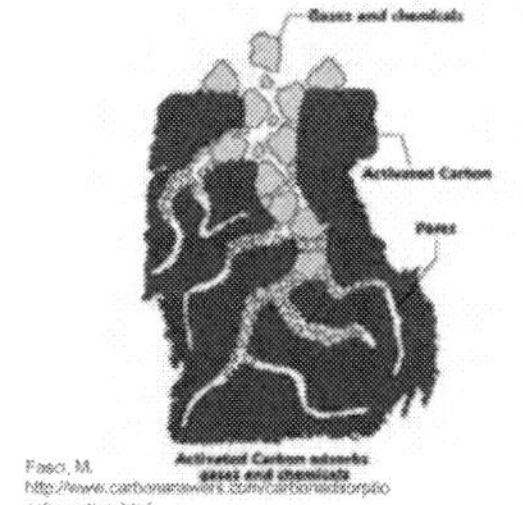

New Mexico State University

X-Ray Difraction Pattern

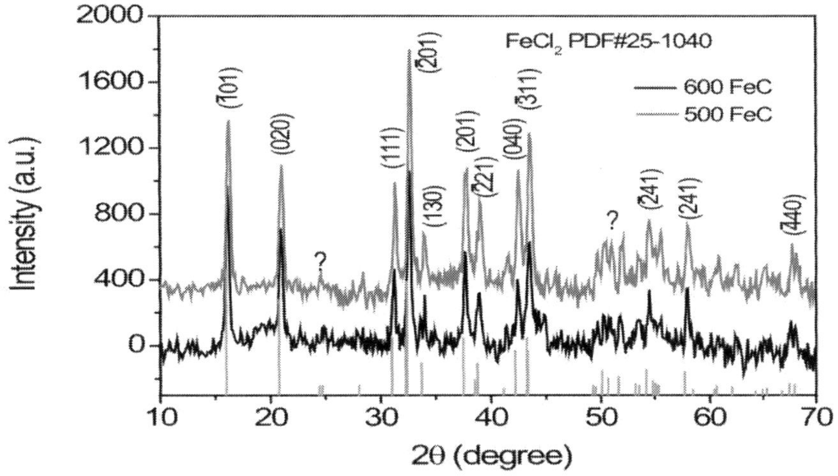

New Mexico State University

Scanning Electron Microscope Imaging

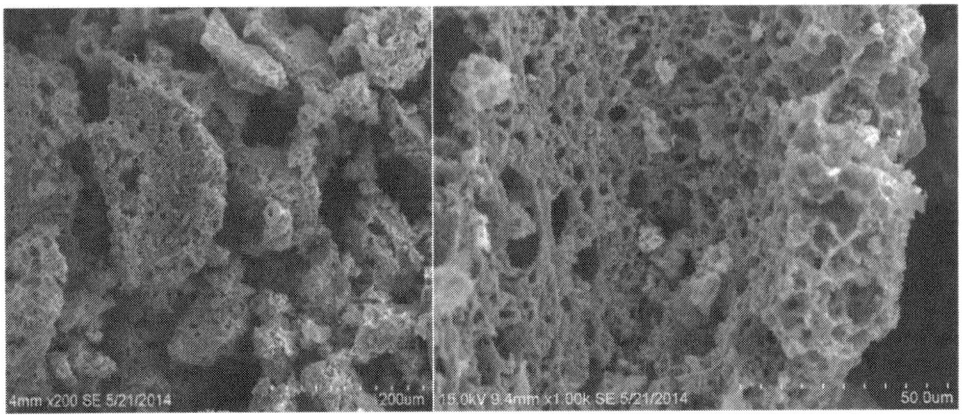

New Mexico State University

Unwashed FeCl(II) Activated Carbon Textural Properties

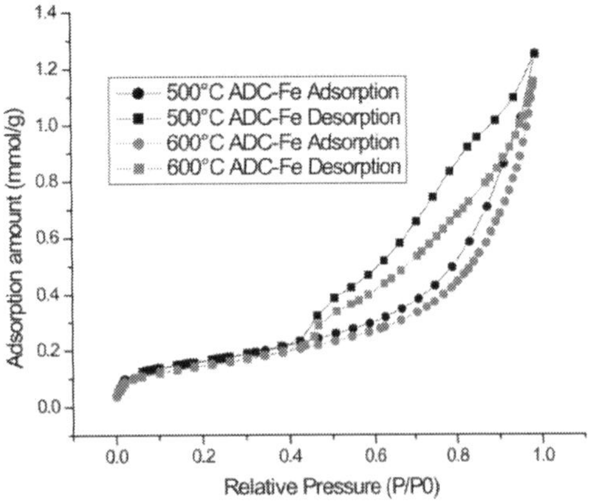

N_2 adsorption and desorption isothermal @ 77K for unwashed samples.

	500 ° C FeCl2/ADC	600 ° C FeCl2/ADC
BET surface area (m^2/g)	12.9893	11.6336
Average pore size (nm)	8.6408	6.5769
Pore volume (cm^3/g)	0.0394	0.0476

New Mexico State University

Washed FeCl(II) Activated Carbon Textural Properties

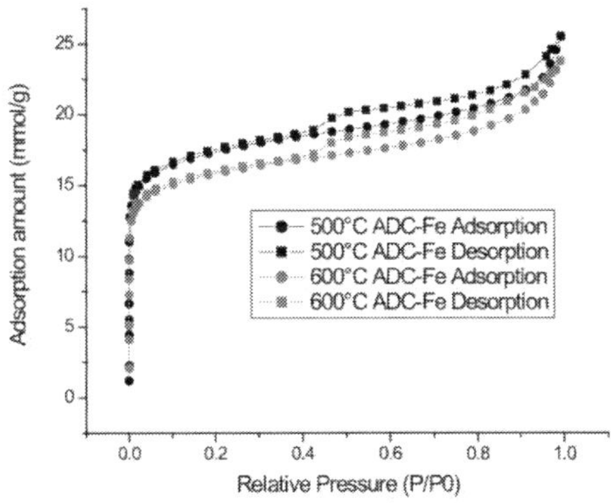

	500 ° C ADC (after washing)	600 ° C ADC (after washing
BET surface area (m^2/g)	1326.0368	1208.1831
Average pore size (nm)	7.8704	7.9340
Pore volume (cm^3/g)	0.3362	0.3305

N2 adsorption and desorption isothermal @ 77K for washed samples.

New Mexico State University

Shaker Experiments

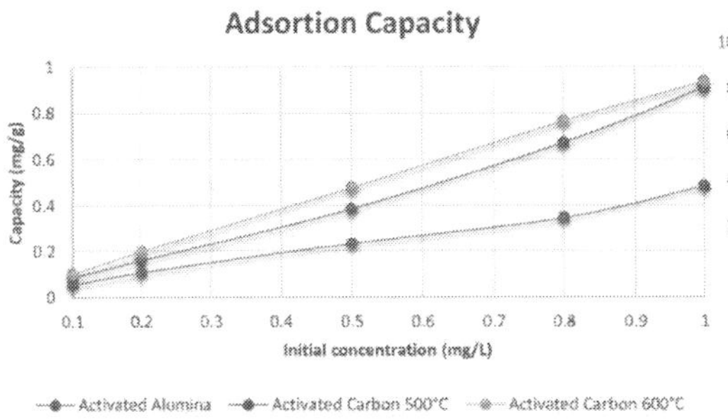

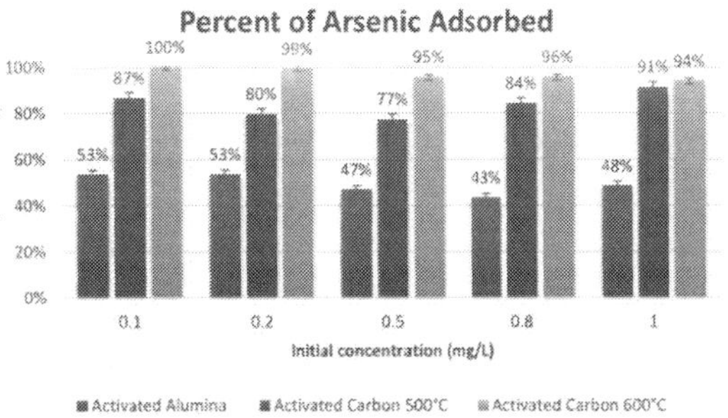

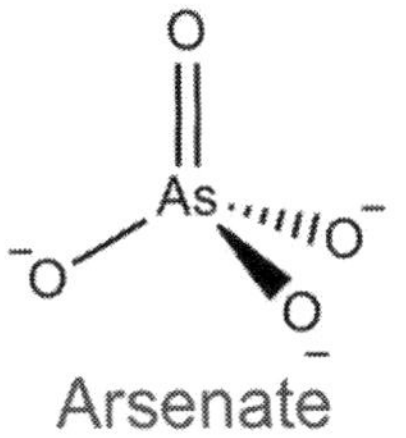

Arsenate

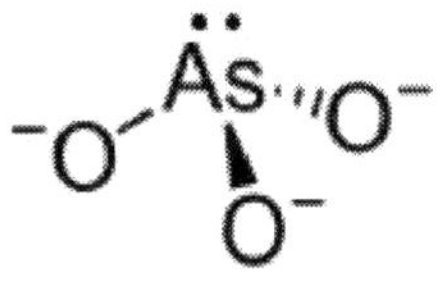

Arsenite

New Mexico State University

Conclusions

- Filters are still being made in Palomas by the teachers today and will continue to be made.
- The results of the shaker experiments and textural property data shows that the carbon activated with Iron (II) Chloride is a viable replacement.

New Mexico State University

Next Steps

- Further Development of the filter.
- Reach out to other communities.
- Advance research with optimized carbon.

New Mexico State University

Acknowledgments

- Dr. Shuguang Deng, Chemical & Materials Engineering dept.
- Meng Zhou, Chemical & Materials Engineering dept.
- Dr. David A. Rockstraw, Chemical & Materials Engineering dept.
- Erin Ward, Water Resources Research Institute.
- Mark Chidester, Chemical & Materials Engineering dept.
- Peter Edmunds, Border Partners.

NSF HRD # 1305011

Deng's Renewable Energy and Advanced Materials

New Mexico State University

Questions?

Email: jgomez38@nmsu.edu

New Mexico State University

394186 Ferroelectric BTO on Si (001) for High-Efficiency Solar Cell Heterostructures

Abstract:

Ferroelectric oxides such as barium titanate (BTO) have recently become a topic of interest in relation to thin-film solar cell production due to the material's stability at a wide range of temperatures, its spontaneous polarization, and the potentially low-cost of production. Integration of BTO on silicon could enable the creation of highly efficient solar cells, as BTO's ferroelectric properties encourage the detachment of photo excited electrons. We have previously demonstrated successful integration of BTO on silicon carbide (SiC) substrates, using a magnesium oxide (MgO) heteroepitaxial layer to enable single-crystal orientation of the BTO film deposited on SiC.[1] BTO has also been successfully grown on silicon substrates, but with the use of strontium titanate (STO) and MgO buffer layers to resist formation of SiO_2.[2] Integration of BTO on Si with a single MgO interface layer would decrease interfacial losses in a final solar cell device. Based on previous work with SiC, we believe a single interface can be possible by controlling the surface interactions during film growth.

In this research, MgO is grown on Si (001) substrates in an ultrahigh vacuum (UHV) environment using a number of growth methods including molecular beam epitaxy (MBE) and sputtering. In comparing differences in surface structure using reflection high-energy electron diffraction (RHEED) and surface chemistry using X-ray photoelectron spectroscopy (XPS), we can understand surface interactions between Si, Mg, and O. By understanding the relationship between atomic level interactions and ultimate film characteristics, we can engineer the most effective process for BTO integration on Si. The substrates are cleaned using wet chemicals to create a hydrogen-terminated surface that resists contamination from laboratory exposure. XPS and RHEED are used to verify the lack of contamination on the surface and proper crystallographic orientation. Contaminant levels of 7.61 % oxygen (0.15 O/Si ratio) and 18.0% carbon (0.31 C/Si ratio) are the most successful results that have thus far been attained. XPS and RHEED will serve to confirm proper growth of MgO (001) on Si and determine thickness of the MgO film. BTO is then grown on the correctly-oriented MgO film using MBE, as has already been demonstrated on MgO/SiC substrates.[1] This presentation will discuss cleaning and growth conditions for successful production of high quality BTO on Si using the single MgO interface layer.

393873 Natural Gas, the Bridge Fuel

Abstract:

Medium and heavy duty trucks and buses produce almost 18 percent of all carbon emissions from transportation. Alternative energy sources need to be used on a larger scale to reduce these emissions, as well as to reduce dependence on foreign oil. Natural gas is the future of fuel in large vehicles. This paper tackles different issues surrounding the usage of natural gas as the main fuel for natural such as public awareness, public policy, existing infrastructure, consumer choice and controversies surrounding hydraulic fracturing. Current pilot programs, in which public city buses are using natural gas as the fuel of their choice, will be analyzed to determine the best way to implement these programs in other areas of the United States as well.

397811 The Federal Role in Fostering an Innovative U.S. Energy Ecosystem

Abstract:

Although renewable sources of energy – solar, wind, hydro, and geothermal – have been around for over one hundred years, they only generate 12.5 percent of electricity in the United States today. Renewable technology has advanced in cost, reliability, performance, and efficiency, but not enough to allow renewables to be a large contributor to electricity generation and widely used in the U.S. Renewables need continued innovation and progression to become more efficient, more reliable, and more cost competitive with fossil fuels, as well as expanding the possibilities of their use. Increased R&D funding for renewable sources is the key to advancing the technology and giving renewables the tools it needs to provide a larger portion of electricity in the U.S.

With renewables generating a larger portion of electricity, the U.S. can reach environmental sustainability, energy security, and a new economic opportunity. In reference to reaching environmental sustainability, renewables do not emit carbon and greenhouse gases, which is important for a cleaner future and mitigating climate change. Because the "fuel" for renewables includes the sun and wind, electricity generation from these sources does not depend on obtaining fossil fuels inside or outside of our country. They also extend the lifetime of these resources to be used for base load power, providing energy security to the U.S. Finally, if the U.S. is able to quickly and effectively develop competitive renewable technologies and deploy them into the electricity sector, it can provide competitive renewable technologies to the global community, boosting the economy and maintaining world leadership.

Although renewables have the capability to provide many benefits for the U.S., they face obstacles in receiving the funding they need to be widely used. These obstacles include climate change skeptics, the fossil fuel industry, the polarized political environment, and having to cut funding for other areas such as education and defense in order to allocate more money to R&D. The demand for renewables has also been relatively weak because electricity generated by renewables is no different than electricity generated by fossil fuels when the consumer uses it. The main way for renewables to be competitive with fossil fuels is through the cost and since wind is the only renewable that is cost competitive with fossil fuels, a strong customer demand has not been present to spur large private sector investment in renewable technologies.

In order for renewables to accomplish necessary advancements and contribute a larger portion of electricity to the U.S., financing to R&D for the technologies needs to be increased through government funding, capital generated through a carbon tax, public-private partnerships, and incentives for investment in the technology and its R&D. A commitment to innovating renewables in order to reach environmental sustainability, energy security, and a new economic opportunity is necessary from governmental officials to make the tough decisions advocating for widespread use of renewables in the U.S.

397856 A Biowall for Improving Indoor Air Quality

Abstract:

The Biowall is a living plant wall that filters air in the HVAC system of a home or small commercial building. This new technology improves indoor air quality be removing Volatile Organic Compounds and has the potential for energy savings by reducing the need for heating and cooling large amounts of outside air in a building. The presentation provides an overview of biofiltration for buildings and discusses the new research being conducted in this emerging field.

Impact of Future Energy on Water-Food-Energy Nexus

AIChE 2014 Annual Meeting, Atlanta, GA

17 November 2014

Joe Powell

Joseph B. Powell, Ph.D.
Shell Chief Scientist -Chemical Engineering

1

DEFINITIONS AND CAUTIONARY NOTE

The companies in which Royal Dutch Shell plc directly and indirectly owns investments are separate entities. In this presentation "Shell", "Shell group" and "Royal Dutch Shell" are sometimes used for convenience where references are made to Royal Dutch Shell plc and its subsidiaries in general. Likewise, the words "we", "us" and "our" are also used to refer to subsidiaries in general or to those who work for them. These expressions are also used where no useful purpose is served by identifying the particular company or companies. "Subsidiaries", "Shell subsidiaries" and "Shell companies" as used in this presentation refer to companies over which Royal Dutch Shell plc either directly or indirectly has control. Companies over which Shell has joint control are generally referred to "joint ventures" and companies over which Shell has significant influence but neither control nor joint control are referred to as "associates". In this presentation, joint ventures and associates may also be referred to as "equity-accounted investments". The term "Shell interest" is used for convenience to indicate the direct and/or indirect (for example, through our 23% shareholding in Woodside Petroleum Ltd.) ownership interest held by Shell in a venture, partnership or company, after exclusion of all third-party interest.

This presentation contains forward-looking statements concerning the financial condition, results of operations and businesses of Royal Dutch Shell. All statements other than statements of historical fact are, or may be deemed to be, forward-looking statements. Forward-looking statements are statements of future expectations that are based on management's current expectations and assumptions and involve known and unknown risks and uncertainties that could cause actual results, performance or events to differ materially from those expressed or implied in these statements. Forward-looking statements include, among other things, statements concerning the potential exposure of Royal Dutch Shell to market risks and statements expressing management's expectations, beliefs, estimates, forecasts, projections and assumptions. These forward-looking statements are identified by their use of terms and phrases such as "anticipate", "believe", "could", "estimate", "expect", "goals", "intend", "may", "objectives", "outlook", "plan", "probably", "project", "risks", "schedule", "seek", "should", "target", "will" and similar terms and phrases. There are a number of factors that could affect the future operations of Royal Dutch Shell and could cause those results to differ materially from those expressed in the forward-looking statements included in this presentation, including (without limitation) (a) price fluctuations in crude oil and natural gas; (b) changes in demand for Shell's products; (c) currency fluctuations, (d) drilling and production results; (e) reserves estimates; (f) loss of market share and industry competition; (g) environmental and physical risks, (h) risks associated with the identification of suitable potential acquisition properties and targets, and successful negotiation and completion of such transactions; (i) the risk of doing business in developing countries and countries subject to international sanctions; (j) legislative, fiscal and regulatory developments including regulatory measures addressing climate change; (k) economic and financial market conditions in various countries and regions; (l) political risks, including the risks of expropriation and renegotiation of the terms of contracts with governmental entities, delays or advancements in the approval of projects and delays in the reimbursement for shared costs; and (m) changes in trading conditions. All forward-looking statements contained in this presentation are expressly qualified in their entirety by the cautionary statements contained or referred to in this section. Readers should not place undue reliance on forward-looking statements. Additional risk factors that may affect future results are contained in Royal Dutch Shell's 20-F for the year ended December 31, 2012 (available at **www.shell.com/investor** and **www.sec.gov**). These risk factors also expressly qualify all forward looking statements contained in this presentation and should be considered by the reader. Each forward-looking statement speaks only as of the date of this presentation, 17 Novembe2014. Neither Royal Dutch Shell plc nor any of its subsidiaries undertake any obligation to publicly update or revise any forward-looking statement as a result of new information, future events or other information. In light of these risks, results could differ materially from those stated, implied or inferred from the forward-looking statements contained in this presentation.

We may have used certain terms, such as resources, in this presentation that United States Securities and Exchange Commission (SEC) strictly prohibits us from including in our filings with the SEC. U.S. Investors are urged to consider closely the disclosure in our Form 20-F, File No 1-32575, available on the SEC website www.sec.gov. You can also obtain these forms from the SEC by calling 1-800-SEC-0330.

 2

FUTURE ENERGY & SUSTAINABILITY

- Global demand for energy will double by 2050.
- Energy for Mobility is a key driver for prosperity.
- Stakeholder needs: more energy, secure supply, responsible stewardship .
- Result: a Mosiac of energy, fuel and vehicle options!

World population

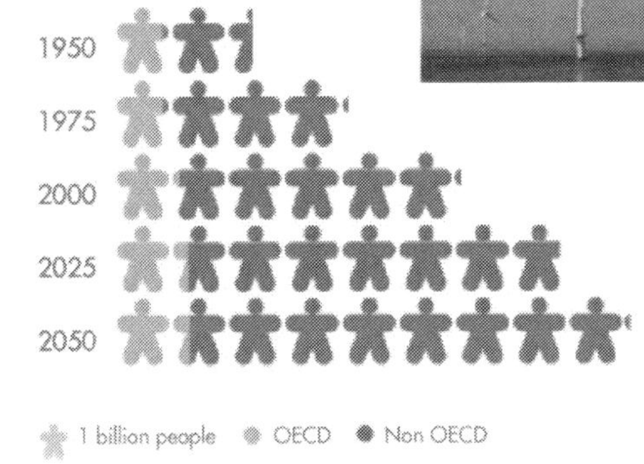

Source: UN Population Division
Copyright of Royal Dutch Shell plc

MOUNTAINS
ENERGY – Policy driven

www.shell.com/scenarios

- Moderate economic growth dampens energy demand and prices
- Supply side investments in big projects with long time horizons
- Urban planning leads to compact cities and fuel-efficient vehicles
- The Gas Backbone – the shale revolution
- Nuclear renaissance
- Renewables sluggish
- 'Clean' energy: CCS takes off by 2030
  - Electricity decarbonised by 2060

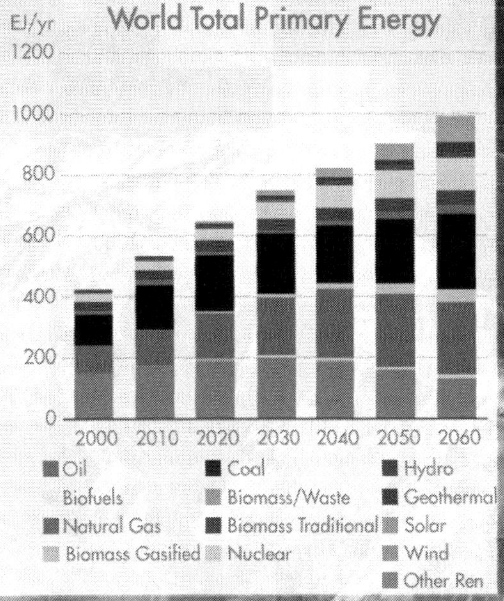

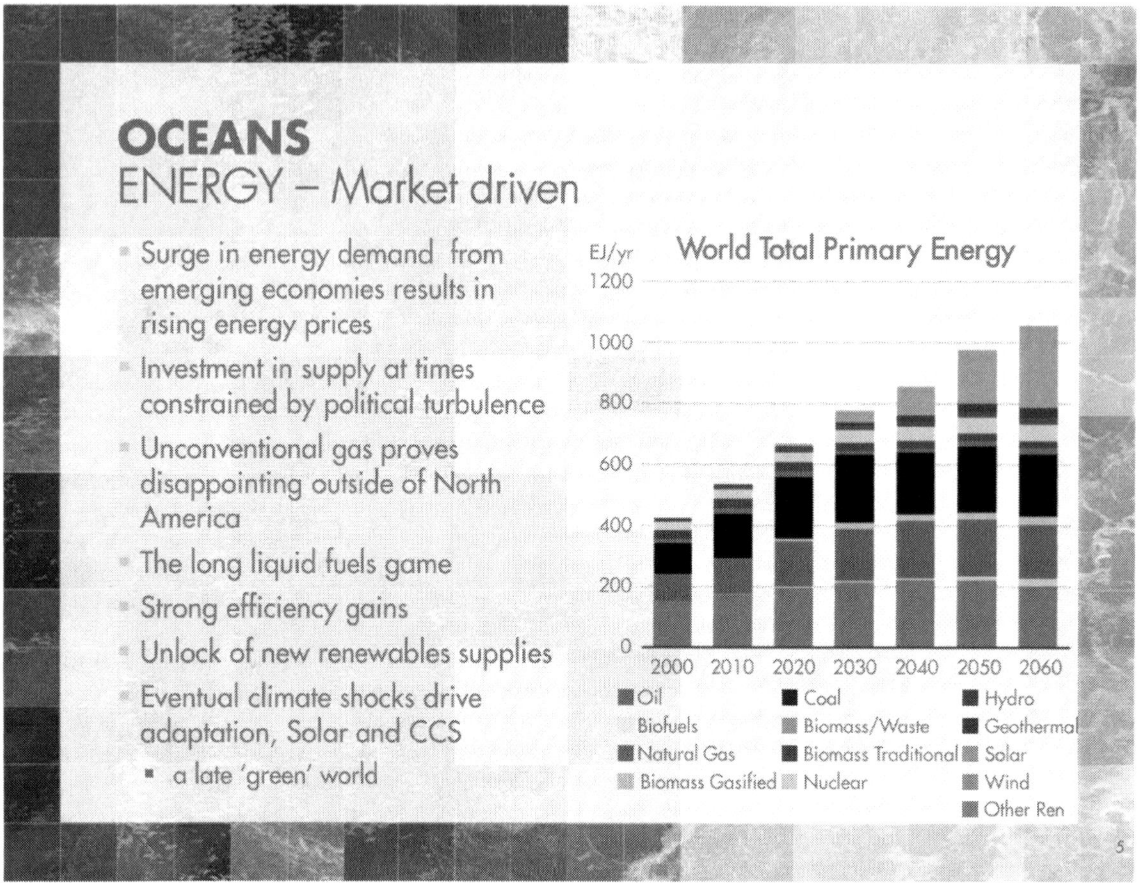
OCEANS
ENERGY – Market driven
Surge in energy demand from emerging economies results in rising energy prices
Investment in supply at times constrained by political turbulence
Unconventional gas proves disappointing outside of North America
The long liquid fuels game
Strong efficiency gains
Unlock of new renewables supplies
Eventual climate shocks drive adaptation, Solar and CCS
a late 'green' world
EJ/yr
World Total Primary Energy
1200
1000
800
600
400
200
0
2000 2010 2020 2030 2040 2050 2060
Oil
Coal
Hydro
Biofuels
Biomass/Waste
Geothermal
Natural Gas
Biomass Traditional
Solar
Biomass Gasified
Nuclear
Wind
Other Ren
5

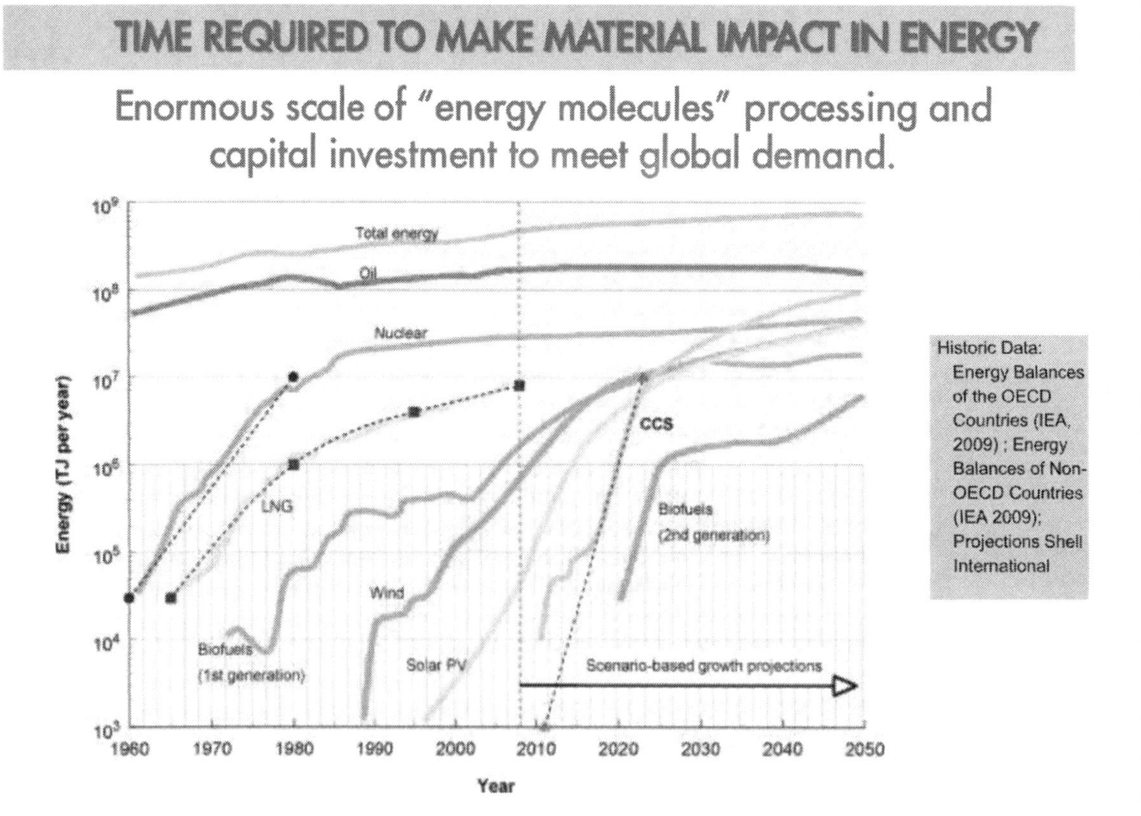
TIME REQUIRED TO MAKE MATERIAL IMPACT IN ENERGY
Enormous scale of "energy molecules" processing and capital investment to meet global demand.
Energy (TJ per year)
Total energy
Oil
Nuclear
LNG
CCS
Wind
Biofuels (1st generation)
Solar PV
Biofuels (2nd generation)
Scenario-based growth projections
1960 1970 1980 1990 2000 2010 2020 2030 2040 2050
Year
Historic Data: Energy Balances of the OECD Countries (IEA, 2009) ; Energy Balances of Non-OECD Countries (IEA 2009); Projections Shell International

SUSTAINABLE DEVELOPMENT: LICENSE TO OPERATE AND A DRIVER FOR INNOVATION

"Meeting the needs of the present generation without compromising the ability of future generations to meet their needs."

People, Planet, and Profits

TECHNOLOGY INNOVATION AND DEVELOPMENT

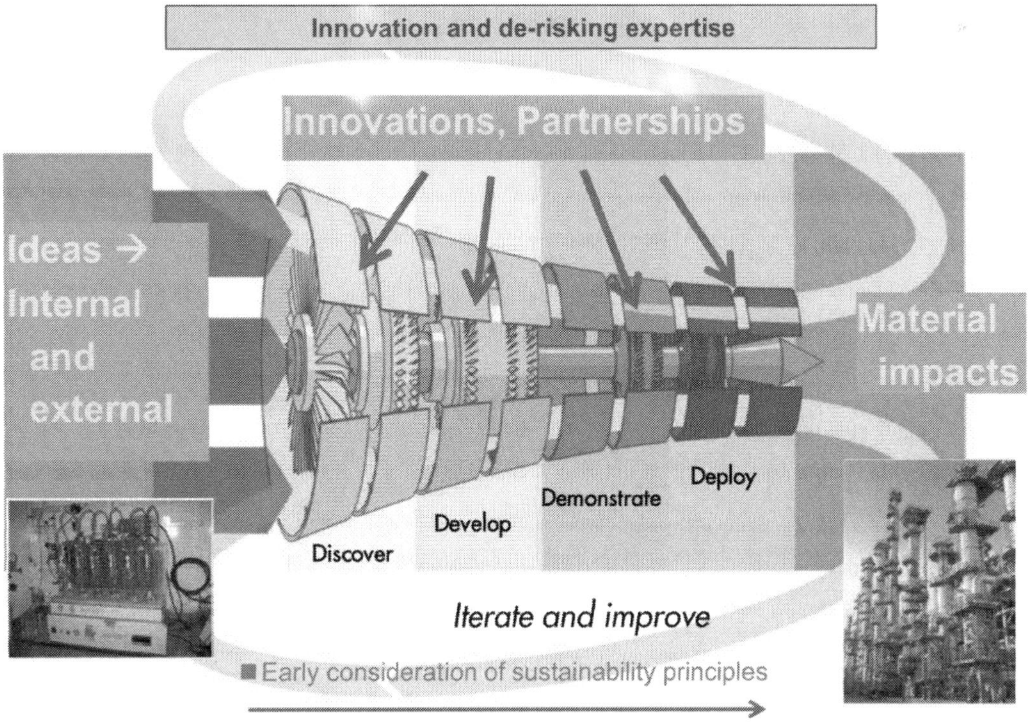

EARLY INTEGRATION OF SD PRINCIPLES

SOONER RATHER THAN LATER

- Integrating social and environmental considerations early – early in the project life or early in the merger, acquisition or divestment process, or early in the development of new products or technology – is critical to success.

MCKINSEY 2030 CO_2 COST ABATEMENT CURVE

Global GHG abatement cost curve beyond business-as-usual – 2030

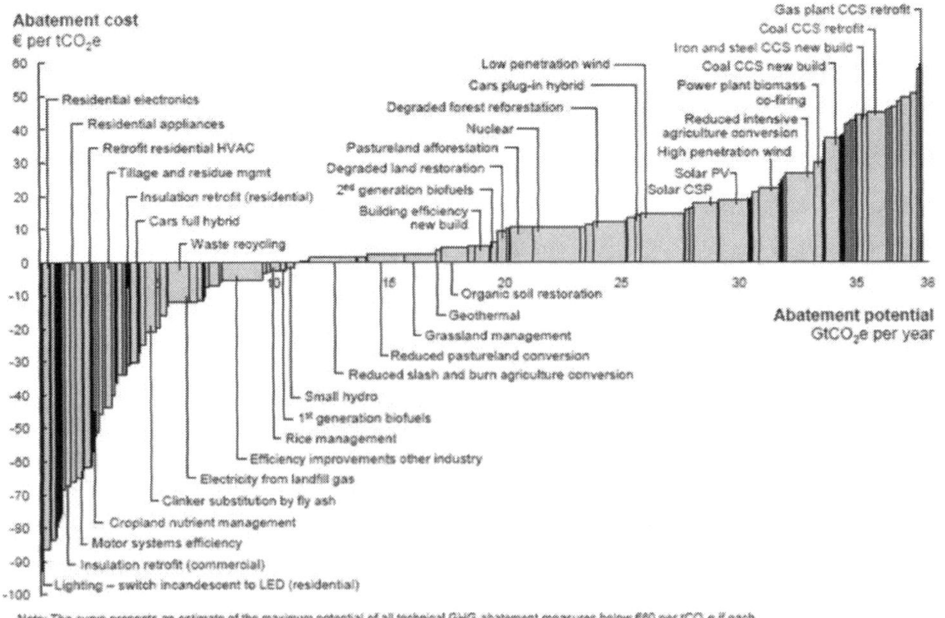

Note: The curve presents an estimate of the maximum potential of all technical GHG abatement measures below €60 per tCO_2e if each lever was pursued aggressively. It is not a forecast of what role different abatement measures and technologies will play.
Source: Global GHG Abatement Cost Curve v2.0

CARBON CAPTURE AND STORAGE

CCS IS A CRITICAL TECHNOLOGY FOR DELIVERING THE CO_2 REDUCTIONS NEEDED BY 2050

WITHOUT CCS THE COST OF MEETING EMISSIONS TARGETS COULD BE SIGNIFICANTLY HIGHER

SHELL IS PROGRESSING CCS TECHNOLOGY

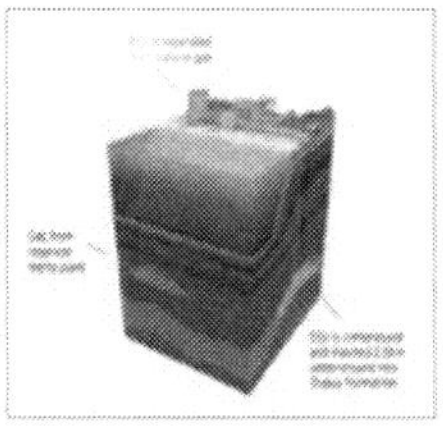

GORGON, AUSTRALIA (PROJECT LEAD – CHEVRON)

- The Gorgon LNG project will include the world's largest CCS project

MONGSTAD, NORWAY (PROJECT LEAD – STATOIL)

- World's largest CCS test facility, demonstrating post-combustion capture

QUEST, CANADA (PROJECT LEAD – SHELL)

- First application of CCS to an oil sands upgrader

Peterhead (UK) → Goldeneye: NG post combustion / amine

Sustainability is more than just CO_2

- Water
- Land Use
- Biodiversity
- Managing Chemicals
- Social performance
- Security / jobs
- Other environmental stressors
- Safety
- Climate and CO_2

THE LINK BETWEEN ENERGY, WATER AND FOOD

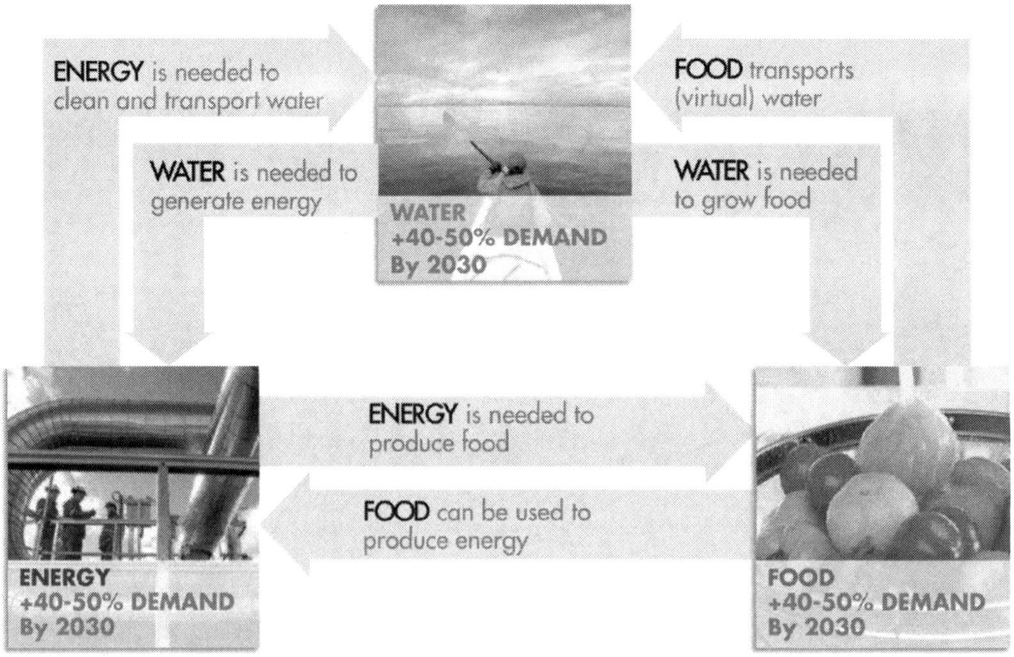

December 3, 2014

BIOFUELS FOR ROAD TRANSPORT

TRANSPORT ACCOUNTS FOR ONE FIFTH OF ENERGY CO_2 EMISSIONS

BIOFUELS OFFER THE MOST PRACTICAL, COMMERCIAL WAY TO REDUCE ROAD TRANSPORT CO_2 EMISSIONS

BIOFUELS vs. Energy Efficiency: REDUCING CO_2 EMISSIONS TODAY

- Today's biofuels are the most realistic commercial solution to take CO_2 out of the transport fuels sector over the next twenty years.

Global CO_2 abatement cost curve for the Road Transport sector - Mix Technology World scenario Societal perspective; 2030

raizen

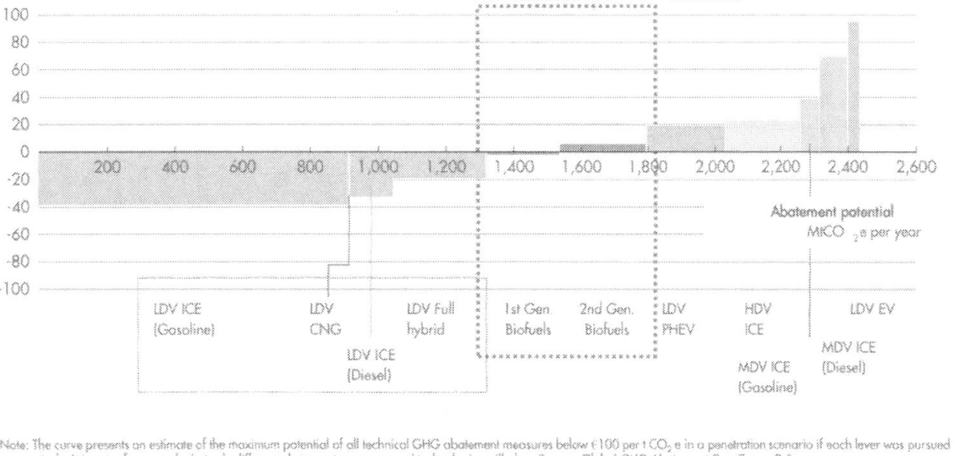

Note: The curve presents an estimate of the maximum potential of all technical GHG abatement measures below €100 per t CO_2 e in a penetration scenario if each lever was pursued aggressively. It is not a forecast of what role different abatement measures and technologies will play. Source: Global GHG Abatement Cost Curve v2.0

WATER CONSUMPTION BY ACTIVITY

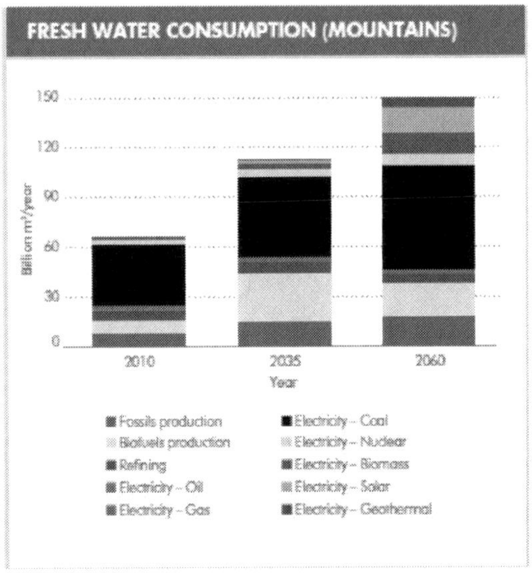

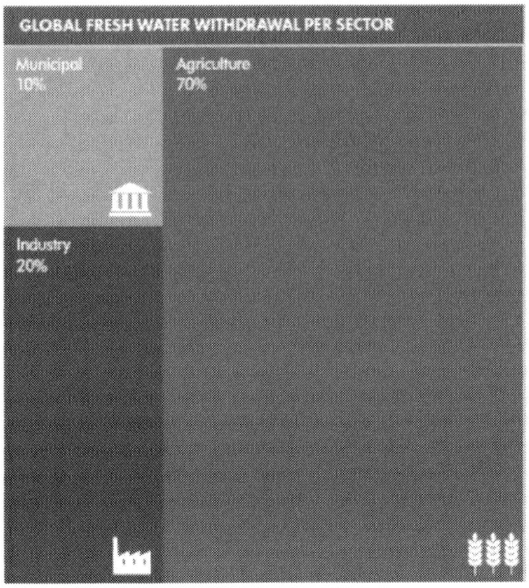

WATER USAGE FOR ENERGY PRODUCTION & TRANSPORTATION

Joost Schornagel*, Frank Niele**, Ernst Worrell**, Maike Böggemann,** Water accounting for (agro)industrial operations and its application to energy pathways, Resources, Conservation and Recycling 61 (2012) 1– 15

A number of charts presented from this reference.

Green Infrastructure

■ www.natureconservancy.org

- Incorporating nature into man-made infrastructure to improve business resilience.
- Examples
  - oyster reefs for coastal protection
  - reed beds that treat industrial waste water:

Petroleum Development Oman LLC (PDO): joint venture with The Shell Petroleum Company Ltd and the Government of Oman (majority)

World's largest commercial wetland covering more than 360 ha and treats more than 95,000 m3 of produced water per day

Thank you!

Joe.Powell@Shell.com

393130 Engineering Global Biological Solutions

Abstract:

The past decades have seen biopharmaceuticals begin to dominate the drug development pathway and already we can see potent biologics bringing benefits to populations on a truly impressive scale. There remains much to do before we can claim however that the benefits of our burgeoning capabilities in the life sciences are fully translated into treatments, delivered globally. That challenge, of enabling the exquisite power of biologically-derived drugs and treatments to benefit world-wide populations, will require significant engineering innovation. This talk will look at some of these development and illustrate potential solutions driven by work from the Department of Biochemical Engineering at University College London (UCL); a pioneer in the field.

Challenges will be used to illustrate the nature of the advances made and of the path ahead. The first is the need to move rapidly from promising drug candidate to a robust and efficient process. Here UCL created the concept of ultra scale-down (USD) which can enable process insights to be gained with a few 10's of mL of material.

Second is the need to make best decisions, be that at the level of technology choice or on a portfolio of drugs for development. So called Decisional Tools have been deployed to address such questions and to provide critical direction to research efforts as drugs move toward manufacture.

Finally we need, as engineers, to understand better the ways in which complete bioprocesses behave so that we can design and operate processes with confidence. A key example is the capacity to gain detailed insights into the processes of chromatographic performance loss. Here the ability to scrutinise at an individual bead level can create rich information for the biochemical engineer.

The talk will be supported with relevant industrial examples to demonstrate how our capacity to engineer global biological solutions continues to advance the translation of exciting life science into commercial outcomes.

393139 Impact of Climate Change on the US Economy

Abstract:

The impact of climate change on the US economy will be presented especially meeting the internationally agreed goal of limiting global temperature rise to 2 degrees Celsius. Global temperature rise beyond the 2 degrees Celsius will lead to many catastrophic consequences including high sea-level rise, extreme disruptive storms, and artic melting. As concluded by the International Energy Agency, meeting the 2 degrees Celsius target will require leaving two-thirds of the earth's known reserves of oil, gas, and coal underground and unburned with economic impact of great magnitude.

393140 Successful Entrepreneurship in the 21st Century

Abstract:

It is well established that the quickest way to become a successful entrepreneur is to get to know and/or spend time with those who are already successful themselves. Tips for success and getting ahead in the 21st Century will be discussed. Strategies that include inviting successful entrepreneurs to university campus to interact and talk to students on how to become successful entrepreneurs will be presented. The presentation will include profiles of prominent entrepreneurs' leaders like Bill Gates (computer science) and Larry Evans (chemical engineering) in their field in today's world.

393143 Space Travel Experience of a Chemical Engineer

Abstract:

Satcher's travel mission experience to the International Space Station to resupply, perform maintenance and conduct tests on how human body reacts in space will be presented. We have learned a lot about what it takes to survive in space. We learned how to start IVs and to make sure that when you push the needle in, you're not also pushing yourself away. Everyday tasks such as personal hygiene were complicated, especially when you had a room full of guys. This included using waterless shampoo and strapping oneself to the toilet, not easy or comfortable adjustments. It took me three days just to get used to sleeping because you're just floating around, even if you're still in the bag.

During the mission, Satcher conducted three space walks to check the station's outside equipment, while employing his finely tuned surgical skills to navigate a robotic arm scanning for shuttle damage. He likened the tedious nature of this task to his experience in the operating room. The fine art of actually getting out of the space suit was an incredibly tricky undertaking which is more difficult than birthing a baby calf.
http://www2.mdanderson.org/cancerwise/2010/05/robert-satcher-md-phd-talks-about-his-nasa-mission.html

393884 A Call for Entrepreneurship Education

Abstract:

From a palate of basic science, and a training in problem identification and formulation, engineers have used the myriad paintbrushes of the engineering discipline to produce some true "masterpieces" of innovative solutions to each generation's most pressing problems. And yet, creativity and innovation are not enough. It is not enough for us to conceive of solutions and walk away after publishing the results in our journals. From their very first day in an engineering program, 21st century students must be made to understand that even the most innovative solutions can only make the world better if they make it into the marketplace. If we are to live up to our reputation as creative problem solvers who make a lasting impact on the world, 21st century engineering education must include entrepreneurial principles and immerse students in an ecosystem that nurtures innovation and entrepreneurship. This presentation is intended to foster a discussion on the need for entrepreneurship education within the engineering and how this might best be achieved concretely.

394222 Slow Growth Increases Myo-Inositol Availability for Glucaric Acid Production in S. Cerevisiae

Abstract:

Glucaric acid has been classified as a top value-added chemical from biomass for potential uses ranging from cancer therapies to sustainable nylons by the Department of Energy. A biosynthetic pathway was constructed in *E. coli* to convert glucose to glucaric acid but activity of the *Mus musculus* myo-inositol oxygenase (MIOX) enzyme was rate-limiting. New expression constructs enabled a one-step knock-in of the four pathway genes to *S. cerevisiae.* Protein sequence similarity mapping paired with simple structure-function hypotheses enabled efficient sampling of over 300 MIOX enzymes. Substituting a plant MIOX in the pathway increased glucaric acid titers detected in the YPD culture broth at 72hrs. Supplementation with 60mM myo-inositol tripled glucaric acid titers for yeast expressing either the plant MIOX (1.5 g/L) or the mouse MIOX (0.94g/L). However, myo-inositol supplementation did not affect glucaric acid titers in yeast expressing the mouse MIOX during slower growth on non-fermentable carbon sources such as glycerol or ethanol. Myo-inositol is a component of yeast membrane phospholipids and carbon-limited chemostat experiments demonstrated that slow growth rates lead to increased myo-inositol availability and glucaric acid titers. We pursued a fed-batch strategy in YPD with yeast expressing the plant MIOX and achieved final titers of 1.2g/L.

Development of Tissue Phantoms for a New Breast Cancer Detection Technique

Matthew A. Conrad
Caitlin B. Douglas
Adam J. Nolte, PhD

Rose-Hulman Institute of Technology | Department of Chemical Engineering

Background Information

- Healthy breast tissue has a Young's modulus of ~3 kPa and tumorous tissue has a Young's modulus of ~20 kPa

⇨ Gelatin hydrogels seem to be acceptably soft

- 250 Bloom B Gelatin was utilized

Rose-Hulman Institute of Technology | Department of Chemical Engineering

Research Goals

- To prepare gelatin tissue phantoms of a Young's modulus of healthy tissue with tumors of various sizes and geometries
- To characterize the mechanical properties such as the effects of aging, multiple tests, strain rate, and sodium alginate as an additive

Rose-Hulman Institute of Technology | Department of Chemical Engineering

Tissue Phantom Preparation

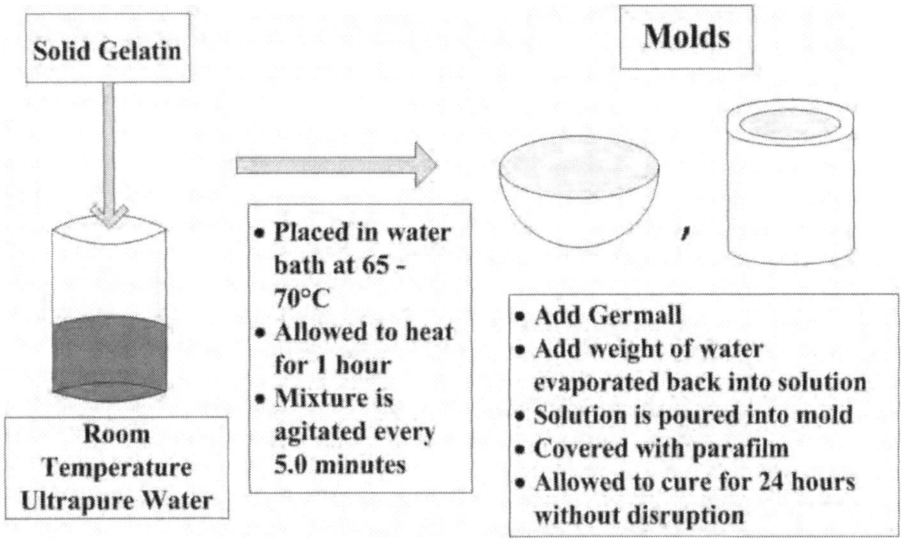

Rose-Hulman Institute of Technology | Department of Chemical Engineering

Tissue Phantom Preparation

Rose-Hulman Institute of Technology | Department of Chemical Engineering

Tumor Inclusion

Two proposed methods:

Liquid Injection via syringe

- 12.0% w/w gelatin liquid solution syringe-injected into congealing 4.0% w/w gelatin solution

⟹ • Not effective

Pre-made inclusions

- 12.0% w/w solid gelatin inclusion placed in congealing 4.0% w/w gelatin solution (T ~ 23.5°C)

⟹ • More effective approach

Rose-Hulman Institute of Technology | Department of Chemical Engineering

Mechanical Testing

Mechanical properties measured:

Young's Modulus
Fracture Strength
Toughness

- Samples tested in compression with an Instron with a 10 kN LC
  - Young's Modulus from strains less than 10%
  - Some samples tested to failure

Rose-Hulman Institute of Technology | Department of Chemical Engineering

Concentration Dependence

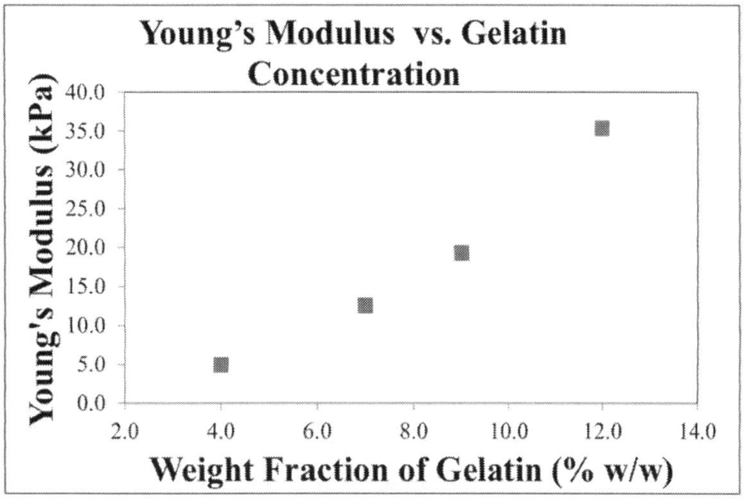

Rose-Hulman Institute of Technology | Department of Chemical Engineering

Aging Effects

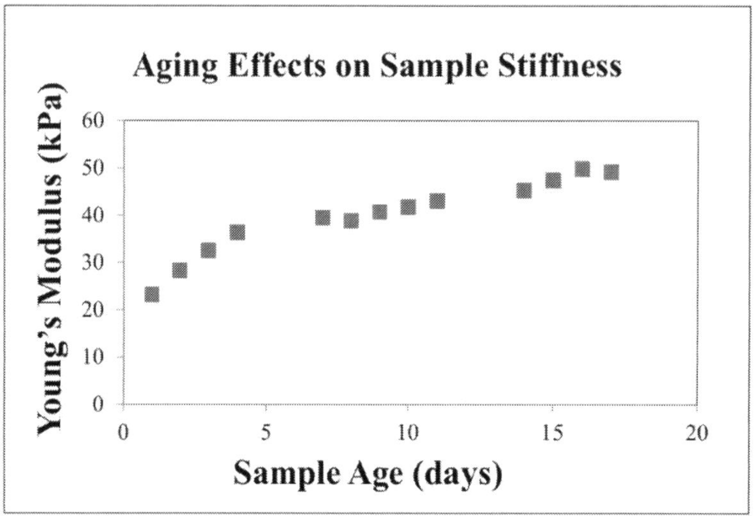

Rose-Hulman Institute of Technology | Department of Chemical Engineering

Testing Speed

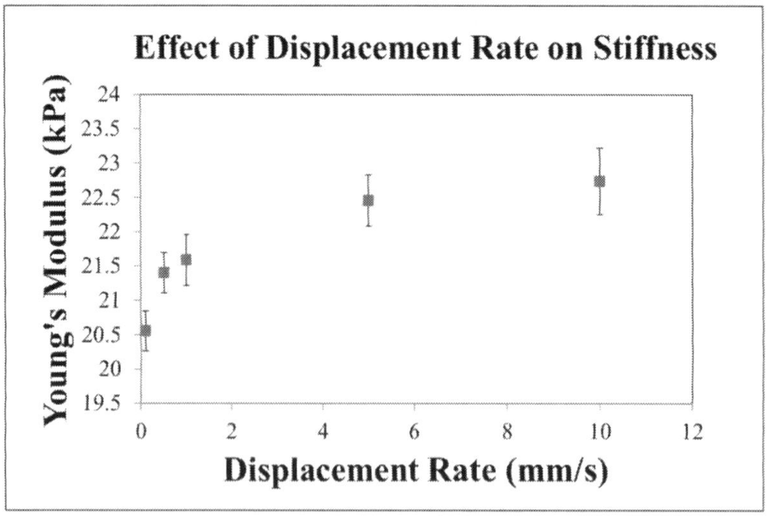

Rose-Hulman Institute of Technology | Department of Chemical Engineering

Multiple Test Effect on Stiffness

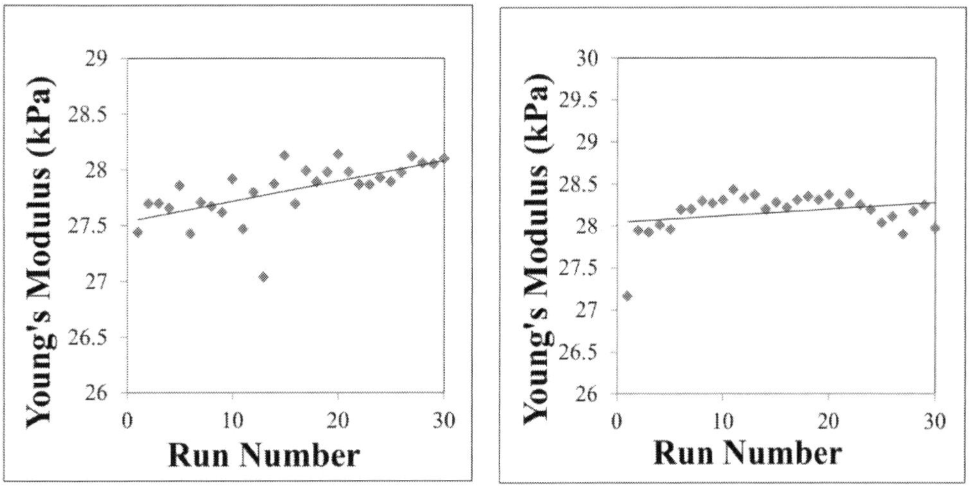

Uncovered during testing

Covered during testing

Rose-Hulman Institute of Technology | Department of Chemical Engineering

Additive Effects: On Stiffness

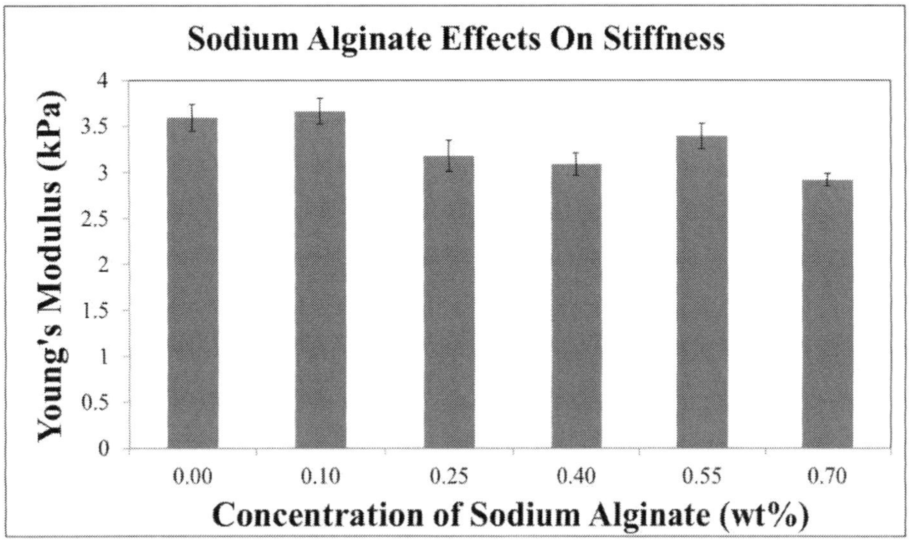

Rose-Hulman Institute of Technology | Department of Chemical Engineering

Additive Effects: On Fracture Strength

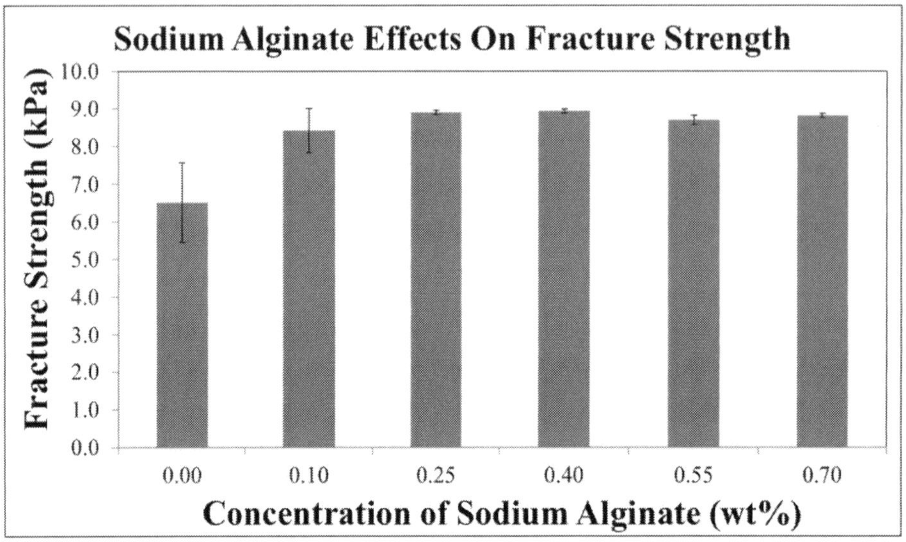

Rose-Hulman Institute of Technology | Department of Chemical Engineering

Additive Effects: On Toughness

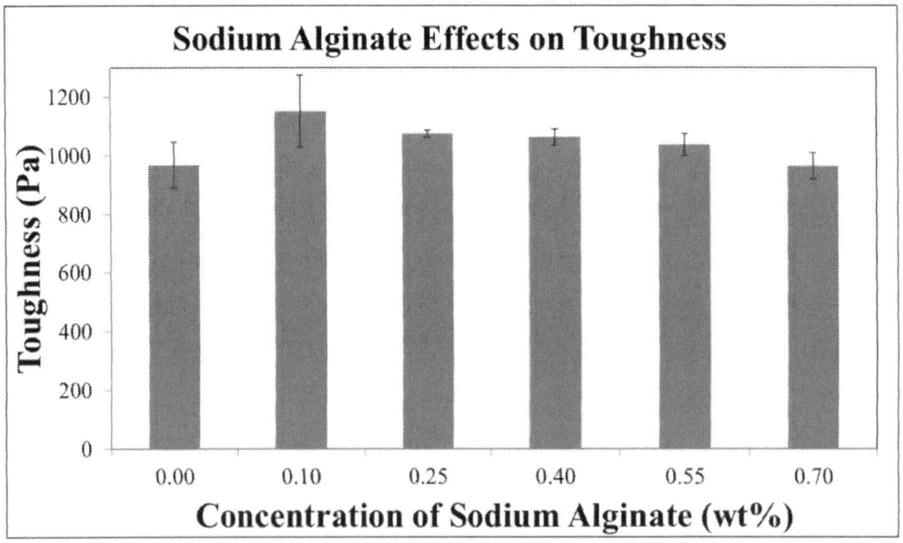

Rose-Hulman Institute of Technology | Department of Chemical Engineering

Conclusion

- Gelatin hemispheres can be made to mimic healthy breast tissue with tumor inclusions by inserting a pre-made tumor into a congealing gelatin solution
- Increasing the sodium alginate concentration appears to improve the fracture strength of gelatin hydrogels while decreasing the stiffness

Rose-Hulman Institute of Technology | Department of Chemical Engineering

Future

- Confirmation of results with texture analyzer
- More accurate relaxation tests
- Optimizing concentrations of additives

Rose-Hulman Institute of Technology | Department of Chemical Engineering

Acknowledgements

- *This research was funded by the National Science Foundation under Grant No. ECSS-1306808*
- *Professor Michael Insana and research group members (UIUC)*
- *Professors Olson and Throne (RHIT)*

Rose-Hulman Institute of Technology | Department of Chemical Engineering

Relaxation Effects

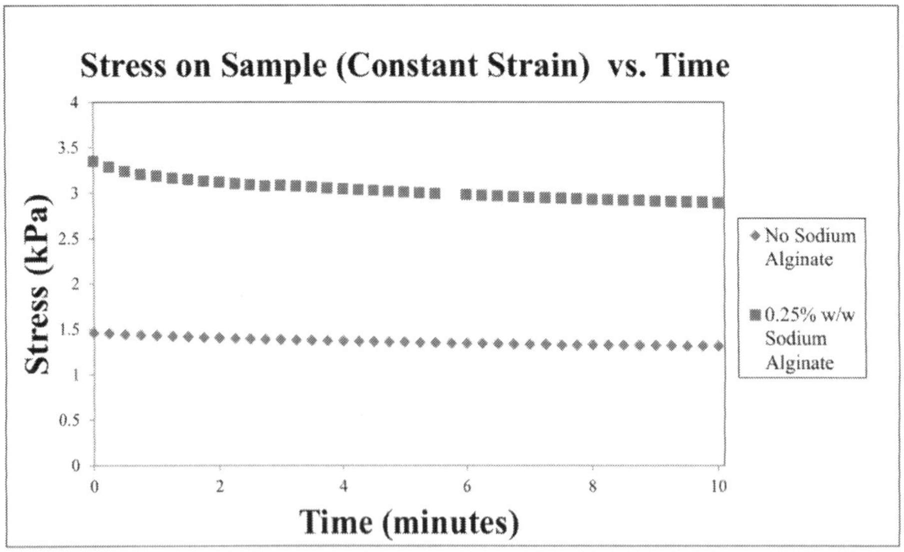

Rose-Hulman Institute of Technology | Department of Chemical Engineering

394192 Effective Antisense Design Using Ensemble of Energetically Sub-Optimal Secondary mRNA Structures

Abstract:

There is a deficit of effective therapies in the current research against bacterial infection. Many therapeutic strategies seek to use small molecules to target the infectious pathogen. One approach to disease inhibition involves direct manipulation of the pathogen at the RNA level. Part of the central dogma, messenger RNA (mRNA) is a transcript of genetic information that encodes the fundamental instruction for protein production. Inhibiting translation of mRNA effectively prevents synthesis of proteins.

In order to effectively block the RNA's message from being read, it must be physically accessed by the therapeutic agent. A field of study has arisen in which a complementary nucleic acid binding strand, called antisense[1], is generated to impede protein synthesis. One of the issues that arises in creating effective antisense RNA is the ability to find appropriate target sites on the mRNA to be inhibited. This problem is in part due to fluctuating secondary structure sequestering target sites. In fact, the kinetics of physical accession of the antisense to the target strand are suggested to be the rate-limiting factor[2] and thus the cause of inefficiency of this method. However, the fluctuations among suboptimal conformations are energetically inherent to the nucleic acid sequence and can be predicted. A software package, GenAVERT, has been developed to thermodynamically determine effective target sites by identifying the ensemble of the most probable energetically suboptimal states and determine which regions of mRNA are most accessible[3]. Once properly identified, a successful and effective antisense can, in theory, be synthesized.

In this work, a proof of principle will be demonstrated in *E. coli* with the downregulation of a short-lived variant of the green fluorescent protein through construction of an inducible antisense molecule. Statistical reduction in the fluorescence upon antisense induction will affirm the efficacy of this model.
Supported by fundamental energetic principles, this method of producing an antisense sequence complement for the most accessible mRNA target region has the potential to greatly reduce the strife of pathogenic infection.

---

[1] Lee, L. and Roth, C. (2003). Antisense technology in molecular and cellular bioengineering. *Current Opinion in Biotechnology*, [online] 14(5), pp.505-511.

[2] Patzel, V. and Sczakiel, G. (2000). In vitro selection supports the view of a kinetic control of antisense RNA-mediated inhibition of gene expression in mammalian cells. *Nucleic Acids Research*, [online] 28(13), pp.2462-2466.

[3] Johnson, E. and Srivastava, R. (2013). Volatility in mRNA secondary structure as a design principle for antisense. *Nucleic Acids Research*, [online] 41(3), pp.e43-e43.

Nanotechnology REU
Summer 2014

Raghav Malik
AIChE National Conference: Undergraduate Research Forum II: Bioengineering, November 17th
Supervising Graduate Student: Giulia Suarato
Principal Investigator: Dr. Yizhi Meng
Department of Materials Science and Engineering

Stony Brook University

Proanthrocyanidin (PAC)

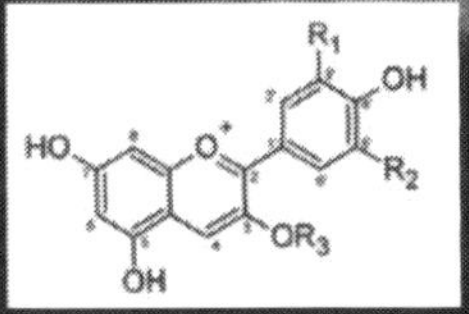

- Organically derived (from fruits and wines)
- Digests in saliva
- Promising chemotherapeutic effects, novel cancer cytotoxicity
- Encapsulation to improve bioavailability

Stony Brook University

Kom Kamonpatana, et. al. Susceptibility of anthocyanins to ex vivo degradation in human saliva, Food Chemistry. 2012: Vol. 135, 738-747

3

Chemotherapeutic Effects of PAC

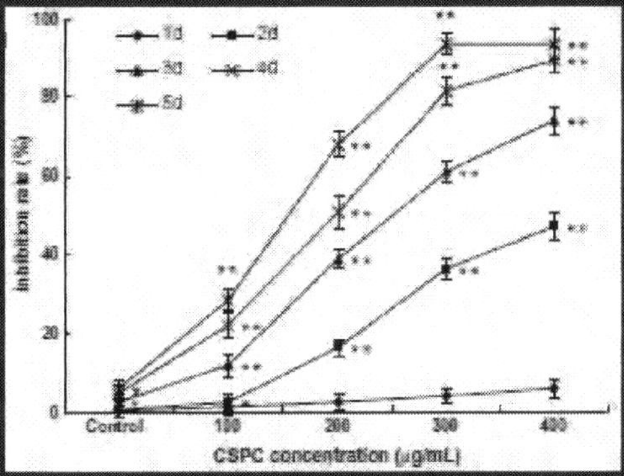

- Human Hepatoma G2 Cells
- Inhibition Rate: (Control – Treated / Control)*100
- Inhibition Rates Dose-Dependent
- Inhibition Time Dependent

NSF

Stony Brook University

H. Zhang, J. Ke, T. Shao, J. Li, Y. Duan, Y. He, C. Zhang, G. Chen, G. Sun, X. Sun. Food and Chemical Toxicology. 2014: Vol. 64, 166-176

4

Objective

Encapsulate PAC using chitosan and characterize the resulting nanoparticle as well as study the effect on MG-63 Osteosarcoma cells *in vitro*.

NSF

Stony Brook University

Why Chitosan?

- High Bioavailability
- Organic, derived from Chitin
- Biodegradable
- Hydrophilic

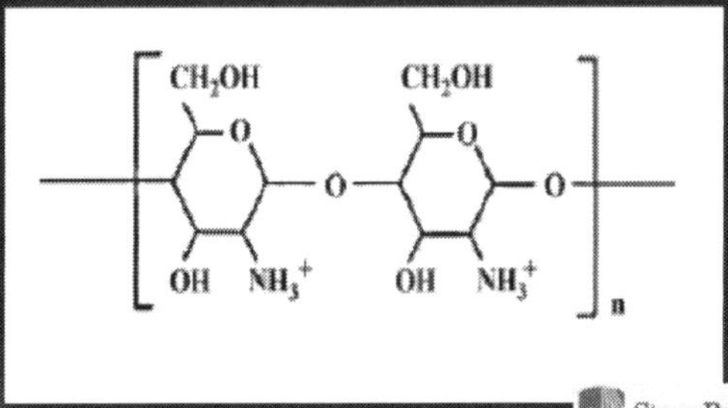

Chitosan

Encapsulation Potential

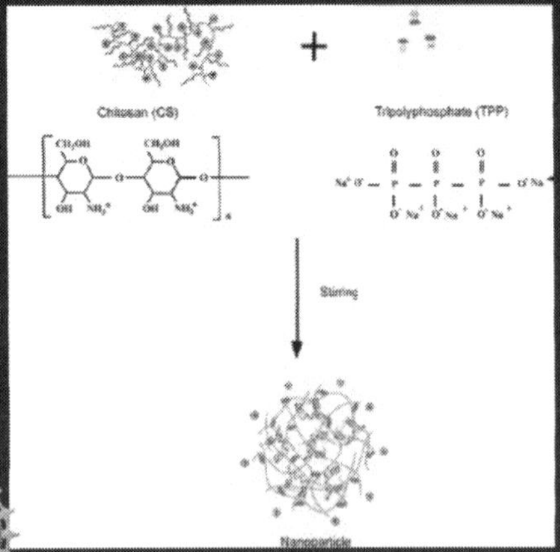

- Ionic Gelation Method
  - Pluronic F68 (detergent)
  - GCS:TPP 8.7:1 and 2.9:1
- Sizes are typically 200-300 nm

Adriana Trapani, Johannes Sitterberg, Udo Bakowsky, Thomas Kissel. International Journal of Pharmaceutics, 2009: Vol. 375, 97-106

Structure of Glycol Chitosan

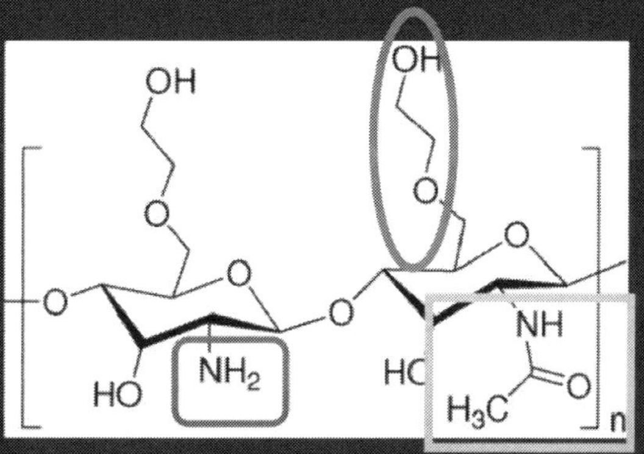

NSF

Stony Brook University

Approach

Synthesis

Characterization

Delivery *in vitro*

NSF

Stony Brook University

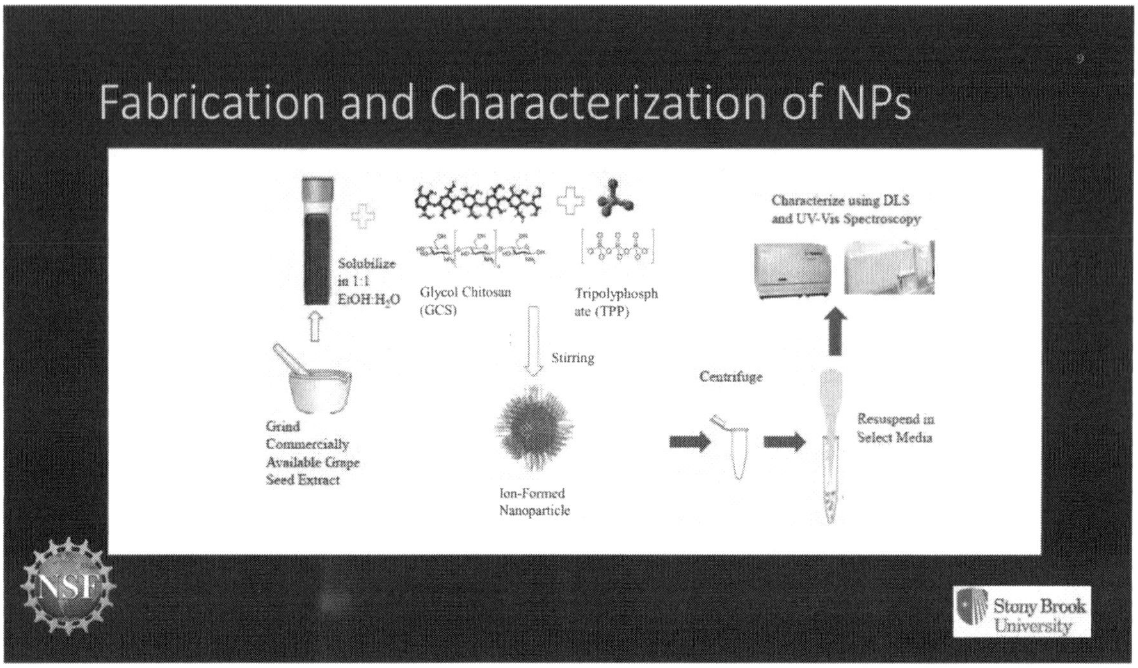
Fabrication and Characterization of NPs
Solubilize in 1:1 EtOH:H2O
Glycol Chitosan (GCS)
Tripolyphosphate (TPP)
Characterize using DLS and UV-Vis Spectroscopy
Stirring
Centrifuge
Grind Commercially Available Grape Seed Extract
Ion-Formed Nanoparticle
Resuspend in Select Media
NSF
Stony Brook University

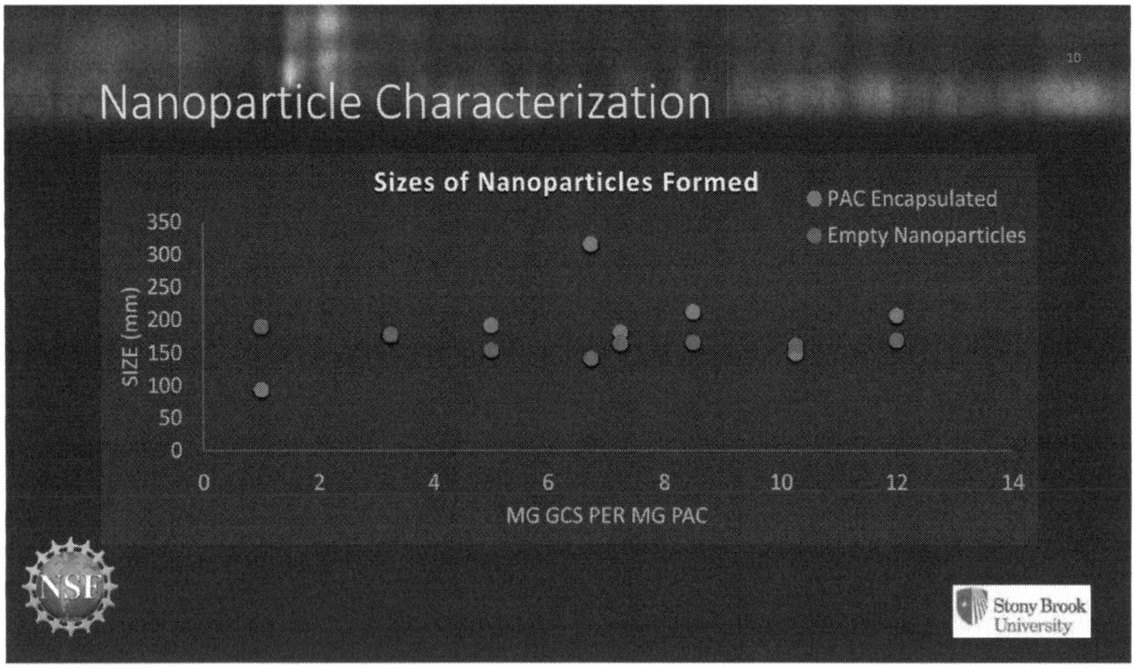
Nanoparticle Characterization
10
Sizes of Nanoparticles Formed
PAC Encapsulated
Empty Nanoparticles
SIZE (mm)
350
300
250
200
150
100
50
0
0 2 4 6 8 10 12 14
MG GCS PER MG PAC
NSF
Stony Brook University

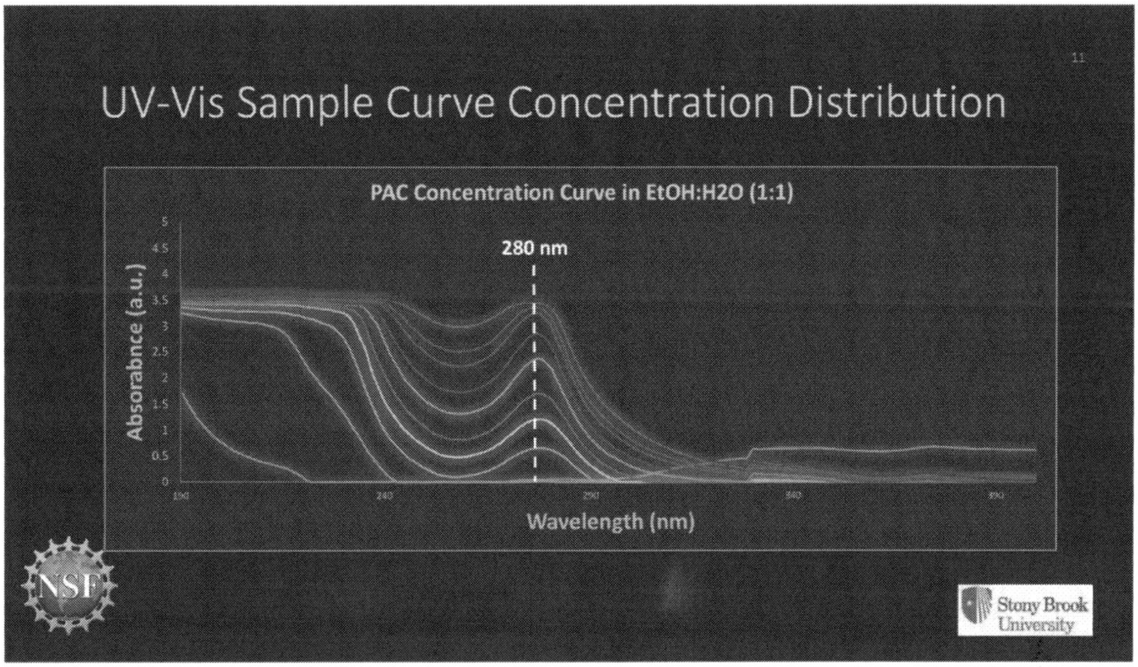
11
UV-Vis Sample Curve Concentration Distribution
PAC Concentration Curve in EtOH:H2O (1:1)
280 nm
Absorabnce (a.u.)
Wavelength (nm)
NSF
Stony Brook University

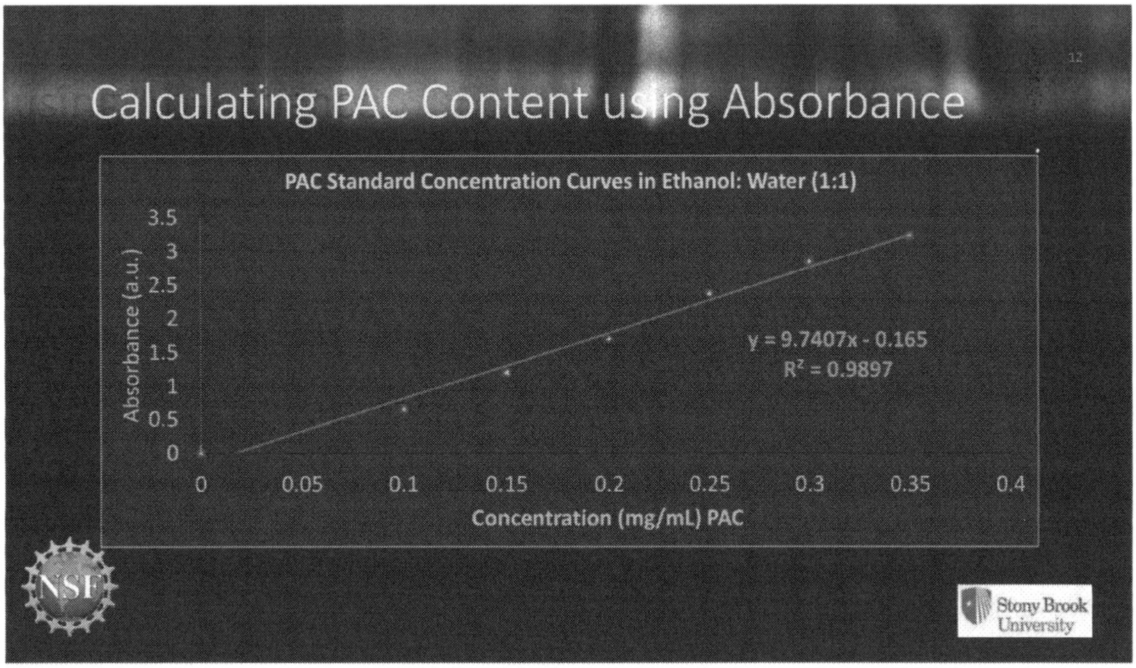
12
Calculating PAC Content using Absorbance
PAC Standard Concentration Curves in Ethanol: Water (1:1)
Absorbance (a.u.)
3.5
3
2.5
2
1.5
1
0.5
0
0 0.05 0.1 0.15 0.2 0.25 0.3 0.35 0.4
Concentration (mg/mL) PAC
y = 9.7407x - 0.165
R² = 0.9897
NSF
Stony Brook University

13

Encapsulation Efficiencies of GCS NPs

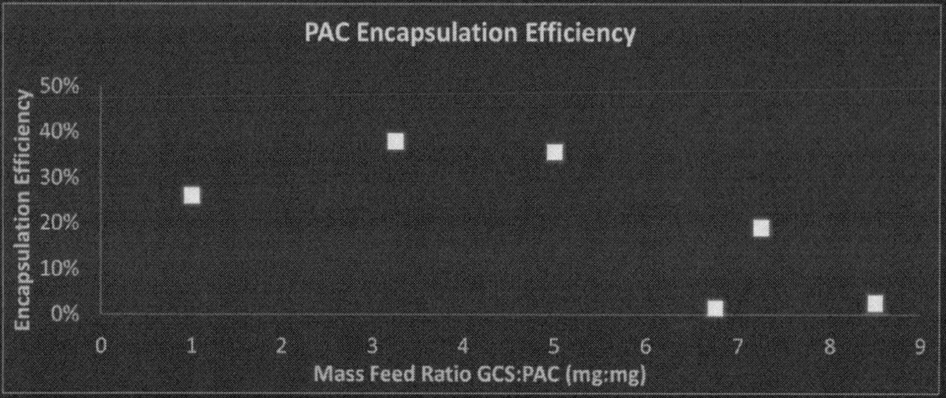

14

Delivery

- Cells used: MG-63 Human Osteosarcoma Cells
- Morphological study of cells after application
  - Comparing free and encapsulated PAC
- Cell Viability Assay done using MTS Assay
  - Comparing free and encapsulated PAC

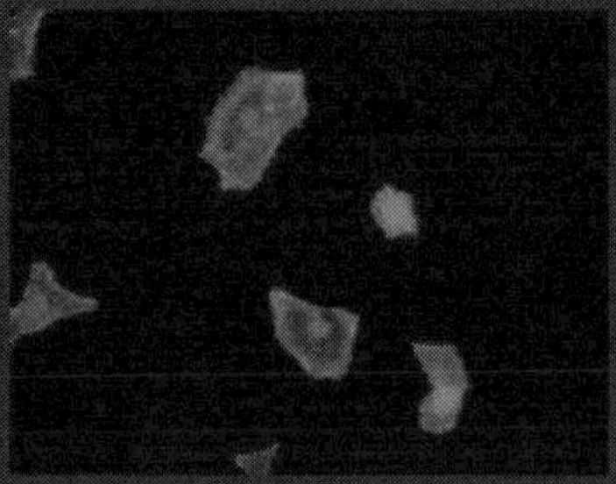

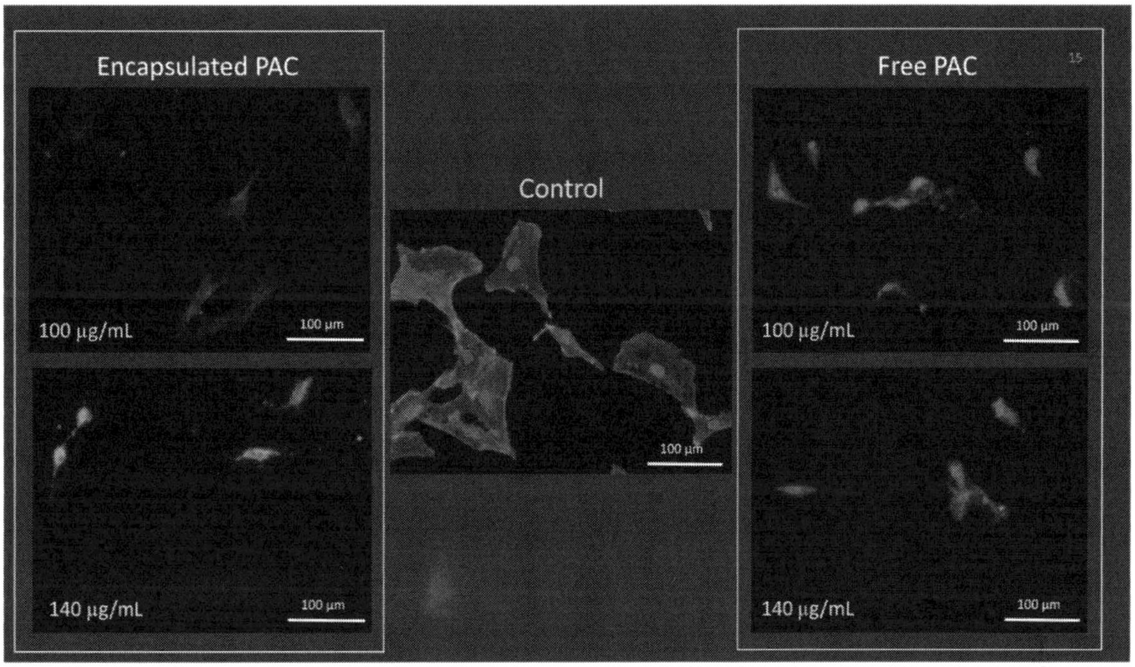

15
Encapsulated PAC
Control
Free PAC
100 μg/mL
100 μm
140 μg/mL
100 μm
100 μm
100 μg/mL
100 μm
140 μg/mL
100 μm

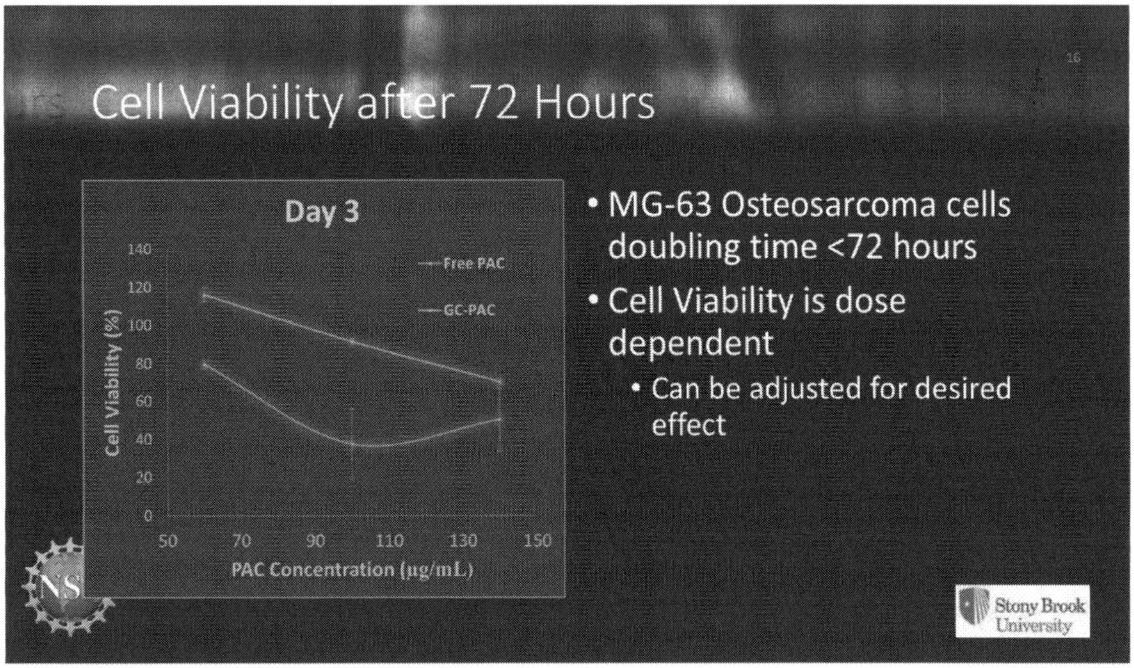

16
Cell Viability after 72 Hours
Day 3
Free PAC
GC-PAC
Cell Viability (%)
PAC Concentration (μg/mL)
• MG-63 Osteosarcoma cells doubling time <72 hours
• Cell Viability is dose dependent
• Can be adjusted for desired effect

17

Percent Cell Viability

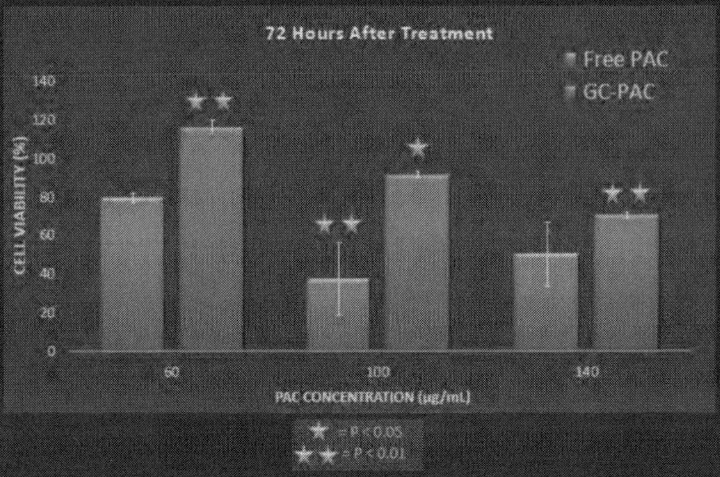

Stony Brook University

18

Conclusions

- Most Efficient Mass Feed Ratios: 5 mg : 1 mg GCS:PAC
- Morphological analysis shows similar effects by both free and encapsulated PAC
- MTS Analysis indicates encapsulated PAC is advantageous over free PAC in its dose-dependency as well as effectiveness with greater amounts of PAC

Future Studies

- Release Kinetics of GCS-PAC NPs
- Amount of GCS in formed NPs

Stony Brook University

19

Acknowledgements

I would like to thank the National Science Foundation, Giulia Suarato, Dr. Yizhi Meng, the Center for Inclusive Education, and all other students and mentors involved in the REU for making this research possible by providing the resources necessary

NSF

Stony Brook University

394191 Single Cell Isolation Via Microfluidic-Based Droplet Delivery

Abstract:

Commercial technologies such as fluorescent-activated cell sorting (FACS) for isolating and delivering single cell samples are often inappropriate for smaller research facilities due to their size and cost. Researcher labs conducting experiments on single cells require an inexpensive and accessible method for creating samples. We demonstrate the practicability of using a gravity-driven microfluidic device under a bench-top optical trap to generate droplets containing isolated single cells for delivery onto a 96 well plate. The microfluidic device dispenses droplets of constant volume at fixed intervals. We establish functionality by delivering droplets containing glutaraldehyde fixed red blood cells (RBCs) onto glass slides. A fluorescent confocal microscope capable of imaging the glutaraldehyde-induced fluorescence of the RBCs is employed to count the number of cells contained per droplet. We attained 64% fidelity of single-cell samples, indicating that delivery from our device is not stochastic. Biomedical researchers often expect droplet delivery methods to be compatible with a 96 well plate standard; to this end we achieved automated delivery into each well of a 96 well plate using a Teensy 3.0 encoded motorized stage. Additionally, we implemented a side and forward scatter detection scheme around the optical trap to show the feasibility of fully automating the cell isolation and droplet delivery process. The entire apparatus is compact and portable enough to fit within a sterile environment to prevent contamination of samples. We propose that our setup is an inexpensive and compact alternative to flow cytometers that is accessible to most labs and can reliably sort single cells into droplets for sterile delivery.

394190 Self-Assembling Zwitterionic Nanogels As Immune Isolating Coatings for Stem Cell Derived Pancreatic Islet Transplantation

Abstract:

The concept of pancreatic islet transplantation to patients with diabetes to restore normoglycemia is a viable clinical strategy. However, widespread clinical application of islet transplantation remains limited because of the harmful side effects of the necessary immunosuppressive therapy to prevent host rejection of transplanted cells. Here we report the fabrication of a superbiocompatible nanogel and a novel coating strategy for protection of stem cell derived pancreatic islets against host rejection in vivo. Zwitterionic nanogels (100-300nm in size) were fabricated using an inverse emulsion polymerization synthesis scheme and further surface modified with CLICK reactive functional groups. Stem cell derived pancreatic islet cells were engineered to express azide functionalities on their cell membranes. Confocal fluorescence microscopy was used to verify covalent reaction and assembly of dual-layered nanogel coatings on engineered islet cell clusters. Cytocompatibility of coating and islet cell functionality were examined using a combination of in vitro glucose tolerance tests, live/dead staining, and in vivo studies on mice treated with streptozocin, a small molecule drug that destroys the beta cells. Our results indicate that the developed nanogel formulation and encapsulation strategy provides a viable strategy for prolonged protection and isolation of transplanted islets against host rejection

394187 Modeling of Selenium Nanoparticle Deposition for Optimized Production of Antibacterial Surfaces

Abstract:

The Webster Nanomedicine lab designs and explores applications of nanomaterials to inhibit bacterial growth and cancer cell responses. Due to its beneficial effects on the body, and known chemopreventive and chemotherapeutic properties[1], selenium has been used in its nano-form to impart nano-roughness and chemical effects in these experiments. To ensure precise and consistent selenium nanoparticle (SeNP) coverage, we aim to tabulate several relevant aspects of the particle. By adjusting the parameters of the reaction producing SeNP, a variety of data can be collected and analyzed. Characteristics that are pertinent include the nanoparticle size, coverage, surface energy, and roughness. Eventually, a cubic response surface model will be developed using a central composite design (CCD), which will be able to predict the interaction of the particles with bacteria and cells.

SeNP are produced through a reaction of glutathione (GSH) with sodium selenite (Na_2SeO_3). The reactants are introduced in a 4:1 molar ratio, respectively. Thin discs of poly(l-lactic acid) (PLLA) are added to the reaction, providing a surface for the cells to grow. In order to precipitate out the SeNP, 2M NaOH is added to the mixture. The reaction is then halted by the addition of DiH_2O. The experimental design used four factors: time in GSH/ Na_2SeO_3, time in NaOH, reaction volume, and the number of samples in the volume. Three results are compared: nanoparticle size, nanoparticle coverage, and surface energy of each sample. Both the size and the coverage of the produced nanoparticles can be measured using scanning electron microscopy. The surface roughness is portrayed through atomic force microscopy. Surface energy can be calculated utilizing the contact angle, which is defined using a Goniometer.

Previous studies show longer time increments (in both cases) increase the deposition of SeNP on the substrate. However, as the increments reach larger values of time (30-60 seconds), there is less of a change in coverage. This suggests the existence of a saturation point. On the PLLA, there is a finite number of binding sites for the nanoparticles. The size of the nanoparticle, with time being the only deviation, remains consistent. It is predicted that development time and reaction volume directly correlate. Recent observations of previous studies have shown that increased coverage correlates with development time in both steps, and therefore, roughness.

Through characterization of the SeNP product, a cubic response surface will be calculated relating the input parameters to expected results, optimizing the coverage of nanoparticles on substrates. These efforts conserve both time and resources because reactions can be designed with focused specifications.

353721 Introduction to Project Management Fundamentals

Abstract:

This presentation introduces the fundamentals of project management. It will be especially useful for new engineers, young professionals, and advanced students, as well as anyone who serves or will served on project teams or leads those teams. We will discuss the importance of people, the importance of setting excellent project objectives, and the importance of good planning. Working with people is perhaps the most difficult. Suggestions will be given on how to better communicate and better achieve unity with project team members. We will discuss the real meaning of the word "Communication" and how that meaning can change the whole approach we take to interacting with people in a productive way. Important ideas and tools for developing excellent project objectives will be presented, including the use of SMART objectives, the triple constraints, and tolerance diagrams. An outline of the project planning process will be reviewed. This planning process includes the use of work breakdown structures, network diagrams, and Gantt charts, as well as resouce histograms. Also, we will briefly discuss doing fatal-flaw analysis.

364220 Entrepreneurial Chemical Mega Projects with Parallel Path Process Development

Abstract:

Entrepreneurial Chemical Mega Projects with Parallel Path Process Development

The start-up world has unleashed the excitement of the American entrepreneurial spirit. The United States leads the world in innovation. The lessons learned that have revolutionized innovation in the manipulation of information can be applied to innovation in chemical mega projects. Manipulating mass, however, is much different than manipulating information. The rules of entrepreneurial efforts have to be applied in a more complex way to insure safety and risk are mitigated within the adventure of changing the world in the chemical industry.

A mega project is a high value, complex, interdependent, multi-organizational undertaking that creates a game changing difference in the landscape of chemical processing. The bonds that are formed, the lessons that are learned and the impact that is felt spans careers, industries and businesses. There is nothing more exciting.

Game changing mega projects are usually based upon some new process development that enables the cost optimization, capacity increase, or yield improvement on some needed chemical product. Feedstock changes, energy efficiency or production routes are usually the chemical engineering technology behind a mega project.

In the old model, process development would be completed, the process work up would be transferred to the process engineering group and the project would move in the serial steps to execution. In today's fast paced market, parallel paths must be considered to accelerate the pace. However, unlike the technology world, the first application out the door, version 1.0, has to be safe and has to work due to the great differences in safety and financial risk between a project that manipulates mass, and a project that manipulates information.

353722 Becoming a Real Team to Accomplish Projects

Abstract:

In project management a major effort goes into becoming a real team! A team is not just a group of individuals! It is a very special group of individuals who have worked and learned to become united in their objectives and willingness to work to accomplish those objectives. This paper will discuss in depth the important principles of people interaction skills needed for people to coalesce into a working unit of synergy. We will discuss communication, what it means, and how to use it. We will talk about leading effective team meetings. We will discuss how to look at issues from other peoples perspectives in order to better understand them and work with them to accomplish objectives. We will discuss listening and barriers to effective listening. All of this will go toward developing a new vision of what it means to be an effective and united team and give important meaning to the word "teamwork."

394223 Exploiting Polymer-Nanoparticle Interactions to Create a Hydrogel with Biomedical Applications

Abstract:

Injectable hydrogels have recently become a relevant material platform in applications such as in vivo drug delivery or as cell carriers because they can be implanted in a minimally invasive manner and closely mimic biological tissue. Attractive properties for such injectable hydrogels are that they easily flow under applied stress (*shear-thinning*) and rapidly heal upon relaxation of the applied stress (*self-healing*). Previously, we developed a shear-thinning and self-healing injectable hydrogel driven by non-covalent interactions between hydrophobically modified cellulose and PEG-PLA nanoparticles. [1] The nature of this gel enables facile loading with a variety of therapeutics as well as tunable release profiles of multiple therapeutics. Specifically, the PEG-PLA nanoparticles allow for loading with hydrophobic therapeutics while a hydrophilic therapeutic can be entrapped within the aqueous phase of the gel.[2] In this work, we have quantified the factors that affect the mechanical properties and release profiles of the gels, demonstrating the range of their tunability. Additionally, we explored other nanoparticles and biopolymers for the fabrication of similar materials. Lastly, we studied the use of this injectable hydrogel as a cell carrier with osteogenic properties. Human mesenchymal stem cells (hMSCs), which are capable of differentiating into bone, cartilage, and fat cells, were encapsulated in the gel and survived injection through a 28 gauge needle into three-dimensional collagen hydrogels. Additionally, PEG-PLA nanoparticles were loaded with dexamethasone to stimulate osteogenic differentiation of the encapsulated hMSCs. We are currently exploring the use of this drug-loaded cell carrier as an injectable system for bone healing. Overall, we present an injectable hydrogel with tunable properties for the controlled delivery of therapeutics with facile synthesis and minimally invasive application *in vivo* that is capable of serving as a stimulatory cell carrier.

---

[1] Appel, E.A., Tibbitt, M.W., Langer, R., Nature Communications., **2014**, *in revision*.

[2] Greer, J.M., Appel, E.A., Tibbitt, M.W., Langer, R., *manuscript in preparation*.

Validation of High Throughput Electrochemical Gas Sensing Screening System

Zixuan Wang

November 17, 2014

UNIVERSITY OF SOUTH CAROLINA

Background

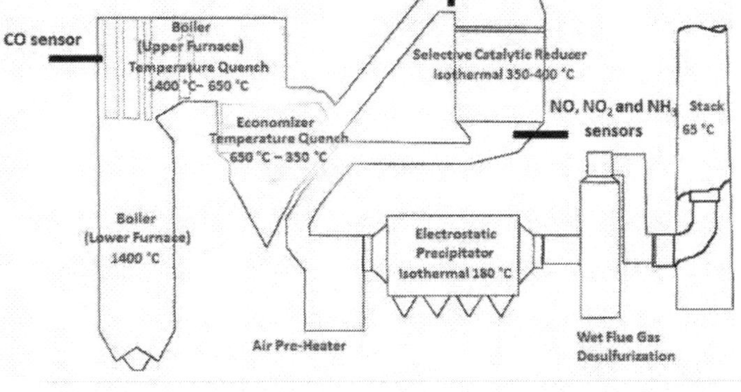

Harsh combustion environment - 800 ° C

Accurate and selective to target gas

UNIVERSITY OF SOUTH CAROLINA

High Throughput Methodology

Subatomic difference affect large scale

Not feasible to individually test material

High throughput synthesis and characterization methods

UNIVERSITY OF SOUTH CAROLINA

High Throughput Methodology

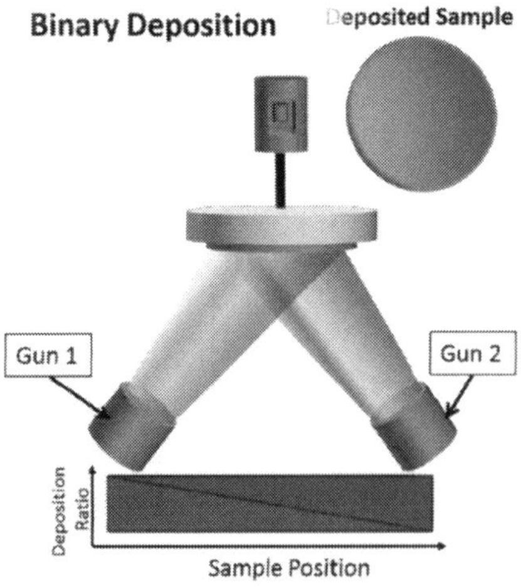

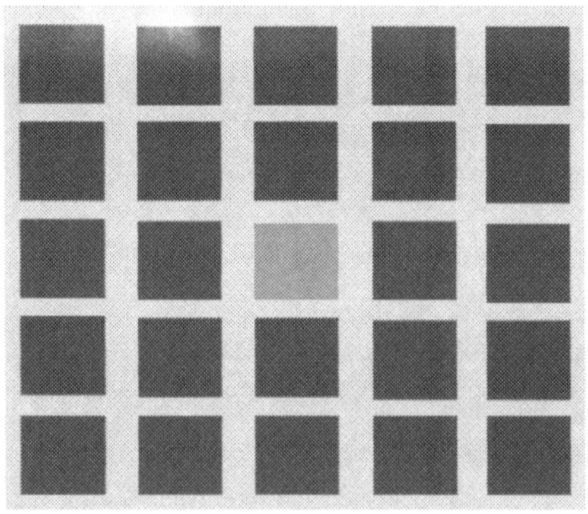

- Normalize affect of distance on the potential

UNIVERSITY OF SOUTH CAROLINA

Development of Uniform Standard Sample

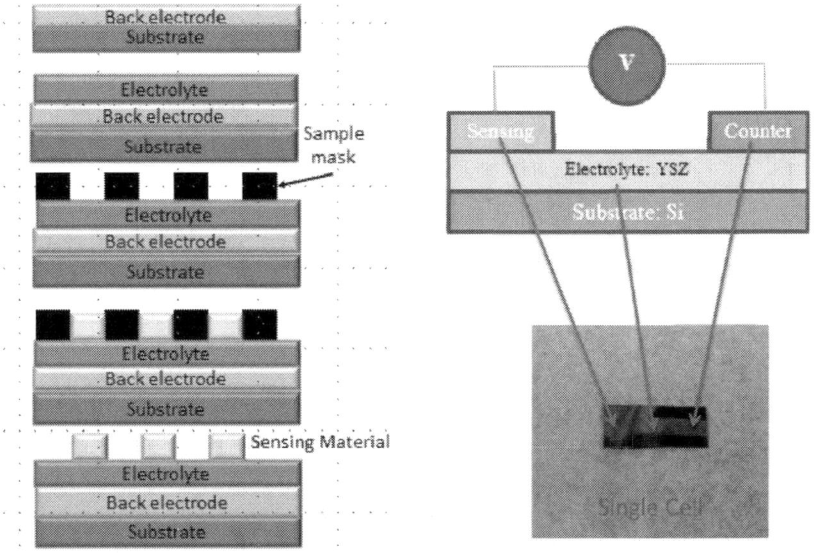

UNIVERSITY OF SOUTH CAROLINA

Surface Mechanisms

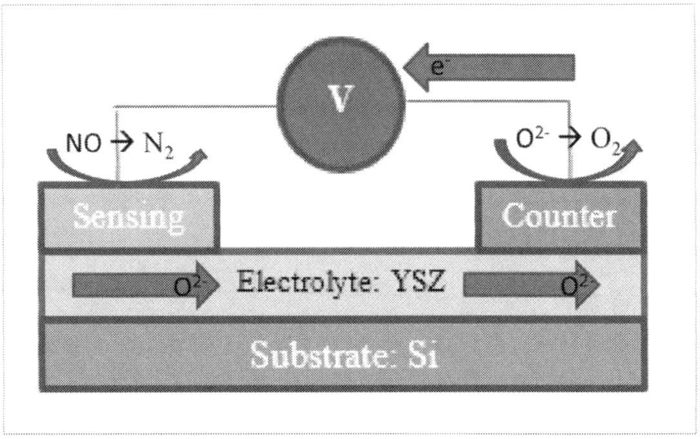

UNIVERSITY OF SOUTH CAROLINA

Sensors and Actuators B 121 (2007) 652-663

Reflectometry Results - YSZ

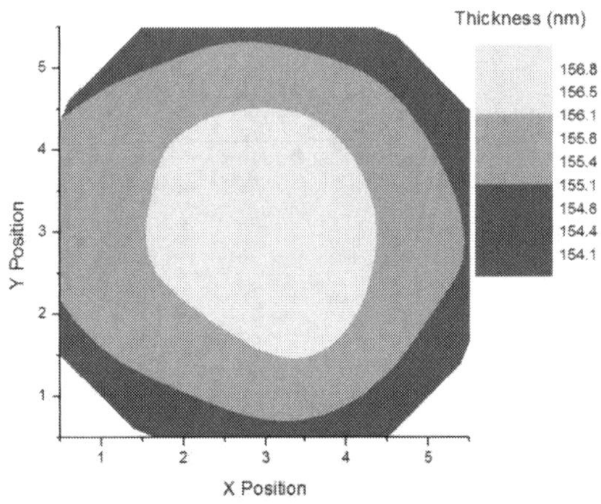

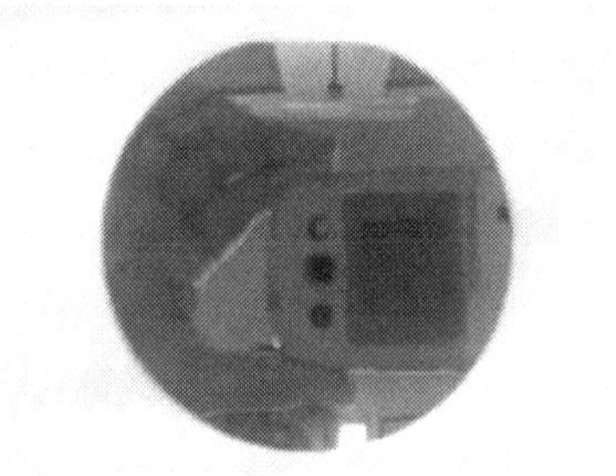

- Results confirm relative uniformity of sample (1.8% difference)

UNIVERSITY OF SOUTH CAROLINA

XRD Results - YSZ

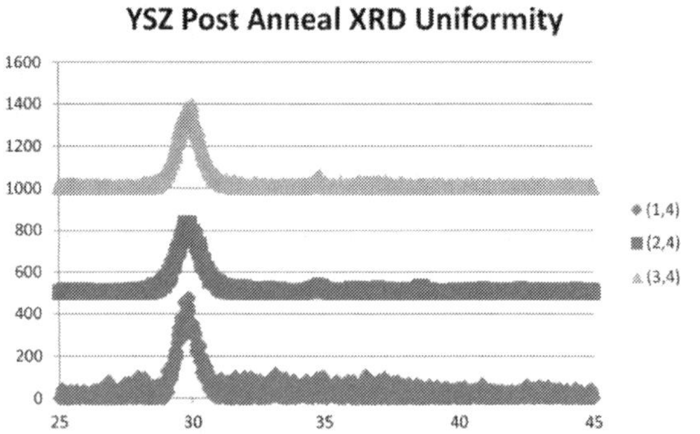

(1,4)	29.745	
(2,4)	29.836	34.848
(3,4)	29.876	34.859

- XRD analysis using Jade
- YSZ Cubic:
  - (111): 29.675°
  - (200): 34.398°
- Confirm identical phases

UNIVERSITY OF SOUTH CAROLINA

Reflectometry Results – WO_3

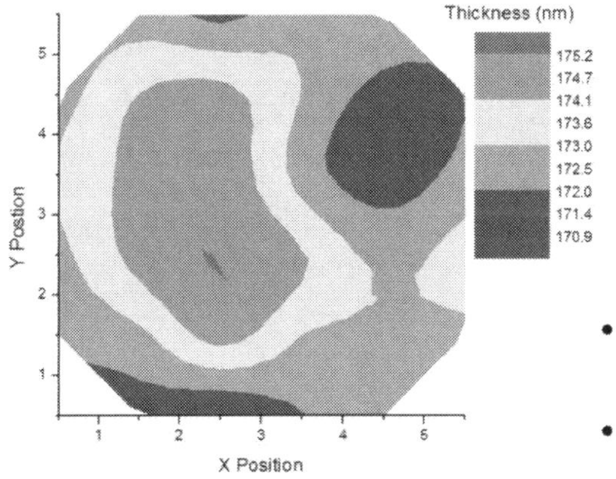

- Confirms relative uniformity (2.5% difference)
- Evidence some parts of the plate that are slightly thicker or thinner

XRD Results – Post Annealed WO_3

(1,4)	23.143	23.653	24.362
(2,4)	23.140	23.671	24.397
(3,4)	23.127	23.641	24.377

- XRD analysis using Jade
- WO_3 monoclinic:
  - (002): 23.118°
  - (020): 23.582°
  - (200): 24.365°
- Confirm identical phases with phase shifts present

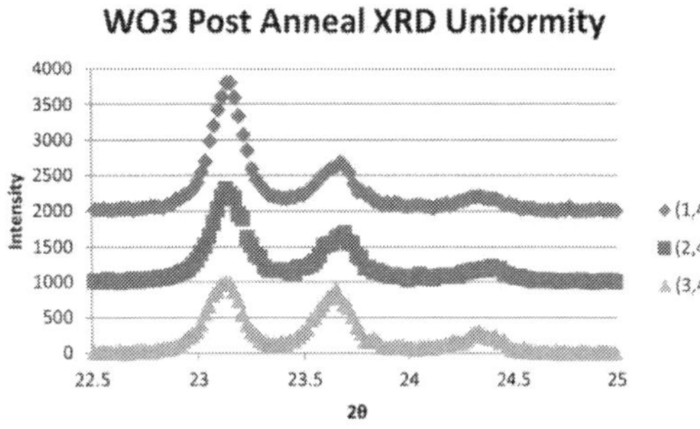

Reflectometry Results - WO_3-YSZ

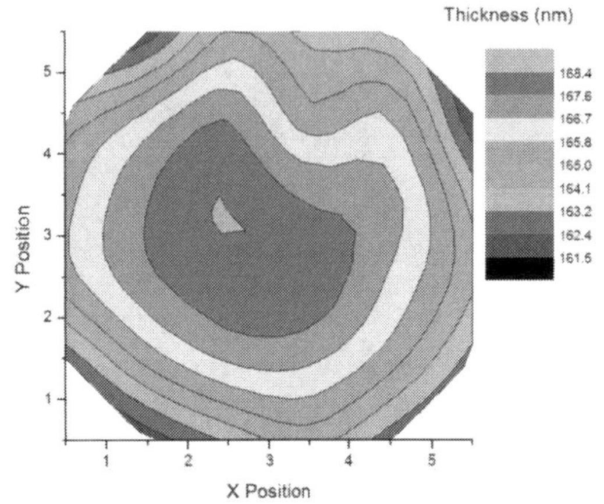

- Sputtering conditions: same
- YSZ held constant
- Confirms relative uniformity (4.3% difference)

XRD Results – WO_3-YSZ Post-annealed

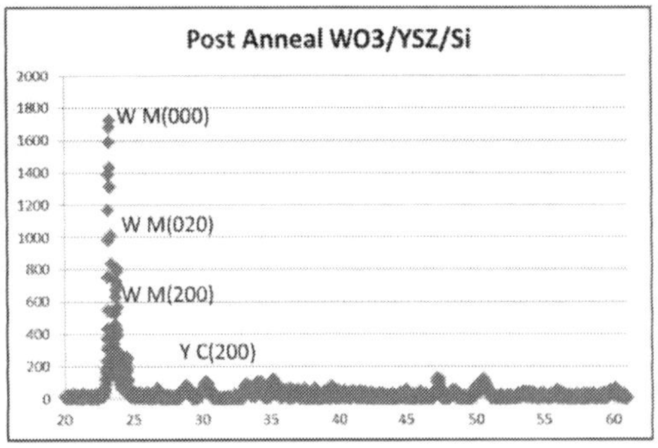

WO_3 monoclinic:
- (002): 23.118°
- (020): 23.582°
- (200): 24.365°

YSZ Cubic:
- (111): 29.675°
- (200): 34.398°

Electrochemical Cell Setup

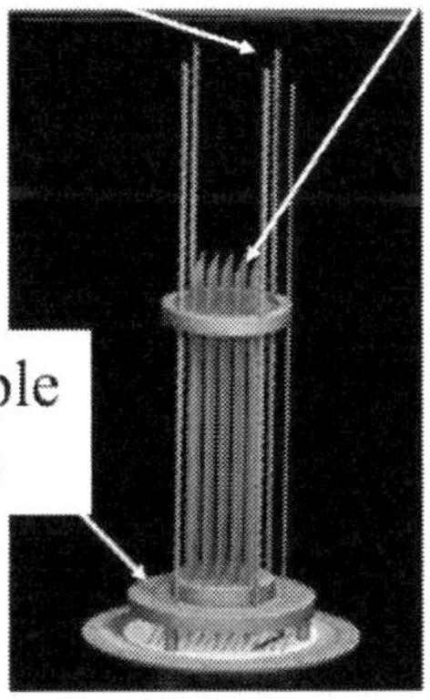

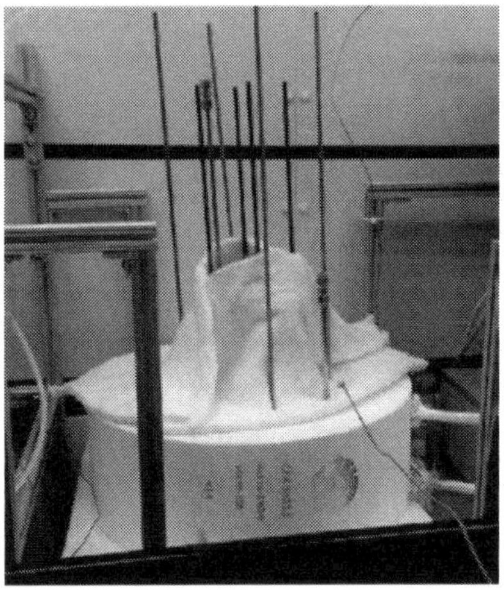

UNIVERSITY OF SOUTH CAROLINA

Single Electrochemical Cell Tests

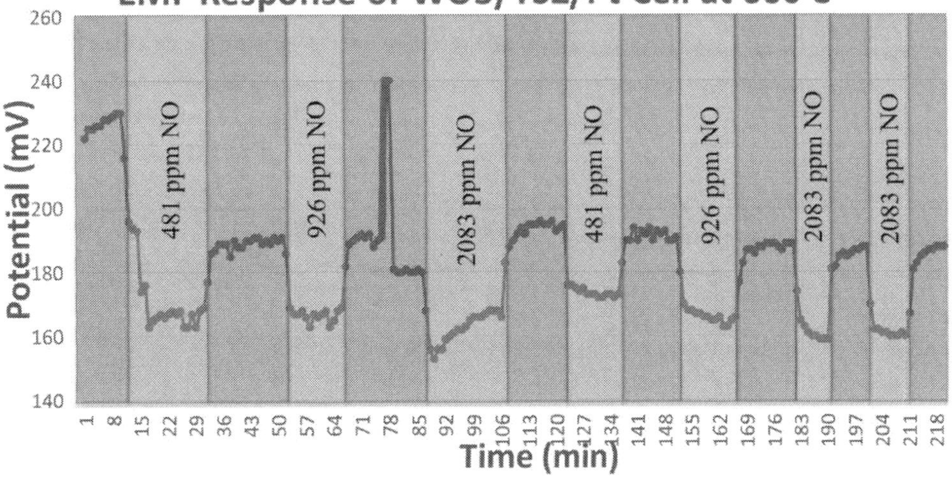

UNIVERSITY OF SOUTH CAROLINA

Electrochemical Cell Tests

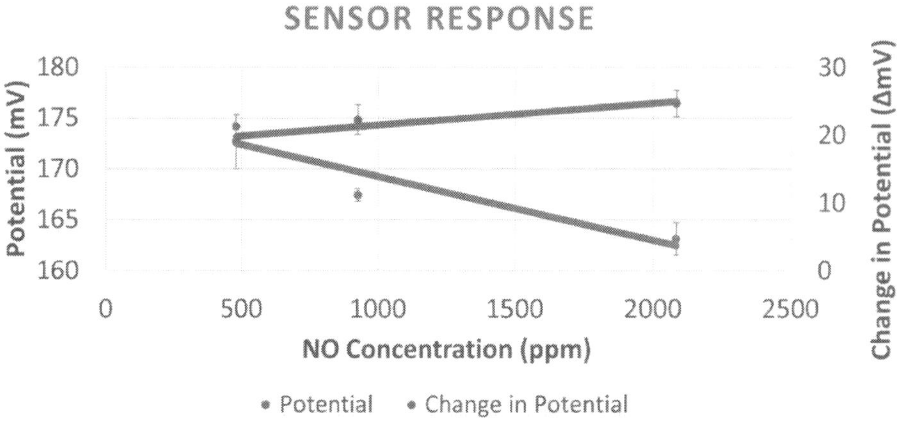

UNIVERSITY OF SOUTH CAROLINA

Multielectrode Cell Tests

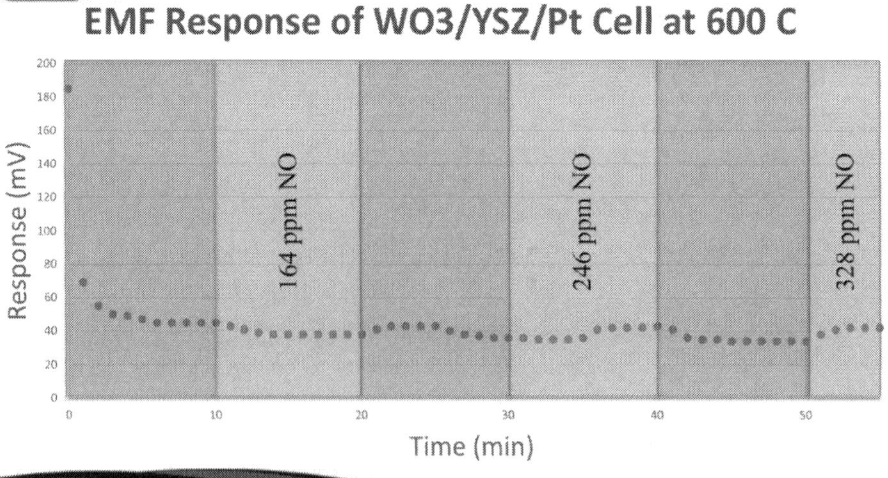

UNIVERSITY OF SOUTH CAROLINA

Conclusions

- Designed planar sensor array
  - Use for combinatorial screening of electrochemical gas sensors
- Confirm deposition uniformity
  - Reflectometry: thickness
  - XRD: phases
- Cell potential
  - Changes valid use of planar sensor design

UNIVERSITY OF SOUTH CAROLINA

Future Work

- Create multielectrode with increased surface area/perimeter of the electrodes
- Normalize potential to distance of electrodes
- Create multielectrode with different composition gradient

UNIVERSITY OF SOUTH CAROLINA

Acknowledgements

- National Science Foundation (NSF)
- University of South Carolina – College of Engineering and Computing
- Strategic Approaches to the Generation of Electricity (SAGE)
- Principal investigators: Dr. Jochen Lauterbach and Dr. Jason Hattrick-Simpers
- Graduate mentor: Ben Ruiz-Yi

UNIVERSITY OF SOUTH CAROLINA

394216 Long-Range Correlations in Liquid Water

Abstract:

The purpose of this research project was to observe water molecules' environments to determine whether or not the system forms identifiable clusters of similar environments. The different clusters aid in proving the hypothesis that different water patches at differing high and low densities values form in a size of 1nm in a system at room and lower temperatures. Molecular dynamics computer simulations were done using the dl_poly classic program with the water model potential TIP4P/2005. The simulations were run at various configurations for density, temperature, and pressure. The results were read by a program, which took the file from the simulations and determined the coordinate positions of the water molecules. Different clusters were found based on the results. These results are indicative of distinctive structures formed in supercooled water that may lead to the water patches of varying densities.

394155 Frictional Study of Polyethylene Glycol Monolayers on Silica Substrate

Abstract:

Lubricating micro- and nano-electromechanical systems (MEMS and NEMS) has become increasingly challenging as devices are continuously miniaturized. As the dimensions of these systems decrease, their surface area to volume ratios increase significantly and frictional forces become dominant where moving parts exist. Solid-state lubrication via self-assembled monolayers presents a promising alternative to traditional macro-scale lubricants which are too bulky to use in such small systems.[1,2] Previous studies have shown that self-assembled monolayers containing alkylsilanes and mixed alkylsilane and perfluoroalkylsilane chains reduce friction coefficients compared to equal length systems. Although, hydrocarbon-based monolayers have been widely studied and shown to improve frictional properties and device life, mixed monolayers impart the beneficial properties of perfluoroalkylsilanes (low critical surface tension and high thermal and mechanical stability) to the film. Furthermore, a mixed length configuration with longer alkylsilane chains relative to the perfluoroalkylsilane chains leads to a liquid-like layer formed by the taller chains that fold over and align with the shearing direction.[3]

In this study, we investigate polyethylene glycol (PEG) monolayers using molecular dynamics simulations as a potential solid-state lubricant for NEMS. The presence of oxygen atoms within the chains render these chains more flexible than their alkyl- and perfluoroalkyl-counterparts. Thus, even sparsely functionalizing a surface leads to a uniform film. We have investigated the tribological properties of such films across a range of surface coverages and chain lengths. Preliminary results show the effects of surface coverage on friction coefficients is diminished as chain length is increased. This suggests that equivalent friction coefficients could be achieved while reducing separation distances by using longer chains on sparser surfaces.

394001 A Machine-Learning Model to Predict Activation Energies of Hydrogenation Reactions

Abstract:

The goal of this research is to develop a machine-learning model to predict activation energies associated to the hydrogenation of organic compounds in the presence of specific catalytic substrates. The current method of predicting activation energies involves a linear relationship with the change in Gibbs free energy of the reaction, which can be very inaccurate. Other methods include experimentation and simulation, both of which have significant cost and/or time drawbacks. This research seeks to improve on these methods to develop a means to predict activation energies quickly, accurately, and with minimal cost.

393764 Microscopic Modeling of the Self Assembly of Poly(ethylene oxide)-Poly(propylene oxide)- Poly(ethylene oxide) (PEO-PPO-PEO) Block Copolymers: Critical Micelle Concentrations

Abstract:

Single chain mean field theory (SCMFT) is applied as a coarse-grain simulation model for Pluronics, a trademarked branch of triblock copolymers. The microscopic model is used to quantify the micellization process of the nonionic class of surfactants at 37 degrees Celsius for drug delivery applications. The simple model with only three interaction parameters is shown to be able to reproduce with high accuracy experimental data for the critical micelle concentrations of different chain lengths of the polymers. These parameters are physically significant and are close to the experimental values from Flory models. In addition, aggregation number of the surfactants and volume fraction profiles are explored. The effects of the chain lengths of hydrophobic tails and hydrophilic heads on CMC values are discussed. SCMFT is transferable in surfactants of different ABA block conformations and other more complex polymer structures where previous simulations have not excelled.

Recommendations for Legislative Actions to Reduce Carbon Emissions in the Electricity Production Sector

Sam White

American Institute of Chemical Engineers
University of Michigan
Washington Internships for Students in Engineering

August 2014

Executive Summary

There is currently a movement in America pushing for environmental effects to be taken into consideration when assessing new and existing business ventures. As emphasized by President Obama"s 2014 State of the Union Address, a major area for this movement to make progress is within the energy sector [1]. Electricity production currently produces the most greenhouse gas emissions of any energy sector in the US, making it the energy sector where the largest environmental impact can be made with proper policy action [2].

Luckily, currently-available, low-emission renewable energy technologies are most well suited for electricity production over any other energy sector application. As a result of recent government regulations combined with high coal prices, coal plants, which currently dominate the electricity production landscape, are being retired at an accelerated rate [8]. This, combined with increased electricity demand, has opened up the door for new electricity production technology to be implemented. However, due to the low price of natural gas resulting mainly from the onset of fracking, efforts to implement new renewable-sourced electricity production facilities have been stalled in favor of natural gas power plants. While natural gas plants produce only half the carbon emissions of coal, and do not produce harmful byproducts such as sulfur dioxide or mercury, they are not as environmentally friendly as the other available alternatives, such as nuclear, solar, or wind [11] [12] [13] [14].

The fact that natural gas dominates the shift away from coal is not surprising, however, given the current economic landscape. The figure below shows the levelized cost of electricity (LCOE) for select available electricity production technologies, giving a $/Mwh cost of production by factoring in all applicable costs (construction, fuel, maintenance, operation, etc.) [23].

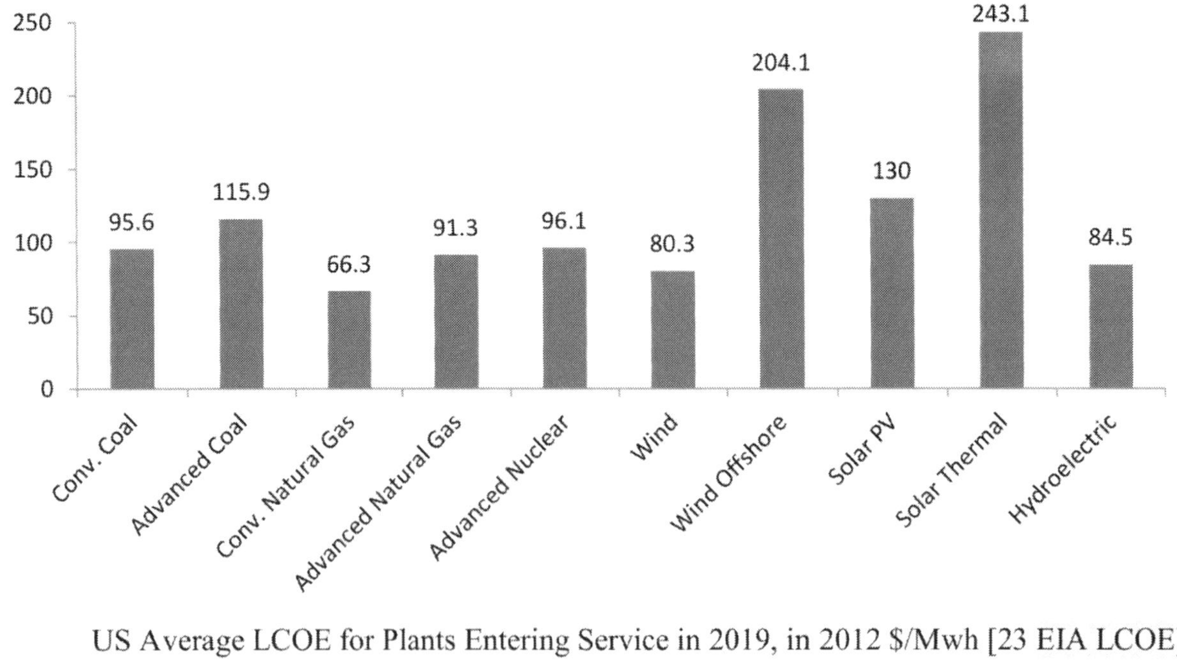

US Average LCOE for Plants Entering Service in 2019, in 2012 $/Mwh [23 EIA LCOE]

As seen in figure above, conventional natural gas currently offers the cheapest form of electricity production. It is important to note, though, that some alternatives like nuclear or wind are already competitive with other conventional electricity production methods like coal. Other alternatives, like solar or offshore wind, are still too expensive to be competitive at the moment, mainly due to the fact that electricity production from renewable sources is still too variable. With the advent of advanced electricity storage technology this issue would be eliminated, but such technology is currently unavailable or too expensive to be implemented economically [19].

Because of their positive environmental impacts, alternative electricity production methods are given some assistance by the federal government in order to make them more economically competitive. These economic aids mainly come in two forms: tax credits and loan programs. On the tax side, an installment tax credit allows solar production facilities a credit equal to 30% of the cost of implementation, while a production tax credit allows wind, hydro, and geothermal plants to receive a .23 cents/kwh credit. However, the installment tax credit is only available until the end of 2016, and the production tax credit already expired at the beginning of 2014 [26]. On the loan program side, the government has extended loans to help both nuclear and renewable power plants, though the renewable plant loan program expired in 2011 [27] [28].

On top of aid being given to alternative electricity production methods, the federal government can also level the playing field by putting a burden on conventional electricity sources. This mainly comes in one of two forms, a carbon tax or a cap-and-trade program, both of which seek to charge electricity producers for the amount of carbon emissions they produce. On top of producing revenue for the federal government, these programs help incentivize electricity producers to innovate new ways to reduce their carbon emissions. Though no such program has ever been implemented, a cap-and-trade program was able to at least pass the House in 2009 [24] [25].

Even small actions by the federal government can have noticeable effects in shifting the production methods used in the electricity production sector. Just a relatively small tax such as $10/metric ton of carbon emissions would cause significant carbon emission reductions compared a scenario in which no government actions are taken [33]. Additional analysis specific to natural gas and renewables shows that any carbon tax would actually increase the amount of natural gas-based electricity production for the next couple of decades, and increase the amount of renewable-based electricity production indefinitely. Additionally, if current favorable tax policies for renewables are renewed long-term, the use of renewables is significantly higher than those projected given current tax policy deadlines, especially after 2030 [34] [35].

As a result, it is recommended that a two-pronged approach be taken to limit the amount of economic disturbance while ensuring the electricity production sector makes a greater shift towards renewables given the reduction of coal use. First, a carbon tax equivalent of $10/metric ton of carbon emissions should be enacted to reflect the social cost of emissions. Each year this tax would increase 5% per year after inflation in order to push companies to innovate new carbon reduction technologies. Second, the production tax credit should be extended for all renewable electricity production methods using the current credit of .23 cents/kwh. This credit would decrease 5% per year after inflation to mirror the carbon tax in order to account for the increased rate of improvement in renewable technologies and the relative advantage given by the increasing carbon tax.

It is important that such a program be long-term in order to allow industry to accurately model the economic landscape their projects will enter. Given the current gridlock in Washington and the long-term nature of implementing new electricity production facilities, a long-term tax plan is the only way to ensure economic stability. Factoring in only revenue from taxes on natural gas use and costs from credits extended to renewables use, such a program would pull in an average surplus of $5.5 billion per year between 2015 and 2040; factoring in coal would increase this figure significantly, particularly in the near future. Feeding at least part of this surplus into loan programs for nuclear plants would ensure that an increased amount of low carbon electricity production facilities are able to come to market.

Table of Contents

Table of Figures

Table of Tables

1.0 Introduction

In his 2014 State of the Union address, President Obama stated that "[O]ne of the biggest factors in bringing more jobs back is our commitment to American energy"[1]. He stressed the importance of "a cleaner energy economy" and acknowledged that "[it] won't happen overnight, and it will require tough choices along the way." Though he touted the success of his "all-of-the-above energy strategy", and highlighted the advantages of natural gas and solar, the President laid out no definitive plans for how the US would approach energy production in the future. However, with the release of recent carbon emission regulations, the executive branch seems to be taking at least some action to direct the country towards a cleaner energy economy. However, if a cleaner energy economy is going to happen in America, the legislative branch needs to get more involved by developing and implementing a long-term energy policy plan.

Of all the areas of energy production in the America, the most promising area for change is the electricity production sector, since a vast majority of alternative energy sources are most viable for electricity production instead of applications like transportation or residential heating. Luckily, this is also the energy sector in which the most impactful changes can be made: in 2012, electricity production was the largest single contributor to greenhouse gas emissions in the US, accounting for 32% of greenhouse gas emissions as seen in Figure 1 below [2].

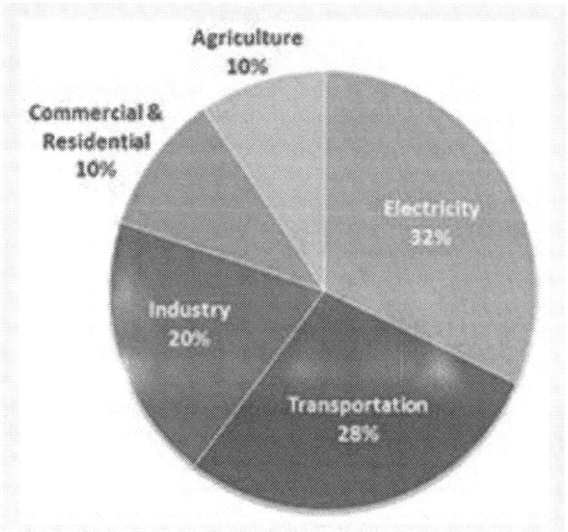

Figure 1. Total Greenhouse Gas Emissions by Economic Sector in 2012 [2]

The implementation of a cleaner energy economy has shown to be a popular policy among voters from both parties. According to a Washington Post – ABC News poll from June 2014, 70 percent of voters say that the federal government should require limits to greenhouse gases from existing power plants. 57 percent of Republicans, 76 percent of Independents, and 79 percent of Democrats support state-level limits on greenhouse gas emissions. Most significantly, voters are willing to pay extra for energy if it means a reduction in greenhouse gas emissions. "Asked whether Washington should still go forward with limits [on energy-sector greenhouse gas emissions] if „they significantly lowered greenhouse gases but raised your monthly energy bill by 20 dollars a month," 63 percent of respondents say yes, including 51 percent of

Republicans, 64 percent of Independents and 71 percent of Democrats" [3]. These results suggest that Americans on both sides of the aisle are in support of implementing a cleaner energy economy.

In addition to the popularity of a cleaner energy economy, it also makes sense from an environmental and health stand point. On top of the fact that the reduction of greenhouse gas emissions reduces the rate of climate change, it brings along health benefits as well. With the release of its Clean Power Plan in June 2014, which seeks to cut carbon pollution from the power sector by 30% by 2030 by imposing stricter emission limits on existing plants, the EPA projects that "annual public health benefits will total somewhere between $55 billion and $93 billion by avoiding up to 100,000 asthma attacks and 2,100 heart attacks each year." By contrast, they predict compliance cost of $7.3 to $8.8 billion per year, significantly lower than the expected public health benefits [4]

However, despite the ongoing executive branch actions regarding energy policy, the legislative branch has taken relatively little action regarding energy policy lately, mainly due to party-politics-driven gridlock. At the beginning of 2014, the Production Tax Credit for renewables was allowed to expire for the first time since 2004, and only the fourth time since its enactment in 1992 [5]. In May of 2014, the Senate allowed a bipartisan energy efficiency bill to die due to disagreement over other energy related issues, such as the fate of the Keystone XL pipeline and possible legislative response to President Obama"s climate change regulations. In fact, it has been 7 years since Congress passed a major energy bill, which means there has been no major legislative action in response to the changing energy landscape as a result of the shale fracking boom [6].

A plan for implementing a cleaner energy economy becomes even more important as energy demands increase, particularly when it comes to electricity production. As electricity demand increases, more electricity production facilities are built, and how these facilities produce electricity greatly affects their environmental impact. According to the EIA"s Annual Energy Outlook 2014 report, US electricity demand will increase from anywhere between 14% and 41% between 2012 and 2040, though their best estimates suggest that electricity demand will increase by 29% over this time period. Additionally, the growth of electricity demand is projected to increase sometime around 2025, marking the beginning of an upwards slope in electricity demand growth for the first time in 75 years [7].

In order to meet this demand and fill the gap left behind by plant retirements, new electricity production facilities need to be built. As a result of recent emission regulations like the Clean Power Plan, coal power plants are being retired at an accelerated rate. However, due to the low price of natural gas in recent years thanks to fracking, much of this gap has been filled by natural gas fired plants instead of alternatives, like nuclear or renewables. This is mainly due to economics: even the cheapest alternative energy sources, which are on the verge of being if not already cost competitive to coal, simply cannot compete with the low price of natural gas. While natural gas is a lower-emissions alternative to coal, it has emissions nowhere near as low as alternative electricity sources.

The onset of natural gas is not a bad thing, however: using natural gas as a bridge fuel to reduce emissions in the immediate future while hoping that renewables become more cost competitive in the future is not unreasonable given the rate of development and investment in renewable technologies. Natural gas also offers the unique advantage of "spinning reserve", a source that can quickly be powered up to meet rapid fluctuations in demand. This means natural gas can offer a flexibility that base load electricity sources like coal or nuclear cannot, while acting as a buffer against the variability of renewables like solar or wind.

Embracing natural gas in the here and now is also economically responsible, as it keeps electricity prices low by offering a cheap alternative to coal and helps support the growing American natural gas production market. However, without government action, this shift towards natural gas will dwarf any efforts to make renewables a larger share of our electricity production sector. If the government is serious about using natural gas as a bridge fuel to a cleaner energy economy in the electricity sector, a long-term policy needs to be enacted to slowly guide the market in the right direction.

2.0 Overview of Electricity Production Methods and Their Environmental Impacts

In order to discuss energy policy, it is first important to establish the electricity needs of the country, as well as the current capabilities of various electricity production methods and their effects of the environment. As seen in Figure 2 below, coal and natural gas currently dominate the electricity production sector, with nuclear and renewables making up the bulk of the remaining production.

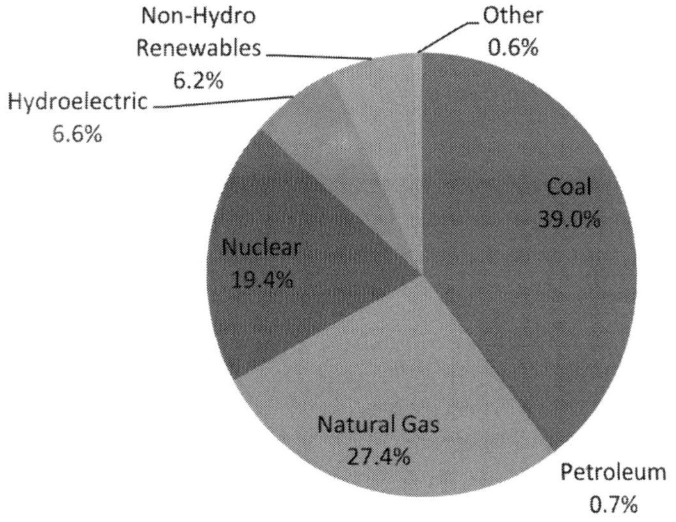

Figure 2. Share of Total US Electricity Production by Source [Adapted from 8]

2.1 Coal and Natural Gas

The two main fossil fuel based / conventional energy producing sources, coal and natural gas power plants are currently the largest source of electricity in America, accounting for nearly two-thirds of the country"s total electricity production [8]. Coal plants are relatively easy to keep in operation over a long time scale, but it is difficult for them to make relatively quick changes to production capacity in response to the changing needs of the grid, particularly during peak-usage hours. As a result, coal plants are very useful in providing a baseline power supply somewhere between peak and off-peak requirements.

Natural gas, on the other hand, allows for a very flexible energy production method. Gas flow rates into a furnace can be controlled with relative precision, unlike solid input streams found in coal plants. As a result, natural gas plants can be very responsive to the changing needs of the energy grid. In fact, natural gas plants are typically used to augment other energy production facilities for this very reason.

From an environmental perspective, the EIA considers conventional electricity sources like coal and natural gas as being the only contributors to CO_2 emissions in the electricity production sector. In 2010, the EIA reported that 1969 million metric tons of CO_2 were released by coal plants nation-wide, while 1285 million metric tons of CO_2 were release by natural gas plants [9].

However, coal tends to have other unique environmental impacts. First, burning coal tends to release nitrogen oxides, which tend to have larger warming potentials than CO_2. For example, N_2O has a SAR 100 year warming potential of 310, meaning that one ton of N_2O has 310 times the impact on global warming compared to one ton of CO_2 over the course of 100 years [10]. From a life cycle assessment standpoint, coal plants produce 974 metric tons of CO_2 equivalent per gigawatt-hour, which accounts for this increased warming potential as well as CO_2 emissions due to non-production factors, such as plant construction or fuel transportation [11]. Coal also tends to produce SO_2, which, along with nitrogen oxides, is the major man-made cause of acid rain. According to the EPA, "In the United States, roughly two-thirds of all SO_2 and one-quarter of all nitrogen oxides come from electric power generation that relies on burning fossil fuels, like coal" [12]. Additionally, coal plants are a major source of mercury, a highly toxic element. According to the EPA, "Coal-burning power plants are the largest human-caused source of mercury emissions to the air in the United States, accounting for over 50 percent of all domestic human-caused mercury emissions" [13].

In the case of natural gas, environmental impacts are lower than that of coal. According to the EPA, "… the burning of natural gas produces nitrogen oxides and carbon dioxides, but in lower quantities than burning coal". In fact, from a life cycle assessment standpoint, natural gas plants produce 469 metric tons of CO_2 equivalent per gigawatt-hour, less than half that of coal [11]. Additionally, the EPA states that "[e]missions of SO_2 and mercury compounds from burning natural gas are negligible". [14]. While the EPA also mentions that unburned natural gas being released either from a power plant or during transportation is a concern, though in 2009 the

EIA stated that "[i]ncreases in energy-related methane emissions [were] largely from underground coal mining", suggesting that methane releases are more prevalent in the procurement of coal than in the use and transport of natural gas [14] [15].

2.2 Nuclear

In the US, nuclear energy makes up nearly one-fifth of all electricity production [8]. According to the World Nuclear Association, "it [nuclear power] is especially suitable for large-scale, continuous electricity demands, which requires reliability (i.e. base load)" [16]. In this way, nuclear power plants are analogous to coal plants in the sense that they are very useful in meeting baseline power needs, but have trouble making quick changes in response to fluctuating peak-hour demand.

From an environmental standpoint, "nuclear power plants do not emit CO_2, SO_2, or nitrogen oxides as part of the power generation process" according to the EPA [17]. However, as with all power production facilities, the construction of nuclear power plants leads to non-negligible greenhouse gas emissions. Based off a compilation of life-cycle emissions studies gathered by the Nuclear Energy Institute (NEI), nuclear energy"s life-cycle emissions are significantly lower than those of coal or natural gas plants, and are in fact comparable to those of renewable like solar, wind, or hydro [18]. From a life cycle assessment standpoint, nuclear produces 15 metric tons of CO_2 equivalent per gigawatt-hour, which is the second-lowest behind wind, and only about 1.6% that of coal plants [11]. Much of these emissions come from the construction of plants, which are significantly larger than other electricity production facilities, and the transportation of fuel.

2.3 Solar and Wind

As shown in Figure 2, only about 6% of electricity production in the US comes from non-hydro renewable sources. Of this 6%, a majority comes from solar (in the form of both photovoltaic cells and solar thermal) and wind [8]. Of all of the available technologies, solar and wind are the most challenging to integrate into the grid due to the fact that they are considered variable sources: if the sun doesn"t shine or the wind doesn"t blow, these technologies are not actively putting energy into the grid, meaning demand has to be met by some other means [19] [20]. In order to offset this, large-scale production facilities are looking into using some form of energy storage system. This would allow excess energy collected during times of sunshine or wind gusts to be used when the sun doesn"t shine or the wind doesn"t blow [19]. However, the storage capacity of current electricity storage technology is limited. As a result, any useful amount of electricity storage is currently too expensive, which limits the viability of renewables.

From an environmental standpoint, neither solar nor wind produce carbon emissions during their operating lifetime, since their energy is derived from the sun and not from burning fossil fuels [19] [20]. However, like nuclear, solar and wind do have some emission impacts due to the processes used to manufacture materials and construct plants: wind is responsible for 14 metrics tons of CO_2-equivalent per gigawatt-hour, while solar photovoltaics are responsible for

39 metric tons of CO_2 equivalent per gigawatt-hour [11]. As with nuclear plants, many of the emissions related to renewables come from construction of production facilities. In the case of solar photovoltaics, higher life-cycle emissions are seen as a result of the emissions that result from the manufacturing process of the solar cells themselves.

2.4 Other Renewables

Solar and wind are not the only renewable sources of energy available to the US. As seen in Figure 2, nearly 6.6% of electricity production comes from hydro plants. The main issue with hydro is that most of the available sites for dams in the US have already been dammed, meaning there is little room for expansion. As a result, the Obama administration has identified a collection of dams that currently do not produce electricity, but outfitting these dams with turbines would only increase the current US hydro capacity by roughly 15% [21]. Another alternative energy source is geothermal, which utilizes naturally-occurring underground sources of heat to power turbines. While this energy is clean, it is only available on a regional basis: Hawaii and the western US have a large geothermal production potential, while areas like New England do not [22]. Since hydropower does not have a lot of potential for growth, and geothermal is regionally specific, they will not be explored as in-depth in this paper.

2.5 Environmental Impact Summary Table

Table 1 below summarizes the life-cycle environmental impact per gigawatt-hour of the main electricity production methods highlighted in this paper.

Technology	Metric Tons of CO_2 released / GWh
Coal	974
Natural Gas	469
Nuclear	15
Solar	39
Wind	14

Table 1. Life-Cycle Environmental Impact Summary of Electricity Production Technologies [Adapted from 11]

3.0 Economics in the Electricity Production Sector

On top of technological limitations and environmental concerns, economics plays a major role in determining which technologies get used for energy production. On the production side, economic factors tend to play a major role in determining which electricity production technologies are utilized by producers. On the consumer side, the electricity production technology used by electricity providers plays a role in the price of electricity sold to consumers.

3.1 Comparing Costs of Electricity Production Technologies

Due the wide range of variables present in energy production, it can be difficult to compare the cost of one electricity production technology to another. However, a useful metric, levelized cost of electricity (LCOE), lays out the cost of given technologies on a per-MWh basis, factoring in such costs as construction, fuel, operating costs, maintenance, and financing costs [23]. Figure 3 below shows the US average LCOE for plants entering service in 2019, in 2012 dollars/MWh, and does not factor in any public financial support (tax credits, subsidies, etc.).

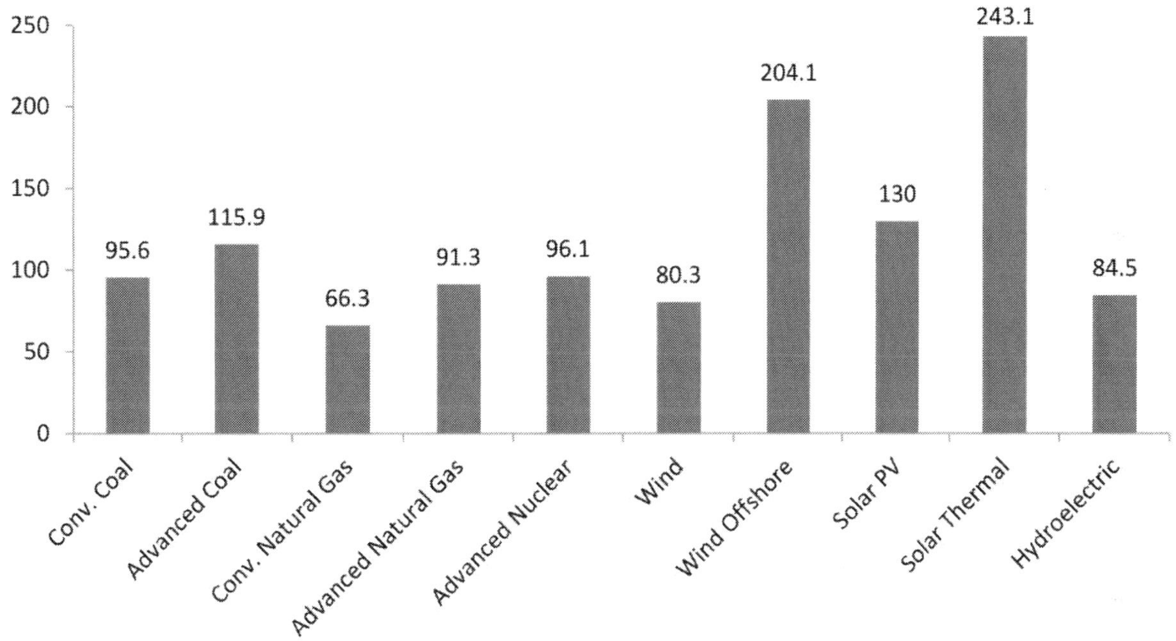

Figure 3. US Average LCOE for Plants Enetering Service in 2019, in 2012$/MWh [23]

As Figure 3 shows, the conventional energy production sources, coal and natural gas, are currently the cheapest available electricity production technologies, which explains why they currently account for such a large portion of US electricity production. Much of this can be contributed to the low price of fuel. Natural gas in particular is benefitting greatly from low fuel prices as a result of the recent American shale fracking boom. However, advanced coal and natural gas facilities, which incorporate carbon capture and sequestration technologies, are significantly more expensive than their conventional counterparts.

While coal and natural gas remain the cheapest electricity production technologies, it is important to note that there are alternatives that are already competitive. For example, the LCOE of nuclear energy is nearly the same as conventional coal and advanced natural gas, and is in fact less expensive than advanced coal. This is particularly noteworthy since nuclear fills the same base load electricity production role as coal, and produces no greenhouse gas emissions during its operational lifetime. It is also important to note that, while many of the renewable energy sources are significantly more expensive than conventional sources or nuclear, wind built on land has the second lowest LCOE, higher than only conventional natural gas.

3.2 Capacity Factor

One factor contributing to the relative expense of renewable sources, such as offshore wind or solar, is their variable nature. Since they are not able to produce electricity at all times of the day, renewable sources suffer from prolonged periods of not operating at peak capacity. The easiest way of quantifying this is through capacity factors, the ratio of actual output to the theoretical output if the plant could operate at full capacity indefinitely. A comparison of capacity factors for each electricity production technology can be seen in Figure 4 below.

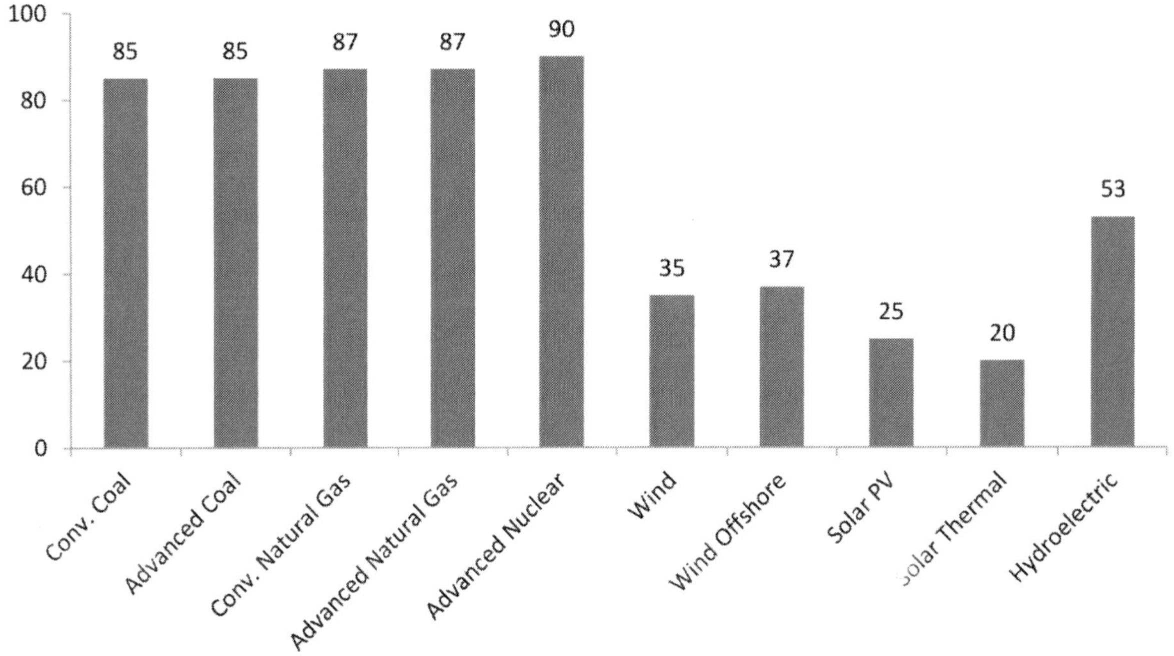

Figure 4. Capacity Factor (%) of Various Electricity Production Technologies [23]

It is important to note that there is no electricity production technology with a capacity factor of 100%: plants do not always run at maximum capacity, and require downtime for things like routine maintenance or system upgrades. However, it is easy to see that renewable sources have significantly lower capacity factors, mainly due to the fact that they cannot produce electricity at all times of the day. Because of this, capital and operating costs of renewable energy plants factor significantly more into their LCOE.

However, with advancements in technology, particularly in the energy storage field, the capacity of renewable sources could significantly increase, and possibly even become comparable with conventional sources. This would cut LCOE by anywhere from 40-70% (depending on current capacity factor). If this electricity storage is cheap enough, even the most currently expensive renewable energy technologies would be competitive with conventional sources.

4.0 Recent Policy Strategies

In recent years, legislators have used a wide variety of strategies to help shape the electricity production sector, most notably by offering some form of economic incentive to desirable technologies. These economic incentives typically come in the form of tax credits, though they may also come in the form of subsidies or guaranteed loan programs. Congress has also made efforts in the past to enact a carbon tax program, but such a program has never been able to make it out of Congress [24] [25].

4.1 Tax Credits for the Electricity Production Sector

The major economic incentive congress has implemented is by extending tax credits to electricity producers, which reduces the amount of taxes they have to pay to the federal government and thus improving their bottom line[1]. Recently, these tax credits have been targeted at renewable electricity resources, most notably solar and wind, though to a lesser extent hydroelectric and geothermal as well [26]. These tax credits have taken on two different forms: an investment tax credit and a production tax credit.

The investment tax credit allows qualifying electricity producers to receive a tax credit equal to a certain percentage of their initial cost of implementation. Solar energy installments are eligible for a 30% credit, while geothermal plants are eligible for a 10% credit. These credits have no limit. As of right now, these tax credits extend to any qualifying facility put in service before January 1, 2017. Wind power facilities put in service between 2009 and 2012 were also eligible for a 30% investment credit, but that has not been renewed [26].

The production tax credit, on the other hand, allows qualifying electricity producers to receive a tax credit based off the amount of electricity produced. As of the end of 2013, this tax credit was 2.3 cents / kWh, and extended to wind, geothermal, and hydropower facilities [5] [26]. If a facility qualifies for both the investment and production tax credit, they are only permitted to claim one of them, at their choosing. The production tax credit extended to any qualifying facility under construction by January 1, 2014, and is available for qualifying facilities for the first 10 years of operation. However, since January 1, 2014, the production tax credit for renewable electricity production facilities has not been extended [26].

4.2 Loan Programs

The federal government also offers loan programs for electricity producers. These loan programs differ from tax cuts mainly in the fact that producers are required to pay back their loan; while there is a cost to the federal government, it is only in the risk of borrowers defaulting,

---

[1] It is important to note that non-profits, such as municipal utilities and utility cooperatives, do not pay taxes, meaning that federal tax cuts do not have an effect on their bottom line. While alternative programs to incentivize non-profits do exist, they are not covered in the scope of this paper.

which has historically been low. Under Title XVII of the Energy Policy Act of 2005, two loan programs were outlined that are still used today: section 1703 and section 1705 [27] [28].

Section 1703 provides loans for "a new or significantly improved technology that is NOT a commercial technology" [27]. Qualifying facilities must also be located in the US. In the current state of the electricity production sector, the only facilities that really meet these requirements are related to nuclear energy; the most recent loans are an $8.3 billion loan for two new nuclear power plants in Georgia and a $2 billion loan for a uranium enrichment plant in Idaho [27]. This program is essential to nuclear power plants, as the most prohibitive cost to entering the market for nuclear plants is upfront capital costs. Due to relatively low operating costs once the plant is built, loan repayment is usually not an issue.

Section 1705, on the other hand, focuses on loan programs for renewable energy production facilities. Unlike section 1703 rules, section 1705 has a time limit, and was only available to projects that had begun construction before September 30, 2011. Since then, this project has not been renewed. Section 1705 loans tend to be lower than their section 1703 counterparts: one of the more expensive section 1705 loans was for $1.3 billion to an Oregon-based wind farm [28].

The major significant difference between section 1703 and 1705 loans is how the Credit Subsidy Cost (CSC) is handled. By law, the CSC, which is "the expected long-term liability to the Federal Government by issuing the loan guarantee", must be paid either by budget appropriations or by the borrower directly. For section 1703, which tend to have more expensive loans and thus larger CSCs, the CSC is paid by the borrower. For section 1705, the CSC was paid for by appropriations [27] [28]. In the American Recovery and Reinvestment Act of 2009, $6 billion was appropriated towards section 1705 CSCs. [29]

4.3 Carbon Tax and Cap-and-Trade Program

On top of offering economic benefits to desirable electricity production methods, Congress can also place economic burdens on undesirable electricity production methods. In the electricity production sector, this can be practically done in one of two ways: either a carbon tax or a cap-trade-program.

A carbon tax works very simply: electricity producers pay an additional tax for each unit of carbon produced during electricity production. This means that carbon-heavy electricity production facilities like coal (and to a lesser extent, natural gas) power plants would pay an additional amount for their electricity production, thus increasing their LCOE. In the same way that tax credits and subsidies seek to level the economic playing field by making renewables cheaper, a carbon tax would seek to level the playing field by increasing the cost of conventional electricity production methods. However, there has never been a carbon tax approved by either chamber of Congress.

An alternative method to a carbon tax is a cap-and-trade program, which seeks to place an artificial limit on the amount of carbon that can be emitted. Once this limit has been determined, electricity producers can buy permits to emit certain amounts of carbon dioxide. The government only sells as many permits as it takes to reach the cap, and then monitors the emissions of each producer to make sure they do not exceed their permitted limit. Producers are able to buy or sell permits to or from other producers as needed.

Cap-and-trade programs create economic incentive to reduce carbon emissions in two ways: first, reducing carbon emissions means that electricity producers need to buy fewer permits, saving them money. Second, if producers reach their permitted cap, their competitors determine the cost of additional permits (since each company now control its own excess permits, not the government). This means exceeding the permitted carbon emission limit can be disproportionately costly, either through the purchase of competitor permits or paying government fines.

In order to ensure long-term carbon emission reduction, cap-and-trade programs also incorporate a permit allowance reduction system. This means that, each year, the amount of permits allowed is reduced, effectively lowering the acceptable amount of carbon dioxide that can be emitted by electricity producers. While a cap-and-trade program did make it out of the House in 2009, the bill was never heard by the Senate, and died when that session of Congress ended [24] [25].

5.0 Projected Electricity Sector Growth: Electricity Sector Capacity Additions for 2013

In order to establish a reference case, it is useful to look at how the electricity production sector has changed recently. Figure 5 below shows the EIA projections for electricity sector capacity additions from 2008.

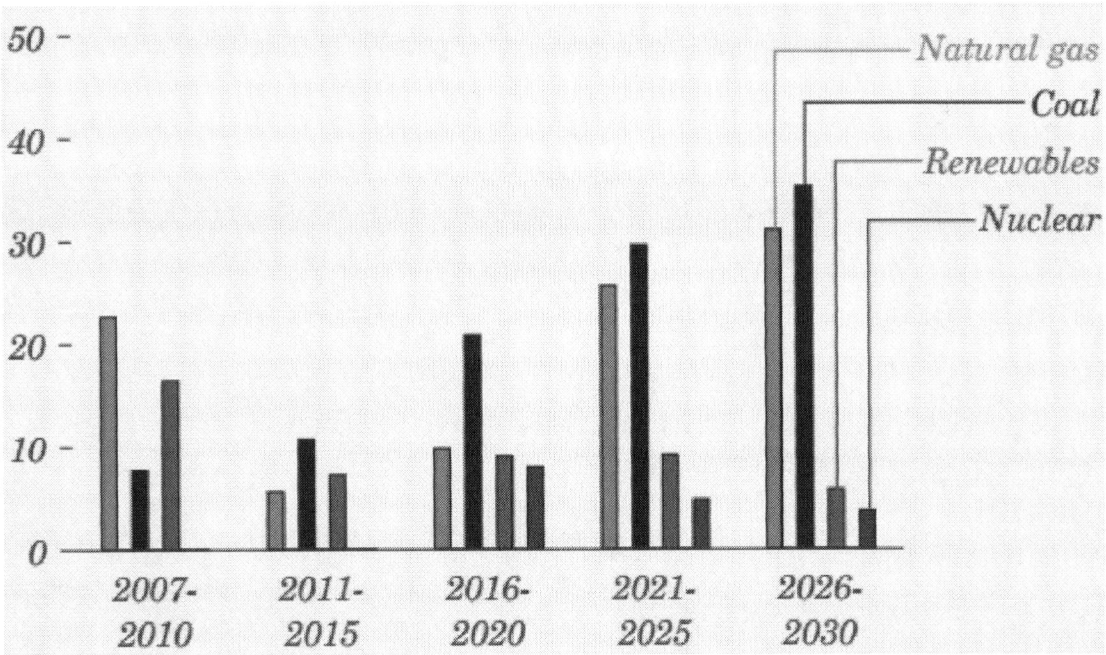

Figure 5. Electricity Generation Capacity Addition Projections by Fuel Type, 2008 (Gw) [30]

As Figure 5 shows, in 2008, it was projected that coal would begin to dominate electricity production additions starting in 2011. Much of this can be attributed to the increasing cost of natural gas, as shale fracking was still in its infancy in 2008. Additionally, renewables, which see a large spike in 2007-2010, were projected to fall by the wayside after 2010 as they begin to be dwarfed by both natural gas and coal.

However, as a result of the shale fracking boom, the reduced cost of renewables, and increased government pressure to reduce carbon emissions, these 2008 projections do not reflect what has actually happened in recent years. Figure 6 below shows the electricity generation capacity additions for 2013.

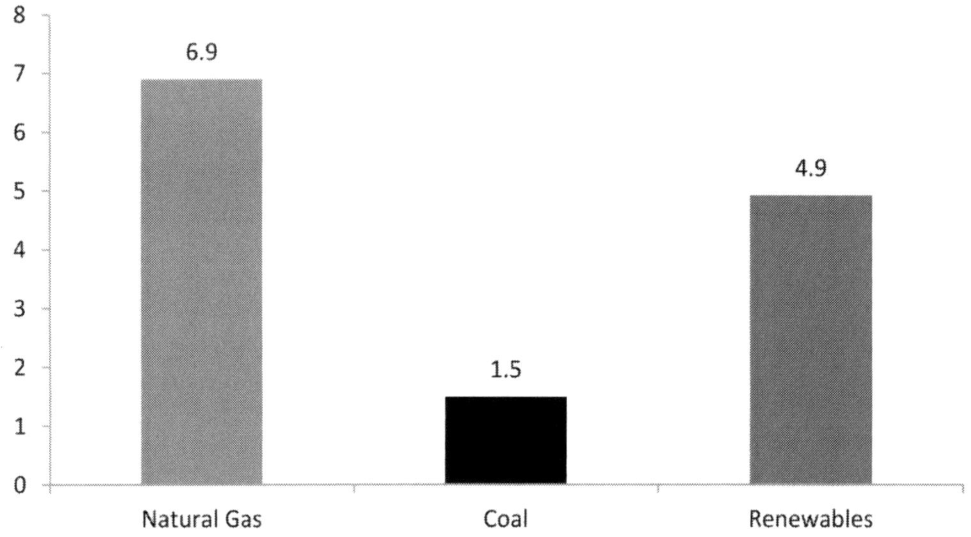

Figure 6. Electricity Generation Capacity Addition by Fuel Type 2013 (GW) [Adapted from 31]

As seen in Figure 6, natural gas dominated new capacity additions, accounting for more than half of the electricity capacity additions in 2013. Renewables made up roughly one-third of new additions, which is attributed to the completion of a number of large scale solar thermal plants in Arizona and California as well as "falling [photovoltaic] costs, aggressive state renewable portfolio standards, and continued federal investment tax credits" [31]. Coal makes up a meager 11% of new capacity additions, in stark contrast to the projections shown in Figure 5.

Long term projections have also changed to reflect the changing economic and political landscape surrounding electricity production. Figure 7 below shows the EIA projections for electricity sector capacity additions from 2014.

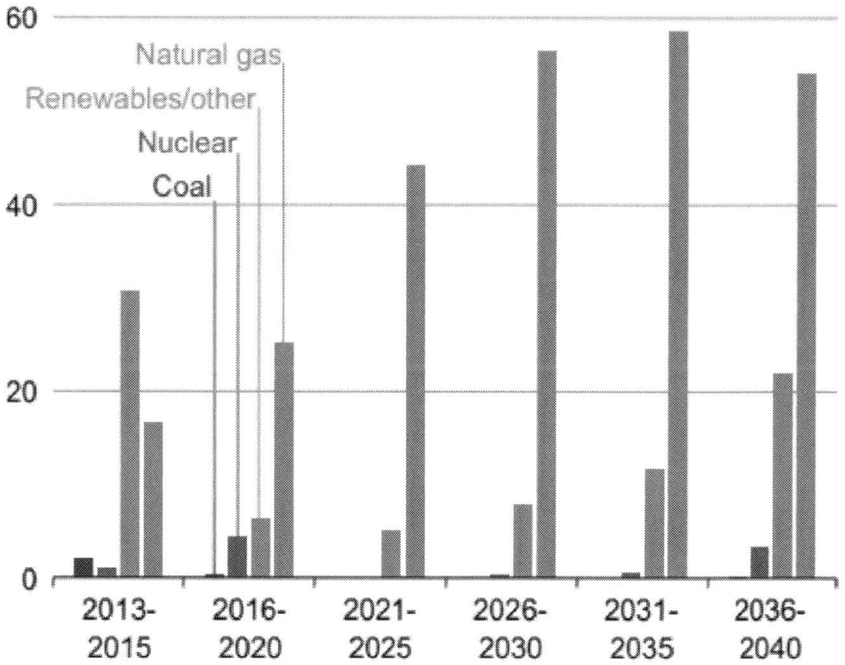

Figure 7. Electricity Generation Addition Projections by Fuel Type 2014 (GW) [32]

As Figure 7 shows, natural gas dominates new additions through 2040, mainly as a result of low gas prices. Coal additions fall virtually to 0 as a result of increased emissions regulations. Most alarmingly, the addition of renewables is actually lower than those projected in 2008 for the period between 2016 and 2035. This is mainly due to the extremely low cost of natural gas relative to any other fuel source, making it take up a larger share of the new additions in the 2014 projections than conventional sources did in the 2008 projections. It is important to note, however, that these projections only show a reference case where no additional government actions are taken.

5.1 Policy Action Effects: Overall Carbon Emissions

Despite the recent transition from coal to natural gas, as well as the increased onset of recently-cheap renewables, carbon emissions from the electricity production sector are projected to increase over the next several decades if current laws and regulations remain in effect. This is mainly attributed to the shift towards natural gas-based electricity production and the shift away from renewables and nuclear as a result of cheap natural gas for the foreseeable future [33]. However, if some form of carbon tax is implemented, carbon emissions could decrease instead increase, as seen in Figure 8 below.

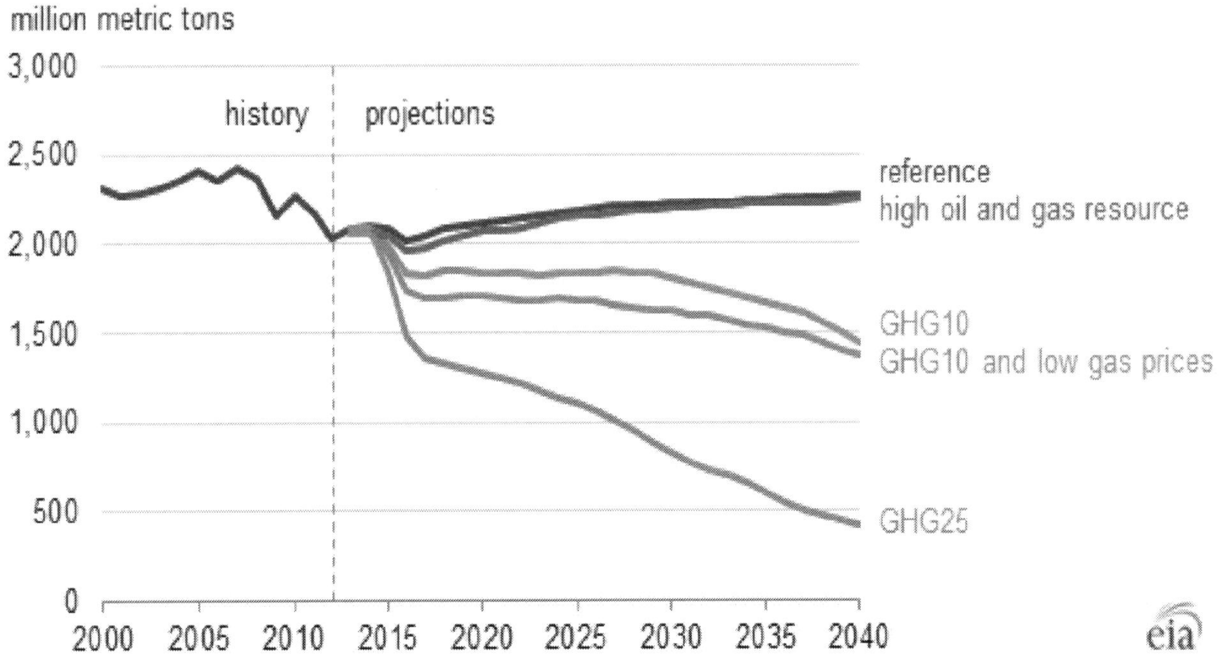

Figure 8. Electricity-Related CO_2 Emissions in Five Cases, 2000-40 [33]

Figure 8 takes into account five different potential cases. The reference case reflects no change in laws or regulations surrounding electricity production and assumes relatively constant economic factors relative to those in 2014. As seen in Figure 8, the reference case shows a 12% increase in emissions up through 2040 [33].

The GHG 10 and GHG 25 cases both refer to the implementation of a carbon tax (or equivalent cap-and-trade permit cost) in 2015 of $10 / metric ton of CO_2 or $25 / metric ton of CO_2, respectively. The GHG 10 case shows a 16% lower CO_2 emission rate in 2025 and a 36% lower CO_2 emission rate in 2040 relative to the reference case. The GHG 25 case shows an even greater decrease in CO_2 emission rates, with a 49% reduction in 2025 and an 82% reduction in 2040. In both cases, there is a large dip in emission rates before 2020, partially due to an accelerated shift away from coal and partially due to an increased implementation of renewables [33].

The High Oil and Gas Resource case assumes the same conditions as the reference case, except that oil and gas resources become more abundant than currently projected. In this case, CO_2 emissions decrease ever so slightly more than in the reference case as a result of increased shift towards natural gas, but by 2040 emission rates are roughly the same as the reference case as a result of producers expressing and increased favor towards natural gas relative to renewables [33].

Finally, the GHG 10 and Low Gas Prices case effectively combines the GHG 10 and High Oil and Gas Resources cases. In this case, reduction of CO_2 emissions before 2020 is greater than those seen in the GHG 10 case, mainly due to an even greater shift from coal to natural gas. However, because a greater share of this shift away from coal is given to natural gas and not renewables, CO_2 emissions do not decrease as drastically after 2030 compared to the

GHG 10 case; it can be extrapolated that CO_2 emissions would actually begin to be higher than the GHG 10 case sometime between 2040 and 2050 [33].

Overall, these cases suggest that some form of action needs to be taken in order to produce a marked reduction in carbon emissions from the electricity production sector. If nothing changes, CO_2 emissions will increase slightly over the next 40 years, in contrast to the significant emission reductions we"ve seen since 2005. Emissions get even worse with an increased gas resource and lower gas prices, which may result with the advancement of fracking and other extraction technologies. Based off of this projection, the only way to continue a trend of CO_2 emission reduction is to accelerate the shift away from coal and ensure that an increased share of that shift is made up of renewables or nuclear.

5.2 Policy Action Effects: Conventional Fuels

When assessing the viability of the cases listed in the previous section, it is important to understand the effects of each action on each specific electricity production method. As discussed previously, there is already a shift towards natural gas and away from coal in response to lower gas prices and increased carbon emission regulations. Figure 9 below shows how possible policy actions may augment that shift by showing the projected natural gas electricity generation in the five previous cases.

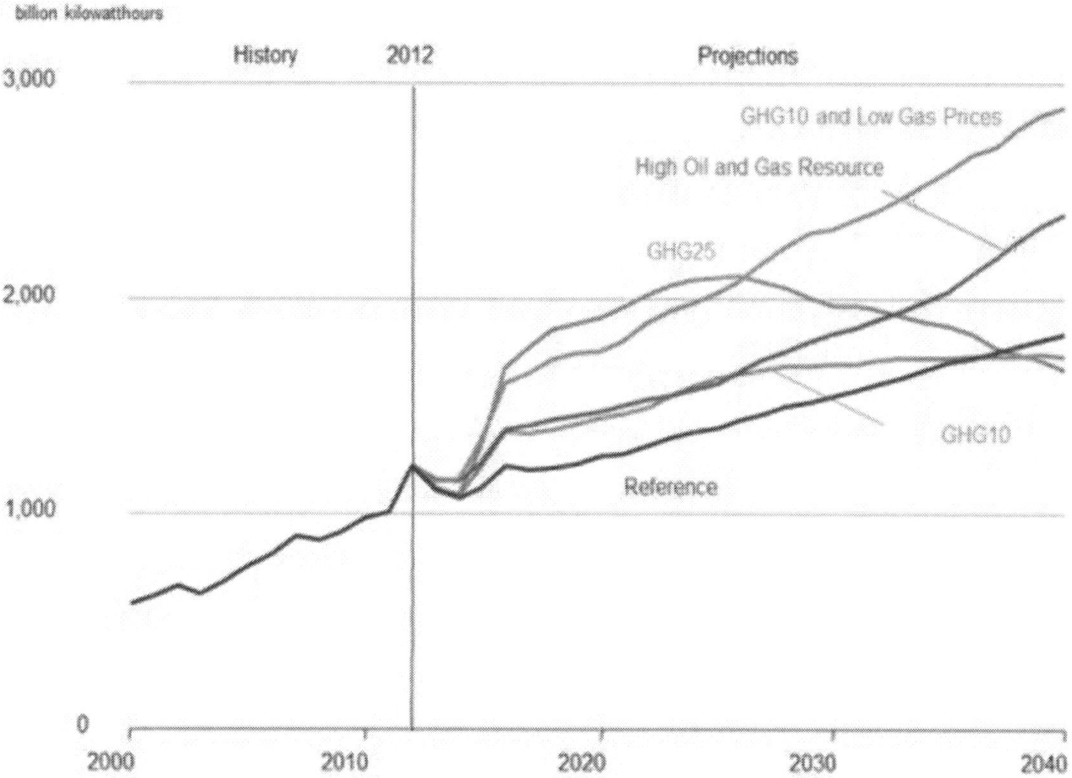

Figure 9. Natural Gas Electricity Generation in Five Cases, 2000-40 [34]

As seen in Figure 9 above, electricity production via natural gas is projected to increase over the next several decades in all cases. In the reference case, we see a relatively stable and constant increase of natural gas fired plants as they replace retiring coal plants; in 2012, 37% of electricity generation comes from coal, and in 2040 that share is 32%, marking a relatively slow shift [34]. With a high gas resource, this shift is more dramatic in the near future, and the long term rate of increase in natural gas plants is also slightly higher than the reference case.

The most interesting part of this projection, however, is the effects of a carbon tax. In all GHG cases, the shift towards natural gas in the near future is increased relative to the reference case. In the GHG 10 case, the share of electricity from coal is 28% in 2025 and 19% in 2040. For the GHG 25 case, the shift is even more drastic, with the coal share being 10% in 2025 and 1% in 2040 [34]. In both cases, the increase in natural gas production is curbed by the increased implementation of nuclear of renewables: in the GHG 10 case, natural gas electricity production levels off around 2030, while in the GHG 25 case, it actually decreases starting in 2025. However, if gas prices remain low, natural gas electricity production continues to increase through 2040, reaching the highest level of any of the five test cases.

5.3 Policy Action Effects: Renewables

Despite the fact that natural gas dominates new electricity production additions, renewables are also seeing an increase in implementation as the electricity production shifts away from coal. However, without some form of government action, the rate of renewable additions will be dwarfed by those of natural gas, leading to the increased carbon emissions seen in Figure 8. Figure 10 below shows the effects of some possible government actions by plotting the projected renewable electricity generation in four cases.

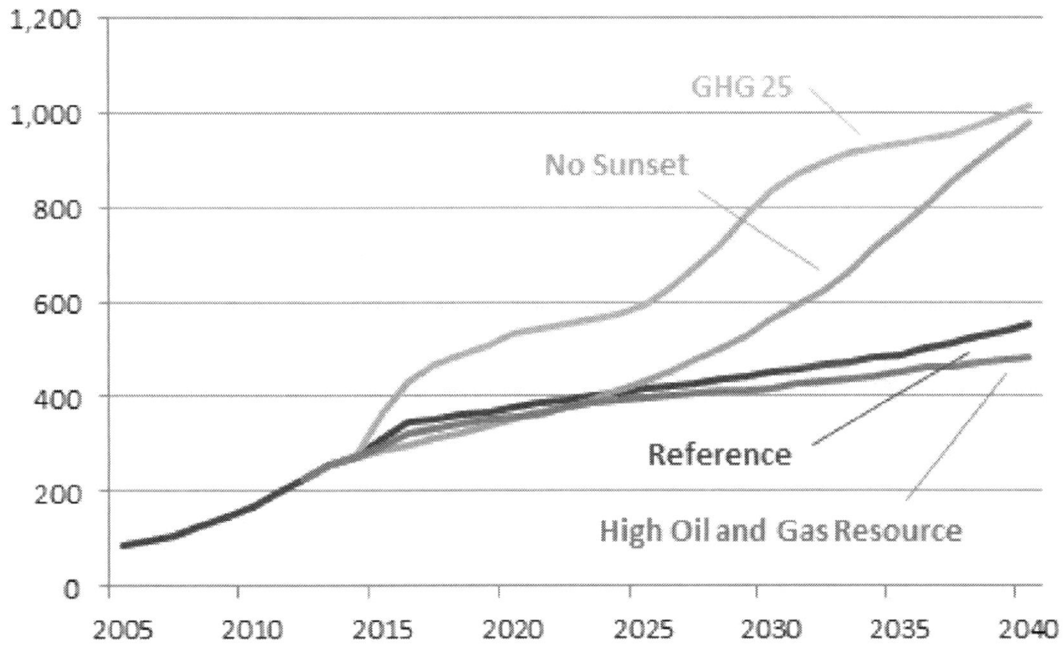

Figure 10. Renewable Electricity Generation in Four Cases, 2000-40 (billion kWh) [Adapted from 35]

In all cases, electricity generation from renewables increases over the next few decades. However, for the reference and high gas resource cases, the rate of increase in renewables is less than that seen between 2000 and 2014, mainly due to the increased implementation of natural gas power plants. The no sunset case shows the effects of extending renewable tax credits (i.e. the installment tax credit and the production tax credit) indefinitely in their current forms. In this case, the growth seen between 2000 and 2014 remains relatively constant through 2025, at which point the rate of implementation of renewables begins to increase steadily. In the GHG 25 case, a large increase is seen just after 2015 in response to the implementation of the carbon tax, after which the rate of implementation of renewables begins to level off. However, around 2025, electricity generation from renewables begins to increase dramatically once again, as more projects come into completion in response to the increased cost of natural gas power plants, whose production begins to decline as seen in Figure 9.

Compared to the reference case, both the no sunset and GHG 25 cases see disproportionate increases in solar and wind relative to other renewables like hydro or geothermal. When it comes to solar, electricity production in both cases is triple that of the reference case in 2040; for wind, it"s double. In both cases, geothermal actually sees slightly lower electricity production in 2040 compared to the reference case, and hydro sees relatively little growth between 2014 and 2040 in all cases [35].

6.0 Policy Recommendations

The electricity production sector in the US is rapidly changing as a result of the shale fracking boom. While the large shift from coal to natural gas will improve carbon emissions on a per kwh basis, the low price of natural gas is also reducing the amount of renewable sources entering the market. If no government action is taken, the growth of electricity demand will still mean an increase in carbon emissions from the electricity production sector over the next 30 years due to the low prices of natural gas. If the government truly wishes to use natural gas as a bridge fuel to a low-carbon electricity sector, a carbon tax needs to be implemented and tax credits for renewables need to be extended.

A carbon tax is necessary to appropriately reflect the societal cost of carbon-emissions-heavy technology. However, a large carbon tax is not necessary, and would cause unnecessary instability as the market acted swiftly to adjust. Instead, a low carbon tax that progressively increases is the most responsible way to implement such a change. Based off the projections in Figures 8 and 9, a carbon tax of \$10 / metric ton of CO_2 would lower carbon emissions in the electricity production sector without drastically changing the economic viability of natural gas facilities. This tax would increase at a rate of 5% per year in order to incentivize investment in further carbon emission reduction technology. For a cap-and-trade program equivalent, this means the acceptable level of carbon emissions each year would be reduced, effectively increasing the cost of emissions a proportional amount.

On the other side of things, renewable technology still needs some assistance to be competitive with conventional fuels, as seen in Figure 3. However, the current split between the investment tax credit and the production tax credit makes the development of new and improved renewable technologies difficult, as it is more difficult to predict the economic environment these technologies will emerge into. A single, long-term tax strategy would offer economic stability and predictability that promotes private investment that has been yet unseen. In this way, a single, long-term tax strategy should be used to incentivize the use of all types of renewables in the electricity sector.

Of the two programs currently available, fully embracing the production tax credit makes the most sense, as it incentivizes the implementation of energy efficient and energy dense technologies[2]. It also incentivizes the development of improvements that reduce the upfront cost of these technologies, which help open them up to small scale residential and commercial consumers. The current rate of 2.3 cents per kwh is an appropriate start-point for such a tax program, but with the implementation of a carbon tax, indirect incentive towards renewables increases proportionately. The increasing nature of the carbon tax combined with the increased rate of improvement for renewable technologies actually means that the production tax credit can be slowly reduced. As such, a 5% reduction per year to mirror the carbon tax rate increase would be appropriate.

In both the carbon tax and the production tax credit, both programs need to be comprehensive and long term in order to allow the electricity producers to accurately model the economic environment that they will be entering with new projects. Due to the long-nature of construction of electricity production facilities, as well as the long operating life-time of these

[2] Again, this credit would not incentivize non-profit utility providers. Alternative programs, not covered in the scope of this paper, could be used to incentivize these utility providers.

facilities, stability in the tax structure from the government is paramount. The unexpected expiration of the production tax credit in 2013 mainly to political gridlock is unacceptable, and should be avoided in the future if at all possible.

All in all, if such a two pronged approach of a $10 / metric ton carbon tax combined with an extension of the production tax credit is used, the government will turn a net profit as a result. Based on the projections in Figures 9 and 10, there will be an average profit of $5.5 billion per year between 2015 and 2040 under such a program, with yearly profits ranging between $4.5 billion and $6.1 billion [adapted from 40 and 41]. Keep in mind that these projections only factor in the revenue from natural gas facilities; factoring in the revenue of coal should increase revenue further, despite the increase in cost from the increased use of renewables.

It is further recommended that the money made from this program be reinvested in the loan programs for electricity producers. While the reintroduction of the section 1705 program may be beneficial, it seems more beneficial to use this money to offset or eliminate the CSC cost burden on borrowers of section 1703 loans. With the accelerated retirement of coal, reduced or eliminated CSC costs would incentivize increased use of nuclear power, which acts as a more practical and carbon neutral base load power supply alternative compared to natural gas.

7.0 References

[1] Obama, B. (2014, January 28). *President Barack Obama's State of the Union Address.* Speech presented at the House Chamber, Capital Building, Washington, DC. Retrieved from http://www.whitehouse.gov

[2] EPA. (2014, June 2). *Learn About Carbon Pollution From Power Plants*. Retrieved from http:// www2.epa.gov

[3] Clement, S. (2014, June 2). A huge majority of American support regulating carbon from power plants. And they're even willing to pay for it. *The Washington Post*. Retrieved from http://www.washingtonpost.com

[4] Koch, W. (2014, June 3). EPA carbon emission limits may shift energy mix. *USA Today*. Retrieved from http://www.usatoday.com

[5] Union of Concerned Scientists. (2014, January 31). Production Tax Credit for Renewable Energy. Retrieved from http://www.ucsusa.org

[6] Davenport, Coral. (2014, May 12). Amid Pipeline and Climate Debate, Energy-Efficiency Bill Is Derailed. *New York Times*. Retrieved from http://www.nytimes.com

[7] EIA. (2014, May 7). *Annual Energy Outlook 2014*. "Market Trends: Electricity demand". Retrieved from http://www.eia.gov

[8] EIA. (2014, May 21). *Electric Power Monthly*. "Table 1.1. Net Generation by Energy Source: Total (All Sectors), 2004 – March 2014". Retrieved from http://www.eia.gov

[9] EIA. (2013, May). *State-Level Energy-Related Carbon Dioxide Emissions, 2000-2010.* "Table 2. 2010 state energy-related carbon dioxide emissions by fuel". Retrieved from http://www.eia.gov

[10] EIA. (2011, February). *Emissions of Greenhouse Gases in the United States 2009*. "Table 5. Greenhouse gases and 100-year net global warming potentials". Retrieved from http://www.eia.gov

[11] Meier, P.J. (2002, August). *Life-Cycle Assessment of Electricity Generation Systems and Applications for Climate Change Policy Analysis*. pg 70. Madison, Wisconsin: University of Wisconsin, Fusion Technology Institute. Retrieved from fti.neep.wisc.edu/pdf/fdm1181.pdf

[12] EPA. (2012, December 4). *What is Acid Rain?*. Retrieved from http://www.epa.gov

[13] EPA. (2014, March 10). *Basic Information | Mercury*. Retrieved from http://www.epa.gov

[14] EPA. (2013, September 25). *Natural Gas | Clean Energy*. Retrieved from http://www.epa.gov

[15] EIA. (2011, March 31). *Emissions of Greenhouse Gases in the US*. Retrieved from http://www.eia.gov

[16] World Nuclear Association. (2014, May). *World Energy Needs and Nuclear Power*. Retrieved from http://www.world-nuclear.org

[17] EPA. (2013, October 22). *Nuclear Energy | Clean Energy*. Retrieved from http://www.epa.gov

[18] Nuclear Energy Institute. (2014). *Life Cycle Emissions Analyses*. Retrieved from http://www.nei.org

[19] EPA. (2012, February 28). *Solar Energy | Energy and Global Climate Change in New England*. Retrieved from http://www.epa.gov

[20] EPA. (2014, May 1). *Wind Energy | Energy and Global Climate Change in New England*. Retrieved from http://www.epa.gov

[21] DOE. (2012, April 17). *Energy Department Report Finds Major Potential to Increase Clean Hydroelectric Power*. Retrieved from http://energy.gov

[22] EPA. (2014, May 1). *Geothermal Energy | Energy and Global Climate Change in New England*. Retrieved from http://www.epa.gov

[23] EIA. (2014, April). *Levelized Cost and Levelized Avoided Cost of New Generation Resources in the Annual Energy Outlook 2014*. Retrieved from http://www.eia.gov

[24] American Clean Energy and Security Act of 2009. H.R. 2454 of the 11th Congress. Retrieved from http://www.opencongress.org/bill/111-h2454/actions_votes

[25] Hulse, Carl. (2010, July 22). Democrats Call Off Climate Effort Bill. *New York Times*. Retrieved from http://www.nytimes.com

[26] 26 U.S.C. §§46, 48

[27] DOE. *Section 1703 Loan Program*. Retrieved from http://energy.gov

[28] DOE. *Section 1705 Loan Program*. Retrieved from http://energy.gov

[29] Wilson Sonsini Goodrich & Rosati. *Status Update – Department of Energy Title XVII Loan Guarantee Program*. Retrieved from http://www.wsgr.com

[30] EIA. (2008, June). *Annual Energy Outlook 2008*. "Early Capacity Additions Use Natural Gas, Coal plants Are Added Later". Retrieved from http://www.eia.gov

[31] EIA. (2014, April 8). *Half of power plant capacity additions in 2013 came from natural gas*. Retrieved from http://www.eia.gov

[32] EIA. (2014, May 7). *Annual Energy Outlook 2014*. "Most new capacity uses natural gas and renewables". Retrieved from http://www.eia.gov

[33] EIA. (2014, June 9). *Power sector CO_2 emissions sensitive to policy changes and natural gas supply*. Retrieved from http://www.eia.gov

[34] EIA. (2014, May 7). *Market Trends: Emissions*. "Energy-related carbon dioxide emissions are sensitive to potential policy changes". Retrieved from http://www.eia.gov

[35] EIA. (2014, April 29). *Renewable electricity projections show growth under alternative assumptions in AEO2014*. Retrieved from http:// www.eia.gov

381133 Membrane-on-Membrane -Development of Liposome-Entrapped Hydrogel

Abstract:

The biological systems have complex environmentally-responsive separation and conversion process. It is therefore necessary to construct a bio-inspired system. If the artificial systems and biological systems are accreted, it is possible to develop new conceptual response system, inspired by biological systems. Liposome membrane is amphipathic molecule and self-assembly organization and metastability condition, and was constructed non-polar and polar membrane interface bilayer, and was changed phase condition with temperature and lipid mixture ratio. Liposome can design the lipid interfacial surface. It was reported that liposome was utilized for Drug Delivery System (DDS) as capsule, in addition, it was reported that misfolded protein was refolded on liposome membrane, liposome was able to control gene expression, and chiral recognition of amino acid and peptide. Liposome was reported close link with vital function expression. However, there are some disadvantages, such as lower immobilized amount of liposome and multi-step for operation immobilization. In this study, development of novel method utilizing liposome membrane entrapped matrix in order to utilize biological membrane function. In order to subject characteristic analysis and functional evaluation, optical resolution of amino acid was investigated. In this study, liposome immobilized method was focused on direct entrapping method of physical containment method.

INTRODUCTION OF PMMA-TAG TO VHH FOR IMPROVING RECOVERY AND IMMOBILIZATION RATE OF VHHS

Dr. Bongmun Kang, Kyoto Institute of Technology, Kyoto-si, Kyoto-hu

Introduction

Camelid antibodies are composed of heavy-chain dimers and are devoid of light chains while nevertheless has an extensive antigen-binding repertoire. Variable domain of heavy chain (VHH) represents the smallest antigen-binding unit with a molecular size of~ 15 kDa in comparison with scFv fragments (30 kDa), Fab fragments (60 kDa), and whole antibodies (150 kDa). They have been expressed in microorganisms and have a high stability and solubility due to simpler structure of beta strand protein with only one disulfide bond. For these reasons, VHH have raised the possibility of various industrial applications such as neutralizing drugs, as well as diagnostic agents for specific detection of biomarkers. Especially, for use of VHH in clinical diagnosis, there is considerable interest in denser immobilization of VHHs on the surfaces of solid materials, such as plastic plates. Michiel et al reported passive adsorption of VHH antibodies onto the surface of polystyrene (PS) plate for use in enzyme-linked immunosorbent assays (ELISA). The highly immobilized VHHs on the plate could offer the potential to increase the sensitivity and specificity of VHHs to various type of antigen. In general, VHHs have been obtained by purification of soluble protein located in periplasmic fraction in a E. coli cell. However, the production yields of soluble VHH are considerably low and it is significantly changed with aminoacid sequences of VHHs that solubility, stability of VHH and protease resistance were highly related. In contrast, larger amounts of insoluble and inactive VHH were obtained from inclusion body. Thus, the refolding of denatured VHHs is necessary to prepare soluble VHH with original antigen-binding affinity and specificity. We recently reported that a newly found affinity peptide retained the ability to interact with the surface of poly(methyl methacrylate) (PMMA-tag) [1]. Consequently, an efficient refolding method for PMMA-tag-fused anti-RNase scFv (scFv-PM) were developed in our lab [2]. However, there is no study about the PMMA-tag-fused VHH for improving refolding and immobilization of denatured VHHs. Thus, this study examined whether PMMA-tag was available for efficient refolding and immobilization of VHHs.

Title

Introduction of PMMA-Tag to VHH for improving recovery and immobilization rate of VHHs

Author/Byline Information

Bongmun Kang, Kyoto Institute of Technology, Kyoto, Kyoto-Fu, Japan
Kagnari Yamakawa, Kyoto Institute of Technology, Kyoto, Kyoto-Fu, Japan
Yoshihisa Hagihara, AIST, Ikeda, Osaka, Japan
Michimasa Kishimoto, Kyoto Institute of Technology, Kyoto, Kyoto-Fu, Japan
Yoichi Kumada, Kyoto Institute of Technology, Kyoto, Kyoto-Fu, Japan

Material and Methods

hCG (Human chorionic gonadotropin) purchased from SIGMAwas biotinylated with biotinamidocaproate N-hydroxysuccinimide ester (Nacalai Tesque Co. Ltd). Biotinylated hCG (biotin-hCG) was used as a model antigen for sandwich ELISA. HRP-labeled streptavidin was purchased from VECTOR Laboratories. As a model, Anti-hCG VHH antibody that the structure was determined by Renisio et al was used in this study. And the gene of anti-hCG VHH was chemically synthesized with optimization of codons to these used in *E. coli*. Other chemicals were of reagent grade unless specified. The VHH gene was amplified by polymerase chain reaction (PCR) with the synthesized DNA as a template and the sense and antisense primers that recognition sitea for restriction enzymes, *Nde* I (catatg) and *Not I* (gcggccgc) were additionally introduced at 5' terminal ends, respectively. . The amplification of VHH gene was confirmed by agarose gel electrophoresis and purified using illustra™ GFX™ PCR DNA and Gel Band0 Purification Kit (GE Healthcare Co. Ltd). The vector pET-PMOMP25 that the nucleotide sequence encoding PMMA-tags (DVEGIGDVDLVNYFEVGATYTFNK) was introduced into the commercially-available expression vector, pET22 (Merck) was constructed in the previous study. The pET22 and pET-PMOMP25 vectors were used for cloning of amplified VHH gene for construction of expression vectors for VHH and PMMA-tag-fused VHH (VHH-PM), respectively. The vectors and DNA fragment of VHH gene were digested with the restriction enzymes *Not* I, and *Nde* I at 37 °C overnight.The digested vectors and DNA fragment of VHH gene were purified, and then they were ligated by addition of Ligation High (TOYOBO Co. Ltd). *E. coli* (HST08 Premium Competent Cell) was transformed with the ligated mixture, and transformant cells were selected on LB-agar plate containing 50 μg/ml ampicilin. The sequences of VHH genes introduced into the vectors were confirmed by DNA sequencing with the primer of T7 promoter. The constructed vectors for VHH and VHH-PM were introduced to *E. coli* Rosetta (DE3) (Merck), and transformants were selected on LB-agar plate containing 50□g/ml ampicillin (Amp) and 34 μg/ml chloramphenicol(Cam). On the basis of the method reported previously, the production and purification of VHH and VHH-PM were performed with some modification. Firstly, single colonies of transformed *E. coli* cells were innoculated in the 200ml Erlenmeyer flask containing 10 ml of LB medium (Difco™ LB Broth, Miller;0.25g per 10ml) supplemented with Amp and Cm, and the flask was shaken at 200 rpm and 37℃ overnight. A portion of the pre-cultured *E. coli* cells was transferred to 500ml shake flask containing 50 ml of Overnight Express™ Instant TB medium (Merck Co. Ltd)containing Amp and Cam, and the final cell density was adjusted to be $OD_{600} = 0.1$. The flasks were shaken at 200 rpm and 37℃ for 24 h. Cells were harvested by centrifugation for 20 min at 10,000 g and 4℃, and then suspended with 10 ml of 1×PBS containing 1 mg/ml Triton X-100 (Nacalai Tesque Co. Ltd) and 1xProtease inhibitor (Nacalai Tesque Co. Ltd.). The sonication of suspension was repeated 4 times by a probe-type sonicator (ULTRASONIC DISRUPTOR UD-211; TOMY Co. Ltd) at 30 W (Duty level: 30%) for 4 min with cooling by iced water. Insoluble aggregates were collected by centrifugation for 15 min at 10,000 g and 4℃ and were washed three times with distilled water followed by lyophilization (FREEZE DRYER FD-1000; EYELA Co. Ltd). They were then dissolved in 5 ml of denaturation buffer (pH 7.5) containing 6 M guanidine-HCl, 5 mM DTT, 300 mM NaCl and 40 mM sodium phosphate. After the removal of cell debris by centrifugation and filtration, the supernatant was directly applied to 5 cm^3 of a HisTrap HP affinity column (GE HealthCare Co. Ltd) equilibrated with buffer A (pH 7.5) containing 8 M urea, 20 mM imidazole and 2xPBS. After washing the column with buffer A, the bound proteins were eluted with buffer B (pH 7.5) containing 8 M urea, 400 mM imidazole and 2xPBS. The flow rate of both buffer A and B was 1 ml/min. The eluate was dialyzed against 1xPBS containing 8 M urea (8 M urea-PBS). Protein concentration in the eluate was determined by DC-protein assay (Bio-rad Co. Ltd), using BSA dissolved in 8 M urea as a standard. Both

VHH and VHH-PM obtained were stored at 4℃ until use. The aggregation properties of VHH, as well as those of VHH-PM, in the presence of 0.5 M urea were evaluated. VHH and VHH-PM were diluted to 24 kinds of refolding buffers that pH and NaCl concentrations were comprehensively different. For adjustment of pH from 6.0 to 8.5, 6 types of good's buffers were adopted. (MES; pH 6.0, ADA; pH 6.5, MOPS; pH 7.0, HEPES; pH 7.5, EPPS; pH 8.0, and TAPS; pH 8.5) Final concentrations of VHH, good's buffer and urea after dilution were 400 □g/ml, 50mM and 0.5M, respectively. Hundred □l portions of the mixture prepared were added to microtiter plate made of PS, the plate was incubated for 6 h at 25°C. Turbidity of mixture was evaluated by measurement of absorbance at 450 nm using microplate reader. A stock solution of VHH and VHH-PM dissolved in 8M urea-PBS was rapidly diluted to 1mg/ml, while urea and TAPS (pH 8.5) were presented at 0.5 M and 50 mM, respectively after dilution. The solutions were dialyzed twice against 1L of 50 mM TAPS (pH 8.5) to remove the residual urea from the solution. The aggregates formed were sedimented by centrifugation for 5 min at 10,000 g and 4℃. The refolding yield (*Y*) was determined by the following Eq (1).

$$Y = (VHH_{\text{conc. after}} / VHH_{\text{conc. before}}) \times 100 \qquad (1)$$

Where, Y is the refolding yield (%); $VHH_{\text{conc. after}}$ and $VHH_{\text{conc. before}}$ are the concentrations of VHHs after and before centrifugation, respectively.

The immobilization conditions of VHHs refolded were evaluated by sandwich ELISA. The microtiter plates made of PMMA was originally-manufactured and used for the immobilization of VHH and VHH-PM. The VHH solutions containing 50mM 6 different good's buffers (MES; pH 6.0, ADA; pH 6.5, MOPS; pH 7.0, HEPES; pH 7.5, EPPS; pH 8.0, and TAPS; pH 8.5) and 4 different concentrations of NaCl (0 ~ 300) were prepared. Final concentrations of VHHs were adjusted to 100 □g/ml. The 100µl solution of each well were transferred to PMMA plate, and incubated at 4□C during overnight. The plates were washed 5 times with PBS containing 0.1% Tween 20 (PBST), and then 300 µl of PBST solution containing 2% BSA (2%-BSA-PBST) were added to each well. The plate was incubated at 25°C for 1 h, followed by washing 5 times with PBST. Hundred µl portions of 0 or100 ng/ml of biotinylated hCG were added, and then the plate was incubated for 1 h at 25℃. After washing 5 times with PBST, 100µl portions of HRP-labeled streptavidin 5,000-fold diluted with 2% BSA-PBST were added to each well. Then the plate was incubated for 1 h at 25℃ followed by washing 5 times with PBST. Hundred µl portions of TMB solution were added to each well followed by incubation for 15 min at 25℃ for color development. The reaction was stopped by the addition of 100 µl of 0.5 M H_2SO_4 and then absorbance at 450 nm (reference at 650 nm) was measured via a microplate reader. The immobilization density of VHH and VHH-PM adsorbed in PMMA plate was measured by Micro BCA™ Protein Assay Kit (Thermo scientific Co. Ltd). Hundred µl portions of 0 ~ 400 □g/ml of VHH and VHH-PM in the optimized buffer for immobilization (VHH: 50mM MES, pH 6.0, VHH-PM: 50mM MES, pH 6.0 containing 50mM NaCl) that were determined by the experiments described in XX, were added onto the PMMA plate and the plate was incubated overnight at 4°C. The plates were washed 5 times with PBST, and 75 □l of PBST was added to each well and, then colored signals for VHHs were developed by addition of 75 □l reagent mixture of microBCA kit (Thermofisher Scientific). The 75 µl of 0 ~ 30 mg/ml BSA in PBST was used as a standard. The plates were incubated at 37°C for 2h and then, absorbance at 562 nm was measured by microplate reader. Hundred µl portions of 100 µg/ml VHHs diluted in the optimized buffer for immobilization (VHH: 50mM MES, pH 6.0, VHH-PM: 50mM MES, pH 6.0 containing 50mM NaCl) were added to each well of a microtiter plate followed by incubation overnight at 4℃. After washing the plate, different concentrations of biotinylated hCG (0 – 1000 ng/ml) were detected by the method described in section.

Results

1. Production and purification of anti-hcG VHH and VHH-PM

In this study, single-domain antibody (VHH) which specifically recognizes -chain of human chorionic gonadotropin (hCG), was used as a model. When both VHH and PMMA-tag-fused VHH (VHH-PM) were produced using *E. coli* Rosetta (DE3), both of them were mainly localized in an insoluble aggregates, inclusion body. Amounts of VHH and VHH-PM recovered from 1L cultures were range of 0.3 g ~ 0.5 g, and more than 10 mg/ml of VHHs were finally obtained after one-step IMAC purification in the presence of 8M urea. They were analyzed by SDS-PAGE under the reduced and non-reduced conditions. As shown in Fig. 1 (a), main protein bands of VHH (lane B) and VHH-PM (lane D) were detected at approximately 20 kDa. Therefore, both the VHHs were highly purified from inclusion bodies by one-step IMAC purification in the presence of 8M urea. Interestingly, positions of protein bands for the VHHs shifted to downwards in the absence of reducing agent (lane A and C), in comparison with those of reduced VHHs (lane B and D), indicating that apparent molecular size of VHHs became larger to some extent under the reduced condition. These results suggested that a disulfide bridge inside a VHH molecule was already formed during purification and dialysis processes in the presence of 8M urea. Oligomer of VHH molecules through intermolecular disulfide bridge was also detected (lane A and C) at the position more than 31kDa. In the purification process, both VHHs was first solubilized with 6M guanidine-HCl, and reduced with 5mM DTT. Therefore, the denatured and reduced forms of VHHs were eluted from the IMAC resin, and their concentration just after purification were significantly more than 10mg/ml. These results suggested that intermolecular disulfide bridges of VHHs were formed after IMAC purification because high concentrations of VHHs as reduced forms would be intramolecularly or intermolecularly oxidized in the absence of reducing agent during dialysis against 8M urea-PBS. It was found that formation of intramolecular disulfide bridge in VHHs was completed within 24h by incubation at 25°C after removal of reducing agent. Furthermore, the oxidization was zero-order reaction that the reaction rate was not changed with protein concentration, while reaction rates of intermolecular polymerization through the disulfide bridge might be a function of VHH concentrations. Consequently, as shown in Figure 1 (b), denatured and monomeric VHHs with an intramolecular disulfide bridge were successfully prepared by repeating reduction and re-oxidation of the VHHs at the concentrations less than 1mg/ml.

2. Effects of refolding conditions on aggregation during the refolding of VHH and VHH-PM

Turbidity induced by aggregation of VHH and VHH-PM in their refolding processes was evaluated using 24 types of refolding buffers with different pH and NaCl concentrations. As shown in Fig.2, aggregation of VHH-PM was considerably suppressed with increase in pH and decrease in NaCl concentration. Especially, the degree of aggregation was hardly detectable when scFv-PM was diluted to 50mM TAPS (pH 8.5) without NaCl. A high degree of aggregation was, however, observed in VHH regardless of the range of pH and NaCl concentrations. These tendencies were consistent with the results obtained in the previous study that scFvs genetically-fused with PMMA-tag (scFv-PM) were refolded. [3] Based on these results, the refolding of VHH-PM was conducted with 50 mM TAPS (pH 8.5) solution without NaCl. As a results, high recovery yield of soluble VHH-PM was shown and significantly higher than that of VHH in this refolding conditions (Fig.3). The genetic fusion of PMMA-tag to C-terminal region of VHHs significantly affect the recovery of their soluble protein during refolding by 50 mM TAPS at pH 8.5. PMMA-tag used in this study consist of 5 acidic amino acids, including Asp (D) and Glu (E), and one basic amino acid (Lys (K)). The apparent differences were observed in isoelectric point (pI)s between VHH and VHH-PM tested in present study (Fig.4). The pI of VHH-PM is 6.03, which is lower than that of VHH (7.83). The pI of VHH-PM was decreased by approximately 25%, compared to that of VHH. Thus, the genetic fusion of the PMMA-tag at the C-terminal region of VHHs lost their positive charge at alkaline pH (approximately 8.5) during refolding. Consequently, there are negative charged at the surface of VHH-PM. As a results, it prefers to interact with water rather than with other protein molecules. Finally, there are significant differences between pI values of VHH-PM compared to that of VHH.

3. Determination of the optimal immobilization condition of VHH-PM on the surfaces of PMMA plate

Firstly, the degree which both of VHH and VHH-PM were immobilized on the surface of PMMA microtiter plate was evaluated at 24 different immobilization conditions of pH (6.0, 6.5, 7.0, 7.5, 8.0 and 8.5) and NaCl concentration (0, 50, 150 and 300 mM), and their antigen-binding abilities in adsorption states were compared by a sandwich ELISA.When the refolded VHHs were immobilized on PMMA plates, significantly higher antigen-binding signals were obtained from the wells on which VHH-PM was immobilized, as shown by a sandwich ELISA (Fig.5C). Furthermore, their S/N ratio, which represent the level of a desired signal to the level of background noise, was larger at pH 6 and low NaCl concentrations (Fig.5D). However, the significant differences of antigen-binding signals and their S/N ratio was not observed with the wild-type VHH without PMMA-tag, regardless of various range of pH and NaCl concentration (Fig.5A and B). Based on these results, we decided to immobilize refolded VHHs on the surface of PMMA plate at pH 6 and 50mM NaCl to minimize the level of background noise.

4. Evaluation of the immobilization density of VHH-PM on the surfaces of PMMA plate.

According to the measurements of immobilization density, scFvs genetically fused with the PMMA-tag highly immobilized on the surface of PMMA plate. Thus, we investigated the VHH-PM could be immobilized to the same extent as scFv-PM, even though different type of antibody. The immobilization density of VHHs on the surface of PMMA plate was measured in Micro BCA™ protein assay (Thermo Scientific Co. Ltd) at pH 6 and 50mM NaCl. As a results, the immobilization density was larger in VHH-PM than those in VHH. Particularly, the VHH-PM showed approximately 0-0.45μg/cm^2, which were about 2-fold greater than those detected with VHH (Fig.6). These data suggest that the immobilization density could be highly improved in VHHs by genetic fusion of PMMA-tag to C-terminal region of VHHs at the same rate as scFvs.

5. Evaluation of the antigen-binding activity of VHH-PM immobilized on the surfaces of plastics

Results presented in 3.4 demonstrated that the VHHs genetically fused with the PMMA-tag lead to increase immobilization density on the surface of PMMA plate. Consequently, we investigated whether the antigen-binding activities in the adsorption state could be affected by the genetic fusion of PMMA-tag in VHH (Fig.7). Obviously, the antigen-binding activities in the adsorption state was significantly high in VHH-PM compared to wild type of VHH. There are approximately 8-fold differences in the antigen-binding activities in the adsorption state between VHH-PM and VHH. These results suggested that the introduction of PMMA-tag on the C-terminal region of VHH lead to high-density immobilization on the surface of a PMMA plate, thereby retaining the antigen-binding activities in the adsorption state. In this study, we practically screened an efficient refolding conditions for PMMA-tag-fused anti-hCG VHH (VHH-PM). It could be refolded with a recovery of more than 95% by dialysis against 50 mM TAPS buffer (pH 8.5). By using this method, the solubility of VHH-PM was significantly increased due to a repulsion of negatively charged VHH-PM molecules. Based on these results, further study is directed toward the genetic fusion of PMMA-tag to another type of single domain antibody, for example, Fab, for efficient recovery of denatured single-domain antibodies during refolding and high immobilization rate of refolded single-domain antibodies.

Figure 1. SDS-PAGE of Anti-hCG VHH (A) and Anti-hCG VHH-PM (C) purified from inclusion bodies. (B) and (D) indicate Anti-hCG VHH and Anti-hCG VHH-PM with reduction under 5mM DTT, respectively.

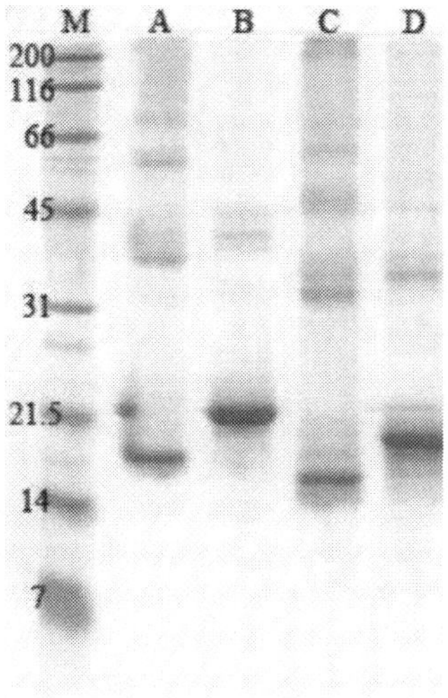

Figure 2. Aggregation test of VHH (A) and VHH-PM (B) diluted in 24 different refolding buffers.

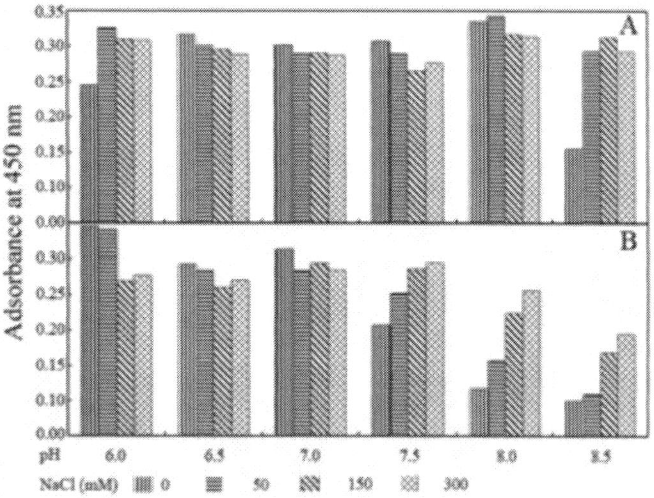

Figure 3. Recoveries of VHH and VHH-PM after dialysis refolding using 50mM TAPS

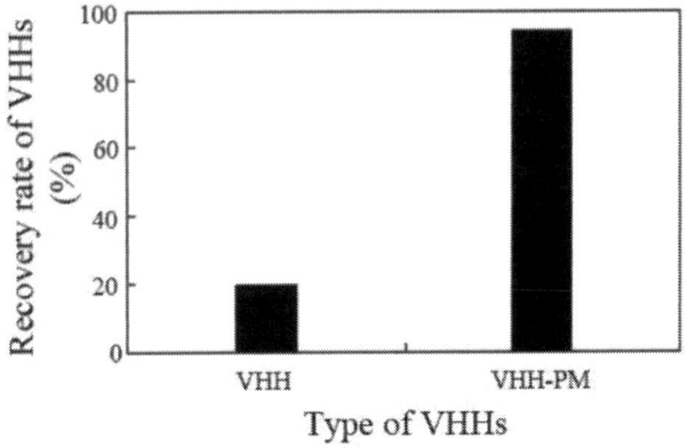

Figure 4. Isoelectric point of VHH and VHH-PM tested in this study

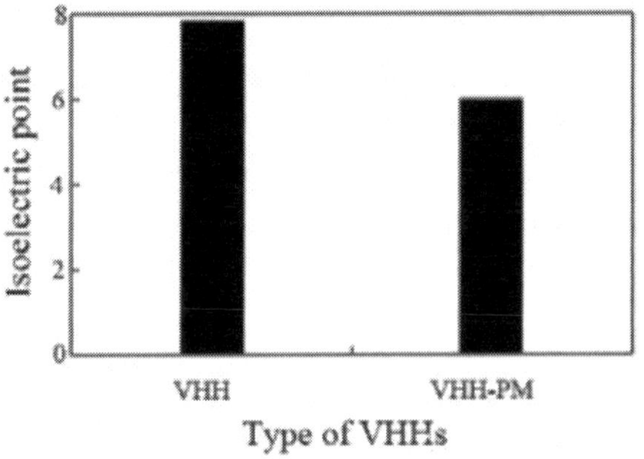

Figure 5. The signal and noise obtained from different immobilization conditions of Anti-hcG VHH after refolding using 50mM TAPS. (A) The signal (S) and noise (N) of Anti-hCG VHH, (B) The S/N ratio of Anti-hCG VHH, (C) The signal (S) and noise (N) of Anti-hCG VHH-PM, (D) The S/N ratio of Anti-hCG VHH-PM.

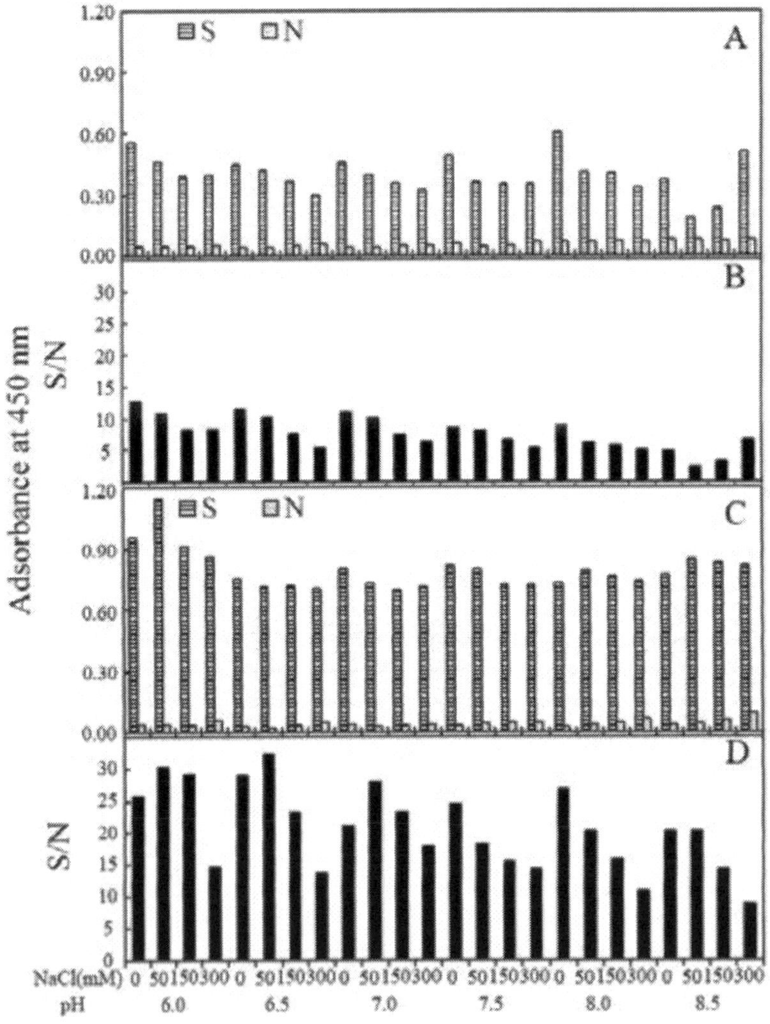

Figure 6. The immobilization density of Anti-hCG VHH and Anti-hCG VHH-PM after dialysis refolding using 50mM TAPS

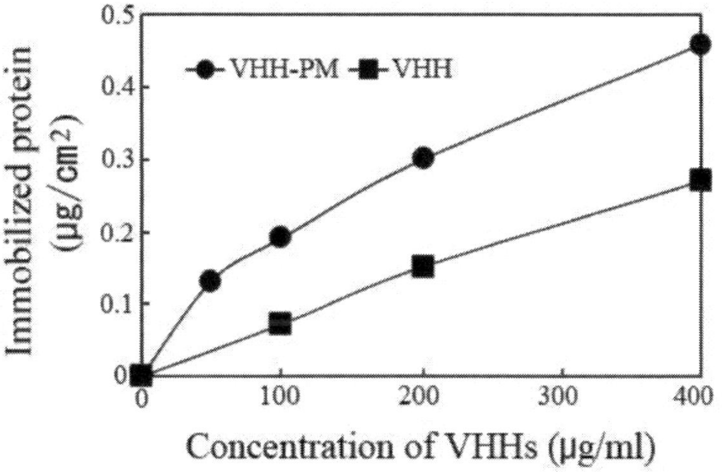

Figure 7. The antigen-binding activity of VHH and VHH-PM after dialysis refolding

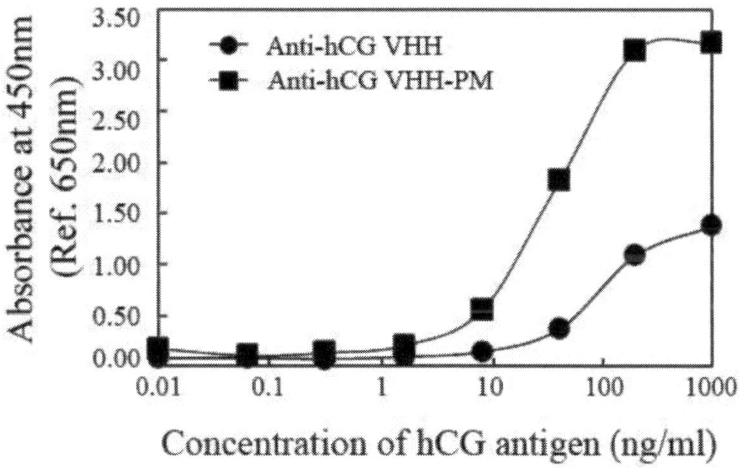

References

1. Kumada, Y, Murata, S., Ishikawa, Y., Nakatsuka, Y. and Kishimoto, M. (2012a) "Screening of PC and PMMA-binding peptidesfor site-specific immobilization of proteins". *Journal of Biotechnology*, 160 , pp. 222-228.
2. Kumada, Y., Shiritani, Y., Hamasaki, K., Nakagawa, A., Sasaki, E. and Kishimoto, M. (2010b) "Novel solid-phase refolding method for preparation of scFv-immobilized polystyrene plates with high-antigen-binding activity". *Analysis and Bioanalysis Chemistry*, 398, pp. 295-303.
3. Ewert, S., Cambillau, C., Conrath, K. and Plückthun, A. (2002) "Biophysical Properties of Camelid VHH Domains Compared to Those of Human VH3 Domains". *Biochemistry*, 41 pp. 3628-3636.

Design and Simulation of an Automated Rare Blood Cell Detector

Zhixi Qian[a], Eugene D. Boland[2], Paul W. Todd[2], and Thomas R. Hanley[1]
[1]Department of Chemical Engineering, Auburn University, AL 36849
[2]Techshot, Inc., 7200 Highway 150, Greenville, IN 47124

Abstract

Today the research of circulating tumor cells (CTC) plays an essential role in cancer's detection and treatment. To detect the cancer at an early stage, a point-of-care diagnostic system for detecting and counting cancer cells in whole blood is required. Our work focus on the simulation of flow pattern in the device, and during the research, computational fluid dynamics (CFD) is used as an effective method for simulating flow in the automated rare blood cell detector. The result showed a well-developed laminar flow inside the magnetic trap, with an experimental work it confirms a flow rate of 1ml/min is best for the magnetic separation. And the simulation of the static mixer showed the mixing pattern and properties of blood and reagents.

Key words: circulating tumor cells; CFD simulation; magnetic separation; point-of-care system; static mixer

Introduction

In recent years there's an increasing demand of using circulating tumor cell (CTC) counts in cancer management. The primary objective of this research is to design, simulate and test an accurate, cost-effective, user-friendly point-of-care CTC detection device. The preliminary design includes three syringes for blood sample, buffer and reagents, a static mixer, a magnetic filter, a microfluidic "chip" and a waste receptor. The device is smaller and cheaper than traditional detection devices. Also, device operation is easier with only one operator required. The device can be used in clinics, nursing homes or residences thus allowing earlier detection of most fatal cancers, including pancreatic cancer, ovarian cancer and melanoma. Our recent research focused on the analysis and simulation of static mixer and the flow cytometer and a further analysis of factors affecting the fluid mixing.

The fluid dynamic analysis of the staged mixer was accomplished using the FLUENT™ analysis package, which performs modeling of fluid flow and tracking particles' movement in complex geometries. The simulation showed that inside the magnetic trap, the fluid performs a well-developed laminar flow. And it's estimated most magnetic particles will be captured at first pair of magnets.

An experiment was set up to test the magnetic trap's separation ability. Various conditions were tested, including flow rate, magnetic size and type. In addition, we tested the blood cell's sedimentation inside the magnetic trap. Then it's possible to optimize the device's flow rate.

Devices and methods

During the research task CFD software simulation and mathematic methods were applied. Computational fluid dynamic analysis of the channel design was accomplished using the FLUENTTM analysis package, which performs modeling of fluid flow and tracking particles' movement in complex geometries. With the help of ICEM CFDTM, a mesh file of the channel geometry was developed, and a laminar flow model was employed for calculation. For the simulation of magnetic particles, the channel was simplified to a 2-D rectangle shape. A discrete model was applied to simulate the particle movement. Tracking of particles movements was performed to show the particle trapping efficiency.

The trajectories of magnetic particles were simulated by choosing a set of coordinates within the channel and adding the flow and magnetophoretic velocity vectors as a function of position in the mesh. At each step in time the velocity is recalculated to account for simultaneous changes in streamline velocity, dy/dt, and magnetophoretic velocity, dx/dt.

To test the magnetic trap's capture efficiency, a hemacytometer is used in counting and calculation of magnetic particle captured. The red blood cell (RBC) and magnetic beads are put in sample tubes. After counting the origin concentration, the sample is diluted and changes its concentration to around 10^7 particles/ml. Then take 5ml of sample, and pump through the magnet trap then collect the outlet. After that, use hemacytometer to count particles' concentration of outlet sample and compare it with original value.

Two types of magnet particle is used, the diameter of magnetic particle include: 2.7µm, 3.13µm and 5.01µm. And the flow rate is changed from 0.25ml/min to 1ml/min.

Results and discussion

Ansys Fluent is used in the simulation of magnetic trap's flow, the result showed the flow's pattern inside the channel, the magnet trap is 15cm long and divided by 2 channels, the size of the channel is 4mm *3mm. The fluid is flowing through the trap at a rate of 1ml/min.

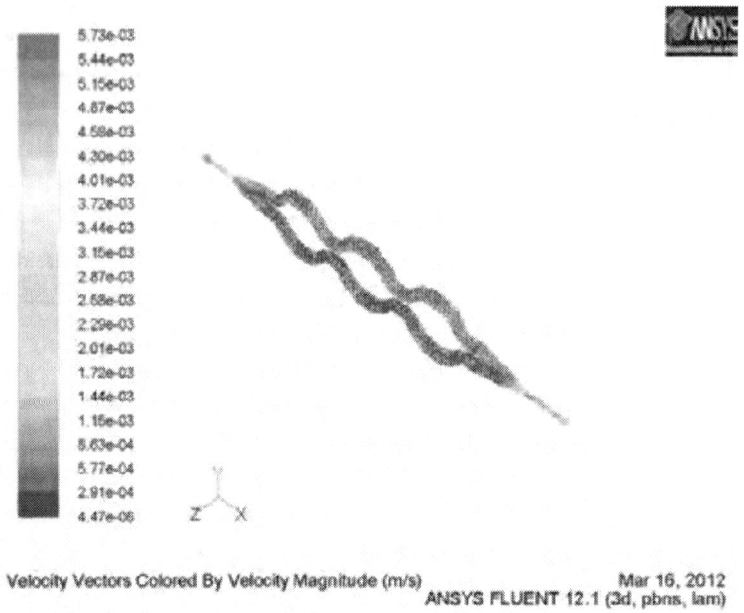

The velocity magnitude of magnetic trap shows that at the middle of static mixer, the velocity is less than 1mm/s, compare with the size of static mixer, it's a rather slow velocity.

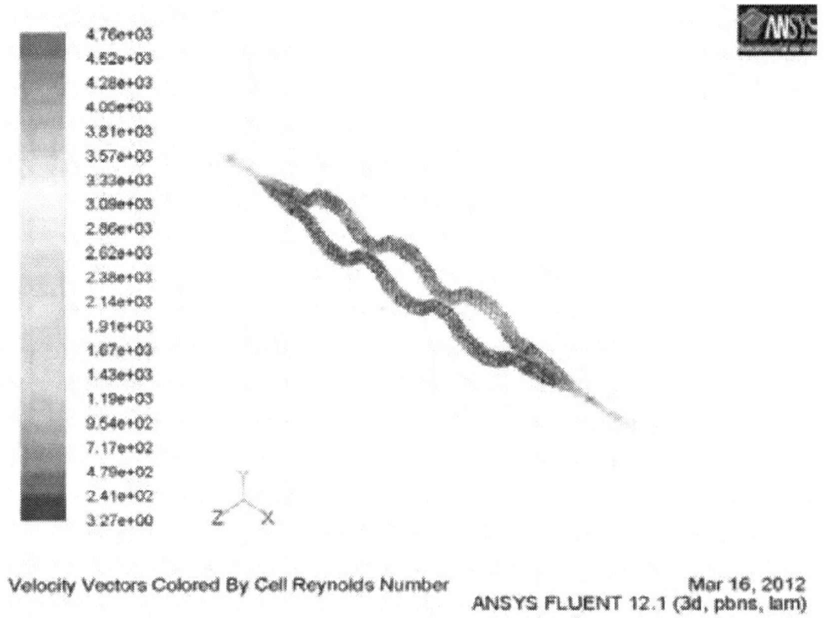

The Reynolds number inside the magnetic trap is generally less than 100, thus it's clear that the fluid inside the magnetic trap is a laminar flow, so the following model set up and calculation can be simplified.

The result of simulation showed the flow pattern of the magnetic trap, figure below showed the velocity vector of partial channel at position of a magnet for magnet particle analysis, to simplify the calculation, the channel is considered as a 2-D model.

It shows the velocity profile of fluid, and inlet average velocity is around 5.32 x 10^{-3}m/s.

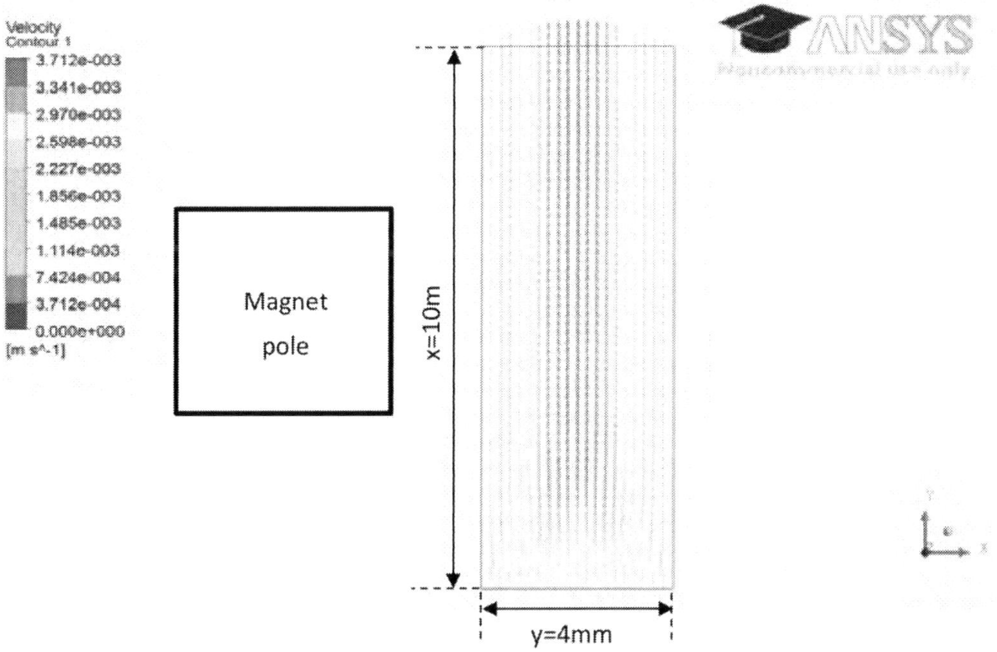

The result of particle track showed the first pair of magnet's able to capture all magnet beads at a flow rate of 1ml/min. However, with a higher flow rate, the flow velocity will dominate the particle movement and most magnetic particles will escape the magnet field.

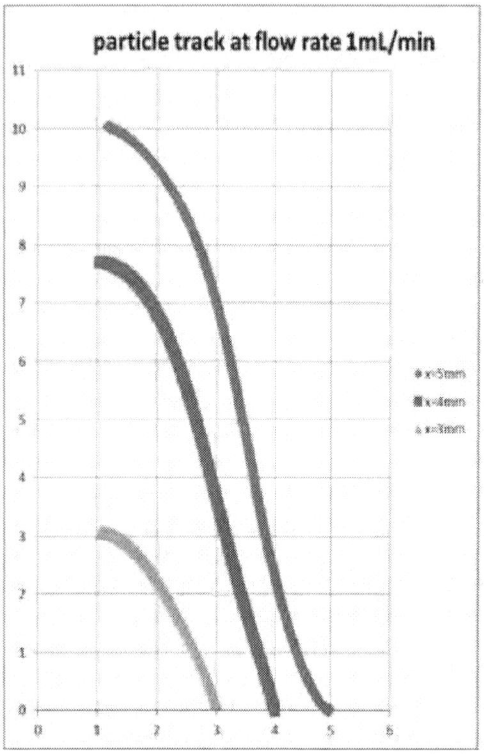

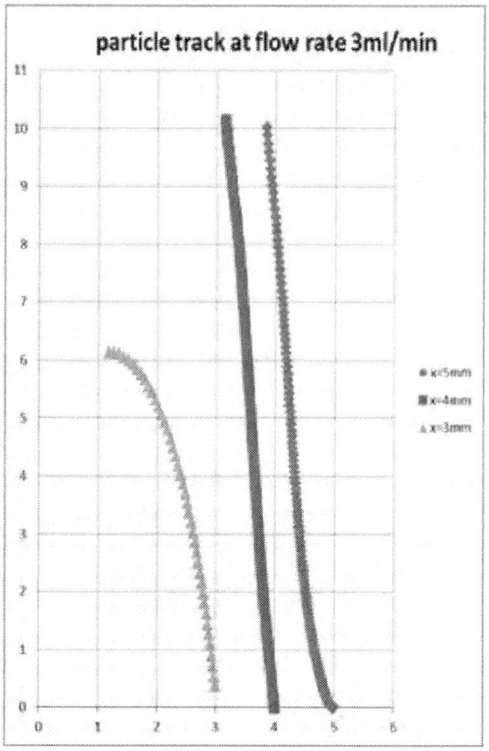

Experiments have been performed to test the magnetic trap's separation ability. The table below showed the result of magnet particle capture.

Magnetic particle size	Flow rate	Initial concentration	Concentration after separation	Capture Ratio
2.6μm(V1)	0.25ml/min	$2.66*10^7$	$8.00*10^5$	96.99%
2.6μm(V1)	0.5ml/min	$2.04*10^7$	$2.80*10^6$	86.28%
2.6μm(V1)	1.0ml/min	$2.04*10^7$	$6.20*10^6$	69.61%
2.7μm(V2)	0.25ml/min	$1.03*10^7$	$1.22*10^4$	99.98%
2.7μm(V2)	0.5ml/min	$1.40*10^7$	$1.00*10^4$	99.93%
2.7μm(V2)	1.0ml/min	$1.67*10^7$	$4.44*10^3$	99.97%
3.13μm (V2)	1.0ml/min	$2.32*10^7$	$5.56*10^3$	99.98%

5.01μm(V2)	1.0ml/min	$1.04*10^7$	$1.11*10^3$	99.99%

As the result of table indicates, the capture ratio is higher with larger magnetic particles and a slower flow speed, while in our magnetic trap, with the designed flow rate and choice of magnetic beads, the capture ratio is high enough to capture more than 99.9% of magnet particles.

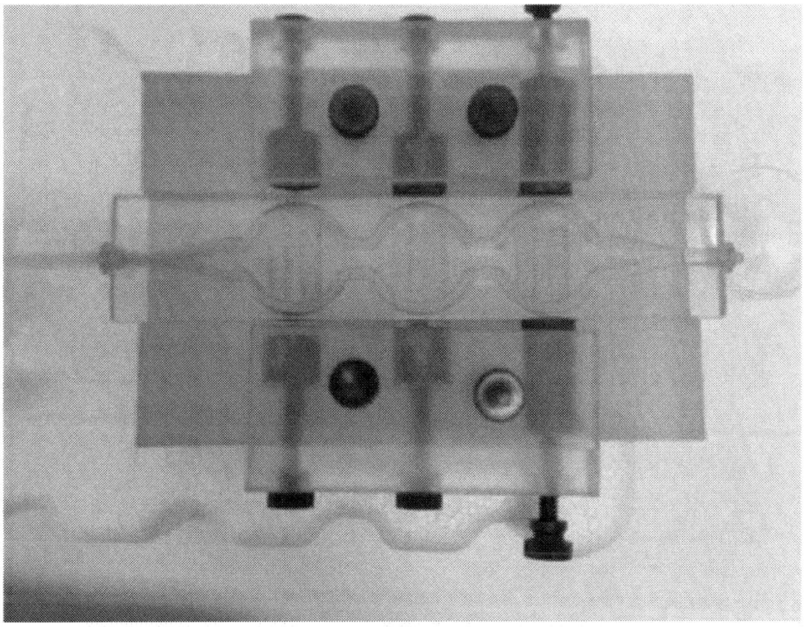

As the photo above shows, most of the inlet magnet particle can be trapped at the first pair of magnet, while the few escaped can be captured successfully in following two pairs of magnet.

The sedimentation is another factor to be considered. For the red blood cells (RBC) left inside the magnet trap will affect following detection, so it's necessary to reduce the sedimentation of unlabeled blood cell inside the mixer.

Type	**Flow rate**	**Initial concentration**	**Concentration after separation**	**Ratio for mass loss**
Magnet particle (5.01μm)	1.0ml/min	$1.39*10^7$	$1.30*10^7$	6.83%
RBC	0.25ml/min	$2.95*10^7$	$1.28*10^7$	56.78%
RBC	0.5ml/min	$2.26*10^7$	$1.89*10^7$	16.37%

RBC	1.0ml/min	$2.14*10^7$	$2.11*10^7$	1.40%

The table showed particle's sedimentation, at a low flow rate (0.25ml/min), almost half of RBCs will stay inside the mixer, which is an unacceptable result, increasing the flow rate will solve the problem. At a flow rate of 1ml/min, the sedimentation rate drops to 1.4%, thus it will not affect the detection of CTCs.

Conclusion

The simulations completed confirm that the proposed magnetic trap design successfully traps positively labeled cells. The magnetophoretic velocity of all beads and cells analyzed to date is much greater than the fluid flow velocity in the magnetic trap and is sufficient to cause cells to be captured by the lowest-strength magnet in the current design. However, a high velocity will reduce the capture ratio considerably thus the flow rate should be no higher than 1ml/min.

The experiment work showed that the magnetic trap has a good performance in particle separation, the first pair of magnets is able to capture most magnetic beads and with a proper choice, the capture rate can be more than 99.9%. The flow rate is also an important factor, when flow rate is less than 1ml/min, the sedimentation will affect the separation process greatly as RBCs will stay inside the trap. So the flow rate should be high enough to avoid particle sedimentation.

Consider both particle sedimentation and residence time, the choice of flow rate should keep a balance between efficiency and sedimentation, so a flow rate of around 1ml/min is a proper choice.

References

Afshar, R., Moser,Y, Lehnert, T and Gijs, M. A. 2011. Three-dimensional magnetic focusing of superparamagnetic beads for on-chip agglutination assays, Annul. Chem., 83(3) : 1022-1029.

Allard, W. J., Matera, J., Miller,M.C., Repollet, M., Connelly, M. C., Rao, C., Tibbe, A. G. J., Uhr J. W and Terstappen, L. W. M., 2004. Tumor cells circulate in the peripheral blood of all major carcinomas but not in healthy subjects or patients with nonmalignant diseases, Clin. Cancer Res.10: 6897-6904.

Cancer Data and Statistics, http://www.cdc.gov/cancer/dcpc/data/

Etzioni, R., Urban N. and Ramsey S., 2003. The case for early detection, Nature Reviews (Cancer), 3: 1-10.

Fan, T., Q. Zhao, J. J. Chen, W-T. Chen and M. L. Pearl, 2009. Clinical significance of circulating tumor cells detected by an invasion assay in peripheral blood of patients with ovarian cancer, *Gynecologic Oncology*, **112**, 185-191

Haun, J. B., Yoon, T-J. , Lee H. and Weissleder, R. 2010. Magnetic nanoparticle biosensors, Nanomedicine and Nanobiotechnology, 2: 291-304.

Hsieh, H. B., Marrinucci, D. , Bethel, K. , Curry, D. N. , Humphrey, M. , Krivacic, R. T. , Kroener, J. , Kroener, L., Ladanyi, A. , Lazarus, N, Kuhn, P. , Bruce R. H. and Nieva, J. 2006. High speed detection of circulating tumor cells, Biosensors and Bioelectronics, 21(10): 1893–1899.

A Split Intein-based Self-cleaving Tag for Recombinant Protein Purification In a Mammalian Cell Expression system

Tzu-Chiang Han

Dr. David W. Wood

Inteins are genetic elements that can self-excise from various host proteins in a post-translational protein splicing process. With a simple mutation of the intein, one can yield isolated, controllable intein cleaving activities at either the N- or C-terminus. By combining self-cleaving inteins with conventional affinity tags, self-cleaving affinity tags have been developed. The self-cleaving tags benefit the tag removal process by a simple, inducible cleaving reaction without any addition of protease. The phenomenal tag removal strategy works successfully in prokaryotic cell systems in our previous studies; however, the *in vivo* premature cleaving of the precursor protein remains the bottleneck of its application in eukaryotic cell systems. Here we present our most recent work in developing a split intein-based recombinant protein purification strategy in a mammalian system. Split intein consists of two complementary fragments of intein that can undergo reconstruction to form a fully-active intein by *in vitro* assembling, while no splicing activity shown individually. The affinity tag-fused intein (N-fragment) was first expressed in *E.coli* and further conjugated with the beads in a typical affinity column. The target protein that fused with the C-fragment of intein was expressed in a mammalian system and then mixed thoroughly with the tagged N-fragments of the intein. After an affinity-based purification, the target protein can be eluted by inducing the self-cleaving reaction of the re-assembled inteins. With the split-intein system, we provide a feasible scheme to overcome the unwanted *in vivo* premature cleaving and generate a potential platform technology for the manufacture of non-antibody glycoprotein therapeutics at large scale.

378777 On-Column (solid-phase) Pegylation and Separation of Pegylated Proteins

Abstract:

PEGylation is known to be an effective method for making a smart protein drug. It prolongs the residence time in the body and makes proteins more stable and soluble. However, producing a PEGylated protein is not an easy process since the reaction yield is not very high and consequently, the reaction mixture contains non-reacted protein and PEG as well as PEGylated proteins. In addition, due to over PEGylation reaction unfavorable, mutli-PEGylated proteins are produced. Separation of a target mono-PEGylated protein from other PEGylated protein forms and native protein is also a difficult process.

Solid phase or on-column PEGylation is claimed to be a promising method for improving the reaction selectivity. We carried out random PEGylation of lysozyme on various cation-exchange chromatography columns. After the PEGylation reaction, linear gradient elution experiments were performed to separate PEGylated protein forms.

Production of multi-PEGylated forms was reduced significantly. However, the yield was relatively low. In order to understand the reason for the low yield, various different chromatography media as well as mobile phase conditions were tested. Porous beads of different particle size (SP-Sepharose FF and HP), hydroxyapatite, Phosphate-Cellufine, Sulfate-Cellufine, SP-Toyopearl650M, Resource S and SO3-monolithic disk were used. Large beads (Sepharose FF) resulted in much lower yield values most likely due to the mass transfer resistance. However, convention-aided chromatography media, monolithic disk did not improve the yield significantly. By choosing a suitable salt concentration of the mobile phase, lysozyme was rather weakly adsorbed onto the disk entirely before PEGylation. This condition increased the yield.

378808 Diffusion Coefficients of DNAs and Pegylated Proteins for Estimating the Performance of Chromatography

Abstract:

Pore diffusion governs the separation or adsorption performance of chromatography. Especially, for large biomolecules such as DNA and PEGylated proteins low pore diffusion coefficients result in low performance of chromatographic separations. Monolithic chromatography, a type of convection-aided chromatography, is best suited for such large biomolecule separations. In order to understand the advantage of monolithic chromatography against porous-bead chromatography quantitatively the diffusion coefficients of large biomolecules are needed. Unfortunately, diffusion data (pore diffusion coefficient D_s and molecular (free) diffusion coefficient D_m) for large biomolecules are not readily available.

In this study, the molecular diffusion coefficient D_m (free diffusion) values were determined by using the Taylor dispersion method for various DNAs and PEGylated proteins as well as standard globular proteins. Necessary conditions for measuring D_m precisely such as tube length, tube diameter, coil diameter and flow velocity were first checked. Single-strand (ss) DNAs showed much lower molecular diffusion coefficients compared with those for globular proteins.

The pore diffusion coefficient D_s values were measured by using isocratic elution at non-binding conditions. D_s of ss-DNA decreased with increasing DNA molecular weight much more sharply compared with D_m. Similar trends were observed for PEGylated proteins. Based on these data we calculated the separation performance of typical porous bead ion-exchange chromatography (IEC) in comparison with that of monolithic IEC.

393632 Turning the Big Idea into a Startup

Abstract:

When a university researcher or garage inventor is thinking about commercializing an invention, there are a number of early considerations that are not always obvious to first time technical entrepreneurs. This talk will focus on some of the early stage commercialization decisions and how those decisions may impact the startup company potential later. Some of the topics to be discussed include fundraising, teambuilding, and intellectual property management.

386885 Starting an Small Engineering Consulting Business – the 10 Year Anniversary of a Startup Company

Abstract:

A small engineering consulting business was started in 2004 with the modest objectives of addressing short-term business opportunities while remaining in a specific geographic location. Full-time commitment to this venture occurred for a seven-year period. Ten years and 154 engineering projects later the business continues to operate. Personal and professional lessons learned on business planning and strategies, accounting and financial tools, fixed asset investments, family involvement, and finally, the pros and cons of this career path versus the traditional non-ownership career path will be discussed.

387532 The Art of the Mass Balance: The Cultural Tools of Entrepreneurial Chemical Startups

Abstract:

The product cycle processes of tech based startups and chemical startups are inherently different. Tech startup thought leaders teach entrepreneurs to hit the ship date even if all the kinks aren't worked out. Later a patch can be uploaded by customers to fix bugs or add functionality at minimal cost to the startup.

Chemical startups differ significantly as processing plants are capital intensive and must incorporate safety, equipment lead times, and construction efforts. Adopting an entrepreneurial approach, chemical startups can deliver against the fast paced timeline while designing the flexibility required to commercialize a first of its kind technology. The tool for success in an entrepreneurial chemical startup is the mass balance.

Keeping score in an industrial startup is dependent upon the value and quality of the product. The disposition of the constituents in the product determines the cash flow possibilities in the operation cost and the needed investment to enable the conversion of the raw material feedstock mass into the valued finished product. The key to a successful mass balance is the validation of the instruments and test methods measure the chemical engineering entrepreneurial effort's success.

The Personal and Family Side of Starting an Independent Business

Starting a business has its special and unique technical challenges relating to deciding on a focus, designing a strategy for communication and outreach, identifying clients, obtaining financing, and many others. But unless you are a single adult, there is another factor and that is your family. If they have gotten used to a regular paycheck, travel being reimbursed, promotions coming on a regular business, and the occasional special recognition, this can all come to a screeching halt. The support and understanding needed from a family and external networks is critical for a new venture to succeed. The level of this support will also depend greatly on how large the imagined enterprise is envisioned to be.

These are some questions to consider:

1. Short term economic survival. How long can you survive without a regular paycheck? Have you built up sufficient savings in the event that your first clients are harder to find than you thought? If your spouse is not working, will they be able to work to cushion the cash flow? Do they want to? What is the market in the local area for the skill set they possess? Will the cost of child care be part of the equation? It is not unusual at all for someone starting an independent business to wait at least six months before receiving their first payment. More companies are moving to 90 day plus payment cycles. Is that tolerable? A discount on pricing and acceptance of a client's corporate credit card are ways of increasing short term cash flow at a small long term cost. There may also be short term training opportunities available through various organizations, but not necessarily at the same rate of pay that may be desired long term.

2. Health insurance. The degree to which this is of concern will depend on a large number of factors including age, state of residence, health history barriers to insurability, and whether a working spouse has insurance to which the new "independent" can be added. Regardless of the specific answers to these questions, the cost of covering health insurance must be considered when starting an independent business. One way of minimizing a major impact is to purchase low cost catastrophic insurance with a very high deductible, usually available in many areas without medical questions. And of course, AIChE offers group health insurance plans, but which come with holding periods before covering pre-existing conditions.

3. Spouses. No two people are created equal and the reaction of family members to such a venture, even when openly discussed ahead of time, can vary and change. What is the personal tolerance for going without a paycheck for an extended period of time (even assuming the points in "short term survival" have been addressed)? Is how the neighbors and your social circle view your situation important? If you are a male and your wife is working or needs to go back to work, are you willing to do household chores you've never done before? Some of these attitudes will depend upon whether your starting of an independent business was totally your choosing or as an indirect result of a reduction in force, a general layoff or closure/relocation of a business, or a personal performance issue. Is your spouse willing to cut back on some of the luxuries they may have enjoyed previously?

4. Children. This is a different issue than spouses and will greatly depend on their ages. What would your child's reaction be to someone in their school, possibly informed by one of your neighbors about your situation, saying something like "Your mommy or daddy is out of work...ha, ha". Will they start a fight? How well have you prepared them?

5. Support structure. In today's world, the stigma of losing a job is far less than it used to be due to numerous factors including mergers and acquisitions, change in business emphasis, and "forced" geographic moves to retain a job. Yes, there are still performance issues, but many times this can be a result of different approaches to a problem or opportunity vs. genuinely poor performance. It is

highly unlikely that anyone today will have only one job for their entire career. Since everyone, over some period of time, will experience the same kind of dislocation, there is virtually no barrier to discussing possible job opportunities with anyone. Today's professional networking tools such as LinkedIn are tremendous sources of connections and opportunities. Local professional engineering chapter meetings are another great source, especially if there is a strong desire to stay in the same geographic area.

It is important to remember that starting such a venture within a family structure has additional needs to consider beyond just hanging out the professional consulting shingle. Consider them all before taking the plunge and your venture has a much higher probability of success.

392019 Aligning Innovation, Patent Strategies, and Business Models

Abstract:

An iterative approach to balancing research efforts, patent strategies, and business models can identify new opportunities as well as provide an attractive package for management or investment support. An overview of the beneficial use of intellectual property is provided and specific examples are shown.

364257 Building an Excellent Entrepreneurial Team

Abstract:

An entrepreneurial team is built on excellence, confidence and passion. All team members must be comfortable with chaos. The cube farm mentality has to be destroyed. The passion to change the world and make a difference has to be unleashed. The recruiting relationship is built on the abnormal. There is no room for conformity on an entrepreneurial team. However, the same traits that make it great also tends to blow it apart.

Entrepreneurial engineering leaders are either superstars or oddballs. Both traits are needed. The team leader must find the mix of innovative off the wall creativity with the ability to drive fast-paced targets and get things done. Decisions must be made on the fly without all the information, but the aspects of safety and ethics can never be compromised. Innovation is unleashed by having mundane tasks taken out of the equation.

An entrepreneurial team is a paradox. Like the grand unified force theory, the moment you think you have handle on how it works, a new opportunity emerges and bang ! pivot to a new direction.

There is nothing more fulfilling, exciting or satisfying.

392020 How to Generate Revenue from Your Technology: Strategies for Licensing Your Intellectual Property

Abstract:

Licensing of intellectual property can be a significant source of revenue for companies. This session will provide practical lessons for generating and increasing value from IP through licensing. Learn how to identify and protect your IP while avoiding common mistakes when monetizing it.

consulting engineers and scientists

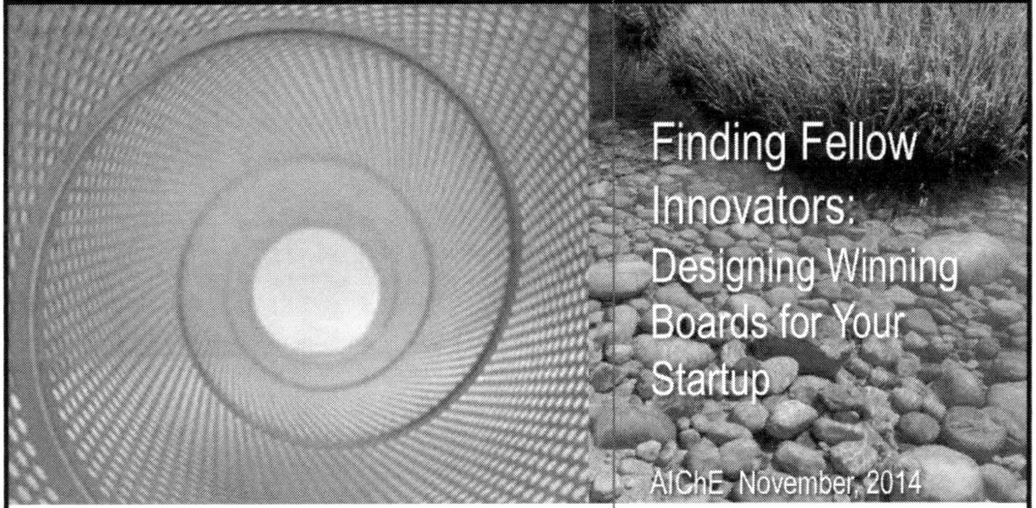

Finding Fellow Innovators: Designing Winning Boards for Your Startup

AIChE November, 2014

John T. Finn, P.E.
GEI Consultants, Inc.

GEI Consultants

Dedication

In Memory of My Father

Robert K. Finn, PhD.
Professor of Chemical Engineering
Cornell University
Member AIChE 1946 - 2012

GEI Consultants

Create Winning Boards for Your Startup

You Want to Be Here

You Are Here

Purpose ?

- What do you want from outside help?
- Depends on:

1. Unique Identity and Circumstances
2. Stage: Idea→ Startup ➡ Growth

Directors, Advisors, and Supporters

- Firm reports to the Directors

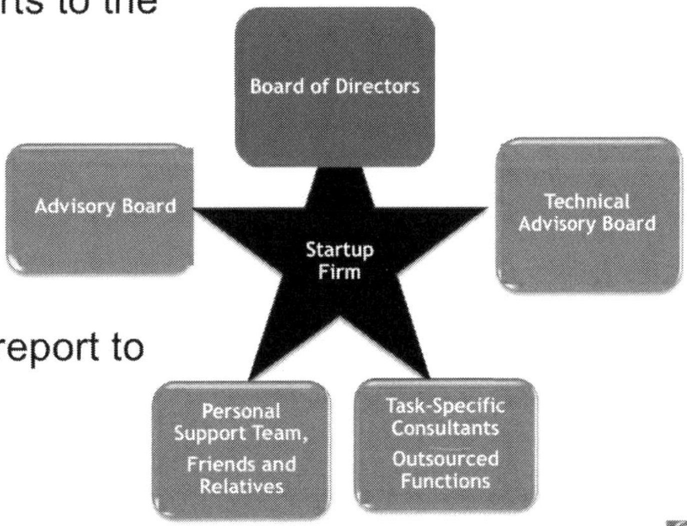

- Advisors report to the Firm

GEI Consultants

Three Types of Boards

Board	Legal Status	Primary Role
Directors	Statutory	Makes Big Decisions, Goals, Selects CEO
Advisory (General)	Informal	Advises on vision, goals, strategy, the market, legal, staffing, financial, governance, etc.
Technical Advisory	Informal	Advises on technical and scientific aspects of the process and product, R&D, external related technologies, etc.

GEI Consultants

Directors

- Usually called "The Board"
- Usually after Idea and Initial Startup
- Founders, VC Investors and Outside Members
- How Many: 7 to 9 to fulfill legally mandated activities
- Commitment: 250 hours/year Chair or active startup 450 hours/ year

GEI

General Advisors

- Provide critical business-related help, introductions to other key supporters
- Focus on your path to the next stage
- 3 to 5 persons, with a range of backgrounds
- For the number of advisors, consider your own bandwidth. 1 hr each per month, min.

GEI

Technical Advisory Board

- Keep it small and focus on your challenges
- Academic and Industry backgrounds
- Include your technical team members
- Assign members 10 minute presentations on an aspect of the field
- Encourage constructive dialogue
- Stress confidentiality of your data

GEI

TAB Special Considerations

- Certainty vs Risk for engineers.
- Audacity (arrogance ?) for Chem Es!
- Humility is the greatest virtue for the TAB

GEI

Finding Fellow Innovators

- Meet a lot of people. Reach outside your circle, referrals from college profs, grad students, conferences, accelerator meetings, AIChE, LinkedIn

...let people know you are looking.

- Board membership benefits the members: Pitch
- Be picky! Proven track record
- Passion about your startup
- Bandwidth to be able to help
  - test this by giving a small assignment
- Good chemistry with you, and board members
- Collaborative, yet challenge assumptions

GEI

Getting the Most from Advisors

- Meet both individually and in groups
- Run the meetings efficiently, with pre-distributed agenda, minutes.
- Give them homework, specific tasks
- Set the next meeting date
- Use calls, email and chat effectively
- Consider a secure site Wiki with company updates, financials, tech developments, marketing tools, Suggestion Box, etc.

GEI

Compensation

Board	Vested Equity	Cash	Meeting Expenses, Lunch
Directors	0.5% to 2% if not a VC or Founder	Not usually	Yes
Advisory (General)	0 to 1%	Small	Usually
Technical Advisory	0 to 1%	Small	Usually

5% Total equity pool set aside for advisors is typical

GEI Consultants

Standard Agreement

- Founder Institute

Founder/Advisor Standard Template "FAST"

- Easily defines

scope, time commitments, compensation

- Proportional to stage: Idea, Start, or Growth
- Term limit, typically 1 year

GEI Consultants

Advisors Compensation

- Wide Range: Lunches at first, to Equity

FAST Scope	Idea Stage	Startup Stage	Growth Stage
Standard: Monthly Meetings	0.25%	0.20%	0.15%
Strategic: Add Recruiting	0.50%	0.40%	0.30%
Expert: Add Contacts & Projects	1.00%	0.80%	0.60%

Source: The Founder Institute

GEI Consultants

Supporters and Consultants

- Asking the Deeper Questions
- Improve Structure
- Project Management

Systems

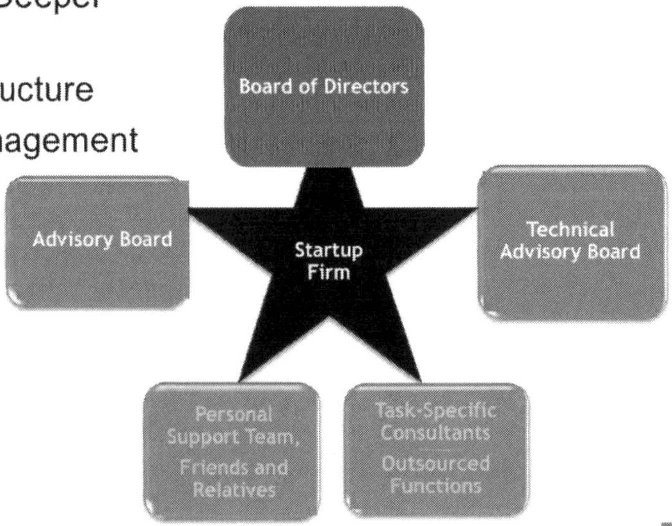

GEI Consultants

Critical Points

- Advisory Boards are essential for most Startups to thrive and progress
- Invest time and attention in setting up and maintaining the board
- Consider general business advisors as well as expert and diverse technical advisors
- Tailor board makeup, scope, compensation to your stage
- Use consultants for specific outsourced tasks
- Friends and family provide personal support

GEI

Thank You

Questions and Discussion ?

GEI

Resources

- Entrepreneur article: www.entrepreneur.com/article/71684
- Ventureneer blog: bloghttp://ventureneer.com/vblog/how-advisory-boards-can-help-your-startup
- RockHealth Video: www.youtube.com/watch?v=sTcpUCmUu4A
- FAST document:

http://fi.co/contents/fast#

- John Finn jfinn@geiconsultants.com

GEI Consultants

392021 The I-Corps Program at NSF (and NIH)

Abstract:

There have been profound recent developments that have changed the way universities and the nation's funding agencies foster the creation of startups. Come learn about this new approach, how the NSF and NIH are implementing it in their new I-Corps program, and what this means to you as you seek to commercialize your innovation

393665 A Personal Perspective on Biotechnology's Progress in a Golden Era (and how Jay Bailey saw it all coming)

Abstract:

Much has changed in biotechnology over the last few decades, but much has remained the same. As a graduate student in Jay Bailey's research group in the early 1980s, I was introduced to a field undergoing vibrant expansion and mentored by one of its most influential and visionary thinkers. As biotechnology was burgeoning from the development of powerful genetic techniques and a renewed integration of engineering principles, Bailey was among the first to recognize the revolutionary potential of exploiting molecular biology in an engineering context, from manipulating heterogeneous populations of protein molecules and whole cells to quantifying and redirecting metabolic pathways. Molecular-level analysis, mechanistic modeling, and a reaction engineering approach to molecular biology were hallmarks of Bailey's early bio program. These signature attributes continue to propel modern research in the field, and comprise the foundation of past and present research in our own program, which has encompassed evolving techniques of biocatalysis, protein engineering, and cellular engineering for applications ranging from drug discovery to biofuel production. This presentation will highlight pertinent accomplishments from our group in these areas and show how many of them can be traced back to the influence and foresight of Jay Bailey.

374209 Hydration of Acrylonitrile to Produce Acrylamide Using Biocatalyst in a Membrane Dispersion Microreactor

Abstract:

In this study, a membrane dispersion microreactor was utilized in acrylonitrile hydration to produce acrylamide. This microreactor can enhance mass and heat transfer rates to improve productivity. The bioconversion of acrylamide is mainly catalyzed by NHase obtained from *R. ruber* TH3 free cells. Bioreaction is generally performed in a stirring tank. However, traditional stirring tanks cannot be used to satisfy the requirements of low energy consumption, bioprocess safety, high purity, and conversion because acrylonitrile and acrylamide elicit double inhibitory effect on enzyme activity and reaction heat is high. Thus, the proposed membrane dispersion microreactor was utilized in this process.

The velocities of two phases and the flow rate ratio were optimized. Heat transfer efficiency was investigated. Acrylonitrile droplets were observed under a microscope. The results showed that the maximum concentration reached 52.5 wt% and the most optimal concentration reached 45.8 wt% in 35 min. By contrast, traditional stirring tanks require 245 min to produce 39.5 wt% acrylamide. Moreover, only a few amounts of acrylonitrile were accumulated in this microreactor system.

Carbon Capture Using Carbonic Anhydrase-Displaying *Escherichia coli* in Biologically Active Foams

Stuart K. Watson, Zhenlin Han, Wei Wen Su, and Eunsung Kan

1. Introduction

In recent years, global climate change as the result of greenhouse gas emissions, especially those of CO_2, has become of great concern. Many technologies have been studied and employed to capture post-combustion CO_2 such as absorption and adsorption (Artanto et al., 2014; Padurean et al., 2012), membrane technology (Low et al., 2013), ionic liquid technology (Albo et al., 2010; Stevanovic et al., 2013), metal-organic frameworks (Cao et al., 2013), and biological methods (Migliardini et al., 2014; Pires et al., 2012). However, current methods have several drawbacks that make them expensive or limit their practical application. Adsorption and absorption have the problems of regeneration of spent media and storage of captured CO_2. Media regeneration often requires temperature or pressure swings, which therefore require additional energy input, and the storage or usage of captured and concentrated CO_2 can be expensive (Mores et al., 2012; Olajire, 2010; Yu, 2012). Furthermore, the commonly used amine solutions can be corrosive, requiring frequent mechanical maintenance, and they are also degraded by O_2, SO_x, and NO_x present in flue gas, leading to overall poor performance in the long term (Dumée et al., 2012). Ionic liquid solutions often must be highly viscous when they are task-specified for carbon capture and the multitude of possible cation-anion combinations makes it difficult to design the optimum formulation (Hasib-ur-Rahman et al., 2010). Membrane separation technologies need pressurization or feed compensation to be able to remove CO_2 at low concentrations such as in flue gas (Low et al., 2013). Furthermore, the large volume of flue gas would require a large membrane for appropriate surface area.

For these reasons, the carbonic anhydrase enzyme (CA hereafter) has been studied for use in carbon capture because of its ability to catalyze the conversion of CO_2 and water into bicarbonate and a proton much faster than the uncatalyzed reaction (Lindskog, 1997; Russo et al., 2013). This process solves the media regeneration and corrosion issues as it creates an environmentally safe and water soluble product at ambient conditions without adding energy to regenerate its carbon capture ability. Bicarbonate, the product from the hydration of CO_2 by CA, can also be harvested as calcium carbonate after reacting with calcium, which can be used in a variety of industrial applications such as fillers for papermaking, coating, and plastics, as well as agricultural applications including odor and pathogen control of animal wastes.

However, current CA technology has several limitations that must be resolved for practical application: the limitations of mass transfer in current bioreactors, the high cost of the purified enzyme, decayed CA activity over time, and CA product inhibition by bicarbonate accumulation. This study seeks to overcome the major limitations of current CA technology by employing recombinant CA-displaying *E. coli* in biologically active foam (using the foam bioreactor, Fig. 1)

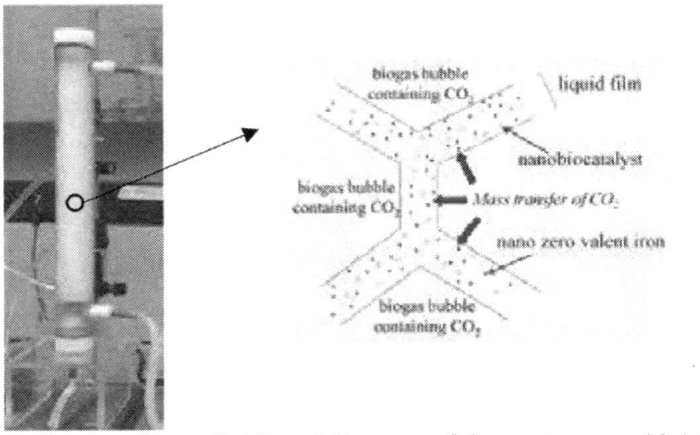

Fig. 1. Foam column (left) and diagram of foam structure (right).

in order to achieve high mass transfer and high CO_2 removal with longer enzymatic stability without the use of expensive purified CA. The primary objective of this study was to investigate CO_2 removal by the CA-displaying *E. coli* in a foam bioreactor. The CO_2 elimination capacity of the foam bioreactor was compared to that of current bioreactors used for carbon capture. This study also attempts to achieve high and stable CO_2 removal in the foam bioreactor by removing bicarbonate (the product from the hydration of CO_2 by CA) semi-continuously to prevent CA activity decay due to bicarbonate accumulation since accumulation of bicarbonate in the bacterial culture showed significant inhibition on CA activity. Overall, the use of whole-cell CA biocatalysts would overcome the problems of the high cost of the enzyme and its gradual decay that current CA technology faces. Because of the large gas-to-liquid surface area available in the foam bioreactor, highly active and dense CA-displaying *E. coli* in the foam effectively converted CO_2 to bicarbonate ions and protons. To the best of our knowledge, this is the first work to demonstrate that CO_2 is captured by biologically active foams containing CA-displaying recombinant bacteria, effectively allowing for higher mass transfer, high removal of CO_2, ease of control, and lower operating costs compared to conventional technologies.

2. Materials and Methods

2.1. Cultivation of carbonic anhydrase-displaying recombinant *E. coli*

Glucose-based auto-inducible medium (glucose AIM, hereafter), which is a modified M9 medium was used for cultivation of the *E. coli* and formulated as follows: 1 g glucose, 4 g lactose, 3.3 g $(NH_4)_2SO_4$, 6.8 g KH_2PO_4, 7.1 g Na_2HPO_4, 0.5 g NaCl, 0.15 g $MgSO_4$, 0.08 g $ZnSO_4$ (0.5 mM), 0.25 g yeast extract and 50 mg ampicillin per liter of DI water. 250 mL of glucose AIM inoculated with the recombinant *E. coli* in a 500 mL flask were incubated under vigorous shaking at room temperature. The 72 h old bacterial culture was directly introduced to the foam bioreactor for testing its CO_2 capturing capacity at various operating conditions.

2.2. Set-up of the Foam Bioreactor and characterization of the foam

The foam bioreactor (Fig. 1) consists of a foam column (0.4 L, an inner diameter of 4 cm, a height of 32 cm), a defoaming chamber (0.5 L), and a cell reservoir (0.5 L, filled with 0.4 L bacterial culture). A metered stream of CO_2 supplemented air was introduced through a fine gas sparger (VWR, Radnor, PA) while the *E. coli* culture was introduced at the bottom of the foam column. Upon feeding to the foam column and contacting with the CO_2-supplemented air at the bottom of the foam column, tiny bubbles (2 – 3.3 mm at gas velocity of 0.48-0.96 m/min) were immediately generated and rose through the column. Note that no additional surfactants were required to make the foam because the proteins in the culture medium assisted in foam generation. After rising through the foam column, the foam leaving through a side port in the column was defoamed in a defoaming chamber by spraying the foam with culture in the reservoir. The defoamed liquid was returned to the cell reservoir in order to be recirculated to the foam column.

2.3. CO_2 Capture Efficiency in the Foam Bioreactor at Various Conditions

To determine the foam bioreactor's performance at various conditions, the effects of cell concentration, gas velocity, inlet CO_2 concentration and bicarbonate removal were tested. All experiments were run with continuously fed CO_2-supplemented air (3.8 – 4.3 % inlet CO_2) with the culture recycled in a closed loop through the system for 1 – 4 hours. Batch operation without wasting any biomass was warranted, as all experiments were relatively short (1–4 h). The pH of the bacterial culture was maintained by using drops of 4.5 M NaOH solution injected directly into the cell reservoir during each experiment. To investigate the effect of gas velocity on the reactor's performance, the reactor was operated with 0.4 L of culture containing 1 g_{dw}/L of the *E. coli* at gas velocities ranging from 0.48 to 0.96 m/min (corresponding to empty bed residence times of 40 s to 20 s, respectively). Similar conditions were used to determine the effect of cell concentration from 0.015 to 4 g_{dw}/L at a gas velocity of 0.8 m/min (corresponding empty bed contact time of 24 s).

3. Results and Discussion

3.1. Effects of Operating Conditions on Removal of CO_2 in the Foam Bioreactor

The effects of the operating conditions including gas velocity, cell concentration and CO_2 inlet concentration on CO_2 removal in the foam bioreactor were investigated. First, the effect of gas velocity on CO_2 removal efficiency (RE%) and elimination capacity (EC) was examined (Fig. 2)

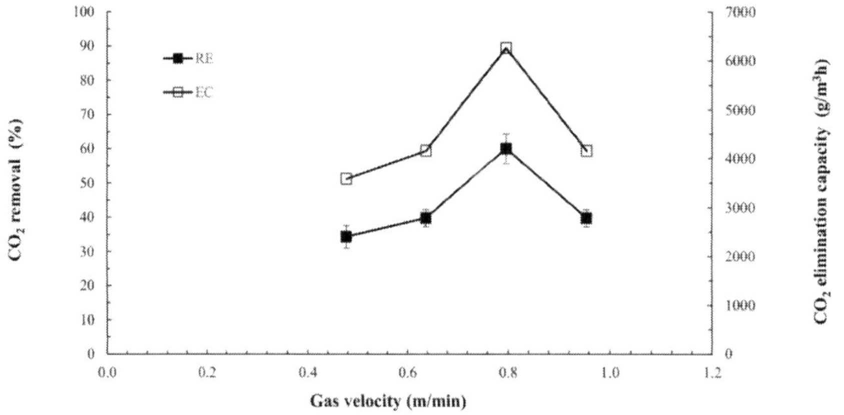

Fig. 2. Effect of gas velocity on CO_2 removal efficiency (RE%) and elimination capacity (EC, g CO_2 removed/m^3–h) in the foam bioreactor. Conditions: CO_2 inlet concentration, 3.8 – 4%; cell concentration, 2.1 g_{dw}/L; total reactor volume, 0.4 L; pH at 10.5-11.

While lower gas velocities (longer gas retention times) generated smaller foam, the lower gas pressure against the liquid held at the bottom of the reactor was not enough for adequate foam generation and stability, leading to poor travel through the foam column. On the other hand, the CO_2 removal in the foam bioreactor decreased at higher gas velocity above 0.96 m/min because high velocities generated larger foam size with low stability. Therefore, the optimum gas velocity was considered to be 0.8 m/min (corresponding gas contact time, 24 s), resulting in stable foam and excellent travel through the foam column, which would support high mass transfer and biological conversion of CO_2 in the bioreactor.

The effects of cell concentration on CO_2 removal were also studied (Fig. 3).

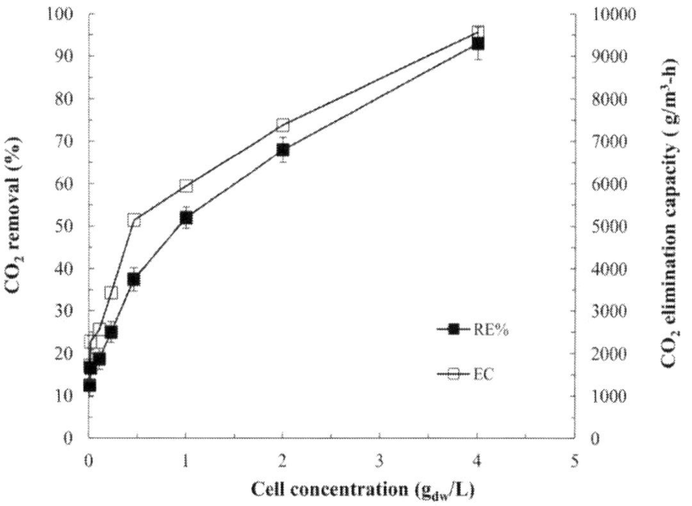

Fig. 3. Effect of cell concentration on CO_2 removal efficiency (RE%) and elimination capacity (EC, g CO_2 removed/m^3–h) in the foam bioreactor. Conditions: CO_2 inlet concentration, 4-4.3%; gas velocity, 0.8 m/min; total reactor volume, 0.4 L; pH at 10.5-11.

When the cell concentration in the foam bioreactor increased up to 0.5 g_{dw}/L, there was a drastic increase in the RE and EC of CO_2 compared to the RE and EC increase using concentrations higher than 0.5 g_{dw}/L, although higher cell concentrations provide greater removal overall. The cell concentrations of 4 g_{dw}/L led to CO_2 removal of 93% and elimination capacity of 9,566 g CO_2 removed/m^3 reactor-h (or 0.036 mmol CO_2/m^2-s based on the gas-liquid area of 1,671 m^2/m^3 in the foam bioreactor). Although a few studies reported CO_2 elimination capacity (or CO_2 mole flux) in bioreactors, the CO_2 elimination in the foam bioreactor was found to be higher than those by the membrane bioreactor using outer membrane-displayed CA in *E. coli* (Fan et al., 2011), but lower than that by the packed bed reactor using CA-accelerated absorption with K_2CO_3 (Zhang et al., 2011). The membrane bioreactor with CA-displaying *E. coli* showed 138 g CO_2/m^3-h at 0.6% CO_2 and 25°C (Fan et al., 2011). The packed bed bioreactor with absorption of CO_2 by K_2CO_3 and immobilized CA exhibited high CO_2 removal (0.08- 0.12 mmol CO_2/m^2-s at 4.6% inlet CO_2 and 50°C), but required higher operating costs than the foam bioreactor in this study due to the chemicals and CA (Zhang et al., 2011). Obviously, high cell density containing high CA loading would help to achieve high CO_2 removal in the foam bioreactor assuming that the foam bioreactor would be (bio)kinetically limited rather than mass transfer limited because of the huge gas-liquid surface area at the optimum gas flow rate. However, the viscosity of highly concentrated *E. coli* (which was found at cell concentration greater than 4 g_{dw}/L) caused excessive foaming and difficult defoaming in the foam bioreactor. Thus, instead of high cell density, the periodic replacement of a portion of the bacterial culture with the same volume of fresh medium would support continuous cell growth and CA induction to maintain high removal of CO_2 in the foam bioreactor over a long term operation as this approach was proven by Kan and Deshusses's previous works (Kan & Deshusses, 2005).

3.2. Effects of Bicarbonate Removal on CO_2 Capture in the Foam Bioreactor

Bicarbonate itself is an inhibitor of CA activity at concentrations around 0.5 – 0.75 mM for various CA enzymes derived from *H. pylori* (De Simone & Supuran, 2012; Maresca et al., 2013). In order to determine the inhibitory effect of bicarbonate concentration on CA activity, Wilbur-Anderson reactions at various bicarbonate concentrations using the CA-displaying *E. coli* were conducted (Fig. 5a).

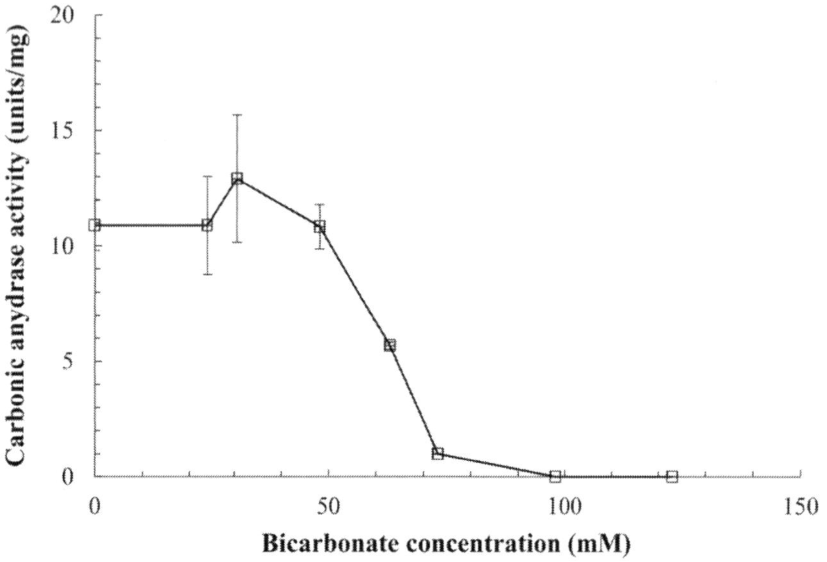

Fig. 5a. Effect of bicarbonate concentration on CA activity using Wilbur-Anderson CA activity test with various initial concentrations of bicarbonate (0-123 mM bicarbonate).

The results indicated that no inhibition of CA activity occurred below 48 mM bicarbonate while a sharp decrease in activity was noted above this concentration. While activity was reduced to half at 63 mM bicarbonate, complete loss of activity was noted at 98 mM bicarbonate. The inhibitory concentration of bicarbonate on the CA displayed at the outer membrane of *E. coli* in this study was much higher than those previously reported by others (De Simone & Supuran, 2012; Maresca et al., 2013) possibly because the CA used in this study was displayed and immobilized on the outer cell membrane and the surrounding lipopolysaccharides possibly protected CA activity against bicarbonate in the medium. Regardless, the rapid production and accumulation of bicarbonate in the culture by the CA reaction would immediately inhibit enzyme activity and subsequently reduce the CO_2 removal in the foam bioreactor. Therefore, bicarbonate removal in the foam bioreactor was conducted to see any enhancement of CO_2 elimination efficiency and stability when the foam bioreactor was operated with and without bicarbonate removal (Fig. 5b).

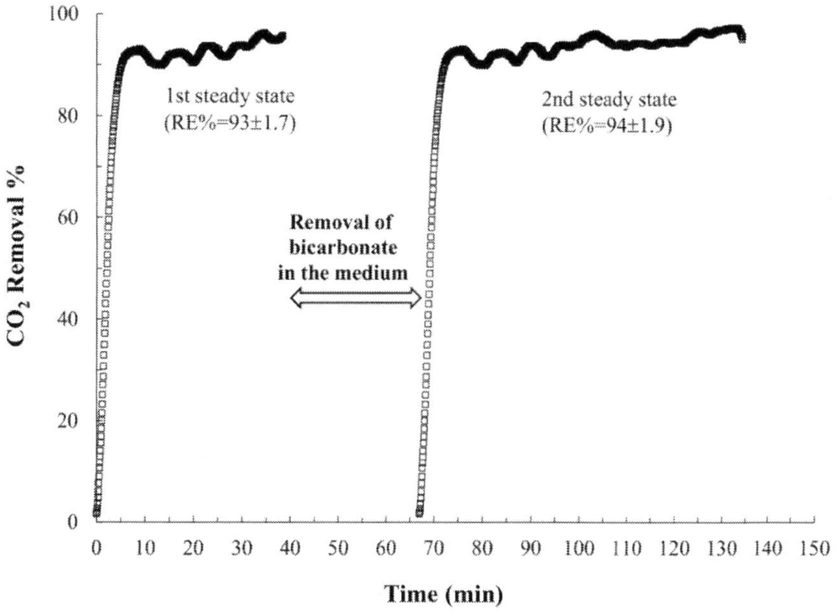

Fig. 5b. Effect of bicarbonate removal on CO_2 elimination in the foam bioreactor. Conditions: CO_2 inlet concentration, 3.8%; cell concentration, 4 g_{dw}/L; gas velocity, 0.8 m/min; total reactor volume, 0.4 L; pH at 10.5-11.

First, the foam bioreactor at the selected condition (3.8% inlet CO_2, 24 s gas contact, 4 g_{dw}/L cells) immediately reached 93±1.7% CO_2 removal at the steady state which lasted about 30 min. After 30 min steady state, the CO_2 removal started to decrease since the measured bicarbonate concentration in the culture (79 mM) inhibited the CA activity leading to reduced CO_2 removal. To completely remove bicarbonate in the culture, *E. coli* cells were suspended in a fresh medium after interrupting the foam bioreactor and separating only the cells via centrifugation, leaving the bicarbonate in the supernatant. When the foam bioreactor was re-operated with fresh medium, it resumed high CO_2 removal efficiency (94±1.9%) at steady state similar to that in the previous operation. Afterwards, the CO_2 removal lasted 55 min at steady state and then decreased over time due to inhibition of CA activity by high concentration of bicarbonate in the culture (~97mM). Therefore, it is evident that complete removal of bicarbonate in the culture resulted in sustaining high CO_2 removal by avoiding inhibition of CA activity. The mineralization percent of CO_2 (as the measured bicarbonate production per the calculated bicarbonate production) during the CO_2 removal by the CA-displaying *E. coli* in the foam bioreactor indicated 78% and 98% at the first and second steady state, respectively. This confirmed that the majority mechanism of CO_2 removal was the CA-driven reaction. Nevertheless, the cyclic or continuous removal of bicarbonate in the medium would be able to sustain high and stable CO_2 removal in the foam bioreactor while avoiding inhibition of CA activity. Therefore, operating strategies for cyclic or continuous removal of bicarbonate in the medium need to be developed.

4. Conclusion

An appropriate auto inducible media was formulated to support cell growth and to be employed directly in the foam bioreactor. Future formulations can be made to support an even higher cell density and to buffer at a more alkaline pH to improve CA function. Direct employment of the inoculated media in the foam bioreactor showed adequate removal and elimination capacity. Variations in gas velocity showed an optimum of 0.8 L/min. Reactions at higher cell concentrations were able to overcome biological limitations of low cell concentrations and achieve long-lasting steady states.

Finally, tests with bicarbonate removal were able to reactivate CA function and point towards the development of continuous bicarbonate removal techniques. Future work will include *in situ* pH monitoring and novel catalytic pH control, development of novel, easily separated amine adsorbents, and adjusted genetic modifications to maximize the bacteria's biological contribution. The work done in this study will be optimized to be applied to continuous operation with product removal, which is more desirable than the batch system used in this study due to the inhibitory effect bicarbonate has on CA.

5. Acknowledgements

This work was supported by U.S. Department of Agriculture Hatch Program (Project number: 5600580).

6. References

Albo, J., Luis, P., Irabien, A. 2010. Carbon Dioxide Capture from Flue Gases Using a Cross-Flow Membrane Contactor and the Ionic Liquid 1-Ethyl-3-methylimidazolium Ethylsulfate. *Ind. Eng. Chem. Res.*, **49**, 11045-11051.

Artanto, Y., Jansen, J., Pearson, P., Puxty, G., Cottrell, A., Meuleman, E., Feron, P. 2014. Pilot-scale evaluation of AMP/PZ to capture CO2 from flue gas of an Australian brown coal–fired power station. *International Journal of Greenhouse Gas Control*, **20**, 189-195.

Cao, Y., Song, F., Zhao, Y., Zhong, Q. 2013. Capture of carbon dioxide from flue gas on TEPA-grafted metal-organic framework Mg2(dobdc). *Journal of Environmental Sciences*, **25**(10), 2081-2087.

De Simone, G., Supuran, C.T. 2012. (In)organic anions as carbonic anhydrase inhibitors. *J Inorg Biochem*, **111**, 117-29.

Dumée, L., Scholes, C., Stevens, G., Kentish, S. 2012. Purification of aqueous amine solvents used in post combustion CO2 capture: A review. *International Journal of Greenhouse Gas Control*, **10**, 443-455.

Fan, L.H., Liu, N., Yu, M.R., Yang, S.T., Chen, H.L. 2011. Cell surface display of carbonic anhydrase on Escherichia coli using ice nucleation protein for CO(2) sequestration. *Biotechnol Bioeng*, **108**(12), 2853-64.

Hasib-ur-Rahman, M., Siaj, M., Larachi, F. 2010. Ionic liquids for CO2 capture—Development and progress. *Chemical Engineering and Processing: Process Intensification*, **49**(4), 313-322.

Kan, E., Deshusses, M.A. 2005. Continuous operation of foamed emulsion bioreactors treating toluene vapors. *Biotechnol Bioeng*, **92**(3), 364-71.

Lindskog, S. 1997. Structure and Mechanism of Carbonic Anhydrase. *Pharmacology and Therapeutics*.

Low, B.T., Zhao, L., Merkel, T.C., Weber, M., Stolten, D. 2013. A parametric study of the impact of membrane materials and process operating conditions on carbon capture from humidified flue gas. *Journal of Membrane Science*, **431**, 139-155.

Maresca, A., Vullo, D., Scozzafava, A., Supuran, C.T. 2013. Inhibition of the alpha- and beta-carbonic anhydrases from the gastric pathogen *Helycobacter pylori* with anions. *Journal of Enzyme Inhibition and Medicinal Chemistry*, **28**(2), 388-391.

Migliardini, F., De Luca, V., Carginale, V., Rossi, M., Corbo, P., Supuran, C., Capasso, C. 2014. Biomimetic CO_2 capture using a highly thermostable bacterial α-carbonic anhydrase immobilied on a polyurethane foam. *Journal of Enzyme Inhibition and Medicinal Chemistry*, **29**(1), 146 - 150.

Mores, P., Scenna, N., Mussati, S. 2012. CO2 capture using monoethanolamine (MEA) aqueous solution: Modeling and optimization of the solvent regeneration and CO2 desorption process. *Energy*, **45**(1), 1042-1058.

Olajire, A.A. 2010. CO2 capture and separation technologies for end-of-pipe applications – A review. *Energy*, **35**(6), 2610-2628.

Padurean, A., Cormos, C.-C., Agachi, P.-S. 2012. Pre-combustion carbon dioxide capture by gas–liquid absorption for Integrated Gasification Combined Cycle power plants. *International Journal of Greenhouse Gas Control*, **7**, 1-11.

Pires, J.C.M., Alvim-Ferraz, M.C.M., Martins, F.G., Simões, M. 2012. Carbon dioxide capture from flue gases using microalgae: Engineering aspects and biorefinery concept. *Renewable and Sustainable Energy Reviews*, **16**(5), 3043-3053.

Russo, M.E., Olivieri, G., Marzocchella, A., Salatino, P., Caramuscio, P., Cavaleiro, C. 2013. Post-combustion carbon capture mediated by carbonic anhydrase. *Separation and Purification Technology*, **107**, 331-339.

Stevanovic, S., Podgorsek, A., Moura, L., Santini, C.C., Padua, A.A.H., Costa Gomes, M.F. 2013. Absorption of carbon dioxide by ionic liquids with carboxylate anions. *International Journal of Greenhouse Gas Control*, **17**, 78-88.

Yu, C.-H. 2012. A Review of CO2 Capture by Absorption and Adsorption. *Aerosol and Air Quality Research*.

Zhang, S., Zhang, Z., Lu, Y., Rostam-Abadi, M., Jones, A. 2011. Activity and stability of immobilized carbonic anhydrase for promoting CO2 absorption into a carbonate solution for post-combustion CO2 capture. *Bioresour Technol*, **102**(22), 10194-201.

378565 PBI Membranes for Pervaporation Dehydration of Phenol Aqueous Solution

Abstract:

Phenol is an important but hazardous commodity in chemical industries. The dewatering is a critical and challenging process for its
application and wasterwater treatment. In this work, polybenzimidazole (PBI) membranes with various morphologies are employed for phenol dehydration via pervaporation, including flat-sheet dense membranes, single-layer and duallayer hollow fiber membranes. The effects of cross-linking modification and post thermal treatment on the pervaporation performance of PBI flatsheet
dense membranes were investigated; effects of operation temperature and feed composition are also studied not only in terms of flux and separation factor, but also of membrane intrinsic permeance and selectivity. In order to achieve a high permeation flux, PBI singlelayer
hollow fiber membranes of thinner selective layers were developed and studied with different spinning parameters. The preliminary study of dual-layer PBI/PBI and PBI/poly(vinylidene fluoride) (PVDF) duallayer
hollow fiber membranes are also carried out to explore the potential of highperformance composite membranes for phenol dehydration. The benchmarking shows the potential of PBI membranes for phenol dehydration with good separation performance. It may open new perspective for highperformance
membranes for the pervaporation dehydration of phenol and other corrosive organics.

378664 Experimental and Mathematical Modelling of Breakthrough Curves of Fe (II) on Immobilized Biomass in a Packed Column

Abstract:

Metal contamination of aquatic ecosystem in and around the Witwatersrand basin has been exacerbated by an overflow of acid mine drainage from spent mines. Physical and chemical treatments are conventional methods widely used in heavy metal removal, however biological methods have proven to be more efficient, cost effective and environmentally conscious. The present work studies the use of fresh-water green microalgae *Desmodesmus sp.* immobilized in a calcium alginate matrix for heavy metal removal. The adsorption performance of immobilized algae beads (IAB) in a continuous flow from aqueous solution was investigated in a packed column. The packed column experiments were carried out as a function of bed height of IAB and alginate beads without algae (ABWA). The total adsorbed quantities, equilibrium uptakes and total removal percentages of Fe (II) ions were determined by evaluating the breakthrough curves obtained at different bed heights. The steepness of the breakthrough curve determined the extent to which the capacity of the adsorbent bed can be used. The total quantity of Fe(II) ions adsorbed in the column as calculated from trapezoidal numerical integration was found to be 1.6 mg; 33.6 mg and 116.1 mg for ABWA (height 26 mm), IAB (height 26 mm) and IAB (height 82 mm), respectively. It was found therefore, that the amount of Fe (II) ions adsorbed in the column was affected by the presence of algae in alginate beads and the increase in bed height of IAB. Thomas; Yoon-Nelson and Adams-Bohart models were used to analyse the experimental data and model parameters were evaluated. Correlation of the experimental breakthrough curves with model predictions was observed. The root mean square error (RMSE) model will be used to measure accuracy of the results.

382308 Effect of Humidity on Bulk Properties and Surface Energetics of Fine Powders

Abstract:

Humid environment often creates various problems in fine powder processing and frequently pose substantial threat to the final product quality.The present study focuses on the effect of humidity on bulk behavior and surface energeticsof fine pharmaceutical excipient powder with and without surface modification. Corn starch powder was surface modified using Aerosil R972 type hydrophobic nano-silica through dry coating. Bulk property characterization for uncoated and surface modified powders was carried out using FT4 powder rheometer. It was found that flow and packing properties of uncoated powder deteriorated with increasing relative humidity (% RH) and prominent changes were observed above 60% RH. In contrast, these bulk properties of surface modified powders were found to be more or less consistent and stable. Further, in order to understand these changes in bulk behavior of powder, surface energy studies of humid powders were carried out using Inverse Gas Chromatography (IGC). IGC studies revealed that both dispersive and specific components of surface energies of uncoated corn starch powder changes with changing humidity conditions from 30% to 90%.On the other hand, surface energies of dry coated powders remained more or less constant at different range of humidity conditions studied. The present study showed that surface modification of powders helped to prevent the detrimental effects of humidity on bulk behavior and surface energetics of fine powders.

386431 Structure and Stability of Layersomes Formed with Dextran Sulfate and Poly-L-Arginine

Abstract:

The self-assembly of polyelectrolyte layers on lipid bilayer surfaces is of great interest in the areas of biological device fabrication, drug delivery, and biomaterial design.The structure, function, and stability of these systems ultimately depend upon the interactions between the polyelectrolytes and lipid bilayers, and how these interactions change as a function of time, temperature, and solvent composition. For example, polyelectrolyte layering on spherical lipid bilayer vesicles or liposomes (i.e. layersomes) can yield stable nanocapsules for drug delivery where the rate of release is dependent upon the permeability of the polyelectrolyte–lipid bilayer shell. The objective of this work was to determine how polyelectrolyte–lipid interactions lead to changes in shell structure and how these changes influence layersome morphology and stability.

Layersomes were prepared in deionized water (DIW) with cationic liposomes (130 nm hydrodynamic diameter) composed of dioleoylphosphatidylcholine (DOPC, zwitterioninc) and dioleoyltrimethylammonium-propane (DOTAP, cationic) at a 1:1 mole ratio. Negatively charged dextran sulfate (6500-10000 g/mol) and positively charged poly-L-arginine (5000-15000 g/mol) were used as FDA approved polyelectrolyte layers. Single (dextran sulfate) and multiple (up to 4) alternating layersomes were prepared using a 'washless' method. Layersome formation was confirmed by zeta potential, dynamic light scattering, and cryogenic transmission electron microscopy (Cryo-TEM) analysis. The layersomes ranged from 170 to 200 nm with zeta potentials alternating between +/-50 mV. There was evidence of polyelectrolyte complexes bound to the layersomes, but these complexes did not affect layersome stability. Long-term studies (30 days) showed that the layersomes were stable in DIW due to electrosteric repulsion. In phosphate buffered saline (PBS), layersomes with outer dextran sulfate layers were stable, but those with outer poly-L-arginine layers were not. Differences in layersome stability can be attributed to differences in the pKa and charge density of the polyelectrolytes, as well as restructuring of the polyelectrolyte shell. Fluidity measurements based on diphenylhexatriene (DPH) anisotropy were used to determine how polyelectrolyte layering affected lipid ordering within the bilayer. Dextran sulfate led to significant lipid ordering (layers 1 and 3), while the addition of poly-L-arginine (layers 2 and 4) partially neutralized dextran sulfate and reduced ordering. Collectively, this study shows that stable layersomes can be prepared using the 'washless' method and that layersome stability is most influenced by the outermost layer.

Minimum reflux and bow tie region in a simple heterogeneous reactive distillation column for esterification reactions

Sánchez, C. A; Rodríguez, G.; Gómez, M. A.
Universidad Nacional de Colombia

Keywords: Heterogeneous Reactive Distillation; Minimum Reflux; Reactive Phase Equilibria; Continuation Methods; Isoamyl Acetate; n-Amyl Acetate.

The basic design of the reactive distillation (RD) columns consists in: 1) establishing the feasibility of the separation, 2) calculating the minimum reflux (minimum energy condition) and 3) calculating the number of theoretical stages and the feed tray for a given reflux higher than minimum. The use of Doherty's transformed compositions allows formulating an analogy with the geometrical methods applied in the heterogeneous azeotropic distillation (HAD) to simplify the solution of the three basic design problems for RD columns in overall equilibrium (simultaneous chemical and phase equilibria). Particularly the minimum reflux problem (MRP) of the sharp reactive separations has the same geometry as the nonreactive case of the HAD when the analysis is performed in transformed molar fractions (Avami et al., 2012).

The solution of the MRP and the localization of the product composition region (bow tie region) for non hybrid columns in overall equilibrium can be made based on the analysis of the reactive pinch points trajectories (RPPT) associated with the distillation products. In relation with these two aspects it is underlined, that there are no numerical details on the calculation of the RPPT known by the authors. With this motivation, this work presents a robust numerical continuation method for the tracking the RPPT and demonstrates its utility solving the minimum reflux problem and locating de bow tie region for two study cases dealing with the synthesis by RD of isoamyl acetate and n-amyl acetate.

In a general perspective, the main contribution consists in adapting the geometrical methods used in the design of the HAD columns with ternary mixtures (Urdaneta et al., 2002; Kraemer et al., 2011) to develop the basic design of RD columns with systems where the transformed compositions space has two dimensions (two degrees of freedom to calculate the reactive phase equilibria). The development of this viewpoint has few previous publications in the open literature (Avami et al., 2012; Sánchez et al., 2013) and contributes to the conceptual design of RD units for the synthesis of acetates.

At the end this work two conclusions are highlighted: 1) the minimum reflux within the best separations of RD columns with an ahead decanter corresponds either with two active pinch points located in different column sections or with a single active pinch point located in the feed tray and 2) the configurations with multiple heterogeneous stages condition the minimum reflux geometry and may reduce the minimum energy consumption.

References

Avami A et al., "Shortcut design of reactive distillation columns". Chem. Eng. Sci., Vol.71, 2012 (166-177).

Kraemer, K.; Harwardt, A.; Skiborowski, M.; Mitra, S. and Marquardt, W. "Shortcut-based Design of Multicomponent Heteroazeotropic Distillation". Chem. Eng. Res. Des., Vol. 89, 2011, pp. 1168-1189.

Sánchez C et al., "Caracterización termodinámica para sistemas de esterificación heterogéneos en destilación reactiva. I. Envolventes reactivas". Revista EIA, año 10, No. 20, 2013 (73-85)

Urdaneta, R. Y.; Bausa, J.; Bruggemann, S. y Marquardt, W. "Analysis and Conceptual Design of Ternary Heterogeneous Azeotropic Distillation Processes". Ind. Eng. Chem. Res., Vol. 41, No. 16, pp. 3849-3866, 2002.

389387 Determining the Effects of Unit Cell Parameters and Solvent Size on the Energy Profile of Organic Semiconducting Crystals

Abstract:

Determining the Effects of Unit Cell Parameters and Solvent Size on the Energy Profile of Organic Semiconducting Crystals

Kristina Lenn, Paulette Clancy Cornell University

Gaurav Giri, Ying Diao and Zhenan Bao Stanford University

Organic semiconducting materials are gaining popularity as solar cell materials, in large part because of their ability to be processed onto flexible substrates and to be manufactured more cheaply. However, due to their "charge-hopping"-like mechanism, organic materials are not nearly as efficient as silicon in regards to charge transport. To circumvent this, the Bao research group at Stanford has initiated a new process called solution shearing on 6,13-bis(triisopropylsilylethynyl) pentacene (6,13-TIPS pentacene) that significantly reduces the pi-orbital overlap, thereby enhancing the charge transport by an 80% increase.

To fully appreciate what is happening at an atomic level and how shearing the unit cell to conform to a configuration that is both energetically favorable and conducive to charge transport, an atomically explicit model of molecules of 6,13-TIPS pentacene were analyzed via *ab initio* and Molecular Dynamics (MD) techniques. *In vacuo*, the parameters of the unit cell [a,b,γ] were varied and subjected to minimization calculations via the MD package TINKER. From these minimizations, an energy profile was generated which allowed us to identify the five lowest energy configurations which we found were well aligned with five polymorphs determined experimentally by the Bao group. Closer examination of the molecular configurations suggested that the fluctuations in the energy profile were due to the bending and twisting of the acene backbone, the wagging of the silylethynyl groups, and the rotating of the methyl groups. Compared to an equilibrium structure, there is a 4.5° deviation in the acene backbone and a nearly 10° difference in the silylethynyl angle across this energy landscape. However, when comparing the changes in these different angles against the energy profile, the minima for both are closely aligned.

6,13-TIPS pentacene was also studied in ten different solvents, ranging in size from tetrahydrofuran (THF) at 3.18 Å to decalin at 3.94 Å. A Lennard-Jones atom was used to represent these solvents, varying only in size. The atom was placed within the unit cell in an energetically favorable location and submitted to energy minimization calculations. From these energies, it can be seen that as the solvent size increases, the stability of the unit cell decreases.

Based on the data gathered, the energy profile of 6,13-TIPS pentacene is significantly affected by the rotating, wagging, and bending of the various side groups and the acene backbone. Whether these deviations occur in sync or are a result of a "domino effect" (i.e., the rotation of the methyl groups leads to the wagging of the silylethynyl groups which in turn leads to the bend in the backbone) is not yet clear. The size of the solvent also impacts the energy profile; as the solvent radius increases, the molecules in the unit cell can be seen to almost push at the solvent in an effort to expel it from the unit cell.

Poly(sulfobetaine methacrylate) (pSBMA) films, known for their antifouling properties, were produced by surface initiated polymerization on a silicon surface via a batch reaction system in a CO_2 expanded liquid reaction medium containing methanol. Atom transfer radical polymerization (ATRP) was carried out using 2-2'-Bipyridyl ligand and CuBr catalyst in a water/methanol mixture with trichloro[4-(chloromethyl)phenyl]silane (CMPS) and ((3-trimethoxysilyl)propyl 2-bromo-2methylpropionate) used as the initiating species. The films were grown in the dense gas environment at a range of conditions and the thickness could be varied up to 10 nm depending on the processing conditions. In contrast to films produced by conventional solvent systems at atmospheric pressure, the polymer films grown by the dense gas process showed uniform thickness and pin-hole free topography. Most importantly, the latter films showed no trace of copper (used as the catalyst) thus obviating the need for post-deposition processing and avoiding adverse effects of the metal leaching during service. The process can easily extended to any surface that can be functionalised.

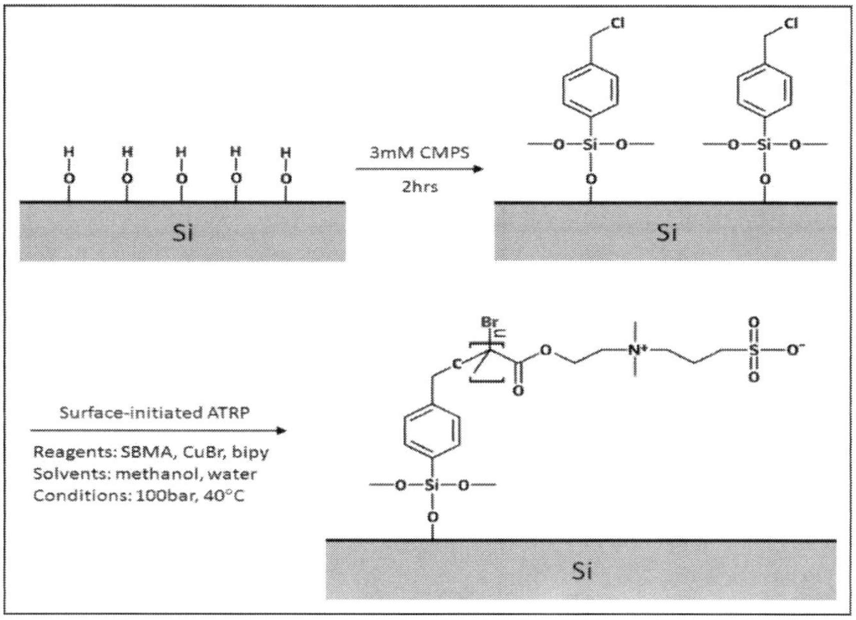

Prediction of the Adsorption Isothermes of n-Aldehyde Mixtures Applying Density Functional Theory

Patrick Zimmermann[1], Thomas Goetsch[2], Tim Zeiner[2], Sabine Enders[1]

[1]Technical University of Berlin, Chair of Thermodynamics, Ernst - Reuter - Platz 1, 10587 Berlin, Germany.

Email: patrick.zimmermann@tu-berlin.de; sabine.enders@tu-berlin.de

[2]Technical University Dortmund, Laboratory of Fluid Separations, Emil-Figge-Straße 70, 44227 Dortmund, Germany.

Email: Thomas.Goetsch@bci.tu-dortmund.de, Tim.Zeiner@bci.tu-dortmund.de

Key words: Adsorption Isothermes of Binary Mixtures, Density Functional Theory, Separation of Close Boiling Mixtures, Separation of n-Aldehydes

Abstract:

The separation of mixtures having components with very similar vapor pressures using distillation is always a very hard task and requires a large amount of invest and energy cost. Mixtures of long-chain n-aldehydes having a similar number of carbon atoms belong to this class of mixtures. An alternative separation method is the adsorption at porous surfaces. For the design and optimization of this separation method the adsorption isotherms are required. In principle, these adsorption isotherms can be measured; however the measurement is very time consuming. Therefore, this contribution deals with the prediction of adsorption isotherms applying density functional theory (DFT) [1,2,3,4] in combination with a suitable equation of state.

Our interest is the understanding of the behavior of pure fluids and binary mixtures in narrow pores. This interest has focused on the density profile and adsorption inside the pore, phase transitions and hysteresis effects, and wetting and prewetting phenomena. The DFT starts with an expression for the grand potential as a functional of the density profile of the fluid, where we assume the fluid is inhomogeneous in only the z direction perpendicular to the wall of the narrow slit pore. The grand potential appropriate the density profile of the fluid when one considers a fluid in an external field at a fixed temperature and chemical potentials of the components in the fluid mixture. At thermodynamic equilibrium the grand potential is in its global minimum with respect to variations of the density profiles and at this minimum its value is equal to the thermodynamic grand potential. The key step in density functional theory is to specify an approximate form for grand potential that is both tractable and accurate.

The properties of the fluid mixture can be described with the inhomogeneous Helmholtz energy, which can be calculated using any equation of state (EOS). In order to model mixtures of n-aldehydes the Peng-Robinson equation of state (PR-EOS) [5] can be applied. The pure-component parameters of this EOS are adjusted to the vapor pressure curves taken from the literature. Unfortunately, no experimental data related to the vapor-liquid equilibria of binary n-aldehyde mixtures could be found in the literature. Therefore, some data points were measured with help of the dynamic method. These new experimental data were used to estimate the binary interaction parameter, k_{ij}. The external potential occurring in grand potential describes the interaction of the fluid with solid wall and can be expressed by the Steele 10-4-3 potential [6]. To solve for the equilibrium density profile, the grand potential functional is minimized with respect to the local density.

Experimental studies of adsorption phenomena of a pure fluid commonly focus on adsorption isotherms; plots of the amount of fluid adsorbed in a porous solid versus the pressure of the coexisting bulk fluid. This information is also obtainable from the density functional theory by integration of the calculated density profile. In the case of fluid mixture the partial density can be predicted using this approach. Additionally, the impact of the porosity on the adsorption isotherm can be studied. The pore distribution function can be obtained by measurement of the nitrogen adsorption isotherm.

We present adsorption isotherms, phase diagrams, and density profiles for pure n-aldehydes as well as binary mixture of n-aldehydes confined to a slit pore. In particular, we concentrate on the gas-liquid transition in the pore (capillary condensation).

Literature:

[1] M.J. de Oliveira, R.B. Griffiths, Surf. Sci. 71 (1978) 687-94.
[2] C. Ebner, Physical Rev. A 22 (1980) 2776-81.

[3] B.K. Peterson, K.E. Gubbins, G.S. Heffelfinger, U.M.B. Marconi, F. van Swol, J. Chem. Phys. 88 (1988) 6487-500.
[4] C. Lastoskie, K.E. Gubbins, N. Quuirke, J. Phys. Chem. 97 (1993) 4786-96.
[5] D.Y. Peng, D.B. Robinson, Ind. Eng. Chem. Fundam. 15 (1976) 59-64.
[6] W. A. Steele, Surf. Sci. 36 (1973) 317-324.

CONTROLLED CRYSTALLINITY OF SILICON POWDERS

Brian S. Holsclaw[1)], Michael J. Molnar[1), 2)], William Herron[1), 2)], Joel P. McDonald[2)], Matthew A. Gave[2)], and Jonathan J. Host[1)]

1) *Hemlock Semiconductor Corporation (Hemlock, Michigan, USA)*
2) *Dow Corning Corporation (Midland, Michigan, USA)*

2014 AIChE Annual Meeting (Atlanta, GA, USA, November 16-21, 2014)

Abstract

Silicon powder generated by the pyrolytic reaction of silane (SiH_4) in a free space reactor has been demonstrated to be produced in varying degrees of crystallinity depending on the temperature and other reactor conditions. The reactor is heated to no more than 650°C while flowing a feed gas comprising silane and a carrier gas into the reactor while maintaining an internal pressure of about 2 atm or less. The non-crystalline (amorphous) silicon is produced in a controlled composition with crystalline silicon. The crystallization composition of the powder is determined by a combination of Powder X-Ray Diffraction and Pair Distribution Function Analysis. Work to produce the silicon powder using a free space reactor was completed by the Institutt for energiteknikk (Kjeller, Norway) on research contract for Hemlock Semiconductor Corporation.

Introduction

The production of silicon powder via chemical vapor deposition and nucleation processes, particularly the deposition of silicon from monosilane (SiH_4), has been discussed in the literature for over a century [1-3]. Elimination of nucleation processes is critical to fabrication of semiconductor devices. Murthy et al. describe the early difficulties of particle management in films of epitaxial silicon derived from monosilane-based deposition and identified a critical concentration of silane precursor to the process [4]. More detailed kinetic mechanisms involving the growth of particles through the consecutive insertion of silylene (SiH_2) into silicon clusters have been proposed by multiple investigators [5-7]. This phenomenon of powder generation extends to silicon precursors other than silane. Herrick and Woodruff describe nucleation of powders in detail for a family of halosilane-based CVD precursors, for which the halogen presence suppresses the spontaneity of the nucleation, but the mechanism only considered routes where the growth of the powder is assumed to occur via clusters of Si(g) vapor adatoms as opposed to silylene-based clustering [8, 9].

On the other extreme, the intentional production of silicon powder as a semiconductor for photovoltaic applications has also been studied. Most notably, work at Union Carbide Corporation examined use of silane in a free space reactor [10-12] and later commercial production of this powder by Joint Solar Silicon GmbH [13]. Co-production of silicon powders in the fluidized bed deposition process for granular silicon has also been well documented [14].

Discussion regarding the structure of these materials has been limited. In the case of a defined film produced from the thermal decomposition of silane, Beers et al. (University of Nijmegen) performed detailed study of the transition of amorphous Si:H materials to amorphous Si to crystalline silicon materials depending on the growth flux and temperature of the growth [15, 16]. For the case of powders grown on seed particles, Iya et al. describe both fibrous growth which was correlated with metallic contamination sites on seed particles and a desired dense phase crystalline growth [17]. Kojima at Seikei University reported similar growth features [18]. Murthy et al. identified crystalline silicon materials in their nucleated powders between 300 Å and 800 Å in size with variants of a face-centered cubic structure (tetrahedral, octahedral, and triangular bipyramidal). In some cases, particularly at nanoparticle scale, incorporation of hydrogen into the structure is clear. For instance, Onischuk et al. describe volatilization of hydrogen from silicon nanoparticles generated during heating plus the presence of polyhydride groups readily detectable by infrared spectroscopy. Kojima at Seikei University reported similar growth features in silane-based deposition to Iya et al. and also showed reduced bulk density, hydrogen content that could be removed with heating, and a smaller atomic spacing prior to heat treatment [17].

Experimental Methods

Silicon powder was generated in a free space reactor located at the Institutt for energiteknikk (IFE) in Kjeller, Norway. This work was performed on research contract by IFE for Hemlock Semiconductor Corporation. For more experimental details, consult the patent application by Dow Corning Corporation [19].

Free Space Reactor

A schematic of the free space reactor used in this study is given in Figure 1. The reaction tube is constructed of alumina, stainless steel or Inconel® with diameter of 78 mm (alumina), 71 mm (stainless steel), or 142 mm (Inconel®) and 1.5 m long. The alumina tube is operable only at or below atmospheric pressure, and the stainless steel tube can be operated at, below or above atmospheric pressure. The flow rates of SiH_4 and the carrier gas may be controlled independently. Prior to introduction of SiH_4 or the carrier gas, the system is evacuated and backfilled with Ar or He one or more times.

The internal reactor volume is heated by either three or four resistive heaters depending on the reactor configuration. The gases fed into the reactor are not preheated. The separate heating zones in the reactor allow for a controlled temperature profile as the gases flow through the tube. The first zone (top zone in Figure 1) is controlled between 200 and 400°C. The second zone is heated to between 300 and 500°C. The third heating zone (reaction zone) is heated between 450 and 650°C. The reaction zone temperature represents the highest temperature of the tube and is the temperature reported for the dust formation. The fourth reactor zone, only used in the Inconel® configuration of the reactor, may be heated between 100 and 650°C but is at a temperature lower than the reaction zone.

Immediately downstream of the heated sections of the reactor, a sintered metal filter traps silicon particles formed in the reactor until a gas back-pulse clears the filter periodically. The powder knocked loose from the filter falls into a stainless steel collection vessel that is removed at the end of the run. The collection vessel is fitted with a valve arrangement to maintain an inert atmosphere over the product during transfer to a glove box, where the vessel is opened and the powder product is removed.

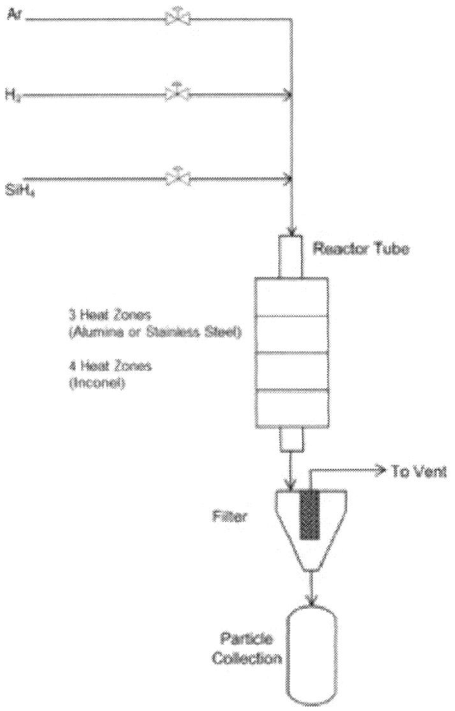

Figure 1. Schematic of the free space reactor employed for synthesis of the silicon powder. SiH_4 and either H_2 or an inert carrier gas (Ar or He) are mixed and fed into the top of the reactor tube. Silicon dust is generated in the heated reactor tube. The dust is then collected in a filter with periodic collection in a stainless steel vessel.

X-Ray Diffraction (XRD)

X-ray diffraction data can provide information about the crystallinity and/or amorphous nature of the specimen. Standard powder diffraction patterns are collected in Bragg-Brentano geometry from 10° to 80° 2θ in 0.02° increments at 2.7°/minute with a Cu anode operating at 40 kV and 44 mA. A 10 mm height limiting slit, 1/2° divergence slit, open scattering slit, and open receiving slit are used, and intensity data are collected with a high speed detector. High temperature powder diffraction patterns were collected for some of the samples. The conversion of non-crystalline Si to crystalline Si was observed with increasing temperature. Temperatures were reached with a heating rate of nominally 10°C/min followed by a brief (~5 min) equilibration time.

Pair Distribution Function (PDF) Analysis

Pair distribution function (PDF) analysis can provide information about both the long-range (>100 Å) and short-range atomic ordering in materials. A method of determining the crystallinity of a silicon powder has been developed. The method entails collecting x-ray diffraction data from a specimen comprising silicon powder, and then performing a Fourier transform of the XRD data to obtain PDF data. The PDF data are fit with a crystallographic model comprising a first unit cell representing crystalline silicon and a second unit cell representing non-crystalline (amorphous) silicon, and then a weight percent crystallinity of the specimen may be determined. A detailed description of the procedure used in the PDF analysis may be found in [19]. To carry out a meaningful PDF analysis, XRD data are typically collected for 24-72 hours using a commercially available lab-scale diffractometer to achieve a high signal-to-noise ratio.

Results

A series of 26 experiments to synthesize silicon powder having a controlled crystallinity was carried out in the free space reactor shown in Figure 1. The process variables included the temperature profile and pressure of the tube, the diluent gas employed and the concentration of silane in the diluent gas, and the total flow rate. The process conditions are summarized in Table 1. It is believed that the primary factors influencing the crystallinity of the resulting silicon powder are the maximum reactor temperature, the internal reactor pressure, and the choice of diluent gas. Residence time in the reactor tube is also important. Examples A-D, 1-12 and 17 were produced in the alumina tube. Examples 13-16 were produced in the stainless steel tube. Examples 18-26 were produced in an Inconel® tube. The total flow rate was measured at standard temperature and pressure (1 atm and 25°C).

Table 1. Compilation of process conditions

	Max. Temp.	**Pressure**	**SiH_4 Conc.**	**Flow Rate**	**Diluent**
Examples	°C	atm	mole fraction	liters/min	gas species
A	580	0.9	0.8	1	H_2
B	600	0.9	0.8	1	H_2
C	620	0.9	0.8	1	H_2
D	750	0.75	0.42	1.2	H_2
1	550	0.9	0.8	1	H_2
2	580	0.5	0.8	1	H_2
3	580	0.5	0.8	2	Argon
4	550	0.9	0.8	2	Argon
5	550	0.5	0.8	2	H_2
6	550	0.9	0.2	2	H_2
7	550	0.9	0.2	1	Argon
8	580	0.9	0.8	2	H_2
9	580	0.9	0.2	1	H_2
10	580	0.5	0.2	2	H_2
11	550	0.5	0.2	1	H_2
12	580	0.9	0.2	2	Argon
13	580	0.5	0.2	1	Argon
14	550	0.5	0.8	1	Argon

15	550	0.5	0.2	2	Argon
16	580	0.9	0.8	1	Argon
17	580	0.9	0.2	2	H_2
18	456	1.0	0.8	3	Argon
19	479	1.0	0.8	3	Argon
20	502	1.0	0.8	3	Argon
21	524	1.0	0.8	3	Argon
22	547	1.0	0.8	3	Argon
23	592	1.0	0.8	3	Argon
24	592	2.0	0.8	3	Argon
25	592	1.0	0.8	3	He
26	592	2.0	0.8	3	He

Prior to the 26 experiments, a set of preliminary powder production runs (Examples A-D) was carried out to determine the transition temperature above which crystalline silicon is produced. The silicon powders obtained from Examples A and B at 580°C and 600°C, respectively, included a significant fraction of non-crystalline silicon, while the silicon powder of Example C appeared to include a larger fraction of crystalline silicon. Example D yielded a conclusively crystalline Si powder. As a result, subsequent runs (Examples 1-26) were carried out at a temperature below 620°C.

Structure Analysis

Powder XRD patterns with very broad peaks are consistent with non-crystalline or amorphous silicon, and patterns with narrow, high intensity peaks are consistent with crystalline silicon. Figure 2 presents the XRD patterns for several silicon powder examples, as described in Table 1. Example 27 refers to ground chemical grade polycrystalline silicon powder obtained from Dow Corning Corporation.

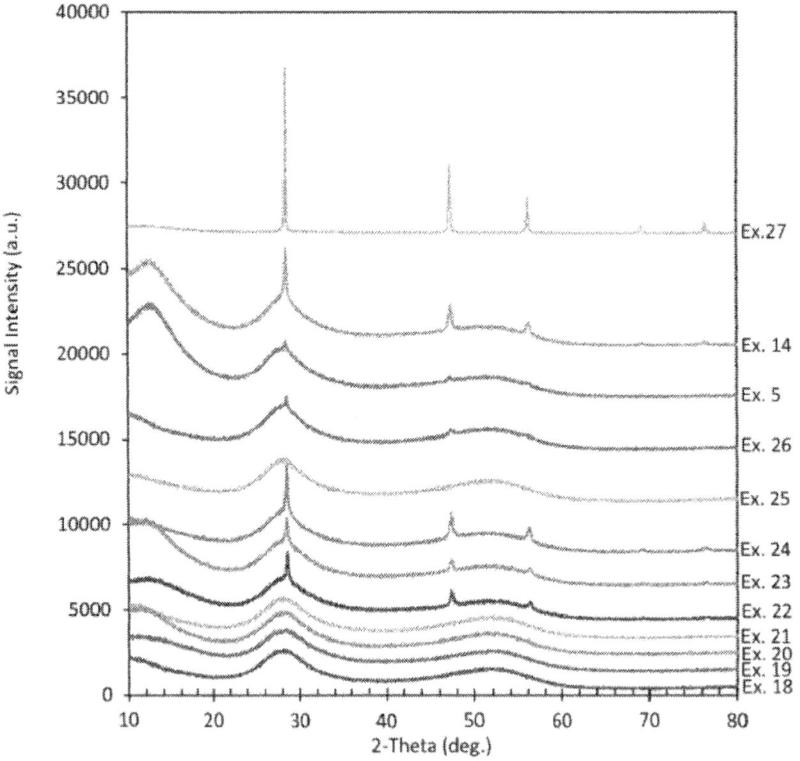

Figure 2. XRD patterns of several silicon powder examples as detailed in Table 1

The XRD pattern of example 5 may be indicative of either an inhomogeneous mixture of non-crystalline and crystalline Si or a homogeneous semi-crystalline Si material. Semi-crystalline Si refers to crystalline Si that may include an appreciable number of defects. Example 5 likely consists of predominately non-crystalline Si with a small amount of crystalline Si which may include defects. The silicon powder of example 14 is composed of both non-crystalline and crystalline Si but with a greater fraction of the crystalline component, as evidenced by the more prominent narrow peaks in the XRD pattern. The silicon powders of examples 18, 19, 20, 21 and 25 comprise non-crystalline Si with little or no (0 wt. %) crystalline Si, while the silicon powder of examples 22, 23, 24 and 26 include non-crystalline Si with some amount (> 0 wt. %) of crystalline Si.

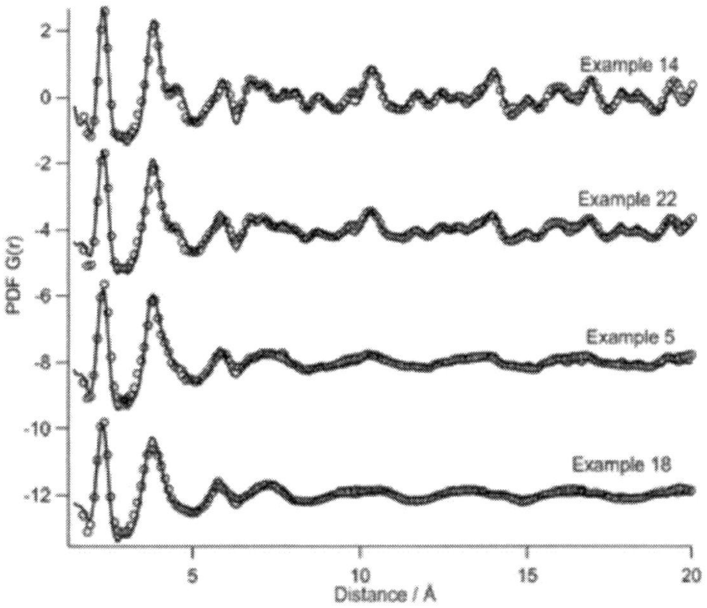

Figure 3. Experimental pair distribution functions G(r) (solid line) for silicon powder examples 14, 22, 5 and 18, with corresponding fits shown as hollow circles, where sparse markers were used for clarity, with 1 hollow circle shown for every 10 data points from the fit. The plots are offset from each other for clarity.

Figure 3 gives the results of PDF analysis for examples 5, 14, 18 and 22. The experimental PDF for the silicon powder of example 22 is very similar to that presented in [20], wherein the material is described as "amorphous silicon." The oscillations up to 20 Å in examples 14 and 22 may be evidence for an admixture of amorphous and semi-crystalline, or crystalline Si. These results suggest that the silicon powders of the examples are composed predominately or entirely of silicon, rather than, for example, SiO, SiO_2, SiC, and others. Furthermore, through a fitting procedure of the PDFs, as described in [19] using models representing the crystalline and non-crystalline or amorphous components, the silicon powder of examples 5, 14, 18 and 22 were found to include about 1%, about 10%, about 0% and about 5% crystalline Si, respectively.

Figure 4 gives the results for high-temperature XRD of example 5 silicon powder. For the silicon powders of both examples 5 and 14 (example 14 results given in [19]), only insignificant changes were observed in the powder patterns up to 550°C, indicating thermal stability of the starting microstructure. At room temperature, XRD data from both examples contained broad features consistent with an amorphous Si component and also narrower peaks assignable to a crystalline Si component, and a single weak reflection that may be assignable to SiO_2 in the cristobalite polymorph. The cristobalite may have been from the underlying SiO_2 example holder or it may have been a part of the example. For examples 5 and 14, a strong (111) peak was observed in the data at 600°C, indicating increased amounts of crystalline Si. At 650°C, the peaks had grown stronger, and beyond that temperature, no additional changes were observed. This behavior is consistent with a sudden enthalpic glass-to-crystalline transition.

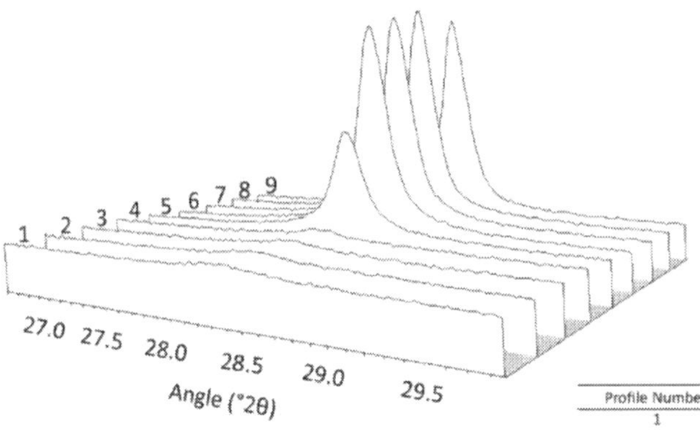

Profile Number	Temperature (°C)
1	30 (before)
2	450
3	500
4	550
5	600
6	650
7	700
8	750
9	30 (after)

Figure 4. XRD data for example 5 comparing changes in the (111) reflection as a function of temperature. A sudden glass-to-crystalline transition is observed.

Additional Analysis

To elucidate the particle morphology, scanning electron microscopy of the silicon powder specimens was conducted using a JEOL 6335 Scanning Electron Microscope (SEM) with a field emission electron source operated at 20 kV and a working distance of between 8 and 20 mm. No conductive coating was needed nor applied for examples 5 and 14, while Pd/Au coating was given in 10 nm thickness for examples 18 and 22. This imaging technique revealed that the silicon powder of both examples 5 and 14 includes spherical particles with a primary size of from about 50-400 nm, with particles around 100-200 nm in size. Figure 5 shows SEM images for silicon powder from example 5. SEM and additional light scattering studies indicate that the silicon powders may include a bimodal or multimodal particle size distribution.

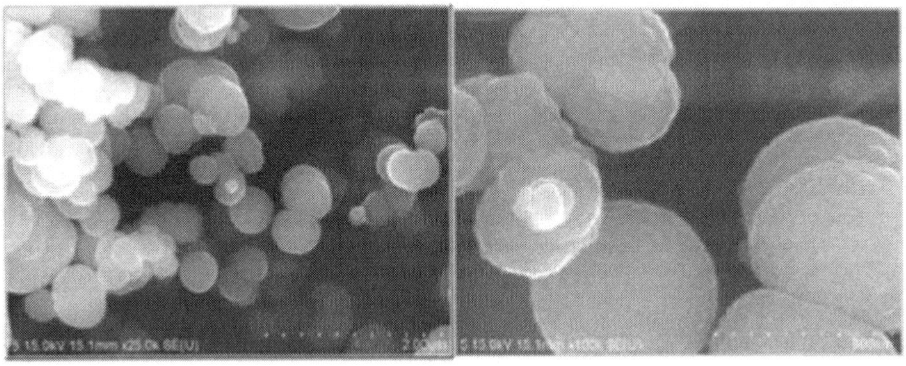

Figure 5. SEM images of silicon powders from example 5.

True density measurements were obtained with a gas pycnometer and represent an average density of individual particles, as opposed to the bulk density of the powder. A true density value of 2.324 g/cm^3 was obtained for the silicon powders of example 14 and a value of 2.309 g/cm^3 was determined for the silicon powders of example 5. The measured density of polycrystalline silicon powder is 2.319 g/cm^3. BET measurements of surface area for examples 5, 14, and 18-26 are given in [19].

X-ray fluorescence analysis using a Rigaku Primus IV gave identical compositions for the silicon powder of examples 5 and 14 of about 98% silicon, 1.5% oxygen, 0.5% carbon, and traces of the main components of stainless steel (Fe, Cr, Ni) as expected from the apparatus used. This procedure has a detection limit below 1 part per thousand by mass for most elements. Pyrolysis gas chromatography mass spectrometry (py-GC-MS) was employed to screen for evolved organic or silicon-containing volatile species that may be present as particle contamination. Thermal extraction was conducted at 300°C for the silicon powder of both examples 5 and 14, and no evolved volatile or semi-volatile compounds were detected.

Thermal desorption spectroscopy (TDS) was performed on four silicon powder examples to evaluate hydrogen content. In a TDS experiment, a known amount of example is heated at a constant heating rate while connected to a vacuum-pumping system equipped with a sensor for monitoring the gas flow. One experiment was performed for each example using a heating rate of 5°C/min to a temperature of 800°C. The integrated amount of hydrogen (assuming all of the desorbed gases are hydrogen) was 0.045 wt. % for example 5, 0.023 wt. % for example 14, 0.025 wt. % for example 16, and 0.032 wt. % for example 13.

Digestion experiments were employed to determine the amount of SiH in the silicon powders. The silicon powders were reacted with KOH in ethanol to convert the SiH to H_2 gas, and the H_2 gas was measured by GC quantitatively to determine the SiH content of the sample. All of the samples examined (the silicon powder of examples 1-17) contained between 400 and 900 ppm of SiH. In particular, the silicon powder of example 5 contained on average 724 ppm of SiH, and the silicon powder of example 14 contained 889 ppm of SiH. The results are based on a 1 hour digestion since it was found that most of the samples showed only a small increase of H_2 (10% or less) with extended digestion times, including setting overnight at room temperature.

Additional Comments

The amorphous-to-crystalline transition temperature of about 620°C for silicon powder, as determined from examples A-D, compares favorably with the temperature found by Bloem and Beers (also 620°C) for amorphous Si films [16]. Bloem and Beers give a plot for the relationship between reaction temperature and growth rate for the crystalline structure of the deposited film. At higher growth rates, the amorphous-to-crystalline transition temperature also becomes higher. In general, there are several conditions that influence the powder crystallinity besides temperature, such as pressure, SiH_4 concentration, diluent gas, and residence time.

Hydrogenated amorphous silicon (a-Si:H) is typically produced in films at temperatures below about 550°C [21]. At temperatures higher than 550°C, little hydrogen is retained in the silicon. Silane can be pyrolytically decomposed at temperatures above about 450°C in the absence of plasma. The silicon powders of examples 1-17 all contained some SiH, as demonstrated by the digestion experiments, and these examples were produced between 550 and 580°C.

Conclusions

Silicon powder generated by the pyrolytic reaction of silane in a free space reactor was demonstrated to be produced in varying degrees of crystallinity. The reactor temperature was the most influential parameter, with additional influence from pressure, silane concentration, gas flow rate (residence time and growth rate), and diluent gas. The non-crystalline (amorphous) silicon was produced in a controlled composition with crystalline silicon. The crystallization composition of the powder was determined by a combination of Powder X-Ray Diffraction and Pair Distribution Function Analysis.

References

1. Buff, H.; Wöhler, F., "Ueber neue Verbindungen des Siliciums", *Justus Liebigs Annalen der Chemie*, **1857**, 104 (1), 94-109.
2. Stock, A.; Somieski, C., "Siliciumwasserstoffe. I. Die aus Magnesiumsilicid und Säuren entstehenden Siliciumwasserstoffe", *Berichte der deutschen chemischen Gesellschaft*, **1916**, 49 (1), 111-157.
3. Hogness, T. R.; Wilson, T. L.; Johnson, W. C., "The Thermal Decomposition of Silane", *Journal of the American Chemical Society*, **1936**, 58 (1), 108-112.
4. Murthy, T. U. M. S.; Miyamoto, N.; Shimbo, M.; Nishizawa, J., "Gas-phase nucleation during the thermal decomposition of silane in hydrogen", *Journal of Crystal Growth*, **1976**, 33 (1), 1-7.
5. Frenklach, M.; Ting, L.; Wang, H.; Rabinowitz, M. J., "Silicon particle formation in pyrolysis of silane and disilane", *Israel Journal of Chemistry*, **1996**, 36 (3), 293-303.
6. Swihart, M. T.; Girshick, S. L., "Thermochemistry and Kinetics of Silicon Hydride Cluster Formation during Thermal Decomposition of Silane", *Journal of Physical Chemistry B*, **1999**, 103 (1), 64-76.

7. Onischuk, A. A.; Strunin, V. P.; Ushakova, M. A.; Panfilov, V. N., "Studying of silane thermal decomposition mechanism", *International Journal of Chemical Kinetics*, **1998**, 30 (2), 99-110.
8. Herrick, C. S.; Woodruff, D. W., "The Homogeneous Nucleation of Condensed Silicon in the Gaseous Si-H-Cl System", *Journal of the Electrochemical Society*, **1984**, 131 (10), 2417-2422.
9. Herrick, C. S. (General Electric Company), "Reducing Powder Formation in the Production of High-Purity Silicon", Patent US 4559219 A (filed April 2, 1984 and issued December 17, 1985).
10. Rogers, L. C., "Polysilicon Preparation", In Handbook of Semiconductor Silicon Technology, O'Mara, W. C.; Herring, R. B.; Hunt, L. P. (Eds.), Noyes Publications, **1990**, 33-93.
11. "Low-Cost Solar Array Project: Feasibility of the Silane Process for Producing Semiconductor-Grade Silicon (Final Report: Oct. 1975 - Mar. 1979, Contract JPL-954334)", Jet Propulsion Laboratory Technical Report No. NASA-CR-162172, **1979**.
12. "Flat-Plate Solar Array Project: Experimental Process System Development Unit for Producing Semiconductor-Grade Silicon Using the Silane-to-Silicon Process (Final Report: Apr. 1979 - Dec. 1981, Contract JPL-954334)", Jet Propulsion Laboratory Technical Report No. NASA-CR-173762, **1983**.
13. Bernreuter, J., "The 2014 Who's Who of Solar Silicon Production: Players, Technologies, Supply/Demand, Prices, and Global Market Perspectives Through 2014", Bernreuter Research Report, **2014**.
14. Filtvedt, W. O.; Javidi, M.; Holt, A.; Melaaen, M. C.; Marstein, E.; Tathgar, H.; Ramachandran, P. A., "Development of fluidized bed reactors for silicon production", *Solar Energy Materials and Solar Cells*, **2010**, 94 (12), 1980-1995.
15. Beers, A. M.; Hintzen, H. T. J. M.; Schaeken, H. G.; Bloem, J., "CVD silicon structures formed by amorphous and crystalline growth", *Journal of Crystal Growth*, **1983**, 64 (3), 563-571.
16. Bloem, J.; Beers, A. M., "In situ observations on amorphous versus crystalline growth of silicon by chemical vapour deposition", *Thin Solid Films*, **1985**, 124 (2), 93-100.
17. Iya, S. K.; Flagella, R. N.; DiPaolo, F. S., "Heterogeneous Decomposition of Silane in a Fixed Bed Reactor", *Journal of the Electrochemical Society*, **1982**, 129 (7), 1531-1535.
18. Kojima, T., "Production of polycrystalline silicon in a fluidized bed", *Trends in Chemical Engineering*, **1994**, 2 (1), 159-173.
19. Dehtiar, M.; Fisher, P.; Gave, M. A.; Herron, W.; Hino, T.; Hwang, B. K.; Larimer, J.; Lee, J. Y.; McDonald, J.; Schrauben, M.; Tabler, R. (Dow Corning Corporation), "Electrode Composition Comprising a Silicon Powder and Method of Controlling the Crystallinity of a Silicon Powder", Patent WO 2013/025707 A1 (filed August 14, 2012).
20. Treacy, M. M. J.; Borisenko, K. B., "The Local Structure of Amorphous Silicon", *Science*, **2012**, 335 (6071), 950-953.
21. Street, R. A., Hydrogenated Amorphous Silicon, Cambridge University Press, **1991**.

*** Formatting Issue - Best Available Paper/Graphic ***

The 14th American Institute of Chemical Engineers, AIChE, Annual General Meeting Atlanta Marriott Marquis & Hilton Atlanta, Atlanta GA, USA November 16 - 21 2014

Recycling Waste-to-Energy (WTE): Studies Carried-out on a Metropolitan City of the North-Eastern Nigeria

Engr. Dr. M. N. Idris* & Ibrahim A.
Energy & Process Refining Technology Group
Department of Chemical Engineering
Faculty of Engineering
University of Maiduguri
Borno State, Nigeria
Tel: +234 (0) 705 877 1096
Email: idrismn@unimaid.edu.ng, muhinu@hotmail.co.uk

ABSTRACT

Studies on municipal solid waste (MSW) were carried-out in some selected locations from the Maiduguri metropolis. These wastes are produced and/or accumulated due to various human activities. Amongst the many factors that these wastes will course to the environment are: environmental degradation, insects, atmosphere (air) pollution, diseases, and disfigure the image of the city. In the effort to meeting the millennium development goals (MDGs) specification on environment, this paper is advocating that an appropriate solid waste management system/mechanism must be instituted. It is reasonable to recover energy from MSW through a variety of a process such as combustion pyrolysis and gasification. In this study, we estimated the energy content of the MSW from the Maiduguri metropolis in order to achieve its optimal system performances.

It was determined that the quantity of solid waste that are been deposited on daily bases at some selected/various locations were respectively 10,000kg/m^2, 20,000kg/m^2 , 40,000kg/m^2, 20,000kg/m^2 , 10,000kg/m^2 from Bagga road, Mairi ward, Monday Market, Post Office, and University of Maiduguri campus. These values determined were compared with the reported data from the literature, were energy values of different materials when incinerated to arrive at the corresponding energy (power) that can be produced from Maiduguri metropolis MSW.

Keywords: *MSW, MDGs, metropolis, environment, degradation.*

Correspondence about this article: *Please for all correspondence on this article, kindly contact the principal author (Engr. Dr. M. N. Idris; MAIChE, MIChemE, MNSE) as stated above. Thank you.*

*** Formatting Issue - Best Available Paper/Graphic ***

1.0 Introduction

Solid waste means any garbage, refuse, sludge from a waste water treatment or air pollution control facility and other discarded materials including solid, liquid semi-solid or contained gaseous material resulting from industrial, commercial, mining and agricultural operations, and from community activities, but does not include solid or dissolved materials in irrigation return flows or industrial discharges that are point source subject (Maduike, 1987).

The problem of solid waste and management is one of the most critical problems in Nigeria. A peculiar case is that of Maiduguri metropolis, where over the years MSW is gradually taken over of virtually every available open space in solid waste. Apart from physically obstructing legitimate human activities, the waste dumps have become fertile grounds for breeding mosquitoes, flies and other pests which have in effect constituted the dumps in to grave health hazard. These wastes, to say the least disfigure the city image and create an eyesore to the people (Maduike, 1987).

There is an obvious need to optimize the generation of waste and to reuse and recycle them, the technology waste to energy (WTE) or energy from waste is the process of creating energy in the form of electricity or heat from the incineration of waste. Most (WTE) process produce electricity directly through combustion, or produce a combustible fuel commodity, such as methane, methanol, ethanol or synthetic fuel which could be use for both domestic and industrial applications.

One event that acted partly to create, or least that has worsened the waste problem in our urban centres is the rapid rate of population growth. Population is not the only affected solid waste by volumes but also made solid waste management strategies incapable of keeping pace with the rate of generation. The consequences to these effects have been disastrous. For example in Maiduguri cities the volume of solid wastes have assumed such alarming proportions that they have constituted blockages across streets and river channels rendering them inaccessible to traffic and water respectively. Indeed it has been opined that the re-occurring flood disaster in the city of Maiduguri and major cities in Nigeria which destroyed lives and properties worth millions of Naira was largely caused by solid waste which choked up channels of rivers and gutters within the city (Kagu, 1996). Waste generation rate in Maiduguri, are affected by socio economic development, degree of industrialization, and climate.

The composition of solid waste in an urban/metropolitan city is described as follows: 51% - municipal solid waste, 38% - agriculture and 11% - industry.

Generally, the greater the economic prosperity and the higher percentage of urban population, the greater the amount of solid waste produced. Reduction in the volume and mass of solid waste is a crucial issue especially in the light of limited availability of final disposal site in many part of the world. Although numerous waste and by product recovery process have been introduced, anaerobic digestion has unique and integrative potentials, simultaneously acting as a waste treatment and recovery process (Maduike, 1987).

This paper presents the possible solution to MSW and harnessing the waste-to-energy, WTE, the studies were carried out at some selected location in Maiduguri metropolis on daily basis. The limitation of the study is to compare the determine quantity of wastes with the standard wastes figures from the literature data.

***** Formatting Issue - Best Available Paper/Graphic *****

1.1 Effects of Solid Waste on Environment

It has been recognized that improper disposal of solid waste has a lot of health hazards and other negative consequences on the people and the environment. Open waste dumps in most areas have become breeding grounds for mosquitoes, flies, rats and other diseases. Some of the major effects of solid waste include the following; Environmental; degradation, insects atmospheric pollution, odour (Oduola, 1986).

1.1.1 Environmental Degradation

Large heaps of waste and the indiscriminate manners in which people dump them have become eyesore to members of the public. It destroys the scenery of the environment. It also becomes a source of psychological disorder. Unclear waste lowers the aesthetic quality of the locality and property values in the affected neighborhood (Oduola, 1986).

1.1.2 Insects

The transmission route of filth induced diseases such as malaria, cough, cholera, dysentery, diarrhea; vomiting and a host of others are largely brought about by insect such as flies, mosquitoes and other insects that use the dumps as their breeding grounds. These diseases are the commonest in almost every urban centre in Nigeria (Oduola, 1986).

1.1.3 Atmospheric Pollution

When waste is burn in the open place, a pool of dense black smoke often covers the site and n neighboring land so that its positions can be located from a distance. Apart from the particular matter that constitutes smoke, gases discharges from incomplete oxide and various other noxious oxide which are dangerous to human health, pollution of water resources is also another inherent character of waste disposal in open areas (Oduola, 1986).

1.1.4 Odour

Waste-constitutes a source of stench and offensive odour to the human being and the environment. These arise as a result of combination of rotten vegetables and other solid wastes that are indiscriminately discarded. When this situation persists all day and nigh. It constitutes a major environmental nuisance. Passer by and indeed people living around such environments find this odour uncomforted and un-attraction (Oduola, 1986)

1.2 Method of Municipal Solid Waste Disposal

The municipal solid waste industry has four components: recycling, composing, land filling, and waste to energy.

1.2.1 Recycling

Recycling is a process of collecting and processing materials that would otherwise be thrown away as trash and turning them into new product. Recycling can benefit your community and the environment (Oyinlola, 2001).

1.2.2 Benefits of Recycling

Reduce the amount of waste sent to land fills and incinerators, conserves natural resources such as timber waste ands minerals, prevents pollution caused by reducing the need to collect new raw materials, saves energy, reduce green house gas emission that contribute to global climate change. Helps sustain the environment for future generations; help create new well paying jobs in the recycling (Oyinlola, 2001). In these methods, efficient recovery of wastes like glass, plastics, metals and cans can be achieved by the separation of the recyclables in Maiduguri (Kagu, 1996). It was observed that plastic waste materials are turned into plastics in the Maiduguri based Mandau plastic company. However, an expert in plastic recycling indicated that there are over 30 different types of plastic in use and it is difficult to differentiate one from the other. A mixture of these therefore produce a products which is brittle (not durable) and has limited useful properties (Kagu, 1996).

1.2.3 Composting

Composting organic materials such as yard trimming, foods scraps, woods waste and paper are helps to manage the largest compound of our trash the garbage waste we dispose of making up more than two-thirds of the solid waste disposal streams. The amount of yard waste and its municipal sold waste disposal market share have declined dramatically in the last four decades which the composting rate has sod red. Backyard compost piles and grass cycling programmed have helped to reduce yard waste generation composting services as and environmentally friendly waste disposal solution by cutting town on the amount of garbage waste headed to landfills. State and local composting requirement have increased the number of commercial composting operation, composting which involves the controlled decomposition of plant remains and other organic materials to male an earthy, dark, crumbly substance that is excellent for enriching soil and preventing soil erosion is a part way to recycle yard and kitchen wastes and reduce the volume of garbage waste sent to land fills or incinerators for disposal (Welsh Assembly, 2005).

1.2.4 Land Filling

Modern land fills are well engineered facilities that are located, designed operated and monitored to ensure compliance with federal regulations. Solid waste land fills must be designed to protect the environment from contaminated which may be present in the solid waste stream. The land fill siting plan which prevents the siting of land fills in environmentally sensitive areas as well as on site environmental monitoring systems which monitor for any sign of ground water contamination and for landfill gas provides additional safeguards. In addition, many new land fills collect potentially harmful land fill gas emission and convert the gas in to energy.

The U.S. environmental protection Agency's landfill methane outreach programmed (LMOP) is a voluntary assistance programme that helps to reduce methane emission from landfill by encouraging the recovery and beneficial use of landfill gas (LFG) as an energy resource. LFG contain methane, a potent greenhouse gas that can be captured and used to fuel power plants, manufacturing facilities vehicles home and more by joining LMOP, companies, state agencies, organization, landfills and communities gain access to a vast network of industry experts and practitioners, as well as to various technical and marketing resources that can help with Landfill Gas LFG energy project development (EPA, 1996 US).

1.2.5 Methane Emission from Landfill

Municipal solid waste (MSW) landfills are the third largest sources of human related methane emissions in the United States, accounting for approximately 17% of these emissions in 2009. At the same times, methane emission from landfills represents a lost opportunity to capture- and use a significant energy resource. LFG is created as solid waste decomposes in a landfill. This gas consist of about 50 percent methane (the primary component of natural gas), about 50 percent carbon dioxide Co_2 and a small amount of non methane organic compounds.

1.2.6 Converting Landfills Gas to Energy (LGE)

Instead of escaping in the air, Landfill Gas (LFG) can be captured converting a used as an energy source. Using Landfill Gas (LFG) helps to reduce odours and other hazards associated with landfill Gas (LFG) emissions and it helps prevent methane from migrating in to the atmosphere and contributing to local smoke and global climate change.

LFG extracted from landfills using a series of wells and a blower/flare (or Vacuum) system. This system directs the collected gas to a central point where it can be processed and treated depending upon the ultimate use for the gas. From this point, the gas can be flared, used to generate electricity, replace fossil fuel in industrial and manufacturing operation, or upgraded to pipeline quality gas where the gas may be used directly or processed in to an alternative vehicle fuel (EPA, 1996 US).

2.0 Waste-to-Energy, WTE

This is the energy recovery from waste in the conversion of non-recyclable waste materials in to usable heat, electricity, or fuel through a variety of processes, including combustion, gasification, pyrolization, anaerobic digestion and landfill gas (LFG) recovery (Herbert, 2007)

Energy recovery from waste is part of the non hazardous waste management hierarchy. Converting non recyclable waste materials into electricity or heat generates a renewable energy source and reduces carbon emissions by offsetting the need for energy from fossil sources and reduces methane generation from landfills. Currently there are 86 facilities in the united stated for combustion of municipal solid waste (MSW) with energy recovery. These facilities are located in 25 States mainly in the Northeast. No new plants have been built in the US since 1995, but some plants have expanded to handle additional waste and create more energy. The 86 facilities have the capacity to produce 2,720 MW of power per year by processing more than 28 million tons of waste in the US. In 2010, combustion of about 29 million tons of MSW (≈ 12 %) for approximately ten percent of the volume remains ash. This ash is generally sent to a landfill (Shi-Ling, 1999)

Waste to energy incinerate is the forth aspect of the EPA's approach to solid waste management. It is the controlled burning of solid waste at extremely high temperature often as high as 2000° F. incineration of all types currently accounts for 14% of solid waste disposal in the U.S. The EPA estimates that 23% of the municipal solid waste stream will be disposed of visa incineration by 1995. WTE incineration is widely used in Japan, some part of Europe, and elsewhere to reduce by as much 80 to 90% the volume of waste that must be land filled. Waste to energy incineration should not be confused with simple open burning of refuse. It is even different from mass bur waste incineration common in the U.S. in the first half of the 20th century. In waste to energy incineration, the heat generated by the process is captured and turned in to useable energy. The energy affiliated organization and contacts industrial classification, and linkages to environmental permit and programmed. (Shi-Ling Hsu, ed, 1999)

2.1 Waste-to- Energy Conversion Path Ways

There are three main path ways for conversion of organic waste material to energy:

2.2 Thermo chemical -Thermo Chemical Conversion:- Characterized by higher temperature and conversion rates, is best suited for lower moisture feedstock and is generally less selective for products. Thermo chemical conversion include; incineration, pyrolysis, and gasification

The incineration technology is the controlled combustion of waste with the recovery of heat to produce steam which in turn produces power through steam turbines.

Pyrolysis and gasification represent refined thermal treatment method as alternative to incineration and are characterized by the transformation of the waste in to product gas as energy carrier for later combustion in for example a boiler or a gas engine. (Rosenthal and Elisabeth, 2010).

2.3 The Biochemical Conversion:- Process which include anaerobic digestion and fermentation, are preferred for wastes having high percentage of organic biodegradable (putrescible) matter and high moisture content. Anaerobic digestion can be used to recover both nutrient and energy contained in organic waste such as animal manure. The process generates gases with a high content of methane (55 – 70%) as well as bio-fertilizer. Alcohol fermentation is the transformation of organic fraction of waste to ethanol by a series of biochemical reaction using specialized microorganisms. (Rosenthal and Elisabeth, 2010).

2.4 The Physiochemical:- Technology involves various processes to improve physical and chemical properties of solid waste. The combustible fraction of the waste is converted into high energy fuel pellets which may be used in steam generation. (Rosenthal, Elisabeth, 2010).

Full pellets have several distinction advantages over coal and because it is cleaner, free from incombustible, has lower ash and moisture contents of uniform size, is cost-effective, and eco-friendly. (Rosenthal and Elisabeth, 2010).

2.5 Factors Affecting Energy Recovery

The two main factors which determine the potential of recovery of energy from wastes are the quantity and quality (Physico-chemical characteristics) of the waste. Some of the important physiochemical parameters requiring consideration include, Size of constituents, density, moisture content, volatile solid/ organic matter, fixed carbon, total inert, calorific value often, can analysis of waste to determine the proportion of carbon, hydrogen oxygen, nitrogen, and sulphur (ultimate analysis) I done to make mass balance calculation for both thermo chemical and biochemical processes. In case of anaerobic digestion, the parameters C/N ratio (a measure of nutrient concentration available for bacteria growth) and toxicity representing the presence of hazardous material which inhibit bacteria growth also require consideration). (Rosenthal and Elisabeth, 2010).

2.6 Significance of Waste to Energy (WTE) Plant

While some still confuse modern waste to energy plants with incineration of the past, the environmental performance of the industry is beyond reproach. Studies have shown that communities that employ waste to energy technology have higher recycling rates than communities that do not utilize waste to energy. The recovery of ferrous and non ferrous metal from waste to energy plants for recycling is strong and growing each year. In addition numerous studies have determines that waste to energy plants actually reduce the amount of green house gases that enter the atmosphere. (Rosenthal, Elisabeth, 2010).

Now a day, waste to energy plants based on combustion techniques are highly efficient power plant that utilize municipal solid waste as their fuel rather than coal, oil or natural gas. For better than expending energy to explore, recover, process and transport the fuel from some distance source, waste to energy plant find value in what other consider garbage. Waste to energy plant recover the thermal energy contained in the trash in highly efficient boilers that generate steam that can then be sold directly to industrial customers or used on-site to drive turbine for electricity production. WTE plants are highly efficient in harnessing the untapped energy potential of organic waste by converting the biodegradable fraction of the waste in to high calorific value gases like methane. The digested portion of the waste is highly rich in nutrients and is widely used as bio-fertilizer in many part of the world. (Themelis, Nickolas J. 2003)

2.6 Waste – to – Energy around the World

To an even greater extent than in the United States, waste to energy has thrived in Europe and Asia as the prominent method of waste disposal. Landing waste to energy for its ability to reduce green house gas emission European nation rely on waste to energy as the European Union has issued as legally binding requirement for its member States to limit the land filling of biodegradable waste (Tangri and Neil, 2003).

The confederation of European waste to energy plant (CEWEP) notes that Europe currently treats 50 million tons of waste at waste to energy plants each year, generating an amount of energy that can supply electricity for 27 million people or heat for 13 million people. Up coming changes to EU legislation will have a profound impact on how much further the technology will help achieve environmental protection goals describing the advances of waste to energy, the German ministry for the environment cities drastic reductions in emission of dioxin, dust and mercury. Twenty years ago, is Swedish waste to energy plant emitted total of about 100grams of dioxin every year. Today the collective dioxin emission from all 29 Swedish waste to energy plants amount of 0.7 of a gram. It is clear that Europe has made similar strides as the United States with respect to emission reduction. (Tangri, Neil, 2003).

*** Formatting Issue - Best Available Paper/Graphic ***

2.7 Combustion with Energy Recovery

2.7.1 Types of MSW Combustion Technologies.

There are three types of technology for the combustion of municipal solid waste.

2.7.2 Mass Burn Facilities:-

Burn facilities are by for the most common types of combustion facilities in the United State. The wastes used to fuel the mass burn facility may or may not be sorted before it enters the combustion chamber. Many advanced municipalities separate the waste on the front end to pull off as many recyclable products as possible. Mass burn units are designed to burn municipal solid waste in a single combustion chamber under condition of excess air. In combustion system, excess air must be used to promote mixing and turbulence to ensure that air can reach all ports of the waste. This is necessary because of their consistent nature of solid waste. Most mass burn facilities burn municipal solid waste on a sloping, moving grate thus vibrated or otherwise moved to agitate the waste and mix it with air.

2.7.3 Modular System

Modular systems are designed to burn up processed mixed MSW they differ from mass burn facilities that they are much smaller and are portable. They can be moved from site to site.

2.7.4 Refuse Derived Fuel System:-

These facilities used mechanical methods to shred incoming municipal solid waste separate out non combustible materials, and produce a combustible mixture suitable as a fuel in a dedicated furnace or as a supplemental fuel in a conventional boiler system.

At a municipal solid waste combustion facility MSW is unloaded from collection trucks and placed in a trash storage bunker. An overhead crane is used to sort the waste and then life it in to a combustion chamber to be burned. The heat released from burning is used to convert water to steam (Michaels and Ted, 2009).

The steam is then sent to a turbine generator to produce electricity. The remaining ash is collected and taken to a land fill. Particulate are capture by a high efficiency bag house (a filtering system). As the gas stream travels through the filters, more than 99 percent of particulate matter is removed. (Captured fly ash particles fall in to hoppers funnel shaped receptacles) and are transported by an enclosed conveyor system to the ash discharger where they are wetted to prevent dust and mixed with the bottom from the grate. The ash residue is the conveyed to an enclosed building where it is loaded in to covered leak proof trucks and taken to a land water contamination. Ash residue from the furnace can be processed for removal of recyclable scrap metals. The figure illustrates how the energy recovery process works (Michaels and Ted, 2009).

*** Formatting Issue - Best Available Paper/Graphic ***

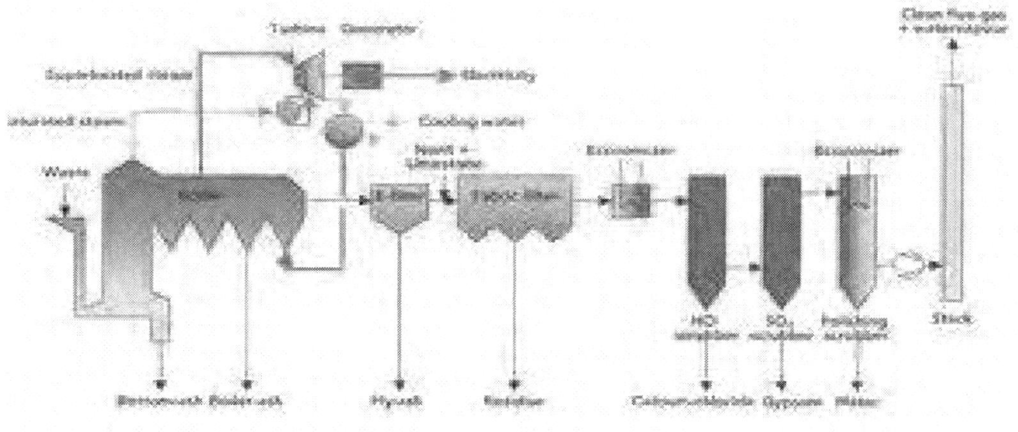

Figure 2.1 Waste –to-energy recovery process
Source: Themelis, Nickolas, J. (2003)

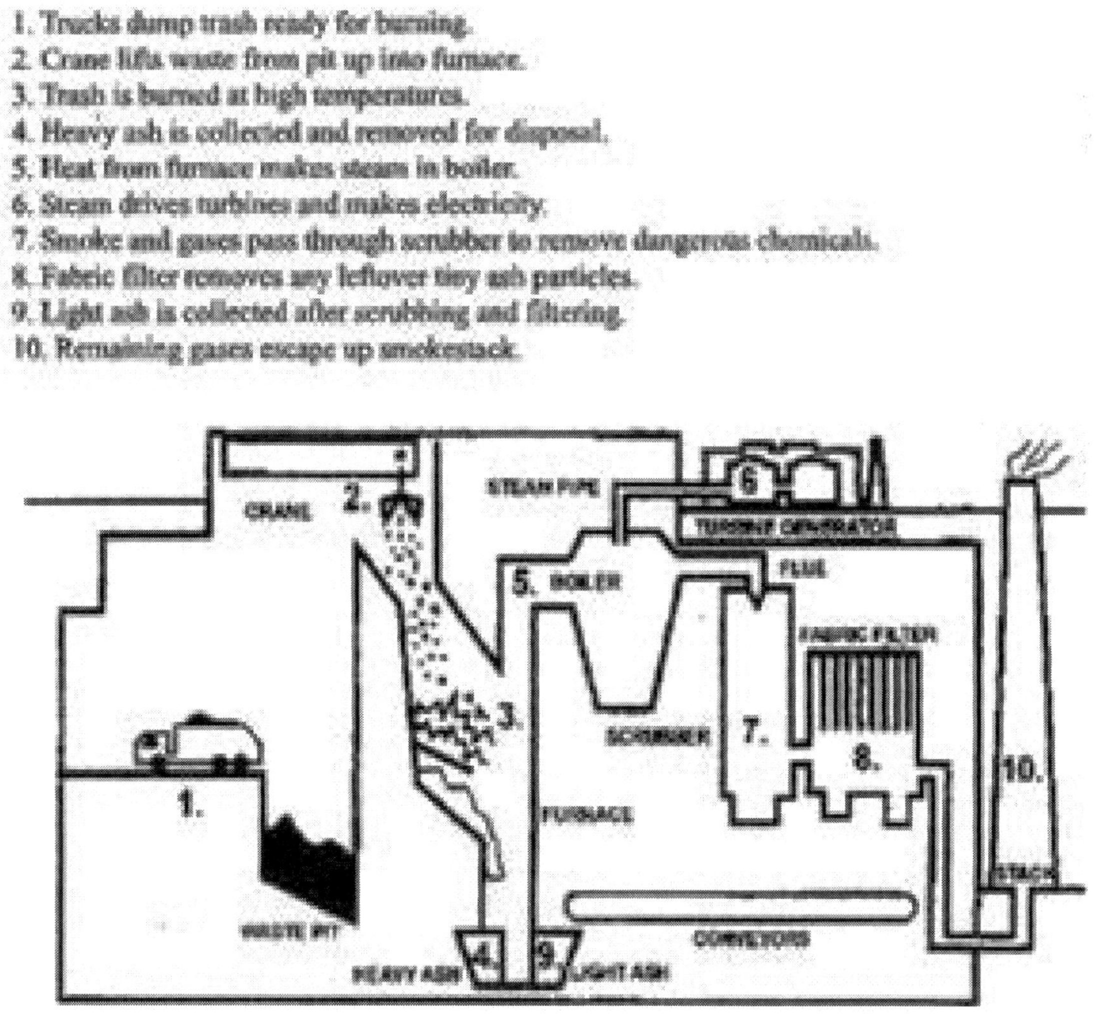

Figure 2.2 Waste –to-Energy ideal process flow Source: (Michaels, 2009)

***** Formatting Issue - Best Available Paper/Graphic *****

Figure 2.3 WTE Control room of a typical moving grate incinerator overseeing two boiler lines
Sources: Themelis, Nickolas, (2003).

Figure 2.4 Picture of the Spittelau incineration plant in Vienna, designed by Friedensreich Hundertwasser
Source: (Themelis Nickolas, 2003)

Figure 2.5 Municipal solid wastes in the furnace of a moving grate incinerator capable of handling 15 metric tons (17 short tons) of waste per hour. The holes in the grate elements supplying the primary combustion air are visible.
Source: Themelis, Nickolas, (2003)

*** Formatting Issue - Best Available Paper/Graphic ***

Figure 2.6 Operation of an incinerator aboard an aircraft carrier, Source: Themelis, Nickolas, (2003)

2.8 Ash Generated From the MSW Combustion Process

The amount of ash generated ranges from 15-25 percent by weight of the MSW processed and from 5 – 15 percent of the volume of the MSW processed. Generally, MSW combustion residues consist of two types of material. Fly ash and bottom; Fly ash refers to the fine particles that are residue from other air pollution control devices such as scrubbers. Fly ash typically amount to 10-20 percent by weight of the total ash (Themelis and Nickolas, 2003)

The rest of the MSW combustion ash is called bottom ash (80-90 percent by weight). The main chemical components of bottom ash are silica (sand and quality) calcium iron oxide and aluminum oxide. Bottom ash usually has a moisture content of 22-62 percent by dry weight. The chemical composition of the ash varies depending on the original MSW feedstock and the combustion process. The ash that remains from the MSW combustion process is sent to land fill (Themelis and Nickolas, 2003).

Table 2.1: Energy values of different materials when incinerated

S/No	Materials	**BTU per pound
1.	Plastics	11.00 – 20,000
2.	Rubber	10,900
3.	Newspaper	8,000
4.	Corrugated Boxes (paper)	7,000
5.	Yard wastes	3,000
6.	Food wastes	2,600
7.	Average of MSW	4,500 – 4,800

Sources: (Michaels, Ted, 2009)

2.0 Methodology of the Study

A careful assessment and quantitative determination of all wastes deposited at some selected location, in the metropolis are namely as follows: Bagga road, Mairi ward, Monday Market, Post Office, Unimaid.

Qualitatively determine the relative weight by the area of each kind of material deposited at the locations on daily basis, such as plastics, rubber, newspapers, corrugated boxes (paper), yard wastes, food wastes at various interval of days as follows, Monday 25 March 2013, Wednesday 27 March 2013, Tuesday 2 April 2013, Friday 5 April 2013, Saturday 7 April 2013, Wednesday 10 April 2013, Friday 12 April 2013.

The determined quantities of all wastes deposited on daily basis in the above locations were then compared with the standard data from the literature. The results help us to know the amount of energy (in terms of power) that can be produce from these wastes.

(a) (b)

Figure 2.8 Picture of combine waste materials at Bagga road (a) on 25th March, 2013, (b) 27th March, 2013

***** Formatting Issue - Best Available Paper/Graphic *****

Figure 2.9 Picture of combine waste materials at Bagga road location on 2nd April, 2013

(a) (b)

Figure 2.10 Picture of combine waste materials at Mairi ward (a) on 25th March, 2013, (b) 27th March, 2013

Figure 2.11 Picture of combine waste materials at Mari Ward location on 2nd April, 2013

(a) (b)

Figure 2.12 Picture of combine waste materials at Unimaid: (a) on 25th March, 2013, (b) 27th March, 2013

Figure 2.13 Picture of combine waste materials at Unimaid location on 2nd April, 2013

(a) (b)

Figure 2.14 Picture of combine waste materials at Monday market: (a) 25th March 2013, (b) 27th March 2013

*** Formatting Issue - Best Available Paper/Graphic ***

Figure 2.15 Picture of combine waste materials at Monday Market location on 2nd April, 2013

(a) (b)

Figure 2.16 Picture of combine waste materials at Post office: (a) on 25th March 2013, (b) 27th March 2013

Figure 2.17 Picture of combine waste materials at Post Office location on 2nd April, 2013

*** Formatting Issue - Best Available Paper/Graphic ***

4.0 Analysis of the Results

The results obtained in this study are presented as follows:

Table 4.1: Bagga road with location area size of $20m^2$

Waste assessment date	Material composition (kg/m²)					
	Plastics	Rubber	Newspapers	Corrugate boxes paper	Yard wastes	Food wastes
25/03/2013	700	70	560	420	490	630
27/03/2013	1050	105	700	700	630	700
02/04/2013	2100	175	840	1400	770	1050
05/04/2013	2800	350	1050	1750	1050	1400
07/04/2013	3500	420	1400	2800	1190	2100
10/04/2013	4200	455	2800	3500	1330	2800
12/04/2013	6650	525	3150	4830	2540	5320
Total average	3000	300	1500	2200	1000	2000

Table 4.2 Amount of Solid Wastes Produce on Daily Basis at Bagga Road Locations

S/No	Materials	Measured weights (Kg)
1,	Plastics	3000
2.	Rubber	300
3.	Newspaper	1500
4.	Corrugated boxes (paper)	2200
5.	Yard wastes	1000
6.	Food wastes	2000
7.	Total	10,000

Table 4.3 Mairi Ward with location Area of $30m^2$

Waste assessment date	Material composition (kg/m²)					
	Plastics	Rubber	Newspapers	Corrugate boxes paper	Yard wastes	Food wastes
25/03/2013	2100	210	350	630	1400	700

*** Formatting Issue - Best Available Paper/Graphic ***

27/03/2013	3500	350	1400	700	2100	1400
02/04/2013	4200	420	2450	1050	2800	2100
05/04/2013	6300	700	2800	1400	3500	2800
07/04/2013	8400	1400	3500	2100	4900	3500
10/04/2013	10500	1820	4200	2800	5600	4200
12/04/2013	14000	2100	6300	5320	7700	6300
Total average	7000	1000	3000	2000	4000	3000

Table 4.4 Amount of Solid Wastes Produce on Daily Basis at Mairi ward Locations

S/No	Materials	Measured weights (Kg)
1,	Plastics	7000
2.	Rubber	1000
3.	Newspaper	3000
4.	Corrugated boxes (paper)	2000
5.	Yard wastes	4000
6.	Food wastes	3000
7.	Total	20,000

Table 4.5 Monday market location with Area of 40m^2

Waste assessment date	Material composition (kg/m^2)					
	Plastics	Rubber	Newspapers	Corrugate boxes paper	Yard wastes	Food wastes
25/03/2013	5600	700	2100	700	2800	2100
27/03/2013	7000	1400	3500	2100	4200	3500
02/04/2013	8400	2100	4200	3500	6300	5600
05/04/2013	9100	2800	4550	4200	9100	8400
07/04/2013	11200	3500	7000	6300	9800	10500
10/04/2013	14000	4200	8400	8400	10500	11900
12/04/2013	14700	6300	12,250	9800	13300	14,000
Total average	10,000	3000	6000	5000	8000	8000

Table 4.6 Amount of Solid Wastes Produce on Daily Basis at Monday market Locations

S/No	Materials	Measured weights (Kg)
1,	Plastics	10,000
2.	Rubber	3,000
3.	Newspaper	6000
4.	Corrugated boxes (paper)	5000
5.	Yard wastes	8000
6.	Food wastes	8000
7.	Total	40,000

Table 4.7 Post Office location with Area of 30m^2

Waste assessment date	Material composition (kg/m^2)					
	Plastics	Rubber	Newspapers	Corrugate boxes paper	Yard wastes	Food wastes
25/03/2013	1400	140	560	490	700	1050
27/03/2013	3500	175	840	630	2100	1400
02/04/2013	4900	280	1190	770	3500	2800
05/04/2013	5600	350	1750	1050	4200	4900

*** Formatting Issue - Best Available Paper/Graphic ***

07/04/2013	7000	490	2800	1190	6300	5600
10/04/2013	9100	700	4200	1330	8400	8400
12/04/2013	10500	1365	6160	1540	9800	10850
Total average	6000	500	2500	1000	5000	5000

Table 4.8 Amount of Solid Wastes Produce on Daily Basis at Post Office Locations

S/No	Materials	Measured weights (Kg)
1,	Plastics	6000
2.	Rubber	500
3.	Newspaper	2500
4.	Corrugated boxes (paper)	1000
5.	Yard wastes	5000
6.	Food wastes	5000
7.	Total	20,000

Table 4.9 Unimaid location with Area of 20m^2

Waste assessment date	Material composition (kg/m^2)					
	Plastics	Rubber	Newspapers	Corrugate boxes paper	Yard wastes	Food wastes
25/03/2013	700	70	490	350	2800	2100
27/03/2013	1400	84	630	420	4200	3500
02/04/2013	2800	112	770	630	6300	5600
05/04/2013	4200	154	1050	700	9100	8400
07/04/2013	5600	224	1190	770	9800	10500
10/04/2013	6300	280	1330	1330	10,500	11,900
12/04/2013	7000	476	1540	1400	13300	14000
Total average	4000	200	1000	800	3000	1000

Table 4.10 Amount of Solid Wastes Produce on Daily Basis at Unimaid Locations

S/No	Materials	Measured weights (Kg)
1,	Plastics	4000
2.	Rubber	200
3.	Newspaper	1000
4.	Corrugated boxes (paper)	8000
5.	Yard wastes	3000
6.	Food wastes	2000
7.	Total	10,0000

Table 4:11 Combined Solid Wastes Data weight of each material at selected locations

S/No	Materials	Weights in Kg
1.	Plastics	30000
2.	Rubber	14000
3.	Newspaper	11000
4.	Corrugated boxes (paper)	21000
5.	Yard wastes	19000
6.	Food wastes	5000

*** Formatting Issue - Best Available Paper/Graphic ***

Table 4:12 Standard data used from literature: The energy values of different materials when incinerated

S/No	Materials	BTU per pound
1.	Plastics	11,000-20,000
2.	Rubber	10,900
3.	Newspaper	8,000
4.	Corrugated boxes (paper)	7,000
5.	Yard wastes	3,000
6.	Food wastes	2600
7.	Average for MSW	4500-4800

Source: Energy recovery council (2009)

Table 4.13 depicts the comparison of Table 4.11 and Table 4.12 to arrive at the energy produced by combined waste materials when incinerated.

Table 4:13 Energy derived from the Comparison of Table 4.11 with 4.12

S/No	Materials	KJ Per KG
1.	Plastics	490575000
2.	Rubber	57497500
3.	Newspaper	118160000
4.	Corrugated boxes (paper)	81235000
5.	Yard wastes	66465000
6.	Food wastes	52117000
7.	Total	866049500

The energy conversions were carried out as depicted below:

3600 KJ	1 KWh
866049500KJ	
= 240,569. 3056 KWh	

5.0 Conclusions

The waste-to-energy (WTE) plants offer two important benefits of environmentally safe waste management and disposal, as well as the generation of clean electric power. WTE facilities produce clean renewable energy through thermo-chemical, biochemical and physiochemical methods. The growing use of WTE as a method of dispose-off solid, liquid wastes and generates power has greatly reduced environmentally impacts of municipal solid waste (MSW). Based on the quantities of MSW produced yearly, their recycle can satisfactorily be used to produce energy that can add immense values (i.e. 240,569. 3056 KWh) to the national power generation, from Maiduguri metropolis alone. WTE can play a vital role in mitigating the problems of disfiguring the city's image and an eyesore to the people.

*** Formatting Issue - Best Available Paper/Graphic ***

Recommendations

1. *The government should introduce this technology of WTE as the best way of MSW management and disposal. Because this project go hand in hand with community and corporate body to cleaner air, renewable energy, economic development which helps to improve public welfare, safety and reduction in green house (global warning) gases. Increase environment protection and better waste management and responsible community planning.*
2. *When these WTE technologies is fully instituted in the Nigeria, the collections of waste should be commercialized at every domestic, industrial etc., this will encourage household to manage their waste for disposal properly.*
3. *Government should, if it has not already done so, enact a law to take care of defaulters of laid down guideline for sanitation. For instance people should be held responsible for their irresponsible waste disposition, either in residential and/or workplace. There should be sanitation inspectors to report cases to law enforcement agencies when the people appear indifferent to the quality of their immediate environment.*

*** Formatting Issue - Best Available Paper/Graphic ***

References

EPA. 1991. U.S Environmental Protection Agency. Air Emission from Municipal Solid waste Landfills.

EPA, 1996 U.S Environmental Protection Agency, Turning a liability in to an asset: a land fill gas to energy project development hand book. September 1996.

Filani M.O dna Abumere, S.I. (1986): Forecasting Solid waste magnitudes for Nigerian cities" in Adeniyi E.O. and Bello Imam, I.B, eds. Development and the Environment. Proceedings of a National Conference of Nigerian Institute of Social and Economic Research (NISER), Ibadan.

Herbert, Lewis (2007). "Centenary History of waste and Waste Managers in London and South East England"(PDF) Chartered Institution of Waste Management.

Kagu, A. (1996): Refuse Generation and Disposed in Maiduguri a paper presented at the 38th conference of the Nigeria Geographical Association at Kashim Ibrahim College of Education (KICOE) Maiduguri.

Maduike, W.I (1987) Solid waste component in Borno State A case study of Bolori Layout Maiduguri. A Paper Presented at the School of Humanities staff Seminar BOCOE Maiduguri.

Michael, Ted (2009). "Letter to Committee on Energy and Commerce" (PDF). Energy Recovery Council.

Oduola S.D (1986): "The Problems and methods of waste Disposal in Nigeria. The case for large scale incineration" in Adeniyi E.O. and Bello Imam, I. B. eds. Development and the Environment proceedings of a National Conference, NISER Ibadan.

Oyinlola, A.K. (2001): Recycling/Resources Recovery Technology option for sustainable Solid Waste Management system in Nigeria" In the Nigeria infrastructural Development, Proceeding of natural Conference and AGM Port Harcourt 5th -9th 2001.

Rosenthal, Elisabeth (2010) "Europe finds clean Energy in Trash, U.S. Lags. The New York Times."

Shi. – Ling Hsu, (1999), "Brown fields and property values", Economic Analysis Council of Alberta Conference 2006.

Tangri., Neil (2003) "Waste Incineration: A Dying technology" GAIA Archived from the original (2007).

Themelis, Nickolas J. (waste – to energy industry, waste management world 2003.

Welsh Assembly (2005) Mechanical Biological Treatment, Environment Country side and Planning Website Welsh Assembly.

381933 High Pressure, High Temperature Wells

Abstract:

On April 20, 2010, BP's Deepwater Horizon oil rig exploded in the Gulf of Mexico. This turned out to be one of the worst environmental disasters in recent history. This high-profile blowout at the Macondo well in the Gulf of Mexico brought the challenges and the perils of drilling into high-pressure, high-temperature (HPHT) fields increasingly into focus. High-pressure, high-temperature fields exist in the Gulf of Mexico, North Sea, Southeast Asia, Africa and the Middle East. Almost a quarter of HPHT operations worldwide is expected to pass on the American continent particularly in North America. There are so many factors that need to be addressed and learned in order to safely overcome the challenges of drilling into and producing from HPHT oil and gas wells.

Drilling into HPHT wells is a new frontier for the petroleum and gas industry. The growing need for oil and gas throughout the world is driving the exploration and production industry to look for raw resources. Some of these resources are located in deeper formations. As we drill into deeper formations we will experience higher pressures and temperatures. Drilling operations in such high pressure and high temperature environments can be very challenging. Accordingly, several companies are compelled to meet or exceed a vast array of technical restrictions as well as environmental, health and safety standards. This report explains the technical challenges in developing HPHT fields, deepwater drilling, completions and production.

387401 Progress Toward Developing Nature-Inspired and Low-Cost Lignocelluloses Processing Technologies

Abstract:

Lowing pretreatment cost of lignocellulosic biomass for obtaining and biochemical conversion of cellulosic sugars requires that plant cell deconstruction occurs under conditions with less severity than that of the currently practiced technologies. Eliminating heat and pressure requirements will allow reactors to be constructed with inexpensive metals. The authors' laboratory has been studying plant cell wall degradation in model natural systems such as that of termite and white-rot fungi. The emerging new knowledge suggests that although termite and fungi have different plant cell deconstruction systems with drastically different efficiencies there are some similarities in their mechanisms for lignin modification. It is believe that radical generation from Fenton type of reactions is the key that turns on the reaction process. Building on these results, different chemical reaction systems have been developed to mimic these reactions using chemical based catalysts. Various degrees of success have been obtained. When ozone was used to modify the lignin structure followed by ammonia soaking to remove the lignin, biomass sacchatification efficiency of 90% was achieved. When hydrogen peroxide was used as the main reaction species for generating radical based reactions, the efficiency was greatly affected by the pH and types of chelators as well as metals employed. The results demonstrate that treatment processes that operate under ambient conditions are possible. Tradeoffs needs to be made between process rate and capital cost.

Utilizatilization of Mixed Culture and Isolated Species for Biomitigation of CO_2

Pratibha Pal[1], Somesh Mishra[2], Suresh Gupta[3], Smita Raghuvanshi[4*]

[2]*Post-Graduate Student, Department of Chemical Engineering*
Birla Institute of Technology and Science (BITS), Pilani- 333031, Rajasthan, India

[1]*Research Scholar, Department of Chemical Engineering*
Birla Institute of Technology and Science (BITS), Pilani- 333031, Rajasthan, India
Email: somesh25162@gmail.com@gmail.com

[2]*Assistant Professor and Head, Department of Chemical Engineering,*
Birla Institute of Technology and Science (BITS) Pilani, -333031, Rajasthan, India
sureshg@pilani.bits-pilani.ac.in

[3*]*Assistant Professor, Department of Chemical Engineering*
Birla Institute of Technology and Science (BITS), Pilani- 333031, Rajasthan, India
Email: smita@pilani.bits-pilani.ac.in

Global warming is 'can of warms' in this century. The roots are many, of which the main crook is the increase in GHG's that is produced by human activities and some natural processes like animal and plant respiration. The main human sources of GHG's emission are: fossil fuel use, deforestation, use of synthetic fertilizer and many industrial processes. Carbon dioxide is one of the GHG's and responsible for Global warming. For CO_2 mitigation biological processes are one of the most understudied methods. In biological process (microalgae, green algae, cyanobacteria) conversion of CO_2 into biomass occurs direct from point source emissions of CO_2 in engineering systems like photo-bioreactors, open pond system etc. On this point biological study is mainly focused on photo-autotrophs and value addition from them. Recent development in CO_2 fixing microorganisms have blurred the distinction between autotrophs and heterotrophs. It was found that facultative microbes are also capable of fixing CO_2 via CO_2 assimilation. The studies also shown the value addition in terms of production of lipids, hydrocarbon from them. The performance of mixed culture was already studied earlier.

The present work is focused on the isolation of microbial species. The culture is obtained from the mixed culture from sewage treatment plant. The comparison was done on the ability of mixed culture and isolated species A in terms of CO_2 fixation. The batch studies was carried out with 15 % inlet CO_2 concentration (V/V %) maintained in gas phase initially. The CO_2 fixation per day basis by mixed culture and isolated species was done on per day basis (Fig. 1 & Fig. 2). The biomass growth rate have been observed on per day basis (Fig. 3). The Gompertz growth model available for without substrate effect was taken into consideration to find the maximum specific growth rate (μ_m). The mixed culture is performed better than isolated species in fixing CO_2. This confirms the use of mixed culture for CO_2 fixation along with value addition. The modeling of growth is necessary to scale up the process.

Keywords: CO_2 fixation, Mixed culture, Isolated species, Gompertz growth model.

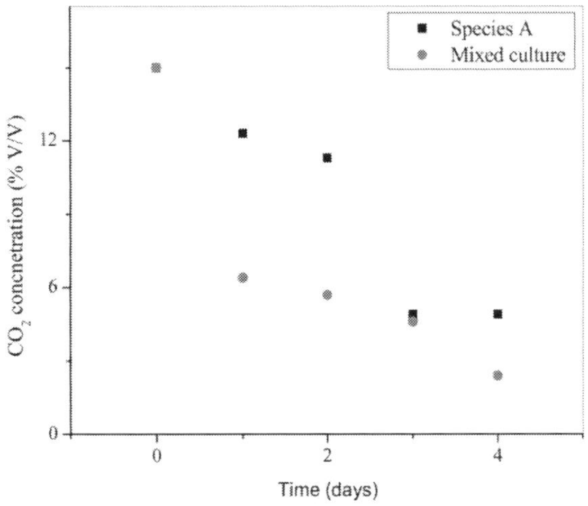

Fig.1 CO_2 fixation on per day basis by mixed culture and isolated species A

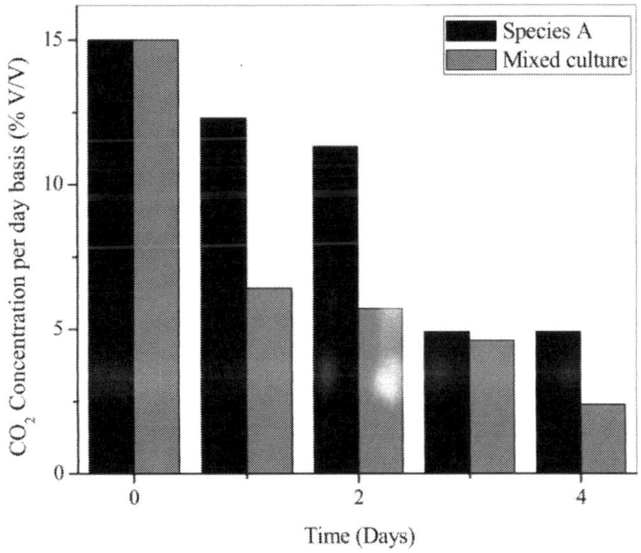

Fig.2 Comparison of CO_2 fixing ability of mixed culture and isolated species A

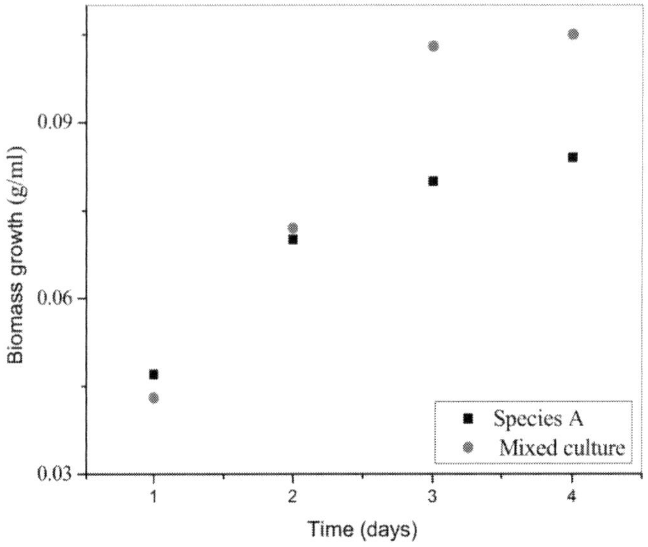

Fig.3 Biomass growth rate of isolated species A and mixed culture per day basis

A New Policy for the Use of Agricultural Residues in Tropical Countries Based on Biorefineries

Daza Laura; Rojas Miguel; Cardona Carlos A.
Research Group in Chemical, Catalytic and Biotechnological Processes
Universidad Nacional de Colombia, Manizales., Colombia
lvdazas@unal.edu.co

14AIChE Annual Meeting, Atlanta, GA

INTRODUCTION

The plantain pseudostem (PP), is the non woody stem generated as residue during the plantain harvesting practices. According to the FAO, Colombia is the second producer of plantain worldwide (Scott Padam, Seng Tin, Yee Chye, & Ismail Abdullah, 2012). The 87% of the total amount harvested is produced under an associated scheme, with other crops such as coffee, cacao and cassava, among others. Mainly by small producers with small lands (less than 5 hectares). The PP is the principal residue representing the 50% of the total biomass associated with plantain crop. The estimated production for 2011 was 1.699.137 ton in 209.931 hectares being the 16.4% of the total area destined to permanent crops. Approximately 268000 people are related to plantain harvesting activities.

Based on a biorefinery scheme, was evaluated the transformation of PP into citric acid, ethanol and xilitol to propose a return profit scheme under public policy scheme. In this work, three scenarios were evaluated in order to find the one that presents the best profit margin. Then, based on the best performance were evaluated two different cases in which the raw material costs were assessed considering the transport cost and an associated cost to transport and a production bonus.

METHODOLOGY

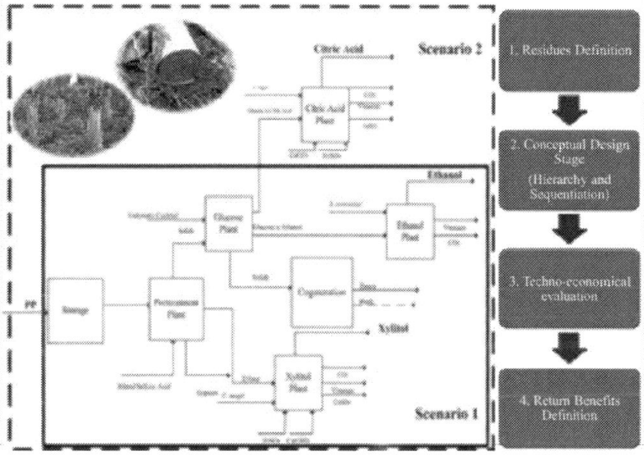

FORMULATED SCENARIOS

First Scenario
Ethanol
Xylitol
Steam
Power

Second Scenario
Citric Acid

Additional Scenario
Citric Acid
Xylitol

Table 1. Chemical Composition PP. Experimental Characterization

Compound	Composition (%)
Cellulose	59.6
Hemicellulose	27.1
Lignin	9.0
Ash	4.3

Policy Guidelines.
The present policy proposal has the aim to contribute to the improvement in life quality of population related with plantain agricultural chain through the following guidelines:
-Scholar Safety: this guideline has the aim to guarantee the material conditions related with school supplies of agricultural Childs, an average kit in Colombian context oscillate between 20-40 USD.
-Technical assistance: this guideline has the aim to enhance the harvesting yields, a percent of total profit should be destined to technical support assistance.
-Market Bonus or Agricultural Inputs Bonus: this guideline has the aim to return the project profitability through food or agricultural input bonus

RESULTS-TECHNO-ECONOMIC RESULTS

Table 2. Evaluation of Profit Margin per Scenario

	Scenario 1	Scenario 2	Scenario 3
Profit Margin	-376%	-82%	118%

Graphics 1-3. Share Cost Distribution third Scenario

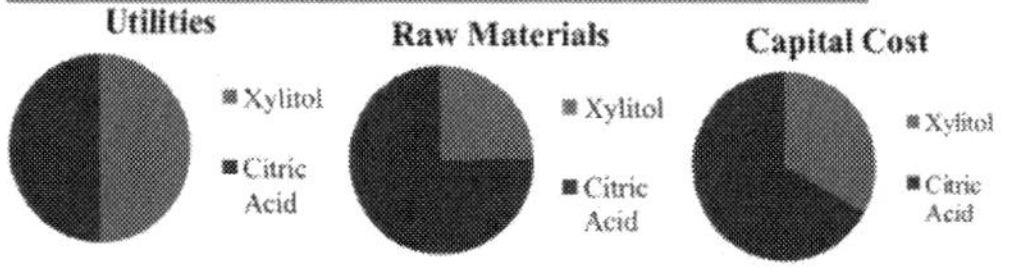

RESULTS- POLICY CONFIGURATION

Study Cases
-First case: Raw material costs associated to transport and basic logistic considerations.
-Second case: Raw material costs associated to transport and bonus production representing a percent of market plantain price.

Case 1. Profit Margin Project 118%
$ 10.295.833 USD/year

Profit Margin returning to Families
$206 USD/year distributed in:
Scholar Safety **70 USD/year**
Market or Agricultural inputs **136 USD/year**
Technical assistance
$514.791 USD/year = 42 professionals

Case 2. Profit Margin Project 67%
$ 5.815.833 USD/year

Profit Margin returning to Families
$116,32 USD/year
Scholar Safety **70 USD/year**
Market or Agricultural inputs **46,32 USD/year**
Raw Material incomes
0,076 USD/kg PP
Technical assistance
$290.791 USD/year = 24 professionals

CONCLUSIONS

-New policies using agricultural residues can impact positively the rural sector through contributions to safety food, scholar safety, technical assistance, and market bonus .
-Strengthening of small producer organizations. Allowing the establishment of profit distribution.
-This type of policy can also contribute to enhance the plantain crop yields through the technical assistance.
-A major impact can be obtained through the second case in which the project profitability does not depend of market price of plantain bunches

ACKNOWLEDGEMENTS

Universidad Nacional de Colombia at Manizales, COLCIENCIAS- Convocatoria 617 Jóvenes Investigadores- for the economical support to realize this research work.

REFERENCES

Mussatto S. I., G. Dragone, I.C. Roberto. (2006a). Brewers' spent grain: generation, characteristics and potential applications. Journal of Cereal Science 43, 1–14.
Niemi Piritta, Craig B. Faulds, Juhani Sibakov, Ulla Holopainen, Kaisa Poutanen, Johanna Buchert. (2012). Effect of a milling pre-treatment on the enzymatic hydrolysis of carbohydrates in brewer's spent grain. Bioresource Technology 116, 155–160.
FAO STAT 2011. Available from: http://faostat.fao.org/site/636/default.aspx#ancor (Accessed 2 August 2013).
MARTINEZ, A., RODRIGUEZ, M. E., WELLS, M. L., YORK, S. W., PRESTON, J. F. & INGRAM, L. O. 2001. Detoxification of dilute acid hydrolysates of lignocellulose with lime. *Biotechnology Progress*, 17, 287-293.
MAZZEO MENESES, M., LEÓN AGATÓN, L., MEJÍA GUTIÉRREZ, L. F., GUERRERO MENDIETA, L. E. & BOTERO LÓPEZ, J. D. 2010. Aprovechamiento Industrial de Residuos de Cosecha y Poscosecha del Plátano en el departamento de Caldas. *Revista Educación En Ingeniería*, 9, 128-139.
RINCÓN, L. E., BECERRA, L. A., MONCADA, J. & CARDONA, C. A. 2013a. Techno-Economic Analysis of the Use of Fired Cogeneration Systems Based on Sugar Cane Bagasse in South Eastern and Mid-Western Regions of Mexico. *Waste and Biomass Valorization*.
RINCÓN, L. E., MONCADA, J. & CARDONA, C. A. 2013b. Analysis of cogeneration as a tool to improve the viability of oilseed based biorefineries. *Catalytic systems for integral transformations of oil plants through biorefinery concept" Catalytic Systems For Integral Transformations Of Oil Plants Through Biorefinery Concept*. Universidad Nacional de Colombia.

There is an urgent need for developing strategies for reducing CO2 emissions and mitigate the climate change. In this sense, valorisation of CO_2 is considered a more suitable option than CO_2 storage since it allows the conversion of captured CO_2 into added value products, such as liquid fuels or raw materials. Electroreduction of CO_2 is particularly interesting because it could be used to store energy from intermittent renewable sources, such as wind or solar energy. Among all the different electroreduction products, one of the most promising is formate. This product is used in several industries as raw material and it has been pointed out as a possible fuel for fuel cells. According with previous work [1], Sn is a very interesting catalytic material for the electrochemical reduction of CO_2 to formate. Recently research efforts on electrochemical reduction of CO_2 are focused on the use of gas diffusion electrodes (GDE)[2]. GDEs are formed by a gas diffusion layer over which particles of metal catalyst are deposited. The aim of this work is to determine the best load and size the of Sn particles, which are key variables to take into account in these types of electrodes. The best GDEs were then used on an experimental system for continuous CO_2 electroreduction to formate and the influence of operational variables such as electrolyte flow and current density were analysed.

Cyclic voltammetries were performed using particles of different size (150 μm, 10 μm and 150 nm) and using different charges (0.1, 0.75, 1.5, 5 mg Sn cm^{-2}). Particles of 150 nm presented the best behaviour, so additional cyclic voltammetries and cronoamperommetries were carried out with different Sn loads (0.1, 0.75, 1.5, 5, 10 mg Sn cm^{-2}) depositing the particles over a carbonaceous support. Physical characterization of these electrodes was performed using Scanning Electron Microscopy (SEM) micrographs. The results obtained in voltammetries and cronoamperommetries highlight that the available Sn surface is not proportional to the total amount of particles deposited. In this way, best use of the catalytic material resulted to be in the electrodes with loads of 0.1 and 0.75 mg Sn cm^{-2}. Therefore, experiments with these optimized GDEs were carried out on an experimental system for continuous electroreduction of CO_2 to formate in a filter-press type cell at ambient temperature and pressure conditions. The faradaic efficiency and the rate of formate production were obtained to assess the performance. Using these electrodes it was possible to achieve rates of formate production around $1.5 \cdot 10^{-3}$ mol m^{-2} s^{-1} and faradaic efficiencies around 70%.

Acknowledgements:

This work was conducted under the framework of the Spanish Ministry of Science and Innovation Project ENE2010-14828

References:

[1] Alvarez-Guerra M, Del Castillo A, Irabien A., Chem. Eng. Res. Des. 2014; 92: 692–701

[2] Prakash G.K.S., Viva F.A., Olah G.A. J. Power Sources 2013; 223: 68-73

Mathematical Modeling of Carbon Capture by Carbonic Anhydrase-Displaying *Escherichia coli* in A Foam Bioreactor

Stuart Watson, Eunsung Kan
Department of Molecular Biosciences and Bioengineering
University of Hawaii at Manoa
Honolulu, HI 96822

Recently, a foam bioreactor for carbon capture has been developed. The foam bioreactor relies on an actively growing culture of carbon anhydrase (CA)-displaying *E. coli*, made into a fine foam with the CO_2-containing air. The foam bioreactor with CA-displaying *E. coli* showed excellent removal of CO_2 at short gas contact time with high stability at mild operating conditions (i.e., 2,800-3,800 g $CO_2/m^3_{reactor}$-h with the removal of 68-91%, at contact time of 30 s and cell density of 2 g_{dwt} L^{-1}). A conceptual mathematical model based on diffusion, mass transfer and biological reaction was developed for the systematic study of mass transfer and biodegradation kinetics for carbon capture by a foam bioreactor which will help reactor design and scale-up. All model parameters including characteristics of foams and biokinetics were determined by independent experiments or taken from the literature and thus the model was fully predictive. The physical characteristics of foam (foam size, liquid film thickness) and gas hold-up in the bioreactor were empirically correlated with gas velocity. The biokinetics for CO_2 consumption was developed using a Michaelis–Menten kinetics with bicarbonate-driven inhibition. The model predictions were a good agreement with experimental data over a wide range of operating conditions (gas velocity, cell density, CO_2 concentration). The model allowed simulation of CO_2 removal at several key operating parameters. The model also simulated CO_2 concentration over the height of the bioreactor, and within the depth of the liquid film of the foam. The model simulations indicated that most of the CO_2 is removed near the inlet port of the foam bioreactor and near the gas-liquid interface, and that bioconversion by carbonic anhydrase in the *E.coli* is the principal mechanism for CO_2 removal. Further studies of the model will include detailed insights on the model parametric sensitivity and on the performance limits of the foam bioreactor.

387482 An Industrial Symbiosis Approach to Produce Bioenergy and Integrate with Urban Areas and Ports

Abstract:

Industrial Symbiosis enhances both environmental and economic efficiency of a certain area, by exchanging industrial flows. These effects are maximized in areas which support a) low transportation costs, b) wide range of industrial activity and c) industrial infrastructure to support future investments. Based on the above, it is apparent that among the most promising areas are the industrial ports, being hubs of agricultural, livestock, industrial and urban material and energy flows. EPIC2020 is a European project that is taking place in four European ports: in Malmoe (Sweden), Wismar (Germany), Mantova (Italy) and Astakos (Greece), in order to create a general methodology on applying Industrial Symbiosis tools and methods for promoting integration of flows and bioenergy production.

Firstly, the ports are interviewed to provide data concerning the energy demands and bioenergy goals of the area, the available infrastructure (e.g. NG, District Heating and steam networks), existing bioenergy actors and stakeholders, ports' geographic boundaries, hinterland and seaward connections and the types of material trade in and out of the ports. Additionally, the ports interview the industries which are present within and around the port area using all the possible communication channels: personal communication, workshops, emails, telephone calls. The industries fill in specially designed templates with as much detailed information as possible for describing the flows including: amount, purity, current solution, supply patterns. Data concerning biomass availability of the greater area are also collected by the ports e.g. manure, straw, branches etc.

Secondly, the collected data are classified based on the details availability: a) statistical - for agricultural and livestock by-products, b) detailed – for industrial flows that are described analytically, c) implied – for industries which gave no feedback or were not directly contacted, d) enablers – new units that produce bioenergy from by-products e) Networks – the final users of the produced bioenergy. Each type of activity is called a "Functional Unit" and it is considered as a "black box", where only inputs and outputs are evaluated.

Finally, ontology engineering was applied for classifying both materials and functional units. This classification enables the automated matching and the proposition of new bioenergy producers. The analysis is refined according to the social guidelines received during the workshops and interviews. The study proves that there is need for new, green and economically sustainable investments and that there is a bioenergy potential of over 2,5 times of the projected goal. A GIS tool is under development to act as a dissemination tool, informing stakeholders of the numerous possibilities for cooperation, establishing energy and material "source to sink" connections and / or selling their by-products to newly established bioenergy units.

387790 Fixation of CO2 into Solid Mineral Matrix Via in-Situ and Ex-Situ Enhanced Weathering

Abstract:

One of the proposed technologies to permanently store CO_2 is to react calcium and magnesium bearing silicate minerals such as olivine ($(Mg,Fe)_2SiO_4$), serpentine ($Mg_3(OH)_4(Si_3O_5)$), wollastonite ($CaSiO_3$), anorthite ($CaAl_2Si_2O_8$), labradorite ($(Ca,Na)(Al,Si)_4O_8$), and basalt (rocks comprising various calcium and magnesium bearing minerals) to form calcium and magnesium carbonates, which are thermodynamically stable and insoluble in water. Permanent carbon storage can occur by directly injecting CO_2 into geologic formations containing these calcium and magnesium bearing silicate minerals and rocks, which is known as *in-situ* carbon mineralization or by mining and crushing these minerals for converting these minerals into carbonates via *ex-situ* carbon mineralization. Regardless of whether carbon storage occurs via *in-situ* or *ex-situ* schemes, it is essential to understand the reaction mechanisms and time scales of CO_2-reaction fluid-mineral interactions which can vary significantly depending on the temperature, partial pressure of CO_2, presence of additives, and pH. The chemical changes in minerals also impact the morphological characteristics such as porosity and surface area. The reaction mechanisms also differ when CO_2 hydration, mineral dissolution, and formation of carbonates occur simultaneously in a single step carbonation process, and when these reactions are decoupled such that mineral dissolution and formation of carbonates are carried out separately at low and high pH, respectively, which favor these individual reaction steps. Therefore, this study aims to provide a comparison of the kinetics of CO_2-reaction fluid-mineral interactions at temperatures as high as 185 °C, pressures as high as 175 atm, in the presence of various additives such as salts (e.g., NaCl, $NaHCO_3$), and chelating agents (e.g., Na-oxalate). In addition, modeling approaches that incorporate coupled chemical and morphological changes from CO_2 interactions with minerals are discussed to predict the fate of CO_2 interactions with minerals over time.

An empirical model to estimate heating value of municipal solid waste (MSW)

Honghong(Crystal) Shi, Nader Mahinpey, Aqsha Aqsha, Rico Silbermann

Energy and Environment Research Group, Department of Chemical and Petroleum Engineering, University of Calgary, 2500 University Drive NW, Calgary, Alberta, Canada T2N 1N4

UNIVERSITY OF CALGARY

Energy & Environment Research Group

Background & Objectives

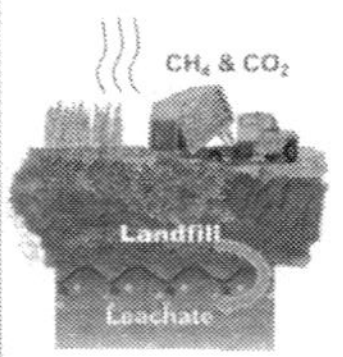

Facts:

- Landfilling causes greenhouse gas and water pollution problems.
- Incineration (controlled combustion) plants produce heat to generate electricity, but with relatively low efficiency.
- Applications of other thermochemical waste-to-energy (WTE) technologies, such as pyrolysis and gasification, are developing slowly and still in the lab research or commercial demonstration stage. **One important reason is the lack of adequate MSW characterization data for the feasibility study and the facility design.**

Why is the heating value (HV) of MSW so important?

Heating value or calorific value, which defines a fuel's energy content, is roughly inversely proportional to the capacity of a WTE furnace/boiler.

Conventional HV determination methods:

Experimental measurements using bomb calorimeter	Obtained from calculations using an empirical model
Advantage: Accurate	Advantage: Fast
Disadvantages: Not all the MSW management institutes are equipped with bomb calorimeters; The experimental process is tedious and requires advanced technical skill in the handling of equipment.	Disadvantage: Various models derived for coal or biomass are not suitable for the HV estimation of MSW. The accuracy of few available MSW models is relatively low, because of limited material types of the samples.

The project objectives are:

- **Fill the current data gap in the characteristics of MSW**
- **Build an empirical model to estimate the heating values of MSW that is both fast and accurate**

Methods

Sample Source and Preparation:

- **14 samples of nonrecyclable landfill wastes**
- Two generation sources (**residential** and **ICI**-industrial, commercial & institutional)
- Seven categories: **1**-Nonrecyclable Paper; **2**-Wood; **3**-Rigid Plastic; **4**-Film & Styrofoam; **5**-Textile; **6**-Carpet; **7**-Rubber
- Dried and ground into small particles (≤ 1mm)

Characterization Technologies:

- Moisture content – oven drying method
- Ultimate analysis – CHNS(O)
- Proximate analysis – volatile matter content, fixed carbon content, ash content
- Heating value – bomb calorimeter

Statistical Analysis:

- Multiple regression analysis
- Best subset method
- Stepwise regression method
- Software – **'MINITAB'**

Results and Discussion

Characteristics Data

Table 1 Characteristics Data of Residential Wastes

Residential MSW (wt%, dry basis)	Paper	Wood	Rigid plastic	Styrofoam & Textile	Textile	Carpet	Rubber	Average
Moisture	22.48	72.57	15.02	0.45	1.07	0.19	5.3	26.27
Proximate Analysis								
Volatile Matter	79	73.46	90.62	98.38	86.12	74.37	90.61	87.17
Fixed Carbon	13.64	21.26	8.28	0.62	12.82	0.89	8.29	8.74
Ash	7.36	5.28	1.1	0	1.06	24.75	1.1	4.09
Ultimate Analysis								
C	41.64	46.75	71.23	90.35	59.37	62.49	43.94	62.76
H	5.62	5.93	7.62	7.93	4.26	8.7	5.13	6.53
N	0.34	0.6	0.08	0.88	0.07	6.36	0.18	0.38
S	1.14	1.12	1.38	1.4	0.87	1.77	1.01	1.24
O	43.9	36.32	18.68	0	34.38	0.98	48.64	26.02
HHV (MJ/kg)								
Weight Percentage	19	10	17	24	11	7	12	-

Table 2 Characteristics Data of ICI Wastes

ICI MSW (wt%, dry basis)	Paper	Wood	Rigid plastic	Styrofoam & film	Textile	Carpet	Rubber	Average
Moisture	24.2	8.4	6.2	0.4	0.2	0.6	0.9	8.8
Proximate Analysis								
Volatile Matter	79.5	77.2	99.7	97.6	89.7	93.8	74.3	86.6
Fixed Carbon	14.9	19.1	0.3	1.7	8.1	3.7	13.6	9.9
Ash	6.7	3.7	0.1	0.6	2.3	2.5	12.1	3.6
Ultimate Analysis								
C	42.9	41.7	85.8	96.4	56.7	61.7	53.4	60.9
H	5.8	5.4	7.2	9.7	6.4	8.7	7.2	7.0
N	0.1	1.5	0.2	3.0	0.4	6.5	3.7	1.7
S	1.1	1.1	1.3	1.8	1.2	1.6	1.4	1.3
O	44.5	46.5	5.5	0.0	33.2	20.1	22.2	25.5
HHV (MJ/kg)								
Weight Percentage	15	25	16	19	5	9	11	-

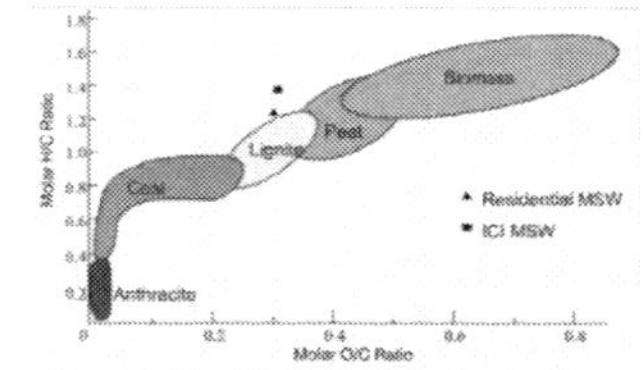

Figure 1 Position of the MSW on the Van Krevelen Diagram (Adapted from Basu, P. 2013, p. 61. [5])

Model Building & Optimization

Input data:

Predictors: Compositions of 5 elements (carbon, hydrogen, nitrogen, sulfur, and oxygen)

Response: Experimental heating values

MINITAB

Multiple regression - initial model:

$$HHV=0.905C-1.63H-0.236N+30.8S+0.499O-68.2\ (MJ/kg)$$

$$R_{adj}^2 = 0.969$$

MINITAB

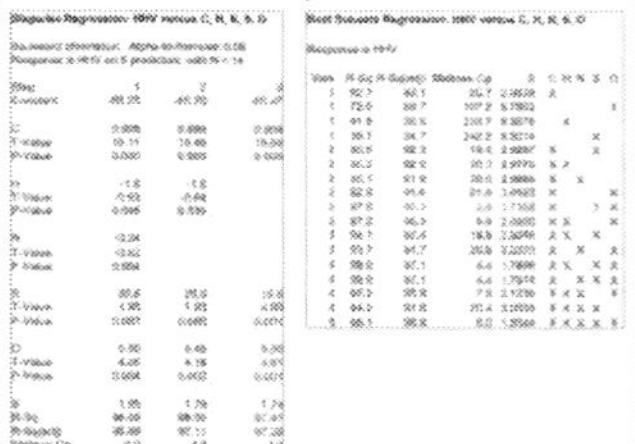

Optimized model:

$$HHV=0.904C+19.8S+0.5O-65.5\ (MJ/kg)$$

$$R_{adj}^2 = 0.973$$

Evaluation of the Model Performance

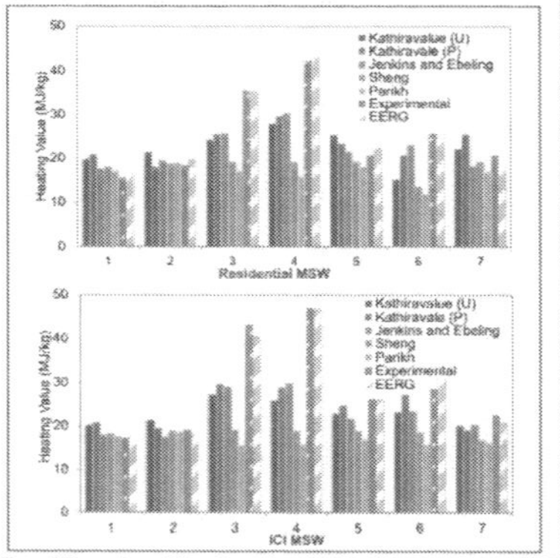

Figure 2 Comparison of Experimental and Predicted Heating Values of MSW

Table 3 Error Analysis Comparison of Different Models

	EERG	Kathiravale(U) [1]	Kathiravale(P) [1]	Jenkins [2]	Sheng [3]	Parikh [4]
Average bias error	-0.3%	-9.1%	-6.8%	-12.4%	-24.5%	-30.6%
Average absolute error	4.9%	23.8%	45.7%	37.8%	14.2%	9.7%
Maximum absolute error	12.0%	44.7%	90.7%	95.1%	24.5%	24.8%

Conclusions

- The characteristics of the MSW samples largely depend on the type of material. The residential wastes are observed to have more moisture than the ICI wastes. However, the dry residential wastes have similar chemical properties to those of the dry ICI wastes, containing high volatile matter content (> 85%) and low fixed carbon content (< 10%).
- The MSW has non-negligible amount of sulfur and nitrogen. Proper De-NOx and De-SOx technologies are essential to reduce the negative environmental or healthy effects of a WTE facility. Meanwhile, our results showed that the MSW has an H/C ratio similar to peat and an O/C ratio similar to lignite. Research on the topic of WTE could expend from the current achievements of utilizing peat or lignite.
- **An empirical model to estimate the heating value of MSW is successfully built.** The optimized model's output is compared with the experimental heating values, as well as with the outputs of other five models in literature. It is shown that the model built in this project has the best predictive performance and the lowest error level among all the models.
- Based on only carbon, sulfur and oxygen data, the model provides a rapid, easy and accurate estimation of the heating value for the MSW. This may be of particular interest in the contexts where experimental measurements are not always available.

Future Work

- Multiple samplings over 12 months are suggested, in order to obtain the year-around moisture content data. It's better to do the sampling in each month or at least in each season because the moisture content depends largely on the weather condition and the season.
- Collect more experimental characteristics data of MSW by laboratory measurements or from literature to add the predictive accuracy of the model built in this project.

Acknowledgements:

- Alberta Innovates – Energy and Environment Solutions
- Waste Management Facility in the City of Red Deer, AB Canada

References:

[1] Kathiravale, S., et al., *Modeling the heating value of Municipal Solid Waste*. Fuel, 2003. 82(9): p. 1119-1125.

[2] Vargas-Moreno, J.M., et al., *A review of the mathematical models for predicting the heating value of biomass materials*. Renewable and Sustainable Energy Reviews, 2012. 16(5): p. 3065-3083.

[3] Sheng, C. and J.L.T. Azevedo, *Estimating the higher heating value of biomass fuels from basic analysis data*. Biomass and Bioenergy, 2005. 28(5): p. 499-507.

[4] Parikh, J., S.A. Channiwala, and G.K. Ghosal, *A correlation for calculating HHV from proximate analysis of solid fuels*. Fuel, 2005. 84(5): p. 487-494.

[5] Basu, P., Biomass gasification and pyrolysis: practical design and theory. Academic press, 2013, p. 61.

Contact Information:

- Dr. Nader Mahinpey, Associate Professor
  Email: nader.mahinpey@ucalgary.ca, Tel: +1(403)210-6503
- Honghong(Crystal) Shi, MSc Candidate
  Email: crystal.shi.calgary@gmail.com, Tel: +1(587)719-6617
- EERG Website: http://www.ucalgary.ca/ENCH/AEG/home.html

370259 Controllable Synthesis of Graphene in Aqueous Phase for Fabrication of Graphene-Based Composites

Abstract:

We report a facile, environment-friendly, and low-cost method for the synthesis of graphene in aqueous phase by the exfoliation of microwave-assisted expanded graphite with the assistance of surfactant cetyltrimethylammoniumbromide (CTAB). Scanning electron microscopy and transmission electron microscopy micrographs reveal that graphene sheets with adjustable number of layers were achieved. Raman and UV-Vis spectra further confirm this result and indicate that the as-prepared graphene sheets are of high quality with low defect density. By measuring the surface energy density of water with and without adding surfactant CTAB, a possible exfoliation mechanism through decreasing the surface energy of water induced by CTAB and subsequent intercalation of CTAB into graphite layers is proposed. By vacuum filtration, the generated graphene sheets can easily form conductive films. Because this method was carried out in aqueous phase, it is convenient to further fabricate the generated graphene sheets into controllable graphene-based composites for various applications, such as photocatalysts, supercpacitors, lithium ion batteries, etc.

388487 Design and Synthesis of Nitrogen-Doped Mesoporous Carbon for Selective CO2 Capture

Abstract:

Solid-state post-combustion CO_2 sorbents have certain advantages over traditional aqueous amine systems, including reduced regeneration energy since vaporization of liquid water is avoided, tunable pore morphology, and greater chemical variability.

Mesoporous carbons are promising for CO_2 capture due to their chemical inertness, low cost, high surface area, and tunable pore structures. Their porous structure and high surface area allow addition of chemical functionality by grafting or impregnation. Nitrogen functionalization plays an important role in surface chemistry to achieve enhanced CO_2 adsorption capacity. We report here an ordered mesoporous nitrogen-doped carbon made using the co-assembly of modified-pyrrole and triblock copolymer through a soft-templating method, which is facile, economical, and fast compared to the hard-template approach. In the synthesis process, the pyrrole precursor acts as both the carbon and nitrogen source. Carbonization of the resultant polymeric assembly generates a graphitic carbon structure and porous network through the removal of the block copolymer template. A high surface area mesoporous carbon was achieved that is comparable to the silica counterpart. The resulting material shows promising CO_2 capture performance, reaching equilibrium adsorption of 1.0 mmol CO_2/g of material at 25 °C and 0.1 bar. Furthermore, the hierarchical macro-meso-microporous structure of this N-doped carbon allows for fast diffusion of CO_2 gas into the adsorption sites and controlled pore condensation within the microporous structure. Another potential benefit is that the thermal conductivity of the mesoporous carbon is higher than its silica counterpart, resulting in faster regeneration with enhanced stability using a temperature swing adsorption process. These properties of mesoporous carbons made from a conducting polymer render these as desirable materials for CO_2 capture.

390736 Mdea As an Effective Solvent for the Natural Gas Sweetening Process

Abstract:

Natural gas has now become a crucial component of world's supply energy. The demand for natural gas has risen drastically over the past few years due to its importance in various industries and also for the domestic purposes. Natural gas need to be purified to meet the quality standards since it contains impurities such as carbon dioxide (CO_2) and hydrogen sulfide (H_2S), which are the main acid gases and can cause corrosion, reduce the heating and sales value of gas. Aqueous amine solutions are proven to be practical solvents for the treatment of natural gas. By simply changing their amine solutions, many inefficient acid gas removal units can be optimized. Tertiary amines like methyl di-ethanol amine (MDEA) are commonly used for the removal of CO_2 from gas mixtures or in gas sweetening processes for the extraction of CO_2 and H_2S. The use of MDEA has been investigated by using a process simulation program called ASPEN Plus. The results show that, at high pressures, amine mixtures have little or no advantage. As the pressure is lowered, it becomes more difficult for MDEA to meet residual gas requirements.

Compatibilized Polyethylene – Graphene Nanocomposites: Component Mapping & Thermal, Rheological and Mechanical Performance

Vikas Mittal[$] and Leslie Krauss[#]

[$]Department of Chemical Engineering, The Petroleum Institute, Abu Dhabi, UAE

[#]CytoViva Inc., Auburn, USA

vmittal@pi.ac.ae

Graphene consists of one atomic thick sheets of covalently sp^2-bonded carbon atoms in a hexagonal arrangement [1]. Due to its superior mechanical and electrical properties, it has generated tremendous research interest and a number of studies on polymer nanocomposites based on graphene as filler have been reported in recent years [1-10]. However, the mismatch of polarity between the polymer and surface of graphene platelets may lead to their poor dispersion in polymers and lower enhancement in polymer properties [1], thus, requiring compatibilization of the phases.

Functionalization of the surface of graphene platelets has been reported to enhance their dispersion potential in polymer matrices, which subsequently results in significant enhancement of the mechanical and electrical properties in polymer nanocomposites [11,12]. Another alternative to enhance the dispersion in polymers is the addition of amphiphilic compatibilizers which due to their polar and non-polar components bridge the polarity gap between the polymer and filler phases [13,14]. Such compatibilizers are of particular benefit for non-polar polymers like polyethylene, where the lack of polar groups in their backbone is a significant hurdle in achieving homogenous dispersion and exfoliation of nanofillers. Schniepp *et al.* used maleic anhydride as a bridge for nanocomposites of low density polyethylene (LDPE) and exfoliated graphite nanoplatelets [15]. Mittal *et al.* [10] in a previous study reported the advantage of introducing different extents of chlorinated polyethylene copolymers of varying chlorination levels on high density polyethylene (HDPE)-graphene nanocomposites. The surface groups on graphene platelets interacted chemically with chlorinated polyethylene compatibilizers by forming ester bonds. Both mechanical and rheological properties were enhanced with higher extent of filler exfoliation achieved for comaptibilizer with higher chlorination level in the structure.

Though different compatibilizers varying in molecular weight and polarity have been reported in the literature to enhance graphene dispersion in the non-polar polymer matrices, but the use of various graphene types with different C/O ratios, different graphene surface modifications and polymer grades as well as different synthesis routes to generate nanocomposites makes the comparison of the resulting nanocomposite properties difficult. Thus, a systematic study looking into the aspect of interfacial interactions of the compatibilizers and filler and resulting impact on nanocomposite properties is needed. Moreover, the mapping of the components in the nanocomposites is also desired in order to confirm the translation of interfacial interactions into the distribution of components. Energy dispersive X-ray analysis would not be suitable for such mapping purposes due to common elements present in all filler, compatibilizer and PE phases. Detection of nanometer thin filler platelets may also be challenging for this method. Raman mapping has also been reported to be a functional method for such mapping [16], but the high energy beam used for scanning damages the polymer samples significantly, thus, requiring the use of methods involving lower energy beams. One such option is hyperspectral imaging and mapping in visible near infra-red (VNIR) range in which polymers can sustain the lower energy beams for longer periods [17-20]. This method also allows the detection of component distribution at nano-scale.

In the current study, the energy dispersive X-ray diffraction analysis revealed an average C/O ratio of 22 in the graphene platelets. In a previous study, graphene platelets produced by similar method and average C/O ratio of 20 were reported to have polar surface groups (hydroxyl, epoxide, carboxyl, etc.), which interacted chemically with chlorinated polyethylene compatibilizers by forming ester bonds [10]. Likewise, the specific choice of the compatibilizers ethylene-co-acrylic acid and ethylene-co-methacrylic acid ionomer used for the current study was based on the possibility of achieving physical/chemical interactions of their polar group with the surface of graphene platelets. As described in Table 1, ethylene-co-acrylic acid (EAA) had 9% acrylic acid content in its structure, whereas the rest was PE. Ionomer ethylene-co-methacrylic acid (with zinc ion) (EMAZ) was chosen to study the effect of ionic species in generating ionic interactions with the filler surface. In this copolymer, methacrylic acid content was 10% and a part of the methacrylic acid along the copolymer backbone was neutralized with zinc ions resulting in polymer structure with three regions: amorphous polymer, crystalline polymer, and ionic clusters. Though the melting point of EMAZ was lower than EAA and overall much lower than the pure PE, but the density of the PE and compatibilizers was similar. EMAZ had significantly low MFI as compared to EAA indicating that the molecular weight of EMAZ was higher than EAA. For nanocomposite generation, the compounding conditions as well as amounts of composite constituents were held constant in order to attribute the observed properties solely to compatibilizers and their interactions with filler surface. The compounding conditions were chosen so as to achieve uniform mixing of the nanocomposites components.

The tensile modulus of PE was enhanced by 13% from 1063 MPa to 1203 MPa for PE-G nanocomposite. For PE-5%EAA-G nanocomposite, the increment in the modulus was 40%, whereas PE-5%EMAZ-G exhibited an increase of 25% in the tensile modulus as compared to PE. Different extents of modulus enhancements revealed different degrees of interactions between the filler and compatibilizers. Without compatibilizer, the filler dispersion did not seem to be optimum as the improvement in modulus was insignificant. Higher extent of modulus enhancement indicated better filler dispersion in the polymer due to the compatibilization effect of EAA and EMAZ. Similar to tensile modulus, PE-5%EAA-G nanocomposite exhibited highest peak stress value of 70 MPa as compared to 62 MPA for PE, whereas a slight decrease was observed for PE-G (61 MPa) and PE-5%EMAZ-G (60 MPa) nanocomposites. The total elongation reduced in PE-G nanocomposite probably due to the presence of aggregates. The PE-5%EAA-G and PE-5%EMAZ-G nanocomposites, on the other hand, exhibited increased total elongation of 4.6 mm and 4.7 mm respectively due to some degree of matrix plasticization by the addition of low molecular weight compatibilizers. Nanocomposites with 7.5% compatibilizer content exhibited further increments in tensile modulus and the extent of increase was significantly higher for EAA (51%) as compared to EMAZ (30%) due to earlier observed effects of better filler dispersion. Comparing the performance with the other filler systems, the polypropylene nanocomposites were observed to have an increment of 26% at 3 vol% (6 wt%) of surface modified clay and 6 wt% polypropylene-g-maleic anhydride compatibilizer [14]. Similarly, polyethylene nanocomposites compatibilized with 7 wt% of polyethylene-b-polyethylene glycol copolymer had an increment of 60% in tensile modulus, but the improvement was attained by using 2.8 vol% (~5.6 wt%) of surface modified clay [13]. In both the cases, the stress at break in the nanocomposites decreased as compared to pure polymer, which, on the other hand, was increased with graphene as filler. These comparisons confirm the potential of graphene as reinforcing filler in nanocomposites.

Figure 1 shows the TEM micrographs of PE-EAA-G and PE-EMAZ-G nanocomposites. Graphene platelets were observed to be flexible, bent and folded. Though both EAA and EMAZ containing nanocomposites exhibited uniform distribution of filler, the composite with EAA copolymer had relatively lower stack thickness of graphene platelets and higher extent of single delaminated platelets. No specific alignment of filler platelets in any particular flow or processing directions was observed. Thus, the shear experienced during compounding was sufficient to cause the filler delamination, but the interactions between the filler and compatibilizers resulted in different microstructures as observed earlier for mechanical performance. Neutralization of acid functional groups in the EMAZ comaptibilizer may have hindered its interaction with the polar functional groups present on the surface of graphene platelets. On the other hand, EAA due to free acrylic acid groups along the main chain would have more efficient interfacial interactions with the filler platelets, thus resulting in improved filler delamination and nanocomposite properties. Occasionally, large aggregates were also observed in PE-EMAZ-G nanocomposites. The PE-G nanocomposites did not show any exfoliation and large filler aggregates were observed in the matrix.

Figure 2 demonstrates the storage modulus of the nanocomposites as a function of angular frequency. The nanocomposites exhibited higher storage modulus than matrix polymer for the whole measurement range, though the differences diminished at higher angular frequencies. Similar to the tensile properties, PE-5%EAA-G nanocomposite showed the highest magnitude of storage modulus. For instance, the *G'* value for pure PE at 1 rad/s was 14600 Pa, which was measured to be 45340 Pa for PE-5%EAA-G nanocomposite and was even higher than 39690 Pa for PE-G nanocomposite, whereas the PE-5%EMAZ-G had a storage modulus of 32020 Pa. This also confirmed the earlier findings from microscopy and mechanical testing that the better dispersion of graphene resulted in the presence of EAA compatibilizer. The intercept of G' and G'' indicating the transition from viscous to elastic behavior occurred at a frequency of 2.5 rad/s for PE-G composite. For PE-5%EAA-G nanocomposite, the transition frequency was unaffected, whereas, the PE-5%EMAZ-G nanocomposite exhibited the transition at 5 rad/s frequency indicating domination of viscous response due to the plasticization effect caused by EMAZ compatibilizer. As shown in inset of Figure 2, the complex viscosity of the nanocomposites also showed similar trends as storage modulus. PE-EAA-G exhibited highest magnitude of viscosity, however, the enhanced viscosity did not pose any difficulty during compounding and subsequent processing operations. It could thus be concluded that the observed changes in the rheological and mechanical properties of nanocomposites resulted from balance between competing factors like matrix plasticization due to comaptibilizer addition thus reducing the properties and filler exfoliation due to better filler-polymer interactions leading to enhanced properties [10,13].

Figure 3 shows the mapping of distribution of graphene and EAA at two different locations in PE-EAA-G nanocomposite. Pixels in red represented the presence of graphene peak, whereas pixels in green represented the EAA peak in the nanocomposite. The dark area in the mocrographs without any colored signal corresponded to PE. Similarly, Figure 4 also shows the mapping of spectral mixing of the graphene and EAA peaks in PE-EAA-G nanocomposite, where pixels in red indicated the presence of graphene peak, pixels in green the EAA peak, and pixels in yellow both peaks. EAA in the PE-EAA-G nanocomposite sample was observed to bond/mix well with the graphene causing spectral shifting and mixing. Several areas were observed to contain characteristic peaks of both EAA and graphene in the nanocomposite. The EAA compatibilizer was thus observed to be effective in dispersing the graphene throughout the PE matrix, which also confirmed the findings of mechanical, rheological and microstructural analysis. In PE-EMAZ-G nanocomposite, no extensive interaction of EMAZ with the graphene platelets was observed and EMAZ was less effective in dispersing graphene in the PE matrix as no evidence of spectral mixing was found in this sample. The components of this nanocomposite were observed to have large domains present in the nanocomposite, also confirming the earlier observed polymer-comaptibilize r immiscibility as well as filler aggregates. Thus, though the IR spectra of the samples did not show any evidence of chemical interaction between graphene and any of the EAA or EMAZ compatibilizers, however, the analysis in VNIR region detected the spectral shifts in VNIR range due to physical interactions between the compatibilizers and filler platelets. Performing the mapping in VNIR is further beneficial as the high wavelength used for the analysis did not damage the sample, whereas extensive sample surface damage was observed while performing the mapping with Raman spectroscopy.

References

1. P. Mukhopadhyay, and R.K. Gupta, *Plast. Eng.*, 32 (2011).
2. H. Kim, A.A. Abdala, and C.W. Macosko, *Macromolecules*, **43**, 6515 (2010).
3. D. Cai, and M. Song, *J. Mater. Chem.*, **20**, 7906 (2010).
4. M.A. Rafiee, J. Rafiee, Z. Wang, H. Song, Z.Z. Yu, and N. Koratkar, *ACS Nano*, **3**, 3884 (2009).
5. P. Steurer, R. Wissert, R. Thomann, and R. Muelhaupt, *Macromol. Rapid Commun.*, **30**, 316 (2009).
6. H. Kim, Y. Miura, and C.W. Macosko, *Chem. Mater.*, **22**, 3441 (2010).
7. D.A. Nguyen, Y.R. Lee, A.V. Raghu, H.M. Jeong, C.M. Shin, and B.K. Kim, *Polym. Int.*, **58**, 412 (2009).

8. M. Fang, K. Wang, H. Lu, Y. Yang, and S. Nutt, *J. Mater. Chem.*, **19**, 7098 (2009).
9. J.R. Potts, D.R. Dreyer, C.W. Bielawski, and R.S. Ruoff, *Polymer*, **52**, 5 (2011).
10. A.U. Chaudhry, and V. Mittal, *Polym. Eng. Sci.*, **53**, 78 (2013).
11. T. Ramanathan, A.A. Abdala, S. Stankovich, D.A. Dikin, M. Herrera-Alonso, R.D. Piner, D.H. Adamson, H.C. Schniepp, X. Chen, R.S. Ruoff, S.T. Nguyen, I.A. Aksay, R.K. Prud'Homme, and L.C. Brinson, *Nature Nanotechnol.*, **3,** 327 (2008).
12. S. Ansari, and E.P. Giannelis, *J. Polym. Sci. Part B*, **47**, 888 (2009).
13. M.A. Osman, J.E.P. Rupp, and U.W. Suter, *Polymer*, **46**, 8202 (2005).
14. V. Mittal, *J. Appl. Polym. Sci.*, **107**, 1350 (2008).
15. H.C. Schniepp, J.-L. Li, M.J. McAllister, H. Sai, M. Herrera-Alonso, D.H. Adamson, R.K. Prud'homme, R. Car, D.A. Saville, and I.A. Aksay, *J. Phys. Chem. B*, **110**, 8535 (2006).
16. N. Furuyama, S. Hasegawa, T. Hamaura, S. Yada, H. Nakagami,, E. Yonemochi, and K. Terada, *Int. J. Pharmaceutics*, **361**, 12 (2008).
17. A. K. Suresh, Y. Weng, Z. Li, R. Zerda, D. Van Haute, J. C. Williams, and J. M. Berlin, *J. Mater. Chem. B*, **1**, 2341 (2013).
18. E. S. Oh, C. Heo, J. S. Kim, M. Suh, Y. H. Lee, and J.-M. Kim, *J. Biomed. Opt.*, **19**, 051207 (2013).
19. E. S. Oh, C. Heo, J. S. Kim, Y. H. Lee, and J. M. Kim, *Proc. SPIE*, **8879** (2013).
20. S. Srinivasan, J. F. Alexander, W. H. Driessen, F. Leonard, H. Ye, X. Liu, W. Arap, R. Pasqualini, M. Ferraria, and B. Godin, *J. Mater. Chem. B*, **1**, 5218 (2013).
21. M.J. McAllister, J.L. Li, D.H. Adamson, H.C. Schniepp, A.A. Abdala, J. Liu, M. Herrera-Alonso, D.L. Milius, R. Car, R.K. Prud'homme, and I.A. Aksay, *Chem. Mater.*, **19**, 4396 (2007).
22. W.S. Hummers, and R.E. Offeman, *J. Am. Chem. Soc.*, **80**, 1339 (1958).
23. V. Mittal, G.E. Luckachan, and N.B. Matsko, *Macromol. Chem. Phys.*, in print (2014).
24. M. Joshi, B.S. Butola, G. Simon, and N. Kukaleva, *Macromolecules*, **39**, 1839 (2006).

Table 1. Specifications of HDPE and compatibilizers as obtained from suppliers

Polymer/ Compatibilizer	Supplier and trade name	Co-monomer content, %	MFI, g/10 min (2.16 kg, 190°C)	Density, g/cm^3	Peak melting temperature, ^{o}C
PE	Borouge BB2581	--	0.35	0.958	147
EAA	Exxon Mobil ESCOR 5050	9% AA	8.4	0.936	97
EMAZ	Dupont Surlyn 9020	10% MA	1	0.96	85

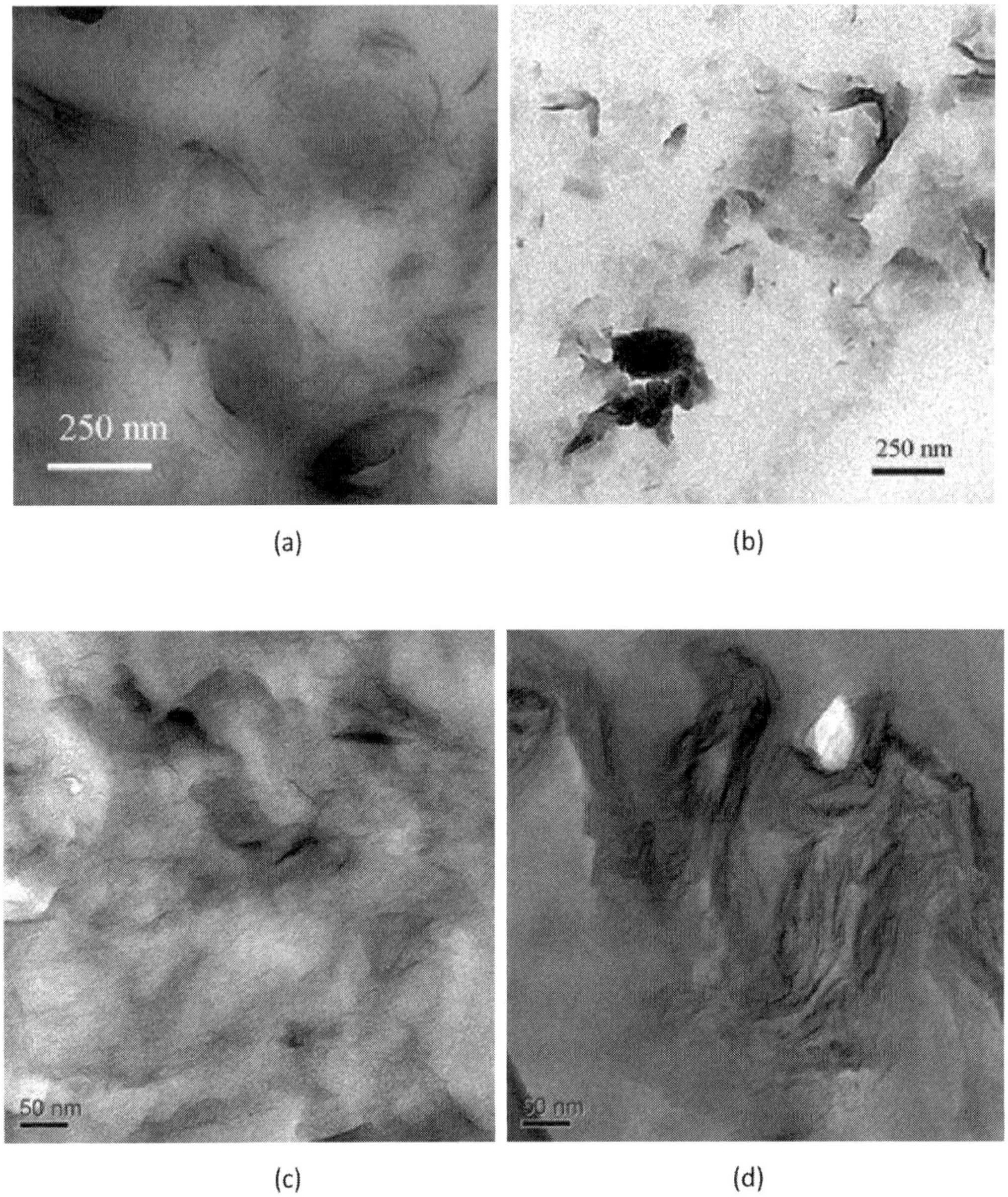

Figure 1. TEM micrographs of (a,c) PE-5%EAA-G and (b,d) PE-5%EMAZ-G nanocomposites. The black lines represent the cross-section of the graphene platelets.

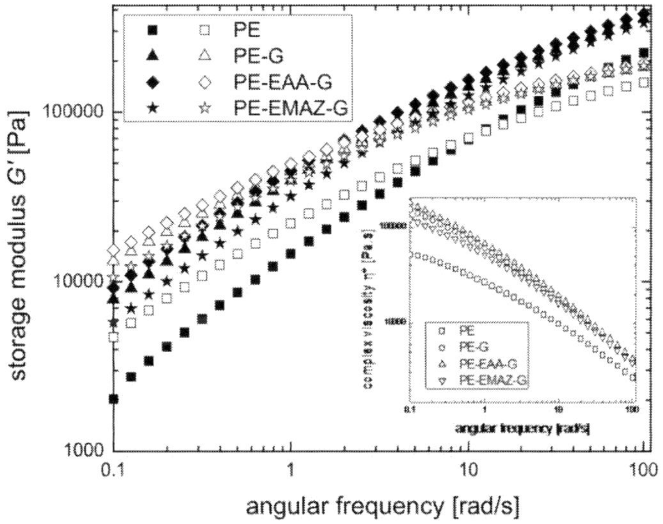

Figure 2. Storage (solid dots) and loss (empty dots) modulus of PE and its composites with and without compatibilizers (5 wt% content of compatibilizers) as a function of angular frequency. The inset shows the complex viscosity as a function of angular frequency.

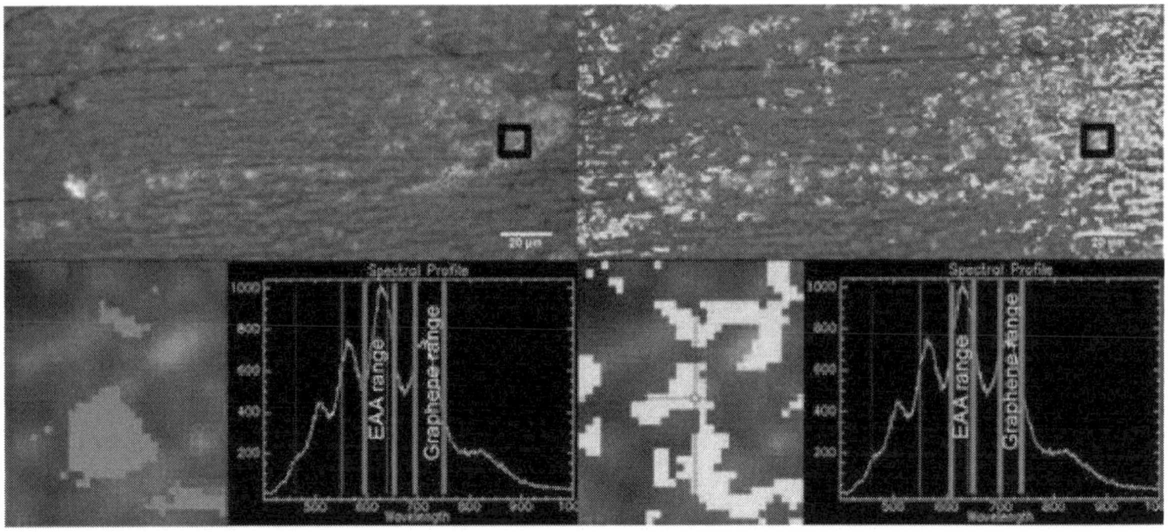

(a)

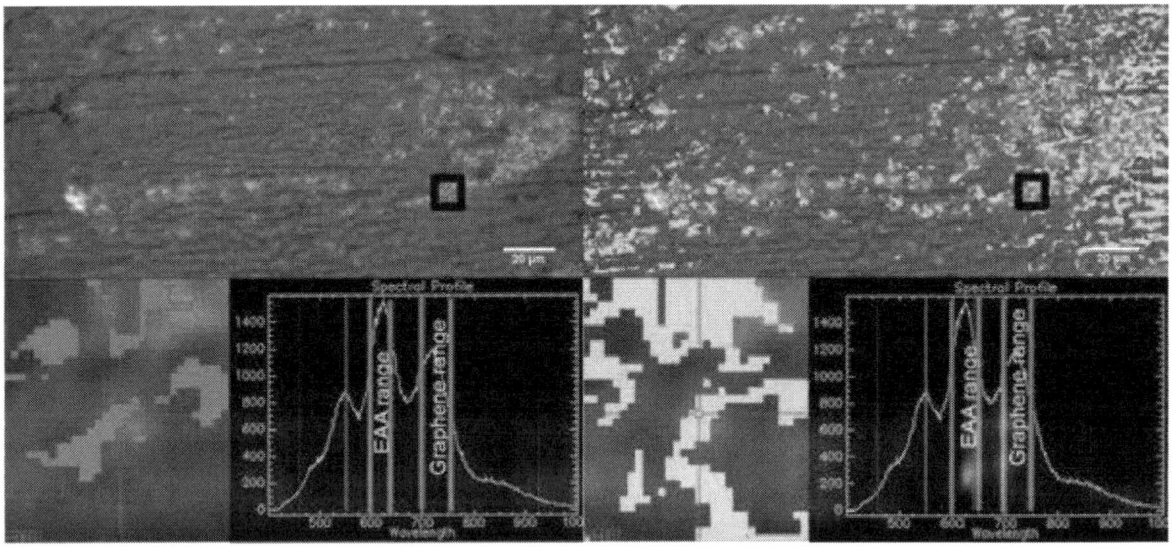

(b)

Figure 3. Mapping of distribution of graphene and EAA at two different locations in PE-5%EAA-G nanocomposite. Pixels in red represent the presence of graphene peak, whereas pixels in green represent the presence of EAA peak in the nanocomposite.

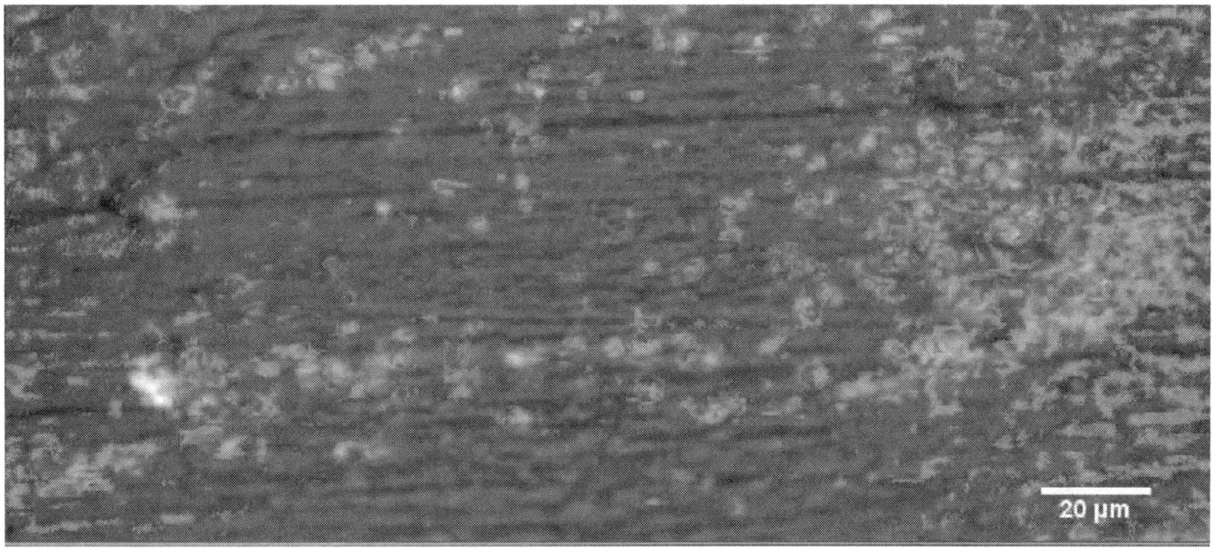

Figure 4. Mapping of spectral mixing of the graphene and EAA peaks in PE-5%EAA-G nanocomposite. Pixels in red indicate the presence of graphene peak, pixels in green indicate the presence of the EAA peak, whereas pixels in yellow indicate the presence of both peaks.

389385 Development of Chemical Heat Transformer for High Temperature Systems

Abstract:

Recently, an energy saving has caught a great attention in chemical industries due to the fuel price rapid increase and green house effect. In fact exothermic reaction is usually carried at low temperatures with cooling and endothermic reaction is carried at high temperatures supplying heat from fuel combustion in many conventional chemical processes. As a result, large exergy destruction is taken places in this process. To provide further energy saving in these processes, we should pay more attention to minimization of this exergy destruction.

One of the ideas to solve this problem is to develop a heat transport system, called heat transformer, which receives heat by an endothermic reaction and releases heat by exothermic reaction. To use this system, the process does not require the fuel combustion and coolers which mainly causes large exergy destruction in the process. However, in many conventional heat transport systems, the unique system for each chemical process has to be designed because all of processes have their own process conditions, leading to increase of investment cost.

To overcome this problem, authors proposed a chemical heat transformer system using a reversible reaction for high temperature range. This chemical heat transformer system transports the heat using chemical reaction. Thus, it can adjust the heat transport condition and amount by controlling conversion ratio and reaction conditions such as reactant composition, temperature and total pressure by following the Le Chatelier's principle.

For this reasons, the chemical heat transformer has been investigated in this research. The target reversible reaction for the representative chemical heat transformer was selected as methane reforming reaction and methanation reaction for this chemical heat transformer because these reactions occur at high temperature range (about 600-800 degree C). The methane reforming reaction and the reverse reaction was examined as the specific chemical heat transformer. At this point, conversion ratio near the turning temperature of the methane steam reforming reaction which is an endothermic reaction of chemical heat transformer was measured by this experiment. Also, it is drawn the optimum reaction conditions by considering efficiency of chemical heat transformer including the reverse reaction from the point of exergy. From these experiment and calculation results, this system has a great potential to contribute the industrial energy saving and to realize the coproduction in chemical industry.

359467 Gelatin Microsphere-Collagen Hydrogel Composite Scaffolds for Tissue Engineering with Human Adipose Derived Stem Cells

Abstract:

Microsphere scaffolds provide a unique three dimensional micro-environment for the cells cultured on their surfaces through their curvature. Many studies have shown previously that the topographical cues like curvature have an effect on the differentiation abilities of the stem cells. Similarly, studies have also shown that mechanical cues like matrix stiffness and various other biomolecular signals can play a crucial role in determining the stem cell fate. In this study, we designed a three dimensional composite scaffold by encapsulating gelatin microspheres in collagen hydrogels, in order to present a combination of topographical, mechanical and biomolecular signals to human adipose derived stem cells (ADSCs) and studied their osteogenic differentiation abilities. Topographical cues are provided in the form of microsphere curvature by culturing ADSCs on gelatin microspheres to form cell-microsphere aggregates. Further, varying amounts of these cell-microsphere aggregates are then encapsulated into collagen hydrogels to fabricate microsphere-hydrogel composite scaffolds with varying mechanical properties. The mechanical properties of such scaffolds were studied using rheometry. ADSCs encapsulated in such composite scaffolds with varying mechanical properties were induced towards osteogenic lineage and further characterized using gene expression studies of osteogenic marker genes and by measuring the alkaline phosphatase activity. We found that encapsulating increased amounts of microspheres in collagen hydrogels increases the storage modulus of the gels and favors the osteogenic differentiation of ADSCs. To further accentuate the osteogenic differentiation of ADSCs, we then provided the biomolecular cues by encapsulating basic fibroblast growth factor (bFGF) into the scaffolds and releasing it in a controlled manner. Gene expression of osteogenic marker genes and alkaline phosphatase activity of ADSCs upon differentiation in the bFGF encapsulated scaffolds seem to be enhanced compared to the scaffolds with bFGF supplementation directly in the media. Overall, this study shows that, osteogenic differentiation of ADSCs can be enhanced by culturing them in microsphere – hydrogel composite scaffolds which can subsequently be used as effective injectable delivery vehicles for ADSCs as well as various biomolecules.

356041 Ferrofluid Flow in a Channel Induced By a Rotating Magnetic Field

Abstract:

Ferrofluids are a clear example of structured fluids; whose rheological behavior is influenced by forces of magnetic polarization. Additionally, these fluids are characterized by an antysimmetric stress tensor when they are under the action of a rotating magnetic field. This work has been motivated by the experimental evidence of the surface flow of a ferrofluid driven by a rotating magnetic field in a circular channel of square cross section. In this work a numerical solution for this particular geometry is obtained, which considers the effect of diffusion of the internal angular momentum of the ferrofluid particles; which is described by the tensor "couple stresses" and quantified by a phenomenological coefficient "spin viscosity". In order to obtain a numerical solution, this study considered the interfacial balances for linear and internal angular momentum obtained for Chaves and Rinaldi as boundary conditions at interface ferrofluid-air. Additionally, a numerical solution for the ferrofluid flow in a duct with a square cross section is also obtained. The numerical solutions obtained showed that the ferrofluid flow in a duct is generated only in case that the existence of the internal angular momentum diffusion is assumed ("spin viscosity"). As for the results of the channel, in the event of "spin viscosity" zero, it is possible to generate ferrofluid flow of superficial character because of discontinuity of "couple stresses" and antisymmetric stresses through the ferrofluid-air interface. Moreover, it was found that the channel flow magnitude is greater when the parameter of flow "spin viscosity" was at nonzero: this can be explained because of the combined action between the mechanism of flow volumetric and surface. Finally, for the same conditions of intensity and frequency of the applied magnetic field, the magnitude of the flow in the channel was found to be higher than the magnitude of the flow in the duct.

Development of a microstructure based model for thixotropy in colloidal dispersions based on population balances

Paul M. Mwasame, Antony N. Beris and Norman J. Wagner

Department of Chemical and Biomolecular Engineering, University of Delaware, Newark, DE 19716

The overall goal of this work is to present a theoretical framework linking fundamental processes in a particle population balance modeling approach to macroscopic suspension rheology. A coarse-grained population balance equation is developed to describe the evolution of microstructure in a colloidal dispersion under shear flow. This provides an alternative derivation of the most general form of the structure kinetics model [1]. The derivation shows that prior models based on intuitive arguments are not formally correct. In particular, the assumption in structure kinetic models that breakage processes only contribute to destruction of structure is not admissible in the population balance approach. The derivation here reveals that the breakage contribution inherently involves fluxes of particles into and out of the various size classes. Therefore break-up must contribute to both structure build up and breakdown simultaneously. Such insight is not readily apparent without an underlying microscopic picture that is provided by the population balance equation in its moment form. For the first time, a microscopically based evolution equation for a scalar structural parameter with close connection to existing empirical structural models is described by applying population balances. This provides some justification for functional form of thixotropic models. The new model developed here is coupled to a constitutive equation for stress and provides a reasonable representation of rheological data [2] when compared to current models in use. Similarities and differences to existing models will be discussed.

The second key result from this work is obtained by the direct identification of the structure kinetic parameter as a physically meaningful parameter related to a linear dimension of the aggregate relative to that of a primary particle. This identification provides a unique opportunity to link independent microstructure measurements from scattering techniques to the rheology of dispersions under flow. The physical identification of the structure parameter provides a crucial link to enable a deeper mechanistic interpretation and analysis of unstable dispersions under flow through simulation of the particle size distribution. Depending on the sophistication of the rate kernels used in the population balance equation, the parameters that emerge in the resulting structure evolution equation have a clear physical meaning and can be related to the characteristics of the system such as fractal dimension and stability ratio among others. In this way, the microstructure based model developed here may further elucidate on key control variables in thixotropic systems and motivate the development of more finely tuned particulate processes. In summary, we show that a simple microscopic picture of aggregation and breakage reveals a deficiency in the current structure kinetic models. Subsequently, the physical identification of the structure parameter emerging from this analysis provides a more physical basis for understanding thixotropic effects in colloidal dispersions.

Acknowledgements: Funding source NSF grant, CBET-1235863

References

[1] C.F. Goodeve, A general theory of thixotropy and viscosity, Transactions of the Faraday Society, 35 (1939) 342-358.

[2] K. Dullaert, J. Mewis, A structural kinetics model for thixotropy, Journal of Non-Newtonian Fluid Mechanics, 139 (2006) 21-30.

14AIChE Annual Meeting, Atlanta, GA

Improving the Sustainability of Antioxidants Production through Biorefinery concept. The Blackberry case

A. A. González, L. V. Daza, C. A. Cardona
[1] Universidad Nacional de Colombia. Instituto de Biotecnología y Agroindustria
Email: aagonzalezvil@unal.edu.co , lvdazas@unal.edu.co , ccardonaal@unal.edu.co

ABSTRACT

The blackberry fruit is a source of different compounds, some of them as anthocyanins and ellagitannins exhibit antioxidant properties that can be extracted with organic solvents. The aim of this study is to present an approach for improving the sustainability of phenolic compounds extraction from blackberry residues through a biorefinery concept in which the phenolic compounds extraction is integrated to the production of bioproducts, biofuels and bioenergy. The Blackberry residues were experimentally characterized. Then the phenolic compounds were extracted at 45°C and 300 bar and the phenolic content was evaluated using the Folin-ciocalteu. The biorefinery from blackberry residues was evaluated using the commercial software Aspen Plus V8.0 (ASPEN TECHNOLOGY USA). In the first scenario the phenolic compounds extraction by supercritical fluids was evaluated. A second scenario where the phenolic compounds extraction by supercritical fluid was integrated to the production of ethanol, lactic acid, xylitol and energy was assessed as well. The economic evaluation was carried out using the commercial software Aspen Process Ecomomic Analyzer V8.0 (ASPEN TECHNOLOGY USA) according to the Colombian. The environmental evaluation was made through the methodology proposed by the Environmental Protection Agency (EPA) Waste Algorithm Reduction (WAR). The obtained results lead to conclude that the biorefinery approach can improve the full utilization of the blackberry residues with economic and environmental competitive advantages.

INTRODUCTION

Castilla blackberry (*Rubbus glaucus benth*) is a fruit from the berry fruits family mainly cultivated in the Colombian with average production yields of 11 ton/ha. The ripe fruit is consumed in fresh and proceed in different forms; it is used for agroindustrial transformation such as juices, jams and cosmetics among other products.

The blackberry fruit is a source of fiber, vitamins, phytosterols, sulphur compounds and carotenoids among others that exhibit antioxidant properties. Different extraction technologies have been employed to obtain antioxidants, especially extraction with organic solvents and supercritical fluid extraction are the widely studied. The operation conditions for solvent extraction are commonly unappropriated because could lead to thermal degradation of the phenolic compounds and it is necessary an additional separation and solvent recovering stage. Besides, some of these solvents have restricted in food, cosmetic and other industries.

Supercritical Fluid Extraction (SFE) technology enhances the extraction yields and could be developed in less time than organic solvent extractions and reduce the recovering and separation costs. The obtained extracts are solvent free which are appropriated in food and cosmetics uses. Usually after SFE of the phenolic compounds the solid matrix is discarded with a lignocellulosic material which could be fully utilized.

METHODOLOGY

Blackberry experimental characterization → Biorefinery desing according potential blackberry products → Techical Evaluation using Aspen Plus V 8.0 software →

Economic evaluation using Aspen Process Economic Analyzer → Environmental evaluation using WAR-GUI software

Figure 2. Methodology steps

RESULTS

Table 1. Blackberry waste chemical composition

Image	Feature	(%)
	Moisture	6,23
	Extractives	12,54
	Hemicellulose	19,14
	Cellulose	40,33
	Lignin	20,21
	Ash	1,54

BIOREFINERY

THE GRAPHIC ON THIS PAGE WAS NOT AVAILABLE ON THE E-MEDIA VERSION OF THIS PAPER.

Figure 3. Blackberry biorefinery flowsheet

ECONOMIC ANALYSIS

Table 2. Mass balance of the proposed scenarios

FLOW	SCENARIO 1	SCENARIO 2
Blackberry (kg/day)	5000	5000
Phenolic compounds (kg/day)	501,32	501,32
Xylitol (kg/day)	-	109,76
Lactic acid (kg/day)	-	804,97
Ethanol (L/day)	-	242,46
Biogas (L/day)	-	276,95
Biofertilizer (kg/day)	-	700,99

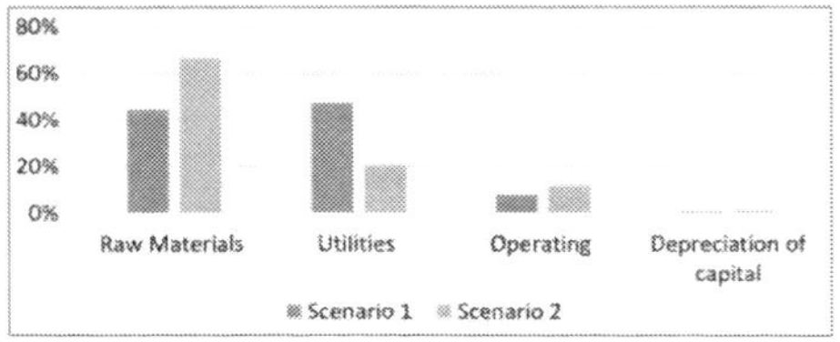

Figure 4. Annual cost distribution

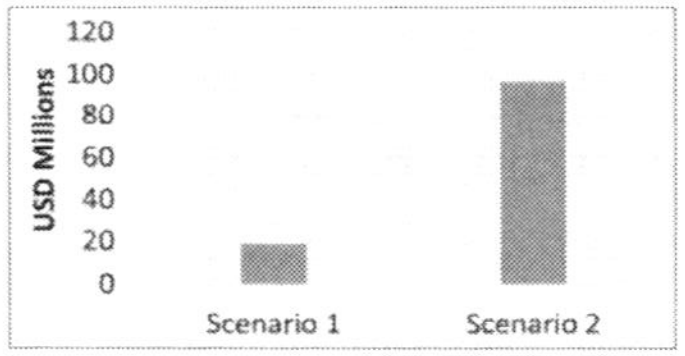

Figure 5. Annual cost

Table 3. Cost production per scenario

SCENARIO	Xylitol	Phen. Compounds	Lactic acid	Ethanol	Biofertilizer	Biogas
	USD/kg	USD/kg	USD/kg	USD/L	USD/kg	USD/L
1	-	18,93	-	-	-	-
2	14,9	17,8	6,30	21,3	9,71	24,6

ENVIRONMENTAL ANALYSIS

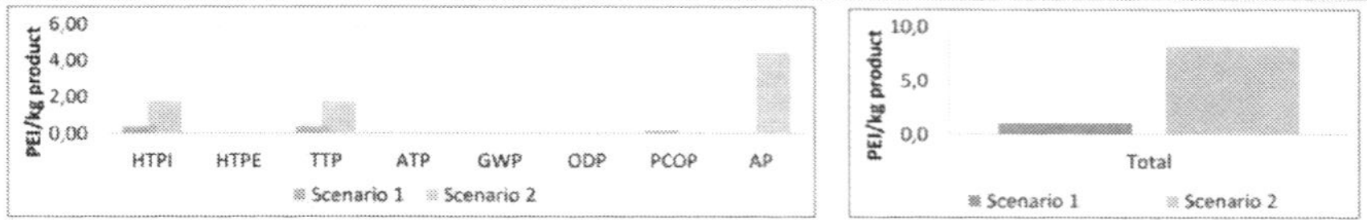

Figure 5. Total environmental impact

CONCLUSION

Blackberry's waste has a potential use to extract valuable products as the phenolic compounds with antioxidant activity. When its use is evaluated under biorefinery concept, this shows the potential for obtaining products as xylitol, ethanol, lactic acid and biofertilizer. It permit to reduce the process waste. However it is necessary to do more study in order to improve the margin economic profile.

REFERENCES

I.X. Cerón, J.C. Higuita, C.A. Cardona. *Design and analysis of antioxidant compounds from Andes Berry fruits (Rubus glaucus Benth) using an enhanced-fluidity liquid extraction process with CO2 and etanol*. J. of Supercritical Fluids 62 (2012) 96-108.
Ó. Acosta-Montoya, F. Vaillant, S. Cozzano, C. Mertz, A. M. Pérez, and M. V. Castro, "Phenolic content and antioxidant capacity of tropical highland blackberry (Rubus adenotrichus Schltdl.) during three edible maturity stages," *Food Chem.*, vol. 119, no. 4, pp. 1497–1501, Apr. 2010.
L. Laroze, C. Soto, and M. E. Zúñiga, "Phenolic antioxidants extraction from raspberry wastes assisted by-enzymes," *Electron. J. Biotechnol.*, vol. 13, no. 6, Nov. 2010.

WebFF: A Smart Force-Field Repository for Polymers and Soft Materials*

Frederick R. Phelan Jr.[†,1] and Huai Sun[2]
[1]Materials Science and Engineering Division, NIST, Gaithersburg, MD 20899
[2]Aeon Technology Inc. and School of Chemistry and Chemical Engineering, Shanghai Jiao Tong University, Shanghai, China

Abstract

The vision of the Materials Genome Initiative (MGI)[1,2] is a new R&D paradigm enabling accelerated materials design via computation leading to a shortening of the time and cost needed to bring new materials to market. This new paradigm will be fostered through the creation of national infrastructure for materials data sharing and analysis that supports integrated multiscale modeling of materials[3] from the quantum or molecular level, upward in both length and time scale to the continuum scale represented by structural components. The NIST role in the MGI is to create the infrastructure for the effort through the development of 1) standards and tools for the representation and interoperability of materials data, and 2) techniques and standards for the interoperation of modeling systems at multiple length and time scales.

In a hierarchical approach to materials modeling, molecular dynamics (MD) calculations are integral to calculating properties and phenomena at length scales ranging from Angstroms to tens of nanometers, on time scales up to tens of microseconds. In addition, these models provide the foundational level upon which progressive, multi-scale models are built. A critical data need required to support MD simulations for materials research are widely available, highly accurate, and validated force field data. In this talk, we will describe WebFF, an open and extensible, smart force field repository based on Amber and CFF type molecular potentials designed to support the MGI for polymers and closely related soft materials. The repository consists of three main components: 1) a web-client interface enabling wide and open access; 2) a software engine to provide robust workflow functionality; and 3) an extensible, multi-table database for the purpose of encouraging and supporting validated data sharing. The repository functions in two main modes of access. The workflow mode supports input of chemical structures, automated force field assignment, and output to support MD calculations. The data sharing mode supports input of new force-field data to the extensible database model to support validated data sharing. The seminar will demonstrate the workflow mode as a means of facilitating the use of accurate force-field data in MD simulations for a number of open source simulation engines. The methods, requirements and goals of force field data sharing for the repository will also be described.

References

1. "Materials Genome Initiative for Global Competitiveness," Office of Science and Technology Policy, Executive Office of the President, Washington, DC, (2012).

† Corresponding author: NIST, 100 Bureau Dr., STOP 8542, Gaithersburg, MD 20899; (301) 975-6761; Fax: (301) 975-4924; frederick.phelan@nist.gov

2. James A. Warren, Boisvert, Ronald F., "Building the Materials Innovation Infrastructure: Data and Standards," NISTIR 7898, National Institute of Standards and Technology, Gaithersburg, MD, (November 2012).
3. "Integrated Computational Materials Engineering," National Research Council, The National Academies Press, Washington, DC, (2008).

Low Cost High Temperature UV-Fused Silica Heated Spectroscopic Cell for Solar Applications

Moisés Romero, Dennis Thomey*, Martin Roeb, Christian Sattler[1]

* Tel.: +49 2203 601-2936, Fax: +49 2203 601-4141, E-Mail: dennis.thomey@dlr.de
[1]Deutsches Zentrum für Luft- und Raumfahrt e.V., Köln, Germany.

Introduction

The European research project SOL2HY2 (Solar to Hydrogen Hybrid Cycles) investigates the Hybrid Sulphur Cycle (HyS) for solar hydrogen production. This process - also known as the Westinghouse Process or "Ispra Mark 11" - is considered one of the most promising approaches to generate hydrogen by the use of thermal energy. Hydrogen, as a potential substitute for fossil fuels, is free of CO_2 emissions if generated from water using renewable energy sources. Conventional water electrolysis, as the state-of-the-art technology, has a low overall efficiency as it demands the generation of electricity. On the other hand, direct thermal splitting of water would require temperatures in excess of 2000°C in order to yield sufficient amounts of hydrogen. Therefore, thermochemical cycles are introduced to lower the process temperature to a technical manageable level of about 1000 °C. The project consortium is led by the Italian scientific consultant company EnginSoft with participation of the DLR Institute of Solar Research, the Finnish university Aalto, the Italian research center ENEA, the Finnish technology company Outotec, the Swiss ceramics company Erbicol and the Finnish gas company Woikoski. The HyS presented in the figure below is a two-step cycle producing hydrogen and oxygen out of water using solar energy [1].

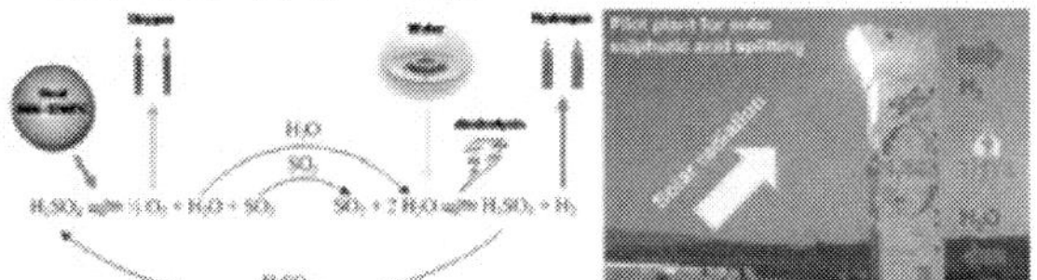

Hybrid-Sulphur-Cycle (HyS) and the proposed Set-up in the Jülich Solar Tower

Technology Challenges

In the first reaction, sulphuric acid is decomposed at high temperature forming sulphur dioxide and oxygen; the latter being separated from the product gas as a by-product. Sulphur dioxide is electrolysed at about 80 °C together with water in the second reaction generating hydrogen and fresh sulphuric acid, which is recycled back to the first reaction. The electrical power required for this electrolysis is only about a fifth of that needed for conventional water electrolysis so that the energy demand for hydrogen production is significantly reduced. This is crucial for an industrial realization of the hydrogen technology [2].

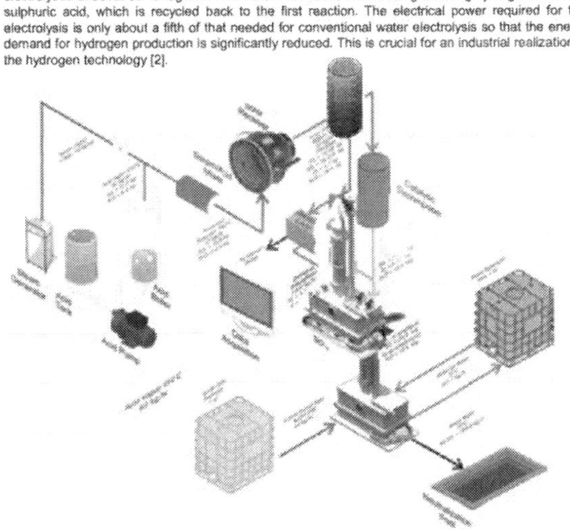

Simplified Diagram of the SOL2HY2 Demo Plant (1 l/min H_2SO_4 >1000°C)

In a **closed volumetric receiver** developed for operation at the Solar Tower in Jülich, evaporated sulphuric acid – consisting of SO_3 and inert H_2O – is superheated to approximately 1000°C. The hot gas will then pass an **adiabatic, catalytically activated reactor** in order to reduce the gases to SO_2 and O_2. The chemical analysis at high temperatures is rather important, as the chemical characterization of the system will determine directly the reaction yield, and therefore the thermodynamic performance of the receiver.

UV-Vis Spectroscopic Cell

A high temperature, low-cost spectroscopic cell that can operate at moderate pressures has been designed for solar applications, in particular to acquire online chemical composition in Solar Receivers that include corrosive media, for instance SO_3, SO_2, O_2 and H_2O. The obtainment of chemical composition is a challenge at high temperatures, which is a common problem for solar processes. Of further complication is the corrosive nature of species within certain process conditions, in this case, corrosive sulphuric acid and sulphur trioxide vapours. To acquire accurate information regarding the yield of a solar receiver reactor that decomposes SO_3 into SO_2 and O_2, a one-phase gas flow cell has been designed. The key aspect of this spectroscopic cell is its cost in comparison to commercial alternatives, its mature technology and ease of operation. It is also resistant to chemical attack at high temperatures if coated with ceramics, such as Al_2O_3 or ZrO_2.

Simulations were performed to determine the transient heat transfer behaviour and the thermal equilibrium properties using two Horst cartridge heaters of 250 W each, achieving thermal equilibrium in approximately 500 seconds. The heaters are 6.5 mm in diameter, and are sufficiently powerful to transfer the necessary heat uniformity throughout the cell. Knowing the extinction coefficient (taken to be 200 l/mol.cm), and taking absorptivity to be ideal between the ranges of 0.1 to 1, a pathlength of 13 mm was chosen to be the ideal for SO_2 concentrations of 25 – 100% at 290 nm. The low cost of the setup has been achieved using standard stock materials, DN40 CF Flanges, a relatively simple in-house design and a robust sealing strategy has been devised for temperatures up to 330°C and pressures up to 5 atm using UV-fused silica windows 2mm diameter and a combination of perfluorinated sealing materials and graphite.

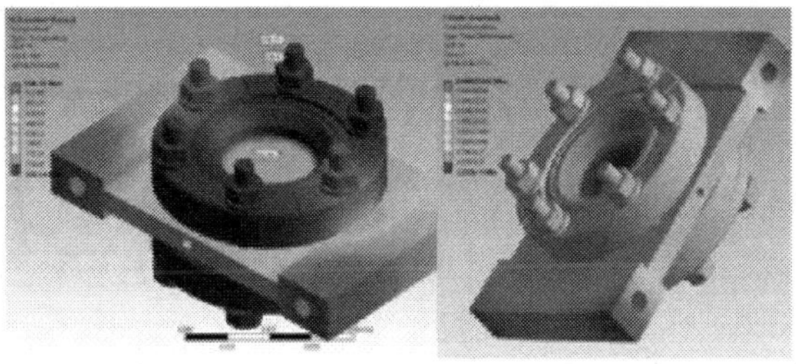

Thermal Transient Simulation and Stress behaviour of In-house UV Cell

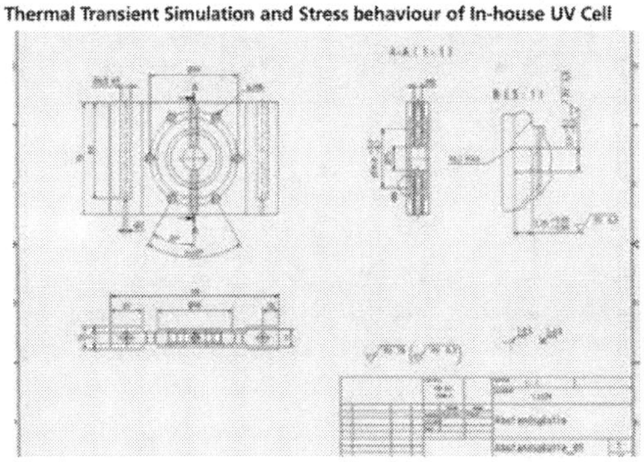

Mechanical Design of Cell, including sections and finishing additions

Outlook

A UV-Vis Spectroscopic cell that operates at high temperature, avoids damage originated from sulphur species corrosion and has an overall low cost in comparison to commercial alternatives has been engineered. The cell was designed to operate with one gaseous phase, therefore mechanical and thermal properties were simulated and tested at 400°C to avoid condensation, including convective cooling from the gaseous analysis. Experimental tests showed comparable s/n ratio to commercial alternatives from major suppliers.

[1] Roeb, M., et al. "Sulphur based thermochemical cycles: Development and assessment of key components of the process." IJHE, 38.14 (2013): 6197-6204.
[2] Wong, B., Thomey, D., Brown, L., Roeb, M., Buckingham, R., & Sattler, C. (2013, July). Sulfur Based Thermochemical Energy Storage for Concentrated Solar Power. In ASME 2013 7th International Conference on Energy Sustainability

The authors thank the European Commission for co-funding the project SOL2HY2 in the 7th framework program (Contract. No.325320)

Knowledge for Tomorrow

Influence of the Selection of Non-conventional Operations on Biorefineries Performance

Daza Laura; Duque Sergio; Gomez Alvaro; Cardona Carlos A.
Research Group in Chemical, Catalytic and Biotechnological Processes
Universidad Nacional de Colombia, Manizales., Colombia
Corresponding author: ccardonaal@unal.edu.co

14AIChE Annual Meeting, Atlanta, GA

INTRODUCTION

The Plantain Empty Fruit Bunches (PEFB) is a lignocellulosic residue generated during the plantain harvesting practices, also known as the plantain flower is classified as an agricultural primary residue due to the nature of the source that is directly related with the seeding, development and harvesting of crops. Lignocellulosic biomass are organized in a compacted structure in which lignin and hemicellulose form a complex through covalent cross-linked bounds encapsulating and giving support to the cellulose. According to the FAO, Colombia is the second producer of plantain worldwide. The 87% of the total amount harvested is produced under an associated scheme, with other crops such as coffee, cacao and cassava, among others. Mainly by small producers with small lands (less than 5 hectares Other type of pretreatments have been explored as an alternative to reduce the operational conditions meaning severity of reactions, residence time, equipment deterioration and its associated costs as well as the environmental effects as formation of inhibitory compounds. Ultrasonic technology is based in acoustic radiation phenomena in which mechanical energy form is transformed to chemical energy generating chemical, physical and biochemical energy forms creating a hydrodynamic shear stress is generated inducing changes in lignocellulosic material structure. Some authors had proposed the use of supercritical carbon dioxide as pretreatment stage for biomass utilization to release and/or removal the lignin and/or hemicellulose through weak acid catalyst formation by reaction with water . In this work were evaluated the production of sugars and inhibitory compounds released using supercritical and ultrasound assisted pretreatments against the base case related to conventional pretreatment.

METHODOLOGY

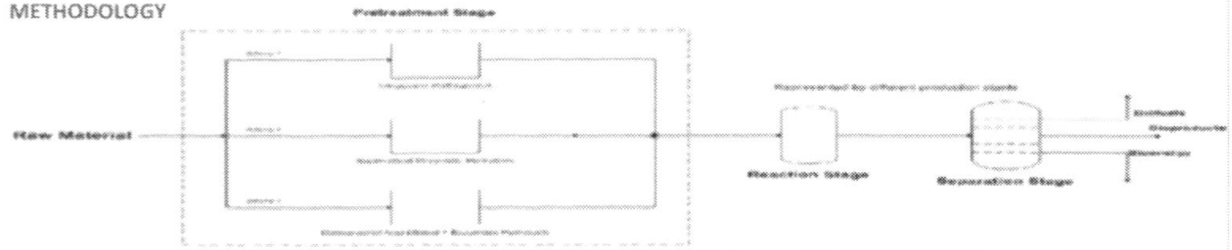

1. PEFB Characterization → 2. Experimental Pretreatment and Hydrolysis. Conventional, Supercritical and Ultrasound Assisted → 2. Conceptual Design Stage (Hierarchy and Sequentiation) → 3. Techno-economical evaluation → 4. Case Comparison

FORMULATED CASES

First Case	Second Case	Third Case
Ultrasonic Assisted	Supercritical Assisted	Conventional
Temperature:70°C	Temperature: 45°C	Temperature:121°C
Sulfuric Diluted Acid 2% v/v	Moisture Content 70%	Sulfuric Diluted Acid 2% v/v
Total Pretreatment time 1 hour	Total Pretreatment time 1 hour	Total Pretreatment time 1 hour

RESULTS

Physicochemical Characterization- Experimental Results D.B

Compound	Composition (%)
Moisture	11.9
Cellulose	14.3
Hemicellulose	3.8
Lignin	23.1
Extractives	45.0
Ash	1.9

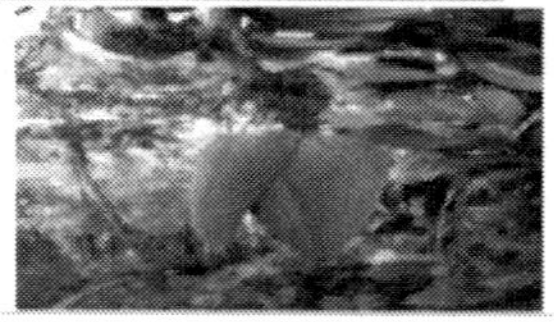

RESULTS-SUGAR CONTENT

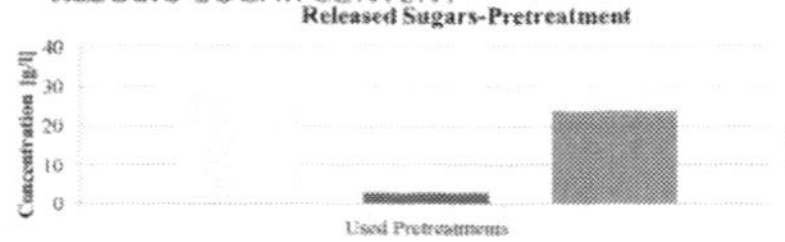

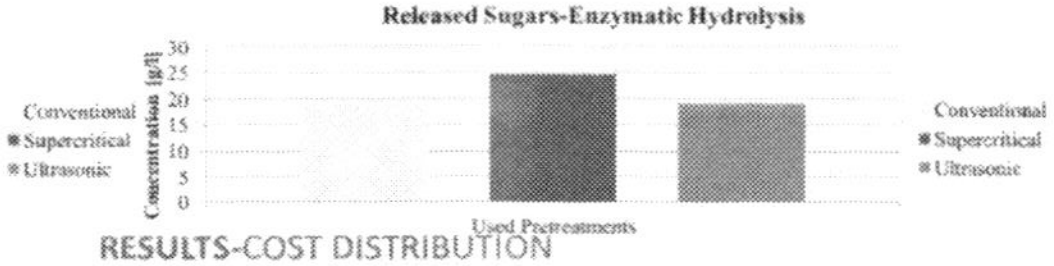

RESULTS- INHIBITORY COMPOUNDS CONTENT

RESULTS-ECONOMIC EVALUATION

Production Cost [USD/kg]	Case 1. Ultrasonic	Case 2. Supercritical	Case 3. Conventional
Ethanol	7.0	4.3	8.1
Xylitol	4.1	6.0	1.35

RESULTS-COST DISTRIBUTION

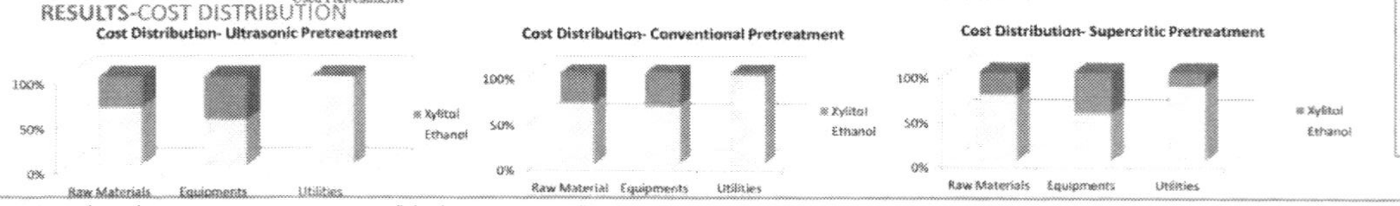

ACKNOWLEDGEMENTS

Universidad Nacional de Colombia at Manizales.
Biotechnology and Agribussiness Institute
COLCIENCIAS- Convocatoria 617 Jóvenes Investigadores- for the economical support to realize this research work.

CONCLUSIONS

The high production costs compared to the market prices in each evaluated case was a consequence of the low content of hemicellulose and cellulose present in PEFB. Despite the low yields of production the use of non-conventional pretreatments can enhance the conversion of lignocellulosic biomass into fermentable sugars reducing the produced amounts of inhibitory compounds, it can be useful for those highly sensitive fermentations due to the low operational conditions compared with the base case. The use of supercritical assisted pretreatments reduce the crystallinity of lignocellulosic material. This pretreatment involves a kind of weak acid hydrolysis due to the reaction between carbon dioxide and water present in the sample. increasing the produced sugars through enzymatic hydrolysis.

REFERENCES

FAO STAT 2011. Available from: http://faostat.fao.org/site/636/default.aspx#ancor (Accessed 2 August 2013).
MAZZEO MENESES, M., LEÓN AGATÓN, L., MEJÍA GUTIÉRREZ, L. F., GUERRERO MENDIETA, L. E. & BOTERO LÓPEZ, J. D. 2010. Aprovechamiento Industrial de Residuos de Cosecha y Poscosecha del Plátano en el departamento de Caldas. *Revista Educación En Ingeniería*, 9, 128-139.
RINCÓN, L. E., BECERRA, L. A., MONCADA, J. & CARDONA, C. A. 2013a. Techno-Economic Analysis of the Use of Fired Cogeneration Systems Based on Sugar Cane Bagasse in South Eastern and Mid-Western Regions of Mexico. *Waste and Biomass Valorization*.
RINCÓN, L. E., MONCADA, J. & CARDONA, C. A. 2013b. Analysis of cogeneration as a tool to improve the viability of oilseed based biorefineries. *Catalytic systems for integral transformations of oil plants through biorefinery concept" Catalytic Systems For Integral Transformations Of Oil Plants Through Biorefinery Concept*. Universidad Nacional de Colombia.

390412 Techno-Economic Analysis of the Antioxidants Production from Fruits Under a Biorefinery Concept

Abstract:

Colombia is a country with a high potential in the production of tropical fruits. The marketing of this kind of products has growing because of the interest on the natural products by the consumers (Moncada, Tamayo, & Cardona, 2014).This fact has led to the increment of the demand of fruits and vegetable species.Among them, the components with high antioxidant capacity, since they delay or prevent different deceases, have gained special interest (Moncada et al., 2014).

The antioxidants are compounds recognized by its ability to delay the oxidation of the molecules present in the cellular structure of the microorganisms.These compounds can be found most of the times in small quantities in natural matrix.Different types of molecules are known by their antioxidant capacities such as the lycopene, the anthocyanins and vitamin C and vitamin E, among others. They can delay and prevent different disease as Alzheimer, cardiovascular and neurological disease, cancer and others. (Anila & Vijayalakshmi, 2003)

The *Theobromagrandiflorum*(copoazu)is an Amazonian fruit used usually to produce juices, desserts and jams pulps. The seeds and skin are residues with no currently use in other different activities. The copoazu pulp is rich in phosphor, pectin and vitamin C. A high amount of lignin is found in the skin and seeds of thecopoazu, other compounds as lipids can be found in the seeds.(HERNÁNDEZ & BARRERA, 2004)

The aim of this work is to propose a biorefinery to obtain antioxidants from Copoazu (*Theobromagrandiflorum*) in order to fully use the raw material. The biorefinery concept consistson obtaining different products from a natural source as a raw material.The raw material was characterized and the supercritical extraction of the antioxidants compounds was experimentally developed at 50-60°C and 300 bar. The extracts were characterized with the FollinCiocalteu and DPPH methods(Brand-Williams, Cuvelier, & Berset, 1995; Cerón, Higuita, & Cardona, 2012)

Furthermore, the biorefinery was designed and evaluated from the techno economic and environmental points of view, using the software Aspen plus v8.0 and the WAR methodology.

The economic and environmental analysis shows that better results are obtained when the production schemes are integrated to the additional products. The best production is achieved when was made an energy and mass integration.

375828 Processing Tissue Engineering Matrix Materials with Supercritical CO_2

Abstract:

Summary

This paper addresses a novel process for preparing tissue engineering scaffolds with the aid of supercritical carbon dioxide (CO_2). Prior research has shown that CO_2 may be useful for sterilizing, removing bioburden, and infusing small molecules, for instance. Specifically, this study addresses how to preserve the volatile content during treatment with CO_2.

Background

According to the Department of Health and Human Services, nearly 7,000 people in the United States die each year while waiting for an organ transplant, and this number is likely to continue to grow. One potential way to address this problem is through the use of artificial organs. An exciting and relatively novel approach to creating artificial organs is the field of tissue engineering. Tissue engineering involves seeding cells onto a biocompatible three-dimensional tissue scaffold and then implanting the tissue scaffold into the body.

There are a number of challenges that have prevented the use of tissue scaffolds in this fashion. The first challenge is the source of the scaffold. Two main approaches are possible: producing scaffolds from synthetic biomaterials, or using natural organs or tissues. Using natural materials is presumed to be advantageous because natural materials should significantly decrease the possibility of bodily rejection and adverse immune response often observed with synthetic biomaterials. Natural organs or tissues contain foreign cellular material, which must be removed prior to seeding the scaffold. The process of removing this material is called decellularization. The objective of any decellularization method is twofold: (1) the removal of all cellular material, and (2) the preservation of the physical and biochemical properties of the extracellular matrix (ECM). Meeting both of these criteria is of utmost importance for scaffold viability [1].

Decellularization is currently accomplished by contacting xenographic tissue with a combination of chemical detergents and biological agents. The main challenge regarding decellularization is finding a balance between preserving the properties of the matrix while effectively removing undesired components. One novel decellularization method involves using a supercritical fluid as a solvent for applying decellularization agents [2]. At temperatures and pressures above the critical point, a substance exists in one homogeneous phase called a supercritical fluid (SCF). SCFs are potentially useful as solvents because they have desirable transport properties and liquid-like densities that are much greater than those of most gases. This combination of properties allows them to penetrate through surfaces easily without damaging them. Upon depressurization, the SCF outgases, leaving no residue in the tissue.

Supercritical carbon dioxide ($scCO_2$) is particularly promising for biomedical applications. It is nontoxic, nonflammable, and inexpensive. It has mild critical conditions of 31.1°C and 7.38 MPa, so processing biological materials can take place at or near body temperature (37°C). Supercritical CO_2 has already been demonstrated other applications, including pasteurization [3], extraction of biological compounds [4], and sterilization, the latter of which our group has published extensively [5-7].

There is currently only one report on using supercritical CO_2 in decellularization: Sawada *et al.* used $scCO_2$ with an ethanol entrainer to decellularize porcine aortas [2]. Sawada reported 100% removal of DNA and 80-90% removal of phospholipids at relatively mild pressures and temperatures. However, considerable tissue dehydration caused by SCF extraction was reported, causing embrittlement of the matrix and a subsequent reduction in mechanical properties. This reported extraction of volatiles during CO_2treatment is not surprising. In fact, this phenomenon is well-known and is the basis of critical point drying, a process commonly used in tissue engineering and other applications. In this case, however, it is desired to prevent drying from occurring.

Hypothesis, Methods, and Results

We hypothesize that the tissue dehydration can be significantly reduced or even eliminated by presaturating $scCO_2$ with water and other volatiles prior to contacting the tissue. If the maximum amount of each volatile substance were dissolved in $scCO_2$to the solubility limit prior to making contact with the tissue, this should prevent volatiles from being extracted.

The objectives of this study were as follows: (1) to construct an apparatus that can presaturate supercritical CO_2 with water and determine the range of CO_2 flow rates at which equilibrium-level presaturation can be attained, (2) to compare the amount of water extracted from model hydrogels and porcine aorta tissue using dry and presaturated CO_2, and (3) to determine if any components of porcine aorta are extracted during CO_2treatment.

Experiments were conducted with an ECM model poly(acrylic *acid-co*-acrylamide) potassium salt, a hydrogel) and with porcine aorta. Both materials were tested by contacting specimens with dry CO_2 and with CO_2 that had been presaturated with water. The conditions used were a pressure of 2000 psi (13.79 MPa), a temperature of 37°C (and also 50°C for hydrogels only), a depressurization rate of 50 psi/min (0.345 MPa/min). Treatment ratios of 30 min/0.1 g gel and 60 min/0.25 g tissue were used so that each specimen would be exposed to the same amount of CO_2.

Presaturation maintained almost all of the volatiles in the matrix (about 99% for hydrogels at both temperatures and 97% for tissue), but both materials lost a significant amount of volatiles (50% and 78% weight retention, respectively) when treated with dry CO_2. These findings confirm the hypothesis and overcome one of the significant barriers to consideration of $scCO_2$ for decellularization of ECM materials. They also confirm the presence of volatiles other than water in the porcine aorta tissue and indicate that further experimentation may be necessary to prevent the extraction of those volatiles during CO_2treatment.

References

1. Keane TJ, Londono R, Turner NJ, Badylak SF: **Consequences of ineffective decellularization of biologic scaffolds on the host response**. *Biomaterials* 2012, **33**(6):1771-1781.

2. Sawada K, Terada D, Yamaoka T, Kitamura S, Fujisato T: **Cell removal with supercritical carbon dioxide for acellular artificial tissue**. *J Chem Technol Biotechnol* 2008, **83**(6):943-949.

3. Spilimbergo S, Ciola L: **Supercritical CO2 and N2O pasteurisation of peach and kiwi juice**. *International Journal of Food Science and Technology* 2010, **45**(8):1619-1625.

4. Bagheri H, Manap MYB, Solati Z: **Response surface methodology applied to supercritical carbon dioxide extraction of Piper nigrum L. essential oil.** *LWT-Food Sci Technol* 2014, **57**(1):149-155.

5. Zhang J, Davis TA, Matthews MA, Drews MJ, LaBerge M, An YHH: **Sterilization using high-pressure carbon dioxide**. *J Supercrit Fluids* 2006, **38**(3):354-372.

6. Tarafa PJ, Jimenez A, Zhang JA, Matthews MA: **Compressed carbon dioxide (CO2) for decontamination of biomaterials and tissue scaffolds**. *J Supercrit Fluids* 2010, **53**(1-3):192-199.

7. Jimenez A, Zhang J, Matthews MA: **Evaluation Of CO2-Based Cold Sterilization of a Model Hydrogel**. *Biotechnol Bioeng* 2008, **101**(6):1344-1352.

381142 Vibrational Spectroscopy Investigation of the Giant Surface Potential of Organic Semiconductors

Abstract:

A phenomenon known as the giant surface potential (GSP), where the surface potential of organic films display linear growth with increasing film thicknesses in the absence of light wasfirst reported by Ito et al (JAP 2002) on (8 hydroxyquinoline)aluminum(Alq_3), a prototypical fluorescent material used in OLEDs. It has been shown that the surface potential of Alq_3 has reached 28 V for a 560 nm thick film by Kelvin probe measurements in vacuum in the absence of light. Since then this phenomenon has been observed for a broad range of molecules thermally evaporated on varying substrates under similar conditions. The effect is independent of the substrate, dependent on the film thickness and decays quickly with illumination at the normal mode of the respective molecule and in elevated temperature. The spontaneous buildup of the GSP cannot be explained by any classical interfacial phenomena. Investigations into the cause of GSP, including the analysis of light and heat on the surface potential are not fully understood.

In this study we use vibrational spectroscopy to understand the nature of the GSP buildup, where we have found a significant change in the vibrational structure of the organic material in thick films where the GSP is present as compared to the thin films. The vibrational spectra of the most commonly studied light-emitting material, Alq_3, on indium tin oxide (ITO) is investigated as a function of thickness using high resolution energy electron loss spectroscopy (HREELS), Raman spectroscopy, high resolution x-ray photoelectron spectroscopy (HR-XPS), attenuated total reflectance infrared spectroscopy (ATR-IR), and density functional theory (DFT) calculations. In order to provide a holistic understanding of the GSP, the results are compared to the vibrational spectra of 1,3,5-tris(N-phenylbenzimiazole-2-yl)benzene (TPBi) on ITO, an electron transporter host material with a measured GSP of 0.07 V/nm, and bis(triphenylsilyl)-dibenzofuran (BTDF) on ITO, a typical electron-conducting host used in combination with hole-conducting deep-blue emitter with a measured GSP of 0.08V/nm. The observed spectra show significant changes with the presence of the GSP in the organic material on ITO, which can be explained in terms of different symmetries of the isomers as well as between complexes and isolated anions. Additionally, it's been found that the surface phase differs from the bulk phase, where a structured layer is evident at the interface of the organic semiconductor, and this layer shifts with increasing thickness and in the presence of the GSP. The present work has provided direct evidence that a different molecular orientation exists at the interface than in the bulk, where the GSP exists.

382469 Polymer Structure and Rheology at Fluid-Fluid Interfaces

Abstract:

Polymer structure and rheology at interfaces have great consequences in biology and chemistry, and there are many examples where our current knowledge of these systems has commercial applications in products like paints, cosmetics, foods, and advanced materials. At fluid-fluid interfaces polymers have found commercial applications as stabilizers in cosmetic and food products, and there are areas of biology where polymers at fluid-fluid interfaces play an important role. An ongoing debate amongst those investigating polymer structures at fluid-fluid interfaces is whether or not the polymer adopts an entangled structure similar to that in bulk solutions of concentrated polymer above a critical molecular weight, or whether the polymer adopts a non-entangled structure analogous to bulk solutions of polymer chains below a critical molecular weight. We have investigated this by measuring the interfacial rheological properties of two similar polymers: poly(tert-butyl methacrylate) (PtBMA), and poly(tert-butyl acrylate) (PtBA), each at an air-water interface. The results we will present do not support the entanglement theory, and additionally indicate that the discussion of polymer structure at an air-water interface need not be restricted to a debate over polymer microstructure. For example, we will show that PtBA forms large heterogeneities at all levels of surface coverage, and that its interfacial rheological behavior is surprisingly different than that of the PtBMA that forms relatively uniform layers at all surface coverages.

* authors affiliated with both the Department of Chemical Engineering at KU Leuven, and the Department of Materials Science at ETH Zurich.

382631 Structure Dynamics and Rheology of Silica-Peg Nanocomposites

Abstract:

Polymer composites are typically composed of inorganic fillers dispersed in a polymeric matrix and have applications in coatings, tires, airplane wings and windmill blades. Composites in which fillers have at least one dimension less than 100 nm are known as nano-composites. Shrinking filler size is associated with enhancements in mechanical, electrical, optical and flow properties. The dispersion of these fillers in the polymer is key to optimizing the benefits derived from the small particle size. Particle dispersion is surprisingly difficult and requires enthalpic gains of polymer adsorption to particle surface to balance the increases in entropy associated with how the particle surface alters polymer configuration. At the same time there is evidence that as the particle size shrinks and the strength of attraction between polymer segments and the particle surface diminishes, the mechanical properties of nanocomposites can be dramatically altered.

In this work we study the structure, dynamics and rheology of bare and surface modified silica particles suspended in polyethylene glycol (PEG) melts. We alter polymer particle interactions by reacting the particle surface with isobutyltrimethoxysilane which will replace some of the surface hydroxyl groups with isobutyl groups in increasing amounts and study changes in microstructure, flow and relaxation properties up to high volume fractions. The reaction technique used here is expected to create silane monolayers on the surface and the coverage is expected to increase monotonically with degree of silanization. We have chosen to work here with PEG with a molecular weight of 400 that corresponds to a degree of polymerization of 8. This molecular weight was chosen because of the success of PRISM (polymer reference interaction site model) at predicting numerous structural properties at low molecular weights. Detailed studies indicate that the polymer adsorbs to bare silica surface producing a layer which increases the particle's hydrodynamic diameter by $\sim 2.9Rg$ (or ~ 2.3 nm). Our studies suggest that at the levels investigated, silanization has a small effect on suspension microstructure and flow properties at the low volume fractions where the composite mechanics are essentially those of particles suspended in a Newtonian fluid and that particle's hydrodynamic diameter is not impacted by surface treatment. At elevated volume fractions, where $h < 6Rg$, the impact of surface treatment becomes evident with higher levels of surface treatment giving rise to faster relaxation of density fluctuations and lower composite viscosities. These studies suggest polymer dynamics near the particle surface are sensitive to particle surface chemistry where increasing the degree of surface silanization results in more rapid polymer relaxation rates.

383100 Biomimetic Topographical Replicas of Small Intestine for Regenerative Medicine and Drug Delivery Platforms

Abstract:

Introduction: Diseases and disorders of the intestinal tract affect millions of people worldwide each year. Inflammatory bowel disease (IBD) represents the leading chronic gastrointestinal disease with increasing healthcare burdens and expenditures [1]. There are over 1.4 million persons living with IBD in the U.S. alone, and as there is no cure, this number increases yearly indicating a need for new therapeutic strategies [2]. Current in vitro models of IBD utilize 2D systems and Caco-2 epithelial cell lines with a limited ability to mimic complex tissue. Native small intestine possesses distinct structures (e.g., crypts-villi) and a heterogeneous cell population (stem, absorptive, goblet, enteroendocrine, and paneth cells). Therefore, we propose to develop functional engineered small intestine with relevant heterogeneous cell populations and spatial organization required for successful regenerative medicine and drug discovery platforms. Recent breakthroughs have enabled long-term primary intestinal culture in the form of enteroids, but size limitations and lack of access to luminal surfaces limits practical applications [3]. Recent work also suggests cell fate is impacted by 3D features and local topography/stiffness [4,5]. Therefore, we hypothesize that incorporating the 3D topography of native intestine into growth substrates with tunable mechanical and adhesion properties will enable dissociated native primary enteroid spatial organization similar to native tissue in a scalable format.

Materials and Methods: To fabricate structurally biomimetic growth substrates, fresh porcine small intestine was rinsed, fixed with 1% GTA: 1% PFA in 0.1M PBS (4°C, 24hrs), sectioned (1cm^2), placed in 0.1% OsO4 in 0.1M PBS (4°C, 72hrs), and agitated to remove the epithelium. Samples were rinsed, ethanol dehydrated, critically point dried, treated with silane (APTS, 12 hrs), and 10:1 v:v PDMS was solution cast for 12hrs followed by baking (60°C, 2hrs). The PDMS negative mold was released from the tissue via soaking in TEA (12hrs, 25°C), and rinsed in 100% ethanol (25°C, 24hrs) and DIH2O (25°C, 24hrs), prior to drying (75°C, 2hrs). Samples were visualized with a Hitachi

S-4800 UHR field emission SEM [6]. Collagen-I:10% Matrigel (v:v) hydrogels were cast into 24-well plates with the PDMS negative replicas or flat surfaces and allowed to gel (2 hrs, 37°C). Primary mouse enteroids from isolated crypts [7] were dissociated and seeded on replica hydrogels or flat controls for 5 days under standard culture conditions. On day 6, samples were incubated with EdU for 24 hrs. Samples were then fixed, stained for alpha-tubulin cytoskeleton and nuclei (Hoechst), and imaged with a Nikon A1R confocal at 20x.

Results and Discussion: Native small intestinal tissue molds were used as templates to fabricate biomimetic PDMS and hydrogel replicas with irregular, but spatially organized multi-scale topographical features. Fixed native tissue and topographical replicas show micro to nano-scale features, containing irregular, but spatially organized crypt-villus structures and smaller features of the epithelial basement membrane (not shown). Preliminary qualitative results indicate that cells positive for EdU (indicating DNA synthesis) are present in higher amounts on topographical replicas compared to flat hydrogel controls (Fig. 1). On topographical replicas, the proliferating cells appear in dense clusters with satellite EdU+ cells throughout the population, whereas on flat substrates there are sparse EdU+ cells throughout the monolayer. Further, microtubules appear to localize to the apical surface, indicating polarity, on replicas but not flat controls. Previous work in our lab has show that crypt-like structures aid Caco-2 differentiation 8,9, indicating the importance of topography in epithelial differentiation and similar effects may be apparent for primary enteroids.

Conclusions: Manipulating primary enteroid spatial organization and differentiation into functional models of native small intestine via presentation of biomimetic topographical cues may provide a better system for regenerative medicine and drug delivery and discovery applications. This work shows the ability to engineer growth substrates that mimic the native intestinal multi-scale topography, and these substrates may impact primary enteroid cell morphology and proliferation. Future work will determine the impact of physical properties (e.g., stiffness), chemical properties (e.g., substrate) and structure (e.g., intestinal topography) on cell behavior (e.g., attachment, signaling, differentiation and spatial organization) for a molecular understanding of how biomimetic topography impacts phenotypic changes.

Acknowledgements: Northeastern STEM Future Faculty Fellowship (ANK), and NIH R21EY021312 (RLK).

References: 1) Park K, Inflamm Bowel Dis. 2011;17(7):1603-1609. 2) Donohoe CL, Reynolds JV. The Surgeon. 2010;8(5):270-279. 3) Sato T, Gastroenterology. 2011;141(5):1762-1772. 4) Evans ND, Eur Cell Mater. 2009;18(1-13):13-14. 5) Reilly GC J Biomech. 2010;43(1):55-62. 6) Pfluger CA, Biomacromol. 2010;11(6):1579–1584. 7) Sato T, Nature. 2009;459(7244):262-265. 8) Wang L, Biomaterials. 2010;31(29):7586-7598. 9) Wang L, Biomaterials. Dec 2009;30(36):6825-6834.

Fig 1. Primary enterocytes, derived from dissociated mouse enteroids, exhibit increased proliferation (green) on hydrogels with topographical features (right) of small intestine compared to flat controls (left). Bar = 10 um, Max projection with orthogonal views. Blue= Hoechst nuclei, Red=alpha-tubulin, Green=EdU.

THE GRAPHIC ON THIS PAGE WAS NOT AVAILABLE ON THE E-MEDIA VERSION OF THIS PAPER.

Assessing the impact of organic loading rate on hydrogen production and mixed anaerobic culture community dynamics

Sathyanarayanan S. Veeravalli[1], Subba Rao Chaganti[2], Saravanan R. Shanmugam[1], Jerald A. Lalman[1,2], and Daniel D. Heath[2,3]

[1]Department of Civil and Environmental Engineering, University of Windsor, Center for Engineering Innovation, 401 Sunset Ave., Windsor, ON, N9B 3P4
[2]Great Lakes Institute for Environmental Research, University of Windsor, 401 Sunset Ave., Windsor, ON, N9B 3P4
[3]Department of Biological Sciences, University of Windsor, 401 Sunset Ave., Windsor, ON, N9B 3P4

Introduction

Depleting fossil fuel supplies coupled with increasing greenhouse gas emissions have become increasingly apparent over the last decade. Developing strategies to combat these issues have forced researchers to develop renewable energy supplies. Hydrogen (H_2), an emerging energy carrier, has a high calorific value and it is considered a renewable energy source because CO_2, a combustion byproduct is carbon neutral (Das and Veziroglu, 2001; Dunn, 2002). Among the biological H_2 production methods, dark fermentation is a promising and preferred route to produce H_2 because of its high productivity, ease and low cost of operation and ability to ferment on wide variety of feedstock (Das and Veziroglu, 2001; Levin et al., 2004).

Hydrogen production via dark fermentation using mixed anaerobic cultures is controlled by the presence of H_2 producers, H_2 consumers and other non-H_2 producers. Changing operational variables pH, temperature, organic loading rate (OLR), hydraulic retention time (HRT) and reactor configuration to shift the metabolic pathways in favor of increasing the H_2 yield is of significant importance (Jung et al., 2011; Li and Fang, 2007). Operating high rate reactors at a pH range of 5.0 - 6.0 and under mesophilic temperature conditions is preferred for optimizing H_2 production (Jung et al., 2011; Lay, 2000; Rittmann and Herwig, 2012). Several studies have examined manipulating the OLR and HRT to suppress methanogenesis and subsequently, increase H_2 production. Establishing a link between data gathered from genomic analysis with metabolic flux data as a means to substantiate the claim that increasing the OLR suppresses H_2-consumption is lacking (Pakarinen et al., 2011; Spagni et al., 2010). Hence, the objectives of this study was to examine the effect of OLR on the microbial fermentation of glucose in continuous upflow anaerobic sludge blanket reactors (UASBRs) and to elucidate the potential link between fermentation type and microbial consortia using 16S rRNA gene based pyrosequencing.

Methodology

The experiments were conducted using duplicate UASBRs (R1 and R2) with a working volume of 8.5 L. The UASBRs were seeded with 10 g L^{-1} VSS of mixed anaerobic cultures which was obtained from a brewery wastewater treatment facility (Guelph, Ontario). Reactor operation and inoculum maintenance was carried out in accordance with Veeravalli et al. (2014). Hydrogen production studies were carried out with cultures operating at 37 □C and at pH 5.0. The influence of the OLR from 2 to 16 g glucose L^{-1} d^{-1} on the H_2 yield, methane suppression and microbial consortia was examined at a 24 h HRT. At each OLR condition, experiments using R1 and R2 were repeated at least 7 times. The gas (H_2 and CH_4) and liquids metabolites comprising volatile fatty acids (VFAs) and alcohols were analyzed according to the methods described by Chowdhury et al. (2007). The flux balance analysis (FBA) was conducted using the data obtained from byproducts of glucose fermentation (Chaganti et al., 2011). The microbial communities were characterized using nested-PCR of16S rRNA gene using 454- pyrosequencing (Andersson et al., 2008).

Results and discussion

A stable H_2 yield of 1.64±0.05 mol mol^{-1} glucose was obtained at OLRs between 8 and 12 g glucose L^{-1} d^{-1} **(Table 1)**. Increasing the OLR resulted in suppressing the methanogenic activity by 95% (approximately) with a simultaneous increase in the H_2 yield. Studies conducted by Spagni et al. (2010) using grass silage suggest that a shift from CH_4 to H_2 production was observed in a CSTR by increasing the OLR from 2 to 10 g VS L^{-1} d^{-1}. Note in the current study, the flux analysis showed that CH_4 production via aceticlastic methanogenic activity contributed to approximately 0.52 and 0.05 mol mol^{-1} glucose at 2 and 16 g L^{-1} d^{-1}, respectively (data not shown). A maximum H_2 consumption of 1.81±0.22 mol mol^{-1} glucose due to methanogensis was observed at an OLR of 2 g L^{-1} d^{-1} while maximum H_2 consumption due to acetogenesis of 0.54±0.03 mol mol^{-1} glucose was observed at an OLR of 16 g L^{-1} d^{-1} **(Table 1)**. The results of the current study suggest that even at pH 5.0 and high OLR (16 g L^{-1} d^{-1}) complete suppression of H_2 consumption was not established. Similarly, Lalman et al. (2013) reported H_2 consumption (mol mol^{-1} hexose) due to an acetogenic flux of 0.45 and 0.12 at high OLRs of 20 and 30 g COD L^{-1} d^{-1}, respectively, for a CSTR operating at pH 5.4 cultures fed sucrose.

Synergesticiae and *Propionibacteriaceae,* which constituted 45% of the microflora, was observed at low OLRs. In addition, the quantity of *Clostridiaceae* and *Ruminococcaceae* relatively observed was 16% and 8%, respectively **(Table 2)**. The fermentation byproducts production with time revealed that mixed VFAs and alcohol production were observed at low OLRs with acetate, butyrate and ethanol as the major by products **(Table 1)**. At elevated OLRs, high levels of *Clostridiaceae* (38%), *Ruminococcaceae* (33%) were detected while low levels of *Synergesticiae*, *Propionibacteriaceae* plus *Bacteroidaceae* (3%) were observed **(Table 2)**. When compared to low OLRs (2 to 6 g glucose L^{1} d^{-1}), an increase in ethanol production under a high loading condition of 16 g glucose L^{-1} d^{-1} was observed. Several studies have shown species belonging to *Clostridiaceae* and *Ruminococcaceae* were responsible for ethanol, VFAs and/or iso-propanol production, in addition to H_2 and CO_2 (Fang et al., 2002; George et al., 1983; Liu et al., 2012)

Conclusions

In summary, increasing the OLR caused an increase in the quantity of H_2 produced and simultaneous suppression of CH_4 production up to 12 g L^{-1} d^{-1} was observed. High OLRs were associated with high homoacetogenic activity and ethanol production. Dominant species belonging to *Clostridiaceae* and *Ruminococcaceae* were detected at high OLR.

Keywords: mixed microbial community, hydrogen production, pyrosequencing, metabolic flux model, organic loading rate

Table 1: Effect of organic loading rate on the hydrogen yield and metabolites

OLR (g L^{-1} d^{-1})	Observed H_2 yield (mol mol^{-1} glucose)	H_2 consumption (mol mol^{-1} glucose)		Acetate	Butyrate	Ethanol
		R29 Flux	R17 Flux	Concentration in mM		
2	0.75±0.09	1.81±0.22	ND	6.6±1.3	3.5±0.2	0.18±0.05
4	1.23±0.15	1.36±0.16	ND	17.1±0.7	7.4±0.4	0.8±0.1
6	1.56±0.12	0.78±0.06	ND	23.0±1.3	11.5±1.0	1.6±0.1
8	1.70±0.05	0.32±0.02	ND	25.3±1.0	14.0±1.4	4.1±1.2
10	1.65±0.04	ND	0.18±0.01	30.4±2.8	18.1±1.5	6.8±1.2
12	1.57±0.05	ND	0.25±0.02	34.9±2.0	21.7±1.5	8.5±1.0
14	1.32±0.04	ND	0.35±0.02	34.5±1.5	26.0±1.0	10.2±0.5
16	1.00±0.04	ND	0.54±0.03	35.7±1.2	24.0±4.3	10.7±2.5

Notes:

1. OLR: organic loading rate and ND: not detected
2. The FBA was conducted to determine H_2 consumption due homoacetogenesis (R17 flux) and methanogeneis (R29) using the model previously developed by Chaganti et. al (2011).

Flux reaction no.	Stoichiometric reaction
17	$4\ H_2 + 2\ CO_2 \rightarrow$ Acetate $+ 2\ H_2O$
29	$4\ H_2 + CO_2 \rightarrow CH_4 + 2\ H_2O$

Table 2: Pyrosequencing results of microbial consortium showing their relative abundance obtained after fermentation at 2 and 16 g L^{-1} d^{-1}

Family	Organic loading rate	
	2 g L^{-1} d^{-1}	16 g L^{-1} d^{-1}
	Percent relative abundance (%)	
Synergistaceae	34.1	1.4
Ruminococcaceae	8.4	33.3
Clostridiaceae	15.7	37.8
Propionibacteriaceae	11.0	0.7
Unclassified Clostridiales	10.4	19.4
Unclassified bacteria	10.1	5.1
Unclassified Bacteroidales	3.7	0.0
Enterobacteriaceae	0.15	1
Others	6.5	1.4

References

Andersson, AF, et al. 2008. Comparative analysis of human gut microbiota by barcoded pyrosequencing. *PLOS One*. **3**(7).

Chaganti, SR, et al. 2011. Flux balance analysis of mixed anaerobic microbial communities: Effects of linoleic acid (LA) and pH on biohydrogen production. *Int J Hydrogen Energy*. **36**(21), 14141-52.

Chowdhury, N, et al. 2007. Biohydrogen production by mesophilic anaerobic fermentation of glucose in the presence of linoleic acid. *J Environ Eng-ASCE*. **133**(12), 1145-52.

Das, D and Veziroglu, TN. 2001. Hydrogen production by biological processes: a survey of literature. *Int J Hydrogen Energy*. **26**(1), 13-28.

Dunn, S. 2002. Hydrogen futures: toward a sustainable energy system. *Int J Hydrogen Energy*. **27**(3), 235-64.

Fang, HH, et al. 2002. Microbial diversity of a mesophilic hydrogen-producing sludge. *Appl Microbiol Biotechnol*. **58**(1), 112-8.

George, HA, et al. 1983. Acetone, isopropanol, and butanol production by *Clostridium beijerinckii* (syn. *Clostridium butylicum*) and *Clostridium aurantibutyricum*. *Appl Environ Microbiol*. **45**(3), 1160-3.

Jung, KW, et al. 2011. Bioreactor design for continuous dark fermentative hydrogen production. *Bioresour Technol.* **102**(18), 8612-20.

Lalman, JA, et al. 2013. Elucidating acetogenic H_2 consumption in dark fermentation using flux balance analysis. *Bioresour Technol.* **146**, 775-78.

Lay, JJ. 2000. Modeling and optimization of anaerobic digested sludge converting starch to hydrogen. *Biotechnol Bioeng.* **68**(3), 269-278.

Levin, DB, et al. 2004. Biohydrogen production: prospects and limitations to practical application. *Int J Hydrogen Energy.* **29**(2), 173-185.

Li, CL and Fang, HHP. 2007. Fermentative hydrogen production from wastewater and solid wastes by mixed cultures. *Crit Rev Env Sci Technol.* **37**(1), 1-39.

Liu, ZD, et al. 2012. Enhanced hydrogen production in a UASB reactor by retaining microbial consortium onto carbon nanotubes (CNTs). *Int J Hydrogen Energy.* **37**, 10619-26.

Pakarinen, O, et al. 2011. The effect of organic loading rate and retention time on hydrogen production from a methanogenic CSTR. *Bioresour Technol.* **102**(19), 8952-57.

Rittmann, S and Herwig, C. 2012. A comprehensive and quantitative review of dark fermentative biohydrogen production. *Microb Cell Fact.* **11**.

Spagni, A, et al. 2010. Effect of the organic loading rate on biogas composition in continuous fermentative hydrogen production. *J Environ Sci Heal Part A.* **45**(12), 1475-81.

Veeravalli, SS, et al. 2014. Optimizing hydrogen production from a switchgrass steam exploded liquor using a mixed anaerobic culture in an upflow anaerobic sludge blanket reactor. *Int J Hydrogen Energy.* **39**(7), 3160-75.

389315 Elucidation of the Reaction Mechanism during Coal Pyrolysis

Abstract:

Recently Integrated coal Gasification Combined Cycle (IGCC) has been attracting great attention as a means of highly efficient power generation. As a previous step of coal gasification, volatiles and char are generated from coal pyrolysis. Tar, one of the volatiles, causes problems such as pipeline plugging, defluidization in fluidized bed reactor, reduction in thermal efficiency, and inhibition of char gasification. Therefore, it is necessary to reduce the production of tar or to separate tar from char in order to overcome these problems. Thus, it is essential to elucidate the reaction mechanism of tar production in coal pyrolysis for improvement in the controllability and thermal efficiency of the coal gasification process.

In many previous studies, both pyrolysis and steam gasification of some kinds of coals, such as bituminous coal or brown coal, were investigated with rapid heating or slow heating using a thermobalance reactor, micro GC and mass spectrometer. It was found that time profiles of weight change of coal samples and the evolution of low molecular weight gases (H_2, CH_4, CO and CO_2). To understand concretely the reaction mechanism of tar production, it is necessary to further investigation about the online time profile of tar evolution and componential analysis.

In this study, the time profiles of gas, tar and char evolution, and tar componential analysis during coal primary pyrolysis and secondary pyrolysis were investigated for elucidating the reaction mechanism of volatiles, in particular tar, production. From this result, tar problems will be overcome and thermal efficiency of IGCC will be improved.

382199 A Computational Model for Microbial Electrosynthesis of Organic Compounds in Reverse Microbial Fuel Cells

Abstract:

Noticeably increased emissions of carbon dioxide as a result of industrial development in recent decades have been attracted more attention to combat and prevent the devastating effects. So to find a way to get rid of harmful greenhouse gases seems necessary. Carbon dioxide is formed the main greenhouse gas. In reverse microbial fuel cells, which compared to microbial fuel cells perform conversely, bacteria, using electrons which are transferred to cathode from anode or from an external energy source, transform CO_2 and H_2O to organic compounds during a process which is like photosynthesis. These organic compounds can be transformed to fuel by themselves. In the present study, modeling of microbial-electrosynthesis to produce organic compounds is presented based on direct conduction of electrons in biofilms. The output of the present model includes the changes of the substrate concentration, electric potential, the distribution of active bacteria in the biofilm, the time variation of the current density, and the thickness of the biofilm. In this model, Coulombic efficiency is a function of substrate concentration and cathode potential. For CO_2 substrate and *Sporomusa ovata* species, the increase of the substrate concentration leads to the decrease of Coulombic efficiency, increase of current density and increase of biofilm thickness and increase of cathodic voltage surface leads to the increase of Coloumbic efficiency and decrease of current density. Because the electrical conductivity of *Sporomusa ovata* biofilm is very high, most part of microbial fuel cell resistances with this microbial community and CO_2 substrate come from the resistance of mass transfer. Coulombic efficiency is short because of ohmic and cathodic resistances in the performance of cell. The maximum Coulombic efficiency was revealed to be %75 in the concentration of 0.025 mmol cm^{-3} and %55 in the cathodic voltage surface of -0.3 V, thus, we can achieve a high range of acetate production by creating an optimal state in this concentration and potential.

390380 Does Saccharomyces Cerevisiea TSH1 Possess the Properties Particularly Suitable for Solid-State Fermentation of Sweet Sorghum Ethanol Compared to Industrial Yeast Strains

Abstract:

With the increasing concerns over cost-effective and environmental friendly production of biofuels, the great potential of solid-state fermentation (SSF) in biofuels area has been recognized recently, due to relatively simple production process and very little disposal of wastewater compared to submerged liquid fermentation (SLF). Over the past several years we developed the advanced solid-state fermentation (ASSF) for fuel ethanol production from sweet sorghum stalks from bench up to industrial scale (500 m3), which overcomes fundamental constraints of traditional SSF ,exhibiting a high ethanol productivity comparable to SLF. Most recently, we have run a 500-m3 continuous ASSF for sweet sorghum ethanol production in China, firstly verifying the feasibility of SSF's application at industrial scale. *Saccharomyces cerevisiae* TSH1 used during the process, is capable of completing sugars utilization within 24 h during the SSF using sweet sorghum containing a total sugar content of 10~14% (w/w), with a 91% of theoretical ethanol yield. In a previous study, we characterized ethanol producer TSH1's physiological and biochemical properties with a lab yeast strain as the reference. The outcome showed that TSH1 exhibits obvious advantages against the lab strain, including higher growth rate, ethanol productivity, and stronger tolerance to inhibitors. These results motivated our interest in evaluating whether TSH1 possesses special properties suitable for SSF compared to industrial yeast strains. In this study, we focused on characterizing TSH1 and several other industrial yeast strains and compared the fermentation performances of these strains under various conditions. This study is aimed to provide an insight into the mechanism of microorganism behavior in SSF, finally harnessing the untapped potentials of SSF for the application in biofuels production.

The use of devices to capture rare circulating cells from the blood under flow is an ongoing challenge for multiple fields within the biomedical sciences. Whether attempting to capture circulating tumor cells for diagnostics or capturing the rare endothelial progenitor cells (EPC) for use in vascular tissue engineering, the target cell populations are incredibly small in comparison to the rest of the cells within the blood. Therefore, there is a great need for the ability to capture rare circulating cells with high specificity, while enabling proliferation of immobilized cells to stimulate tissue regeneration. Vascular endothelial growth factor, VEGF, is an extensively studied mitogen for its role in angiogenesis in both healthy tissue and tumor microenvironments. VEGF binds with high affinity to its respective receptors and in turn induces proliferation and differentiation as well as serving as a chemo-attractant to nearby cells. Therefore, the high affinity binding between VEGF and its receptors could be utilized to selectively capture cells expressing VEGF receptors with high specificity.

We hypothesized that immobilized VEGF might capture EPC and/or other VEGFR expressing cells under in-vivo like conditions such as physiological shear.

In this study we pursued this hypothesis using a genetically engineered VEGF that was immobilized on the surface of a microfluidic device and assessed its ability to capture cells under flow. Our work may have implications in tissue engineering, diagnostics and development of personalized medicine strategies. Recombinant human VEGF-165 was produced in large quantities and immobilized by the naturally occurring heparin binding domain of VEGF to surfaces treated with heparin. Surfaces functionalized with Heparin and VEGF were placed under a microfluidic device designed to modulate flow with adjustable shear strengths by differing flow rates, from low shear to physiological shear. Human umbilical vein endothelial cells (HUVECs) and ovine pulmonary artery endothelial cells (OPAECs) were used as representative endothelial cells. Mouse fibroblasts (NIH-3T3), human hair-follicle mesenchymal stem cells (HF-MSC), and human neonatal dermal fibroblasts (hd-FB) were used as non-endothelial control cells. In addition, runs were performed using HUVEC supplemented human whole blood, whole blood, and isolated peripheral blood mononuclear cells (PBMC). Surfaces functionalized with PLL-heparin-VEGF were shown to capture endothelial cells under static and varying flow conditions. Immobilized VEGF retained the ability to induce proliferation HUVECs. Under flow conditions, endothelial cells were selectively captured with high efficiency under a range of shear strengths. Remarkably, VEGF mediated capture was highly specific, capturing only endothelial cells with high efficiency under low shear. VEGF mediated capture is physiologically and clinically relevant due to the ability to capture spiked endothelial cells (HUVECs) from whole blood in addition to the capture of the rare endothelial progenitor cell from runs of whole blood, as identified through immunocytochemistry.

Immobilized VEGF is capable of selectively capturing endothelial cells under a range of shear conditions as well as in homogenous or heterogeneous populations, including EPCs from whole blood. These results are very promising and could be implemented in multiple ways such as tissue engineered vessels capable of regenerating an endothelial lumen through VEGF mediated capture of EPCs and subsequent VEGF induced proliferation, migration, and differentiation towards a mature endothelial phenotype or for use in capturing circulating tumor endothelial cells for cancer diagnostics.

Introduction: The shortage of donor islets for treating diabetes highlights the urgent need for generating insulin-producing β-cell from human embryonic stem cells (hESCs). Despite the improvement in β-cell differentiation in in-vitro cultures, a major gap remains in the lack of cellular maturity. This maturation defect may be partly due to the lack of 3D structure and interactions with stromal cells, both of which influence embryonic pancreatic development in vivo. We attempt to re-engineer this organogenetic interaction by repopulating decellularized pancreas with hESC derived pancreatic progenitors and endothelial cells. We hypothesize that 3D micro-architecture from decellularized pancreas and co-culture interactions with endothelial cells would provide the optimal microstructure and microenvironment niche respectively to guide organogenesis and promote β-cell functionality.

Materials and Method: Cadaveric pancreata were isolated from adult mice (n=8) and decellularized via detergent (2% Triton-X, 1% DOC) perfusion. Resulting decellularized matrices were characterized for acellularity, preservation of ECM and microstructures. Pancreatic progenitor (PP) cells were generated from step-wise differentiation of hESC into definitive endoderm (Activin and Wnt for 4 days) followed by sonic hedgehog inhibition (cyclopamine) and retinoic acid induction for another 4 days. For recellularization, hESC-derived PP cells ($3x10^6$ cells) were combined with HUVEC ($3x10^6$ cells) and introduced into the acellular 3D pancreatic scaffold via portal vein in 3 steps, with 30 min interval between each step. The recellularized 3D construct was cultured on a perfusion bioreactor and evaluated via immunostaining, TUNEL, and qRT-PCR after 7 days.

Results and Discussion: Perfusion-decellularization of pancreas with detergent resulted in complete removal of cells. Characterization of the decellularized pancreas revealed preservation of ECM and 3D architecture. At day 7, qRT-PCR analysisrevealed that recellularized pancreatic construct(n=3) had significantly increased expression of early pancreatic marker genes (PDX1, NKX6.1, NGN3, ISL1) and maturation markers (MAFA, INSULIN) compared to cells seeded in matrigel (n=6) and control (n=6), $P<0.05$ (**Figure 1**). Consistent with this, immunostaining result showed significantly higher number of PDX1+ cells ($P<0.023$) were found in recellularized pancreas construct compared to matrigel. In addition, immunostaining result also showed that majority of PDX1+ cells (n=10, 82±11%) were detected in close proximity with endothelial cells (CD31+ cells), suggesting the involvement of stromal-cell-dependent factors in pancreatic maturation.

Conclusion: Perfusion-decellularization of pancreas efficiently removes cells while retaining ECM protein and microstructure. This perfusable 3D whole organ scaffold yields complex3D tissue construct that favored pancreatic differentiation when seeded with hESC-PP and HUVEC. This is a first step proof of concept to show that the 3D pancreatic maturation can be recapitulated in vitro by co-culturing pancreatic progenitors with endothelial cells in decellularized pancreas.

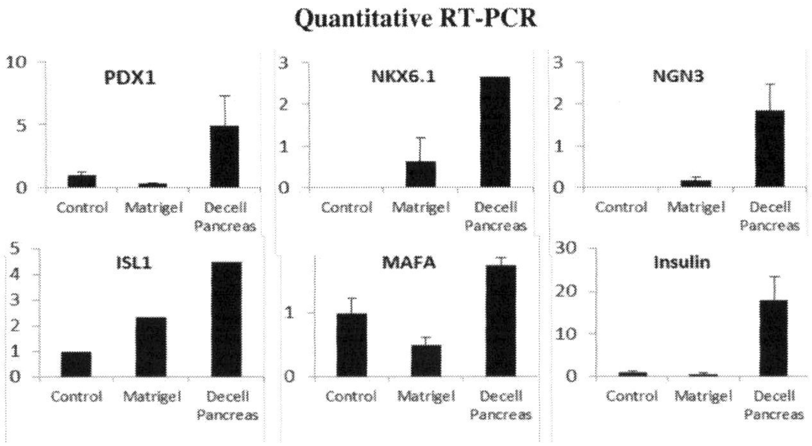

Figure 1.Quantitative PCR analysis of pancreatic marker gene expression in recellularized pancreas with hESC-PP and endothelial cells at day 7 of culture. Control samples were hESC:HUVEC that had been grown as separate cell types and mixed together at the same ratio of cell types as in recellularized pancreas construct.

374473 Facile Access to Cation Functional Biodegradable Microparticles

Abstract:

Biodegradable polylactide(PLA), poly(lactic-co-glycolic acid)(PLGA) and poly(ε-carpolactone)(PCL) microparticles bearing a cationic surface have been investigated for mucoadhesive sustained release of therapeutic proteins, peptides, gene and vaccines. In the past years, different approaches have been developed to achieve a positively charged surface by either adsorbing cationic agents and surfactants or coating with polyelectrolyte, which is not very stable [1]. Here, we report on facile preparation of robust cationic biodegradable microparticles by treating poly(L-lactide-co-vinyl sulfone carbonate) (P(LLA-co-VSC)) microparticles with cationic mucoadhesive agents such as cysteamine and glycol chitosan.

P(LLA-co-VSC) random copolymers were synthesized by ring-opening copolymerization of L-lactic acid and vinyl sulfone carbonate (VSC) [2]. The microspheres based on P(LLA-co-VSC) were prepared by water-in-oil-in-water (W/O/W) emulsion process, followed by Michael-type conjugate addition with the thiol group of 2-mercaptoethylamine and the amine group of glycol chitosan, respectively. The results showed the P(LLA-co-VSC)$_{3.57\%}$ microparticles had an average size of ca. 4 μm and a narrow particle size distribution. ^{1}HNMR and XPS results displayed successful modification of microparticles with cationic agents under mild aqueous condition. SEM showed that cationic microparticles had a smooth surface morphorgy and a slightly increased particle size. Notably, the immobilization of glycol chitosan on the microspheres resulted in an increase of zeta potential to ca. +13 mV. The fluorescence images demonstrated obvious surface conjugation of amine-containing molecules. This gentle surface functionalization is important to maintain the bioactivity of proteins/peptides, gene and vaccines.

Facile Functionalization and Assembly of Live Cells with Microcontact Printed Polymeric Biomaterials

*Zhibin Wang,† Junfei Xia, † Yuanwei Yan,† Ang-Chen Tsai,† Yan Li,† Teng Ma,† ‡ Jingjiao Guan†,‡,**

†Department of Chemical and Biomedical Engineering, FAMU-FSU College of Engineering, Florida State University, Tallahassee, Florida 32310

‡Integrative NanoScience Institute, Florida State University, Tallahassee, Florida 32306

Live cells are commonly functionalized with biomaterials for various biomedical applications such as drug delivery, in vivo cell tracking, sensing and immunological modulation. On the other hand, assembling live cells into functional constructs is central to tissue engineering and biomaterials are frequently employed to assist the assembly. Among various biomaterials used to functionalize and assemble live cells, microfabricated biomaterials, which are characterized by precisely controllable dimensions, shape, structure and composition, attract widespread interests. We here reports development of a novel and facile method for functionalizing and assembling live cells with microfabricated biomaterials by integrating microcontact printing of polyelectrolyte multilayers and poly(lactic-co-glycolic acid) with spin coating of poly(N-isopropylacrylamide) for preparing a temperature-sensitive sacrificial layer. It allows immobilizing cells or cell aggregates on the printed biomaterials at 37 °C and releasing the cell-biomaterial complexes at the room temperature.(Figure 1) The printed biomaterial can be made with any two-dimensional shapes ranging from micrometer to millimeter in size and with biodegradable and biocompatible polymers. Three types of cells including a human cancer cell line, mouse embryonic stem cells and human mesenchymal stem cells in the forms of single cells and aggregates have been functionalized and assembled using this method. In addition, this method is facile and inexpensive compared to the conventional methods. It thus holds potential for various applications including drug delivery, cell tracking and tissue engineering.

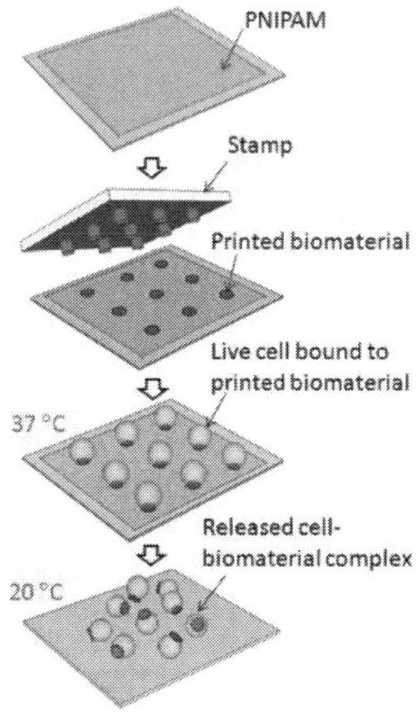

Figure 1. Procedure of functionalizing or assembling live cells with microcontact printed biomaterials using spin-coated PNIPAM as the sacrificial layer. Microparticles and single cells are used here as a model system.

Preparation of Zeolite Coatings on Glass

Melkon Tatlier, Çiğdem Atalay-Oral, Department of Chemical Engineering, Istanbul Technical University, Maslak, 34469 Istanbul, Turkey

Zeolite coatings may be useful for applications related to separation, catalysis, sensing, heating/cooling, etc. Coatings have been grown on different substrates, including metals, silicon wafers, carbon fibers and ceramic surfaces. Zeolite coatings on glass may also be favorable for some applications, such as dehumidification, deodorization and provision of heating-cooling, especially in houses. However, there are only a few studies in the literature on this issue, mainly related to the ready dissolution of glass in the generally rather alkali zeolite synthesis mixtures. ZSM-5 was deposited on glass fibrous supports via synthesis in fluoride-containing media at neutral pH [1]. Zeolite Beta coatings with 1–2 μm thickness were synthesized on borosilicate glass, previously coated by two thin films of zirconia and titania by atomic layer deposition [2]. Highly *b*-oriented silicalite-1 thin films were obtained on the surface of glass slides modified with Ti–OH groups by a surface sol–gel process [3]. In this study, coatings of zeolite X were prepared on soda-lime type glass plates of different sizes by conventional synthesis. The coatings were characterized by X-ray diffraction, scanning electron microscopy and thermogravimetry.

Hydrothermal synthesis was performed in an oven kept at different temperatures. A clear solution composition of 70 Na_2O:Al_2O_3:15 SiO_2:2100 H_2O previously shown to yield zeolite X [4] was taken as a basis and various reaction mixture compositions were tested with molar concentration ratios in the range H_2O/Na_2O=30-60, SiO_2/Al_2O_3=3-26 and H_2O/SiO_2=80-700. It was observed that coatings of zeolites, such as X, A, P, hydroxysodalite could be obtained on glass plates. The glass dissolved to some extent during the experiments which affected the zeolite phase obtained. The thickness and stability of the coatings depended on the synthesis conditions. The thermogravimetric analyses indicated good water adsorption/desorption properties for the coatings. This study indicated that zeolite coatings of even low-silica zeolites, which are generally synthesized in quite highly alkaline reaction mixtures, can be prepared on glass substrates. Such zeolitic materials may be useful for different applications requiring the use of glass.

References

[1] B. Louis, C. Tezel, L. Kiwi-Minsker, A. Renken, Catal. Today 69 (2001) 365.

[2] O. Muraza, E.V. Rebrov, J. Chen, M. Putkonen, L. Niinistö, M.H.J.M. de Croon, J.C. Schouten, Chem. Eng. J. 135 (2008) S117.

[3] J. Di, C. Zhang, W. Yan, X. Wang, J. Yu, R. Xu, Micropor. Mesopor. Mater. 145 (2011) 104.

[4] M. Tatlier, K.B. Cigizoglu, H.K. Cigizoglu, A. Erdem-Senatalar, J. Porous Mat. 15 (2008) 389.

381616 The Role of Alpha-Mangostin on Cellular and Intracellular Level of Melanin Bio-Synthesis and Tyrosinase Activity of B16F1 Melanoma Cells

Abstract:

In the present study an inhibitory effect of melanin synthesis and tyrosinase activity of α-mangostin from mangosteen (*Garcinia mangostana*) pericarp extract has been investigated and verify on B16F1 melanoma cell line. Mangosteen pericarp contains mangostins of which a major constituent is α-mangostin. Alpha-mangostin has been known to possess antibacterial causing acne. Our investigation focused upon identification of anti-melanogenic efficacy of alpha-mangostin since it has been known to possess strong anti-oxidant activities. Determination of alpha-mangostin was carried out by using UV-spectrophotometric method and NMR analysis. Anti-melanogenic effect of alpha-mangostin was analyzed using cultured B16F1 melanoma cells. Cytotoxicity of that compound was measured using 3-(4, 5-dimethylthiazol-2-yl)-2,5-diphenyltetrazolium bromide (MTT) assay; the highest concentration of the extract that did not affect cell viability was 5 μg/mL. Formation of melanin from cultured B16F1 melanoma was estimated using spectrophotometry. α-MSH-induced melanin synthesis was significantly inhibited with dose-dependent manner by treatment of alpha-mangostin, which was more powerful than kojic acid and arbutin. α-mangostin also directly inhibited intracellular melanin formation as well as intracellular tyrosinase activity of B16F1. The inhibition of intracellular tyrosinase activity was found to be exerted at the protein expression level when analyzed by immunoblot and tyrosinase zymography. The expression level of microphthalmia-associated transcription factor (MITF) was also reduced by treatment of alpha-mangostin. Down-regulation of intracellular tyrosinase expression seemed to correlate with decrease in MITF (microphtalmia-associated transcription factor) protein level since MITF is the key factor for the genes involved in melanogenesis. The overall results suggest that α-mangostin has strong and potent anti-melanogenic activity that is exerted by direct inhibition of tyrosinase enzyme activity and by down-regulation of the genes expression involved in the melanogenesis pathways. These finding will lead to the development of a safe pharmacological or cosmetic agent based on α-mangostin.

Liquid Extrusion 3D Printing and its Application to Tissue Fabrication:
3D Culture of HFF-1 cells in a Photopolymerized PEGDA Hydrogel Scaffold

Brent Hukill[a]
Dr. David B. Henthorn[a]
Dr. Jennifer O'Connor[b]

Rose-Hulman Institute of Technology
[a]Department of Chemical Engineering
[b]Department of Biology & Biomedical Eng.

Objectives
- Learn to culture HFF-1 cell line
- Perform cytotoxicity tests to determine material effects on cell viability
- Analyze the effects of shear stress and UV irradiation on Hff-1 cells

Overall Concept & Goal of the Printer
Modern tissue engineering scaffolding techniques struggle to provide a highly specific three-dimensional architecture that is needed for successful autografting procedures. 3D printing technologies are being investigated as a potential solution to this challenge; however, novel 3D printing methods must be developed that are better suited for cell printing.

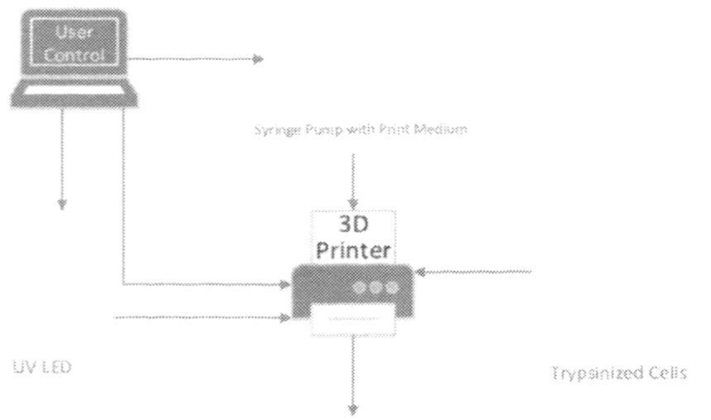

Figure 1: Flow of information and material into and out of the 3D printer. The UV LED (365 ± 5 nm) and the mechanical syringe pump can be computer controlled to modulate UV intensity or pulse pattern and print medium flow rate, respectively. Cells are also fed to the printer, where they are combined with the print medium. The medium—now containing cells—is extruded and immediately irradiated with UV light producing cell-laden hydrogels.

Abstract
Much of modern tissue engineering research aims to develop methods for 3D cell culture and scaffold design[1-3]. Our team strives to extend the versatility of 3D printing technology to such tissue engineering applications. Most commonly available 3D printers operate at elevated temperatures (typically above 100°C) in order to melt a polymer filament into a moldable state, precluding their use with live cells. In this work, we propose an alternative method – namely, the printing of a viscous reactive solution that is polymerized (gelled) as it emerges from the extruder. In this study, polymer solutions are prepared with 20 wt% PEGDA and 3 wt% photoinitiator (Irgacure 2959). Small drops of this solution are dispensed from a mechanical syringe pump and exposed to high-intensity UV light (λ=365±5 nm) to create a hydrogel stream. Also, in order to create a design suited for printing living cells, a liquid extruder is also designed to replace the heated filament extruder included with the printer.

Cell Culturing of the HFF-1 Cells
Human foreskin fibroblasts (HFF-1) were used in this study. These cells are prevalent in the literature and are known for their relative resilience in moderate to harsh conditions. As a result, the cell culturing conditions of these cells is well documented and the cells are readily purchased.

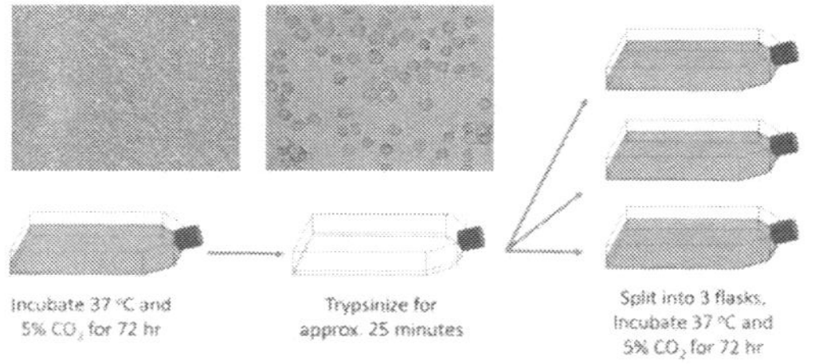

Figure 2: After 3 days of incubation at 37 °C and 5% CO_2, the HFF-1 cells form a confluent monolayer in the flask. The cells are trypsinized briefly then passed into three new flasks. The process is repeated.

Cytotoxicity Test

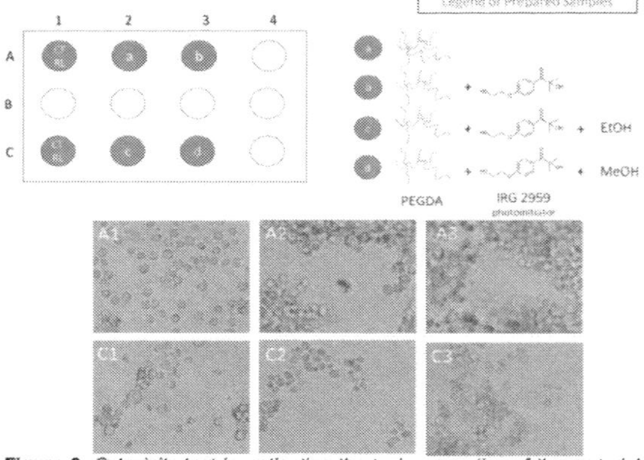

Figure 3: Cytoxicity test investigating the toxic properties of the materials used in the print medium. I) Well plate in which the colored well represent cell-containing wells, II) shows the samples that were prepared and placed in the corresponding well, and III) shows the micrographs of the corresponding wells. Notice that samples A and B cause cell clustering while the other sample did not show this response.

Effects of UV Irradiation and Shear During Exttrusion

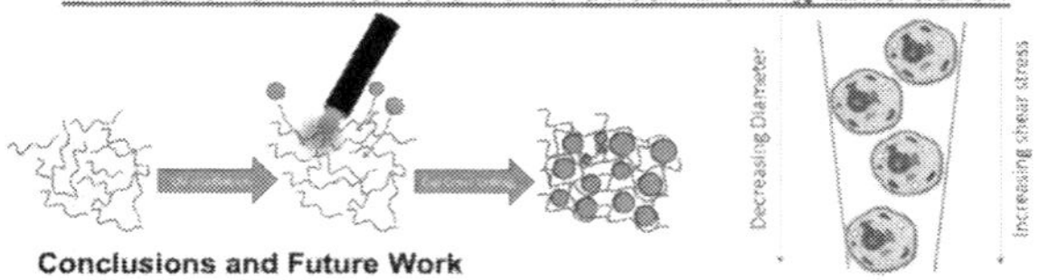

Conclusions and Future Work
The HFF-1 cells exposed to mixtures a and b clustered together. This behavior was not observed in the other mixtures. Replicates still need to be performed, and live/dead cell staining assays need to be performed. Furthermore, the effects of shear stress through extrusion and high-intensity UV irradiation need to be studied.

References
1. K. T. Nguyen and J. L. West, "Photopolymerizable hydrogels for tissue engineering applications", *Biomaterials*, **23** (2002) 4307–4314.
2. N. A. Peppas, J. Z. Hilt, A. Khademhosseini, and R. Langer, "Hydrogels in Biology and Medicine: From Molecular Principles to Bionanotechnology", *Adv. Mater.*, **18** (2006), 1345–1360.
3. S. Lee and D. B. Henthorn, *Materials in Biology and Medicine*, CRC Press, 2012.

Acknowledgements
The authors would like to thank:
1. ArcelorMittal and IPROP for funding this project
2. Dr. Paul Leisher for his contribution to project

ROSE-HULMAN INSTITUTE OF TECHNOLOGY

Independent Project / Research Opportunities Program
End of Quarter Symposium, November 14th, 2014

ArcelorMittal

Stability analysis of two phase stratified Poiseuille flow in a rectangular channel

Dinesh N. V. S. S. R. Bhagavatula, J. R. Picardo and S. Pushpavanam [*]

Department of Chemical Engineering, Indian Institute of Technology Madras, Chennai, India-600036.

(Corresponding authors E mail address: spush@iitm.ac.in)*

The study of stability of two phase stratified flow in a rectangular channel is important from both theoretical as well as practical perspectives. Two phase stratified flows arise in extraction operations in microfluidic systems. In this work we present the stability analysis of stratified two phase Poiseuille flow in a rectangular channel. A two-dimensional model for flow through the rectangular channel is considered. Shear and interfacial modes of instability are analyzed. Linear stability analysis of a base laminar profile is carried out. The linearized equations along with the boundary conditions are numerically solved using Chebyshev collocation method. This leads to eigen value problem which is solved using QR decomposition. The role of different parameters such as density ratio, viscosity ratio on the stability characteristics is analyzed.

Introduction

The stability of stratified flow in a rectangular duct (Figure 1) is often encountered in engineering practices. The aspect ratio of the rectangular channel may vary from one to infinity. In the limit of aspect ratio equal to one, we have a flow in the square duct which is a stable configuration in similarity to Hagen-Poiseuille flow in a circular tube. In the limit of aspect ratio tending to infinity the flow through the duct tends to stratified Poiseuille flow between two parallel plates. Thus there must be a critical aspect ratio for which stable flow transits to an unstable flow pattern.

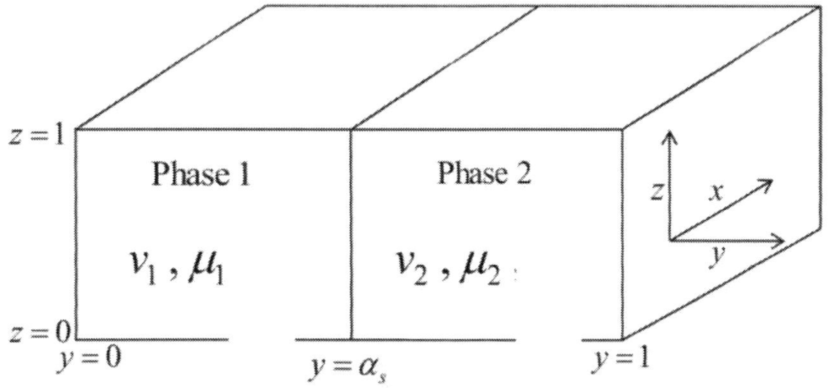

Figure 1 Schematic of stratified in a finite geometry ngular micro-channel (2-D Model)

The base state velocity is a func of y and z. The b tate is perturbed by giving a small disturbance to velocity and press he disturbance fi re decomposed into Fourier modes with respect to x. We obtain set o linear equations f disturbance amplitudes of velocity and pressure. Since we are deali ...th small disturbances we neglect the nonlinear terms with respect to the disturbance amplitudes and obtain a set of linear equations.

For single phase stratified flow between parallel plates (Figure 2) the critical Reynolds number is found to be 5772.22. The flow becomes unstable for Reynolds number greater than 57772.22.

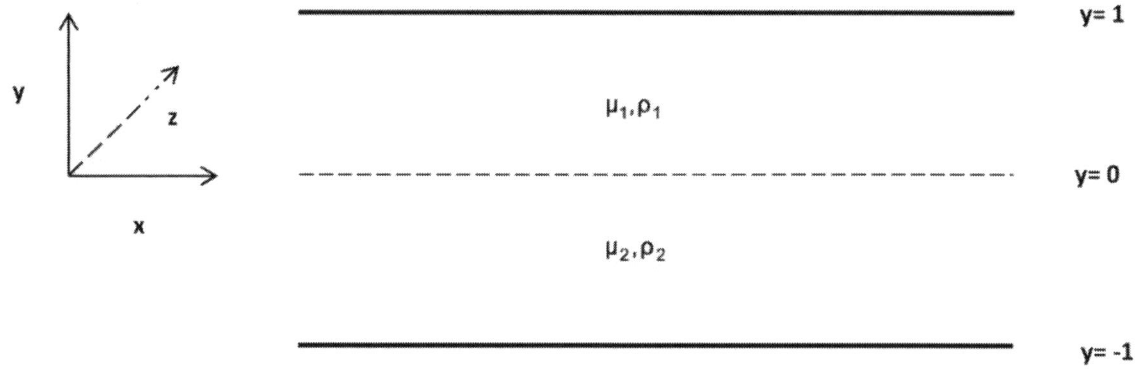

Figure 2: Flow configuration for plane Poiseuille flow of two superposed fluids

Numerical Procedure

We present a Chebyshev collocation method for solving the eigenvalue problem that defines the stability of the two phase stratified flow in a rectangular channel. The governing equations are solved using expansions in Chebyshev polynomials. The stability of stratified flow through a rectangular duct of arbitrary aspect ratio is investigated numerically by expanding the flow fields into a series of Chebyshev polynomials in both y and z directions. The derivatives involved in the governing equations are evaluated at the Gauss-Lobatto grid points. A linear transformation is used to scale the width and height of the rectangular channel such that the Chebyshev expansions used are orthogonal in the limit [-1, 1]. The resulting eigen value problem is solved using QR decomposition.

Key Results

The spectral Chebyshev collocation method is applied to two phase stratified flow between parallel plates extending to infinity. The single phase dispersion curve (Figure 3) is obtained by choosing the properties of both the fluids similar. we can conclude that the flow becomes unstable for Reynolds number exceeding the critical value 5772.22 and is stable for Reynolds number less than the critical value 5772.22.

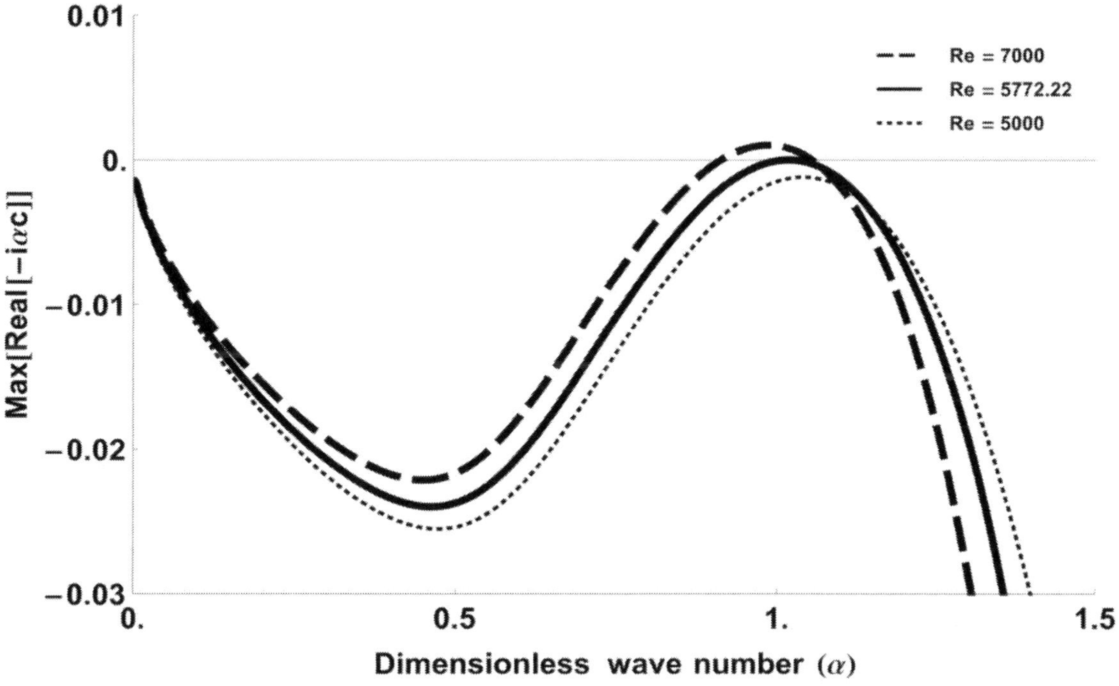

Figure 3: Growth rate vs. Dimensionless wave number (α)

386372 Biocompatibility Evaluation of Skin Wearable Biosensor Systems

Abstract:

Recent advances *in vitro* diagnostics on human skin or organ surfaces rely on the use of sensors. The innovative body-powered nano-enabled wearable health monitoring systems has gained considerable interest recently owing to the advantages of low cost on energy harvesting and convenience on long-term wear. However, engineered materials may display unique properties that have impact on human health. Lack of biocompatibility in a skin-attachable sensor, can result in complications including: cell cytotoxicity, skin irritation, skin sensitization, skin chronic inflammation, and even the recalls on commercial products due to the reported dermatitis. Biocompatibility is the ability of a biomaterial to perform its desired function, without eliciting any undesirable local or systemic effects in the recipients. An evaluation of biocompatibility is one part of the overall safety assessment of wearable biosensors since it will protect human safety. Assessment of the materials biocompatibilities on skin is currently performed with the use of animals. Alternative non-animal related *in vitro* methods are demanded by Animal welfare and EU legislation. Here, we performed standardized *in vitro* screening of skin-wearable sensor materials. Nano-materials and other sensor components were assessed *in vitro* with two skin-related cell lines to characterize and evaluate their biocompatibility. The Food and Drug Administration (FDA) G95-1 and International Standard ISO 10993 serve as a frameworks for selecting tests to evaluate the biocompatibility of medical devices.

We performed a screening of engineered materials and nanomaterials used for skin-wearable biosensors using two cell lines and 3-dimensional tissues. Three characterized in vitro assays were used to measure the cytotoxicity, irritation and sensitization of the skin-attachable sensor materials. These *in vitro* assays could meet the FDA requirement, as well as minimize the usage of live animals.

Continuous Spray Synthesis of Doped-Pyrochlore Oxide Catalysts for Hydrocarbon Reforming

Jonathan Yancey[a,b], Mark Smith[a],
David Berry[a], Dan Haynes[a], Dushyant Shekhawat[a], Edward M. Sabolsky[a,b]

[a] National Energy Technology Laboratory, U.S. Department of Energy, Morgantown, West Virginia 26507, USA

[b] Department of Mechanical and Aerospace Engineering, West Virginia University Morgantown, West Virginia 26506, USA

Improved performance of nano-catalysts with complex stoichiometry and morphologies will increase the demand for continuous, high-yield synthesis methods. Doped oxide catalysts with pyrochlore and perovskite structures are of particular interest in catalysis due to their ability to be modified by many metal dopants, which permit direct modification of the electrocatalytic properties. The purpose of this study is the synthesis of complex pyrochlore nano-catalysts using a continuous spray synthesis process. This spray synthesis method permits control of catalyst properties with continuous production. An aqueous solution containing various metal salts and organic additives was utilized to control the final nano-catalyst morphology and phase purity. The key variables identified include precursor composition, residence time, in-situ sintering temperature, and concentration. The surface area, crystallite size, and morphology of the catalyst were measured using BET, XRD, and SEM, and the performance of the catalyst was tested in a hydrocarbon reforming reactor. The results were confirmed by a series of experiments predicting and verifying catalyst reforming characteristics.

388901 Understanding Structure and Dynamics of Single Stranded DNA: Effect of Charges, Backbone, Length and Bases

Abstract:

Single stranded DNA (ssDNA) has been known as an important intermediate material in many key biochemical process, as well as novel materials in nano-biotechnology, such as biosensors and drug delivery vectors. It has been reported that structural flexibility of ssDNA plays an important role in understanding biochemical process in the cell and their applications. To elucidate the role of the ssDNA constituents including backbone, charges, length and bases on the structure and dynamics of ssDNA, we performed molecular dynamics (MD) simulation on regular ssDNA, neutralized ssDNA where all negative charges were removed, and backbone where all bases were truncated. In this study, we demonstrated different folding pathways of each case showing that regular ssDNA can form loop-like structures via internal non-bonded interactions, but neutralized ssDNA and backbone fold into entangled structure. By comparing folding pathway with persistence length and radius of gyration of each case, we found the importance of negative charges on stiffness of ssDNA. Moreover, it is shown that internal base pairing and pi-pi stacking between bases play a pivotal role in stiffness of ssDNA, which can't be fully explained by conventional polyelectrolyte theories. Overall, our study suggests the pivotal factors which can affect the structure and dynamics of ssDNA and explains the discrepancies between estimations by MD simulations and theories, which can give an insight on understanding of self-assembly of nucleic acid based materials and designing its applications.

Stability of Molten Liquid Jets Undergoing Solidification

Monisha Natchiar[1] S.R., D.S. Pillai[2], S. Pushpavanam[3]

The breakup of a capillary jet into drops is mainly attributed to instabilities driven by surface tension. Capillary breakup of jets under isothermal conditions, wherein there is no heat transfer between the jet and the ambient fluid have been extensively studied [1]. In this work, we study the scenario wherein the jet loses heat to the ambient fluid and thereby in the process undergoes solidification. The motivation for this work rests in its application for safety design of pool-type Fast Breeder nuclear reactors. In the event of a core disruptive accident in a pool-type nuclear reactor, the nuclear core material melts and falls as a jet into a pool of coolant. As the molten core descends, it loses heat to the coolant, solidifies and breaks into drops [2]. Solidification of the molten jet during its fall significantly affects its breakup dynamics. We develop a simple mathematical model to capture the effect of solidification of jet on its breakup dynamics.

In the present study, we assume the jet to be slender, i.e., the length of the jet is much larger compared to its radius. Thus, the problem consists of a small parameter called the slenderness ratio (ratio of the radius of jet to its length). We then proceed by expanding the dependent variables as a power series in terms of the slenderness ratio and obtain simplified equations to the lowest order called the *slender jet equations*. These one-dimensional coupled nonlinear partial differential equations govern the evolution of the interface position and the axial velocity of the jet. These equations are solved using method of lines on a staggered grid.

The effect of solidification is included by assuming the presence of a solid crust at the interface of the jet and the ambient fluid. The normal stress condition at the interface between the jet and the ambient fluid is then suitably modified to accommodate for the additional stress generated due to deformation of this solid cylindrical crust. The additional stress generated due to the crust formation however depends on the thickness of the solid crust, which is not known a priori. The thickness of the crust depends on the total time taken for the jet to breakup. Thus an iterative scheme is used to carry out an overall heat balance on the jet and determine the thickness of the crust and thereby predict the breakup dynamics.

[1] monishanatchiar@yahoo.co.in (Corresponding author)
[2] dipinsp@gmail.com
[3] spush@iitm.ac.in

This work thus aims to present a simple mathematical model to take into account complex effects of solidification on the breakup dynamics of the jet. The main advantage of this model is its computational elegance and the physical insight that it provides as compared to a conventional CFD simulation.

References

[1] J. Eggers and E. Villermaux, "Physics of liquid jets," *Rep. Prog. Phys.*, vol. 71, no. 036601, 2008.

[2] D.S. Pillai et al., "A nonlinear analysis of the effect of heat transfer on capillary jet instability," *Phys. Fluids*, vol. 24, no. 124106, 2012.

389789 Synthesis of Homogeneous and Mixed Ligand Alkanethiol Functionalized Silver Nanoparticles Via DLVO Modeling of Nucleation and Growth

Abstract:

Classical nucleation theory and Derjaguin, Landau, Verwey, Overbeek (DLVO) theory for colloidal stability were applied to gain insight into the synthesis of alkanethiol functionalized silver nanoparticles (NPs) by reduction of silver nitrate with sodium borohydride in ethanol. This analysis indicated the importance of quickly establishing a dense alkanethiol ligand brush on inherently unstable primary particles to achieve colloidal stability. The DLVO calculations also indicated that the electrostatic potential was a minor contributor to repulsive interactions, signifying that it would be possible to control NP size and uniformity in solutions with high ionic strength, as long as sufficient alkanethiol ligand was available to form a densely packed self-assembled monolayer (SAM) on the NPs. These insights were applied to design a new straightforward, one-step, one-phase synthesis for the production of a wide variety of alkanethiol-functionalized silver NPs. To test the insights from DLVO theory, 16 docanethiol samples were synthesized in the parameter space R=3-12, S=1-12 where R=[$NaBH_4$]/[$AgNO_3$], S=[DDT]/[$AgNO_3$], and [$AgNO_3$]=10mM. In general, samples with R=3 or S=1 were polydisperse, however, samples in the R=6-12 and S=3-12 range had uniform particle sizes with average diameters between 3.5-4.7 nm. Additionally, samples with R=72-108 and S=12 were synthesized to test particle stability at high ionic strength; again uniform NPs with average diameters from 3.5-3.8 nm were produced. Finally, the synthetic methods developed with dodecanethiol were extended to silver NPs having SAMs composed of single ligands from butanethiol through undecanethiol, as well as mixed SAMs composed of a combination of dodecanethiol and a ligand from the series butanethiol-undecanethiol. All syntheses were conducted with R=12, S=12, [$AgNO_3$] = 0.5 mM and resulted in similar quality of NPs to pure dodecanethiol functional NPs. NP size did not change appreciably from dodecanethiol as ligand length was shortened, however, polydispersity was observed to increase as the average ligand length of the SAM decreases. This result follows directly from our understanding of the growth processes of this system through DLVO, as the steric barrier to aggregation was calculated to be smaller for shorter ligand lengths, resulting in more NP aggregation events. Ultimately, the insights gained from DLVO theory successfully guided the development of a one-step, one-phase technique for the synthesis of uniform, spherical alkanethiol-functionalized silver NPs with both single-ligand and mixed SAMs composed of ligands from butanethiol through dodecanethiol.

394217 Size Dependent Interactions of Gold Nanoparticles with a Supported Lipid Bilayer

Abstract:

With their small size and unique surface area properties, nanoparticles are being investigated for applications spanning fields from biomedical applications to uses in consumer products. While the use of NPs in applications appears promising, the consequence of their disposal is unknown. This study serves to understand how NPs of varying sizes will interact in environmental and biological systems. A supported lipid bilayer (SLB) was used as a model surface for NPs to interact with. The changes on the bilayer were monitored by Quartz Crystal Microbalance with Dissipation, which is able to measure frequency (inverse of mass) and dissipation (surface rigidity) changes at the nano-level in real time. Using this novel technique, we studied gold NPs of sizes 2, 5, 10, and 40 nm in water and poly(methacrylic acid) (PMA). Water simulated a clean lab condition while the poly(methacrylic acid) modelled an environment more typical of soil. In water, all particles behaved similarly with some lipid perturbation causing small amounts of mass loss from the SLB. However, when PMA was introduced onto the bilayer and coated the NPs, the interaction mechanism was altered. In PMA, the SLB-NP interaction with smaller particles (2 nm and 5 nm) resulted in mass addition onto the bilayer surface, which is due to the particles remaining in the bilayer during interaction. The 10 nm particles behaved similarly in PMA as they did in water with small amounts of lipid removal. The largest particles studied, 40 nm NPs, resulted in large mass loss (~ 60 ng) from the lipid bilayer. Simulating a typical environment in which NPs are disposed of is crucial to understand the interaction mechanisms. Our results demonstrate the importance of both NP core size and bulk solution for their fate within our environment.

394188 Greer

Abstract:

Injectable hydrogels have recently become a relevant material platform in applications such as in vivo drug delivery or as cell carriers because they can be implanted in a minimally invasive manner and closely mimic biological tissue. Attractive properties for such injectable hydrogels are that they easily flow under applied stress (*shear-thinning*) and rapidly heal upon relaxation of the applied stress (*self-healing*). Previously, we developed a shear-thinning and self-healing injectable hydrogel driven by non-covalent interactions between hydrophobically modified cellulose and PEG-PLA nanoparticles. [1] The nature of this gel enables facile loading with a variety of therapeutics as well as tunable release profiles of multiple therapeutics. Specifically, the PEG-PLA nanoparticles allow for loading with hydrophobic therapeutics while a hydrophilic therapeutic can be entrapped within the aqueous phase of the gel.[2] In this work, we have quantified the factors that affect the mechanical properties and release profiles of the gels, demonstrating the range of their tunability. Additionally, we explored other nanoparticles and biopolymers for the fabrication of similar materials. Lastly, we studied the use of this injectable hydrogel as a cell carrier with osteogenic properties. Human mesenchymal stem cells (hMSCs), which are capable of differentiating into bone, cartilage, and fat cells, were encapsulated in the gel and survived injection through a 28 gauge needle into three-dimensional collagen hydrogels. Additionally, PEG-PLA nanoparticles were loaded with dexamethasone to stimulate osteogenic differentiation of the encapsulated hMSCs. We are currently exploring the use of this drug-loaded cell carrier as an injectable system for bone healing. Overall, we present an injectable hydrogel with tunable properties for the controlled delivery of therapeutics with facile synthesis and minimally invasive application *in vivo* that is capable of serving as a stimulatory cell carrier.

---

[1] Appel, E.A., Tibbitt, M.W., Langer, R., Nature Communications., **2014**, *in revision*.

[2] Greer, J.M., Appel, E.A., Tibbitt, M.W., Langer, R., *manuscript in preparation*.

Microdosing of Pharmaceutical Products into Capsules

E. Faulhammer[1, 2], *M. Besenhard*[1, 2], *S. Fathollahi*[1, 2], *M. Llusa*[2], *S. Lawrence*[3], *S. Biserni*[4], *V. Calzolari*[4], *J.G. Khinast*[1, 2]

1. Technical University, Inffeldgasse 13, 8010, Graz, Austria. Email: khinast@tugraz.at
2. Research Center Pharmaceutical Engineering, Inffeldgasse 13, 8010, Graz, Austria
3. GlaxoSmithKline (GSK), New Frontiers Science Park, Harlow, Essex CM19 5AW, UK
4. MG2, Via del Savena, 18, I-40065 Pian di Macina di Pianoro, Bologna, Italy

Abstract

Introduction

In the pharmaceutical industry a wide range of capsule filling systems, which employ different technologies and principles, are used. Hard capsules can be filled in several ways from manual preparation to fully automated industrial production (Podczeck and Jones 2004). The current trend in pharmaceutical industry is to manufacture small doses (< 50mg) of potent active pharmaceutical ingredient (API) for early research clinical trials using drug in capsule approach and as an inhalation formulation (Ashurst et al., 2000). These low doses pose challenges during manufacturing (Edwards 2010, Eskandar et al., 2011, Llusa et al., 2013). One opportunity to fill very low doses is to use capsule fillers with vibration assisted gravimetric feeding (i.e. MG2 Microdose, Capsugel Xcelodose®S, and 3P Innovation Fill2weight). These filling systems are of special interest for their capability of filling with doses in the range between 0.1 mg and 100 mg in capsule. Therefore, these systems need in-process control to confirm the accuracy and uniformity of the filled weight.

The MG2 Microdose system can be operated in several ways: as stand-alone unit (speed depending on operator), on the lab scale machine (Labby) with a production speed up to 700 capsules per hour (CPH) and installed on the high end industrial machine (Planeta) with an output of approximately 3000 CPH, depending on product type.

The present study aims to investigate the feasibility of filling microdoses of diverse APIs and excipients (single or multi dose) using the MG2 Microdose System.

Materials and Methods

For the present study, the MG2 Microdose System was used as a stand-alone unit. In this dosing principle, the capsule body is filled directly via a sieve merged with a vibrating plate ("pepper shaker principle"). Figure 1 shows the operating principle of the system. Hard gelatin capsules (size 3, Capsugel) were filled up to target weights ≤ 5mg. The amount of powder dispensed is a function of different material attributes and process parameters. The vibration amplitude as well as its frequency can be adjusted to achieve the desired fill weight. This is done either manually or by using a PID controller. Further the diameter and number of holes in the sieve have major influence on capsule fill weight. The latter is measured during each dosing event by a capacitance system, allowing full traceability of the samples. Model powders with different material attributes were selected and analyzed regarding their vibration dependent filling attributes. With the knowledge of the filling behavior of every single model substance, combinations of the latter were successfully dosed into single capsules. All experiments were performed under controlled temperature and humidity to avoid moisture uptake.

Results

The controlled parameter in our study is the amplitude of vibration during feeding a target weight of 5 mg. When the target is reached the vibration stops and amplitude goes to zero (Figure 1). Our results demonstrate that highly accurate dosing can be achieved using adequate vibration settings. It was possible to prevent overdosing, non-uniform capsule fillings and varying filling times originating from inhomogeneous powder beds on the vibration plate. We plan to characterize flow behavior of powders as a function of amplitude and frequency.

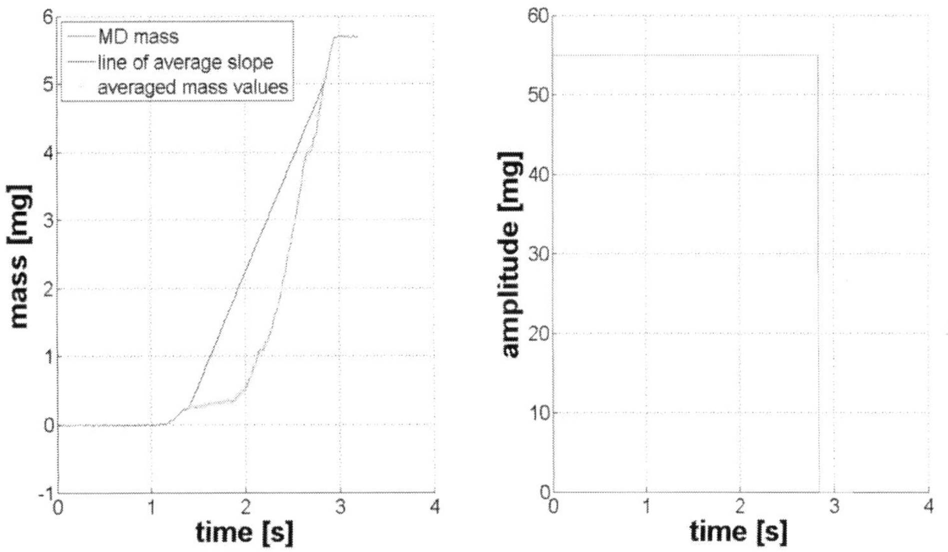

Figure 1: Controlled parameters in mcirodose system

Conclusion

The filling behavior of the model substances could be related to their critical material attributes (CMAs) and to the critical process parameters (CPPs). Furthermore, we demonstrated how the model powders can be filled most precisely (single and multidose). Accurate filling of several substances into single capsules with the Microdose system seems highly promising from the perspective of personalized medicine.

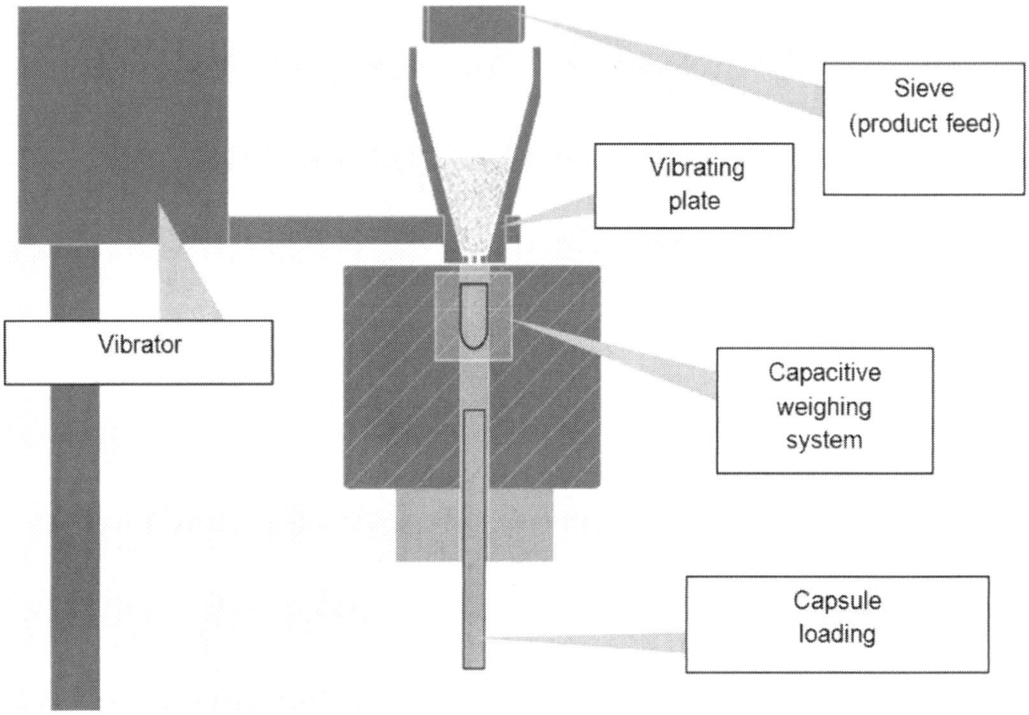

Figure2: Microdose Operation Principle

References:

Edwards, D. (2010). Applications of capsule dosing techniques for use in dry powder inhalers. *Therapeutic delivery*, *1*(1), 195–201.

Eskandar, F., Lejeune, M., and Edge, S. (2011). Low powder mass filling of dry powder inhalation formulations. *Drug Development and Industrial Pharmacy*, *37*(1), 24–32.

Llusa, M., Faulhammer, E., Fink, M., Scharrer, G., Calzolari, V., Biserni, S., Lawrence, S., Bresciani, M., Khinast, J.G.. https://aiche.confex.com/aiche/2013/webprogram/Paper306359.html

Podczeck, F., and Jones, B. E. (2004). Dry filling of hard capsules. In F. Podczeck and B. E. Jones (Eds.), *Pharmaceutical Capsules* (2nd ed., pp. 119–138). London: Pharmaceutical Press.

393303 Numerical Simulation on the Feasibility of Bulk CO¬2 Separation in Natural Gas via Supersonic Nozzle Expansion System

Abstract:

The feasibility of separating CO_2 from natural gas through a convergence and divergence nozzle in supersonic condition has been studied. Due to the decrement of pressure and temperature at the throat, separation of denser phase i.e. liquid CO_2 is made possible at supersonic, compressible flow condition. A full hydrodynamic CFD simulation in Fluent was performed on an axisymmetric nozzle to assess the pressure, temperature, velocity and the turbulence profile. To evaluate the separation of CO_2 from the natural gas, a selection of thermodynamic and physical properties has been requisite and assessed numerically as well as through experimental means. The combinations of these physical properties; nucleation, growth, coalescence and breakage of CO_2 liquid are combined in the Population Balance Model (PBM) to attain the amount of CO_2 liquid formed in the nozzle.

The non-equilibrium condensation is modelled by combinations of Classical Nucleation Theory and other empirical formulas coupled with SRK equation of state. A computational code written in C language is linked with Fluent as User-defined Function (UDF) to generate the particle size distribution of CO_2 droplets as well as the number density of the droplets.

Results generated by the full coupling exercise shows clear formation of CO_2 liquid droplets in the divergence section of the nozzle. The amount of droplets produced increases after the throat indicating that the balancing of each model had been achieved successfully in determining the separation efficiency of CO_2 in natural gas containing CH_4. The vapour CO_2 mass fraction is found to be decreasing as expansion grows whereas the distribution of droplet number density is mainly concentrated at the wall of the nozzle after the throat. The results indicate that separation of CO_2 through convergence-divergence nozzle is possible with further efficiency of separation can be done via optimum geometrical design of the nozzle.

366038 Proteomics Analysis of Extracellular Matrix Derived from Pluripotent Stem Cell Aggregates

Abstract:

The characterization of proteins in the extracellular matrix of pluripotent stem cells is of value in analyzing the paracrine signaling in which the cultured cells are engaged in. Proteomics analysis based on liquid chromatography–mass spectrometry was used to analyze the proteins secreted in the extracellular matrix by three different types of murine embryonic stem cell cultures: undifferentiated aggregates, embryoid bodies, and neural progenitor cells derived from the murine stem cells. Undifferentiated aggregates are able to proliferate extensively and maintain the pluripotency. Embryoid bodies are aggregate-like structures that contain the cells from three germ layers. Neural progenitor cells were derived through the embryoid body formation treated by retinoic acid. These different cellular systems secrete distinct composition of extracellular matrix shown in the proteomics analysis. Data analysis was performed using Scaffold proteome software and the Ensemble database. Data sets were filtered using gene ontology classification for cellular components based on the terms "extracellular proteins". Different proteins were identified in various subsets of the three groups which may be responsible for the tissue development of different cellular systems of pluripotent stem cells. The proteomics analysis performed in this study provides the knowledge and the understanding on the differentially expressed proteins found in the extracellular matrix of pluripotent stem cells which experienced different developmental paths.

379086 Mathematical Modeling of the Physicochemical Evolution of Biomass Particles during Pyrolysis and Gasification

Abstract:

Gasification has been identified as an energy-efficient, environmentally-friendly and economically-feasible technology to partially oxidize biomass into a gaseous mixture of syngas consisting of H_2, CO, CH_4 and CO_2. High-quality syngas can be further used to catalytically synthesize liquid fuels and produce hydrogen. Multi-scale physical and chemical changes at both particle and reactor levels occur during biomass pyrolysis and gasification. At a biomass particle level, the intra-particle reactions, heat and mass transfer, and structural evolution occur during pyrolysis and gasification when the volatiles are released. The yield and composition of the products from a reactor depend on these physicochemical changes of the biomass particles. This research is to develop a transient mathematical model to describe the physical and chemical changes of biomass particles during pyrolysis and gasification at different reaction environments. The mathematical model consists of four sub-models to describe heat transfer, mass transfer, reactions and structural evolution, respectively. A heat transfer sub-model is used to determine the time-temperature profile of a biomass particle with varying physicochemical properties under different reaction environments. A temperature-dependent reaction kinetics sub-model is then used to determine the volatilization rates and predict the dynamic elemental composition of the particle. A mass transfer sub-model is used to describe the intra-particle transport of the volatiles. A structural evolution sub-model is used to predict the changes of porosity, surface area and particle size of the biomass particle that are caused by the volatilization during pyrolysis and gasification and significantly affect the heat and mass transfer of the particle. The model will be validated by the experimental data of dynamic mass, elemental composition and physical properties of biomass particles during pyrolysis and gasification. The dynamic mass of the particles during pyrolysis and gasification are measured by a thermogravimeter (TGA). The elemental compositions of the biomass samples collected at different stages of pyrolysis and gasification are analyzed by an elemental analyser. The porous structure of the biomass particles including pore size, pore volume and surface area is studied by using a Micromeritics ASAP 2020 analyzer. Direct observation of the porous structure is performed using a Carl Zeiss Auriga-BU FIB FESEM microscope and Carl Zeiss Libra 120 plus TEM microscope.

381410 A Comparision of Biocrude-Oil Characteristics from Coffee Ground and Sawdust

Abstract:

Biomass conversion to biocrude-oil has been extensively studied worldwide as a renewable energy technology and an answer to global warming from fossil fuel, since it has little carbon emission with respect to fossil fuels. Biocrude-oil also has logistical advantages such as storage and convenience of transport than other renewable sources of energy. Fast Pyrolysis Biocrude-oil (FPBO) is an effective technology used in producing biocrude-oil but production costs and unstable supply of feedstock are the main drawbacks. Even though many studies have been done to minimize these problems, there are little studies done on abundantly found coffee ground based biocrude-oil. This study focuses on biocrude-oil products from coffee ground in comparison with sawdust. The characteristics of biocrude-oil production by coffee ground are compared with biocrude-oil from sawdust in a bench scale bubbling-fluidized-bed reactor for fast pyrolysis. Product yields and oil characteristics such as higher heating value, water content, viscosity, pH, density, ash, solid residue and functional group composition were also investigated at various reaction temperatures ranging from 400°C to 600°C.

389965 Analysis of the pH and Thermo-Responsive Behavior of a Series of Amino (Meth)Acrylate Polymer Brushes on Silicon Substrates By in-Situ Ellipsometry and AFM Measurements

Abstract:

Stimuli responsive polymers have potential use in a wide range of applications, from nanometer-scale drug delivery to flow control in microfluidic devices. In this study, surface-confined atom transfer radical polymerization (ATRP) or single electron transfer living radical polymerization (SET-LRP) were used to synthesize a series of surface-grafted poly(amino (meth)acrylate) brushes from silicon/gold substrates that were pre-modified with silane self-assembled monolayers (SAMs). The following polymers were synthesized and characterized: poly(2-(dimethylamino)ethyl methacrylate) (PDMAEMA), poly(2-(dimethylamino)ethyl acrylate) (PDMAEA), poly(2-(diethylamino)ethyl methacrylate) (PDEAEMA), and poly(2-(tert-butylamino)ethyl methacrylate) (PTBAEMA). *In-situ* spectroscopic ellipsometry studies were performed using a temperature-controlled liquid cell to collect dynamic and equilibrium data at different pH/ionic strength conditions. Multilayer models were generated based on a general-oscillator (Genosc) dispersion model that accounted for the polymer, initiator, and silane SAM layers. In-situ atomic force microscopy (AFM) measurements were collected at different pH and temperature conditions using a liquid cell. AFM scans were used to determine any phase transition or aggregation behavior of the grafted polymers in response to pH and temperature changes. In addition to examining the structure-property relationships for a series of surface-tethered pH and temperature responsive poly(amino (meth)acrylates) brushes, a novel Genosc model for evaluating optical constants of surface-tethered polymer brushes is described.

368941 The Correlation Between Dioxin Level of Flue Gas, Ambient Air, Vegetation and Soil Nearby Mswi

Abstract:

We set up information about Polychlorinated dibenzo-p-dioxins and polychlorinated dibenzofurans (PCDD/F) level in flue gas, ambient air, banyan leaves and soil nearby a municipal solid waste incinerator(MSWI). The International Toxicity Equivalency Factors (ITEQ) concentrations of total PCDD/F in flue gas are calculated, respectively, in spring and autumn. The mean PCDD/F concentrations in ambient air and banyan leaves nearby MSWI are also given, respectively at four seasons. The correlation between theses level is discussed in this paper.

CFD Modeling of Commercial Scale Thawing Time from Small Scale Bags

Wei Chen, Mita Maity, Vikram Sadineni and Rao Mantri

Research and Development
Bristol-Myers Squibb Co.
AIChE Annual Meeting 2014

Background

1. It is imperative that the parameters required for freezing and thawing of the frozen formulated drug substance (FFDS) be thoroughly evaluated and optimized before transferring the process to commercial DP manufacturing site.
2. This unit operation contributes the most to the process parameter, time out of refrigeration (TOR), which is a CPP considering the inherent instability of biopharmaceuticals upon exposure to ambient temperatures

Objectives

1. Understand protein thawing behavior and identify key impacting parameters
2. Support protein thawing scale up from 30ml to 6L and 12L bag

Freezing and Thawing Theory

Schematic of energy balance for a small thawing element

$$\frac{\partial}{\partial t}(\rho H) + \nabla \cdot (\rho \vec{v} H) = \nabla \cdot (k \nabla T) + S$$

Accumulation **Convection** **Conduction** **Source**

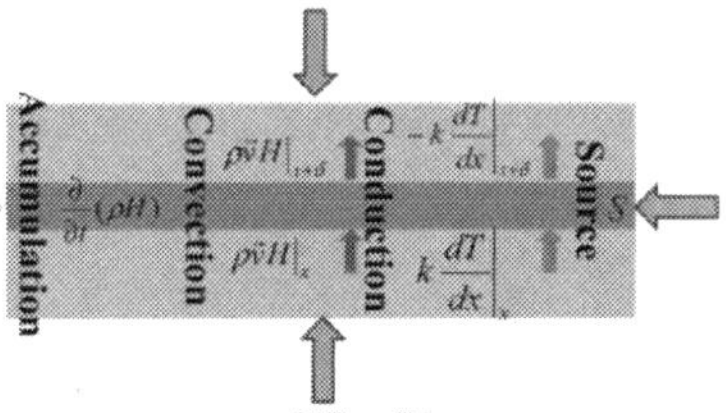

$H = h + \Delta H$

H Enthalpy, ρ *Density*, *S Source term*, $\vec{v}$ *Fluid velocity*

$\Delta H = \beta L$ *Latent heat*, *L Latent heat of materials (Liquid)*

$h = h_{ref} + \int_{T_{ref}}^{T} c_p dT$, *Sensible enthalpy*

$\beta = \dfrac{T - T_{solidus}}{T_{liquidus} - T_{solidus}}$ *if* $T_{solidus} \le T \le T_{liquidus}$ *Liquid fraction*

Thawing Model Verified by 30 ml Bag Experimental Data

Model Parameters & Boundary Condition

Bag (30ml)

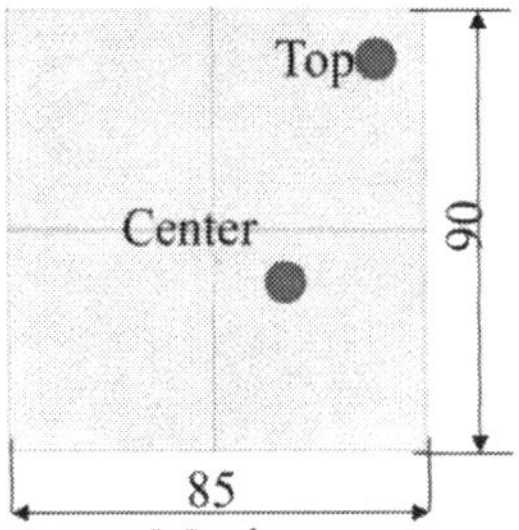

Mesh

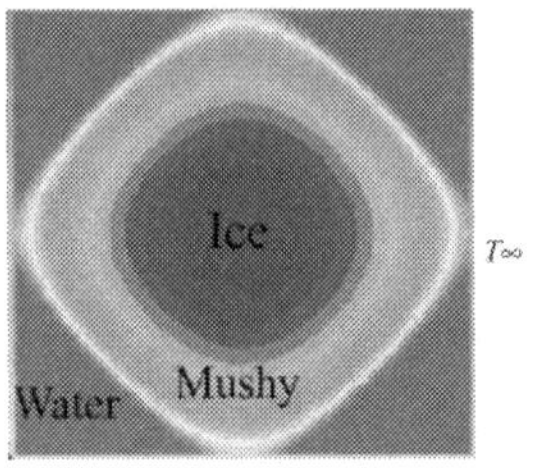

Water Volume Fraction

1. Environmental Temperature : 20 C
2. Heat convention coefficient ; Static: 5w/m^{2}K & Agitated: 8w/m^{2}K
3. Heat capacity from 2040(Ice)~4182(Liquid) ,J/KG
4. Conductivity change from 1.88(Ice)~0.6(liquid) ,w/m^{2}K
5. Thawing heat change : 92082 J/kg
6. Tsolidus :267 , Tliquidus: 273

Thawing Experimental Setup

Placement of Probes

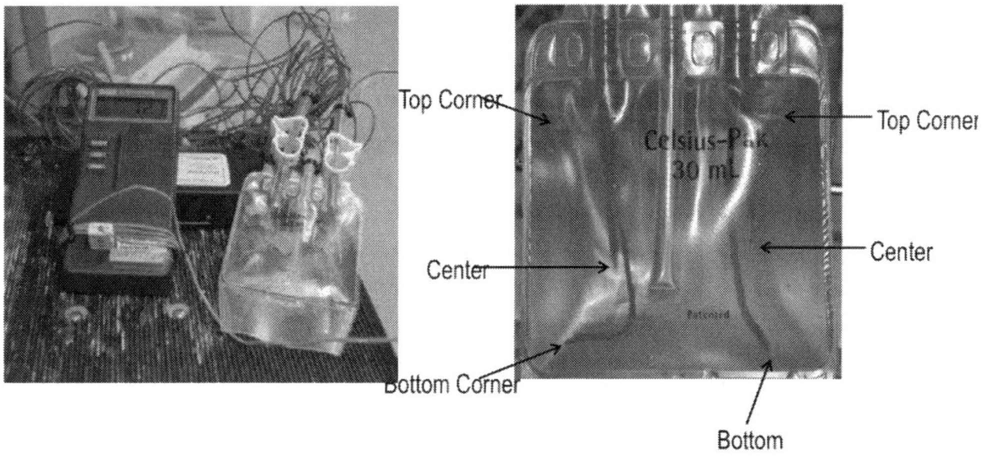

Thawing Model Verified by 30 ml Bag Experimental Data

Temperature Comparisons between experiment and prediction during Thawing

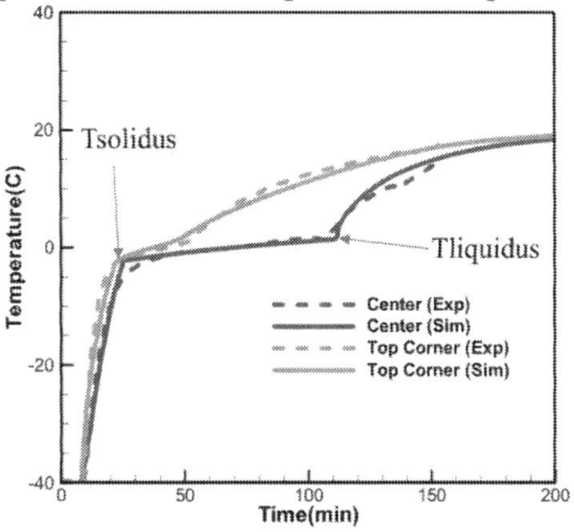

1. Different location temperatures predicted agree well with experimental data
2. The time for the temperature to rise to the thawing point (sensible heat)is relatively shorter than the "thawing" time (latent heat) , the deep freezing does not matter the thawing time.
3. The shape of the bag play an important role in the thawing time
4. Tliquidus and Tsolidus depend on protein formulation(composition)

Thawing Model Verified by 30 ml Bag Experimental Data

Water Volume Fraction Change with Time

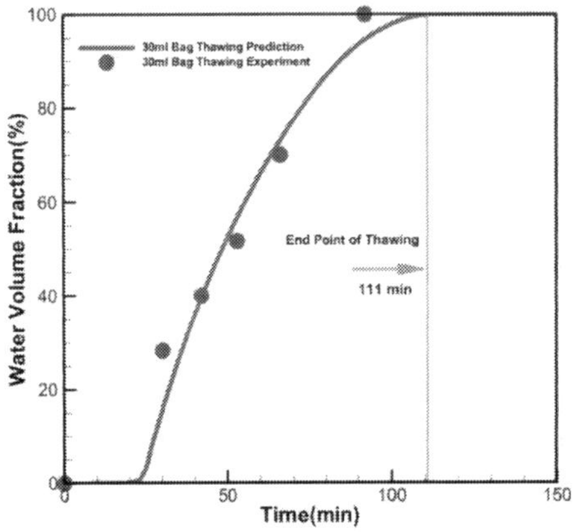

Thawing Dynamic

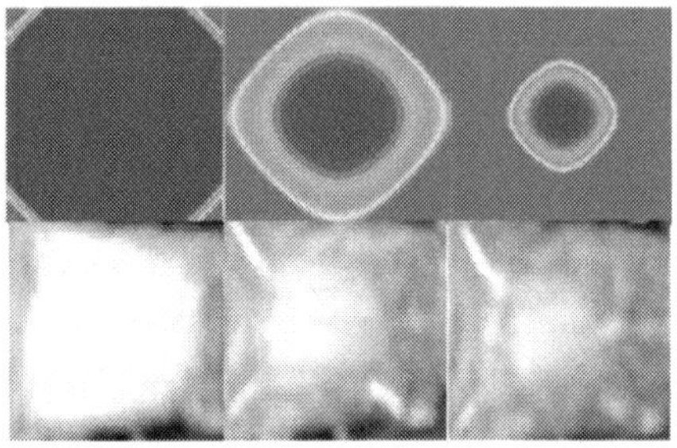

t 12min 48min 84min

1. The predicted thawing time agree well with experimental data
2. Mushy zone size is important for thawing process design can be predicted by Tliquidus and Tsolidus

Verified Thawing Model Predicts 6 L Bag Thawing Behavior

Thermocouple Point

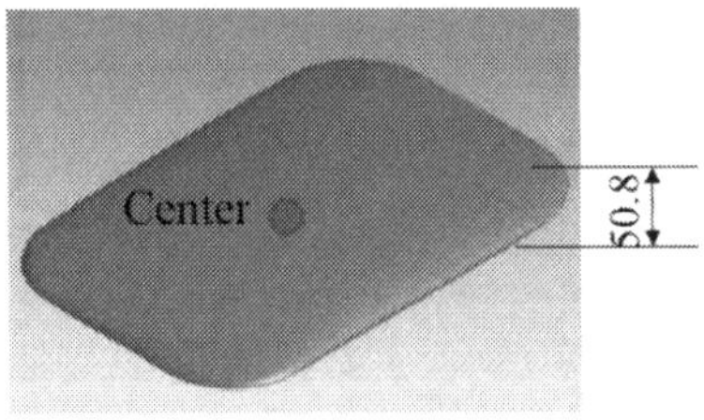

6 L Bag

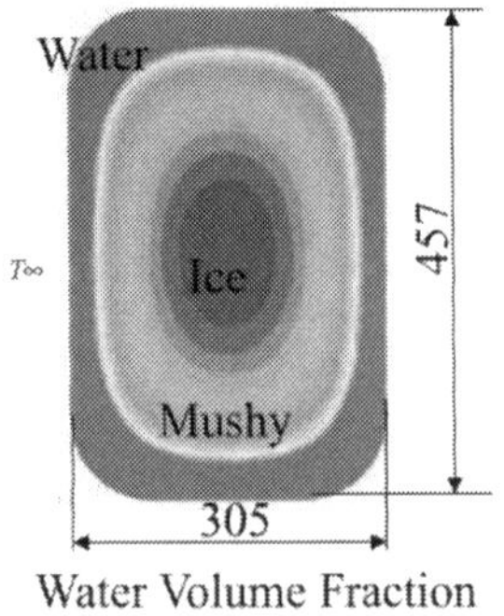

Water Volume Fraction

Apply model parameters of 30ml into 6 L Bag

1. Environmental Temperature : 16 C
2. Heat convention coefficient ; Static: 5w/m^{2}K & Agitated: 8w/m^{2}K
3. Heat capacity from 2040(Ice)~4182(Liquid) ,J/KG
4. Conductivity change from 1.88(Ice)~0.6(liquid) ,w/m^{2}K
5. Thawing heat change : 92082 J/kg
6. Tsolidus :267 , Tliquidus: 273

Verified Thawing Model Predicts 6 L Bag Thawing Behavior

Water Volume Fraction Change with Time

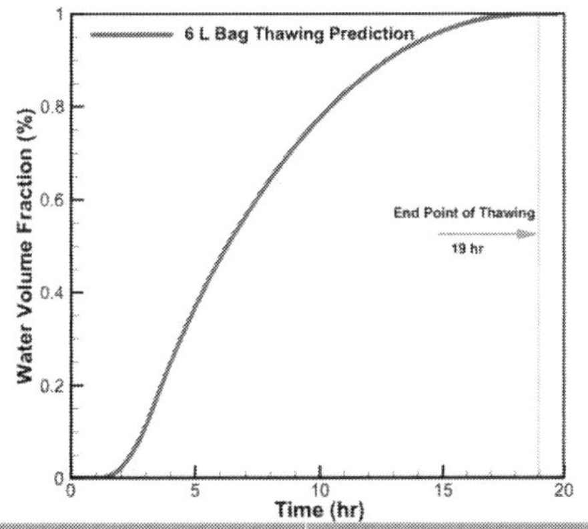

	6L-Static Cycle-Exp-1	6L-Static Cycle-Exp-2	6L-Static Cycle-Pred.
Thawing Time(min)	1265	1139	1140

The predicted thawing time (19hr) agree well with experimental data

Sensitivities Study: Heat Convection Coefficient vs Thawing Time

Natural Convection

$$Ra = Gr\,\mathrm{Pr} = \frac{g\beta(T_s - T_\infty)\delta^3}{\nu^2}\mathrm{Pr}$$

$$Nu = \frac{h\delta}{k} = CRa^n$$

Forced Convection

$$\overline{Nu}_D = \frac{\bar{h}D}{k} = C\,\mathrm{Re}_D^m\,\mathrm{Pr}^{1/3}$$

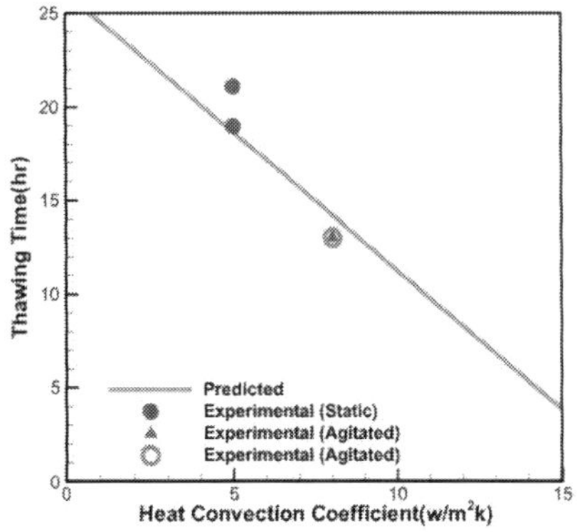

Any process parameters , which make Heat Convection Coefficient increase such as agitated and forced air flow , decrease the thawing time

Effect of Sucrose on Thawing Behaviors

5mg/mL protein in 3.0%/8.5% Sucrose

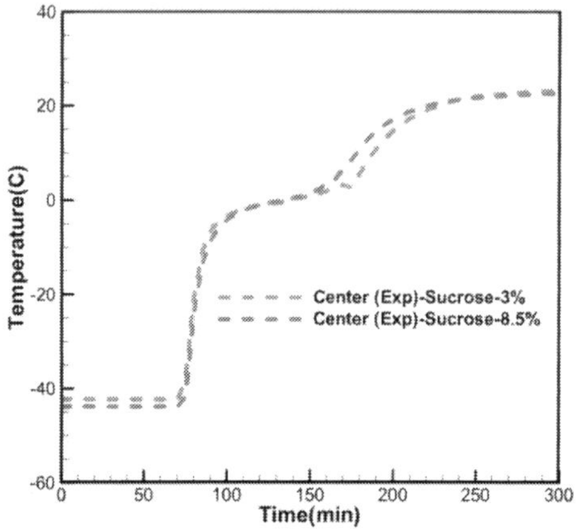

1. Latent heat of fusion and Tliquidus slightly reduces with increase of Sucrose concentration
2. Thaw is much easier with adding more Sucrose

Effect of Shape of Bag on Temperature

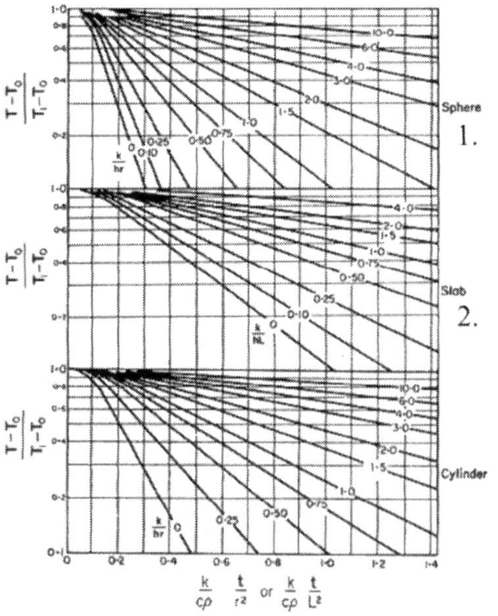

1. The graph shows the temperature of the center of the sphere, slab and cylinder changes with time(without phase change), generally the time to reach the same temperature , Slab>Cylinder>Sphere
2. The thawing time is expected : Slab >Cylinder>Sphere

http://www.nzifst.org.nz/unitoperations/httrtheory4.htm

Conclusion

1. A mechanistic protein freezing and thawing model using CFD was developed to predict thawing behavior
2. The model was verified and its parameters were obtained through 30ml bag experiments, The verified model successfully predict 6 L bag thawing time.
3. The protein composition specially excipients affects the Tliquidus ,Tsolidus and latent heat of fusion
4. The thawing time is mainly controlled by latent heat (thawing) not by sensible heat. The deep freezing does not matter the thawing time

367926 Bioelectrical Impedance Measurements to Detect Changes in Tight Junction Expression at Cell Interfaces

Abstract:

Precisely determining barrier function of endothelial and epithelial cells to ion passage requires advanced electrical techniques. Impedance Spectroscopy, a non-invasive measurement system that can quantify the barrier integrity of a cell monolayer in real time, provides the best opportunity for doing so.We analyzed the effect of phenylarsine oxide (PAO)-induced proteolysis of occludins on Madin-Darby canine kidney (MDCK) cell monolayer impedance as a function of time using this method. Occludins are tight junctional proteins that have been shown to regulate the transport of ions, water and other molecules through cell monolayers. The cell monolayer is modeled as a parallel RC circuit, and the cell resistance and capacitance are recorded before and after the addition of the PAO containing media, providing information on the effect of occludin proteolysis to overall cell behavior. Impedance spectroscopy of the MDCK monolayers before and after treatment show a clear response of the cells both resistively and capacitively to the PAO containing media indicating that occludin proteolysis effects both the junctions and the cells. Further investigations are needed to determine how to modulate tight junction proteolysis in a manner that minimizes the effect on the cell capacitance and only affects the junctions.

Nanostructure modified Ni-YSZ anode by atomic layer deposition for solid oxide fuel cells

Xiaodan Cui[1], Alan D. Zdunek[1] and Christos G. Takoudis[1,2]

[1]Department of Chemical Engineering, University of Illinois at Chicago, Illinois 60607, USA

[2] Department of Bioengineering, University of Illinois at Chicago, Illinois 60607, USA

E-mail address: xcui8@uic.edu

Abstract

Solid oxide fuel cells (SOFCs) are energy devices that directly convert chemical energy to electrical energy through electrochemical reactions. They attract a lot of interest due to their high energy efficiency and environmental safety.[1] Ni-YSZ (yttria-stabilized zirconia) is a common anode for SOFCs. The performance of Ni-YSZ anode is critically dependent on its nanostructure.[2] Under SOFC operating conditions, conventional Ni-YSZ anodes suffer from vapor loss, sintering and coking problems of the nickel phase due to poor coverage of the metal phase by the YSZ. Development of an ideal Ni-YSZ anode nanostructure requires optimizing the ceramic phase coverage protecting the metal phase.

In this work, a novel method combining glancing angle deposition (GLAD) and atomic layer deposition (ALD) was applied to fabricate Ni-YSZ anodes with a unique nanostructure. The fabrication process of the modified nanostructured Ni-YSZ anode is shown in Fig.1. The nickel nanorod backbone was deposited by GLAD. ALD was used to coat a uniform YSZ layer over the surface of nickel. Finally, the anode was annealed to open gas pathways for gaseous fuel. Electrical conductivity measurements were performed using a 4-point probe method and the morphology of the nanostructured Ni-YSZ anode was examined with SEM (Fig. 2). By coating YSZ film on nickel nanorods, a continuous metal phase in contact with the ceramic phase is achieved and the metal phase is stabilized. Consequently, not only is the triple phase boundary (TPB) extended to the entire nickel surface, but the catalyst activity and electrical conductivity of nickel are also maintained.

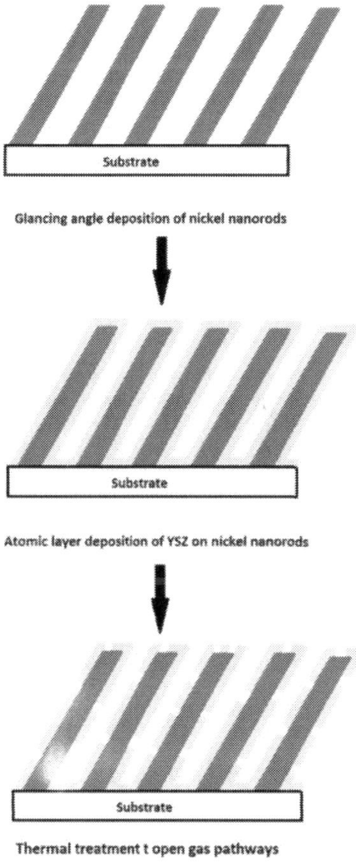

Figure 1 fabrication process of nanostructure modified Ni-YSZ anode

Figure 2 SEM image of nanostructure modified Ni-YSZ anode

References

1. Brett, D. J. L.; Atkinson, A.; Brandon, N. P.; Skinner, S. J., Intermediate temperature solid oxide fuel cells. *Chem. Soc. Rev.* **2008,** *37* (8), 1568-1578.

2. Jiang, S. P.; Chan, S. H., Development of Ni/Y2O3-ZrO2 cermet anodes for solid oxide fuel cells. *Mater. Sci. Technol.* **2004,** *20* (9), 1109-1118

381615 Lithium Chloride and Expanded Graphite-Mediated Hydration Performance Enhancement of a Novel Chemical Heat Storage Material for Lioh/Lioh•H2O Reaction

Abstract:

Low temperature exhaust heat below 373K is enormously unused and discharged from various fields in the world due to the difficulty in utilization caused by its low exergy; e.g. it accounts for about 65% of waste heat discharged from industrial sector over the world. Besides, combined heat and power is recently expected to play an important role in improving total energy efficiency. The key point of this system is to store and utilize heat below 373 K with high efficiency generated from power generator, such as gas engine and fuel cell.

Reversible chemical reactions are appropriate for thermal energy storage because of their high energy densities compared to conventional sensible or latent heat storage systems. Chemical heat storage for such low temperature thermal energy is expected to play a key role to establish a highly energy utilization system. Therefore, in this study, we have first tried to find the promising compounds for low-temperature heat storage by measuring endothermic heat and temperature with a differential scanning calorimetry. As a result, it was found that lithium hydroxide monohydrate dehydrated at 337 K with endothermic heat of 1,440 kJ/kg. However, a serious problem needing to be settled is that hydration of lithium hydroxide proceeds very slowly. The conversion of LiOH hydration was only 21% at 2h, which severely reduce the cycle efficiency of heat storage in the reaction system.

The hydration performance enhancement for $LiOH/LiOH{\cdot}H_2O$ reversible chemical heat storage reaction was discussed. A novel composite, denoted as ELL, was developed by mixing pure lithium hydroxide with lithium chloride and expanded graphite, which were employed as hygroscopic material and heat transfer enhancers, respectively. A high-yield impregnation method was used to prepare the composite as follows: a specified amount of LiCl and expanded graphite were mixed with 100 mL deionized water, followed by addition of pure LiOH powder into the solution. Next, the mixture was sonicated for 40 min to generate a sufficient dispersion. Subsequently, the excess water in the solution was removed by centrifugation, and the remaining wet product was dried at 110 in an oven overnight.

Fundamental study of dehydration behavior of ELL was also performed with a thermo gravimetric analyzer. With respect to pure lithium hydroxide powder, ELL showed much higher reaction rate for hydration transformation without a great loss of energy storage density, the heat output capacity after 30min hydration was nearly 1,200 kJ/kg. The dehydration kinetics of the ELL composite was well fitted with a first-order rate equation in the temperature range between 336K and 373K, addition of LiCl and expanded graphite was found to decrease the estimated activation energy in the dehydration. In conclusion, it was established that the newly developed ELL composite is a promising candidate for novel chemical heat storage materials for chemical heat pumps at working temperatures below 373K.

Reaction Studies on Na_2WO_4-Mn/SiO_2 Catalyst for Oxidative Coupling of Methane

Aseem[a], Geofrey Goldwin[b], Marlon Conato[a], Jeffrey D. Rimer[a], Michael P. Harold[a] and Ken Blackmon[b]

[a] *University of Houston, Department of Chemical and Biomolecular Engineering, Houston, TX 77204*
[b] *Total Petrochemicals & Refining USA, Inc., LaPorte, TX 77571*

The oxidative coupling of methane (OCM) to higher molecular weight hydrocarbons has great appeal as an alternative route to chemicals and polymers. The reaction system is challenging because of large differences in the reactivity of the reactant methane and desired products (e.g., ethylene, ethane, etc.) at the temperatures needed to activate methane. Moreover, the reaction system is highly exothermic and is noted for significant parametric sensitivity.

A large number of systems have been studied over the past few decades, but a catalyst and reactor concept has not been found that give sufficiently high C_{2+} yield. The most studied catalysts to date for the OCM reaction system have been alkali-promoted alkaline earth metal oxides, transition metal oxides and rare earth metal oxides. Na_2WO_4-Mn/SiO_2 catalyst has attracted a significant amount of attention because of its favorable catalytic performance. In this study we evaluate the activity of the catalyst in particulate and washcoat form using both fixed-bed and monolith reactor configurations. We also carry out a parametric study to quantify the effects of residence time, reaction temperature, oxygen concentration, and dilution.

The catalyst was prepared by the incipient wetness impregnation method and its structural properties were characterized by a variety of analytical techniques (e.g., XRD, BET, ICP and SEM). The prepared catalyst has a surface area of 2.6 m^2/g and was sieved to a size of 50-70 mesh. Residence time variations were accomplished by changing the weight of catalyst while keeping the flow rate constant, thereby avoiding changes in flow patterns inside the reactor. Figure 1 shows the dependencies of methane conversion, C_2 selectivity, and C_2 yield on residence time at 800°C. The conversion of methane increased with increase in residence time but the C_2 selectivity decreased due to deep oxidation in the reactor with prolonged exposure to the catalyst.

We observed that, with higher CH_4/O_2 ratios and concentrated feed, the reactor system becomes quite sensitive to the operating conditions. In particular, the steep axial temperature limits the maximum oxygen concentration that can be used. By diluting the feed, the reaction can be conducted at higher CH_4/O_2 ratios. We will also show the effect of dilution of catalyst with inert and catalyst structure (i.e. particles vs. washcoat) on the product compositions. Washcoating is of particular interest since it reduces the contact time of C_2 hydrocarbons within the internal porous structure of the catalyst, thereby inhibiting deep oxidation.

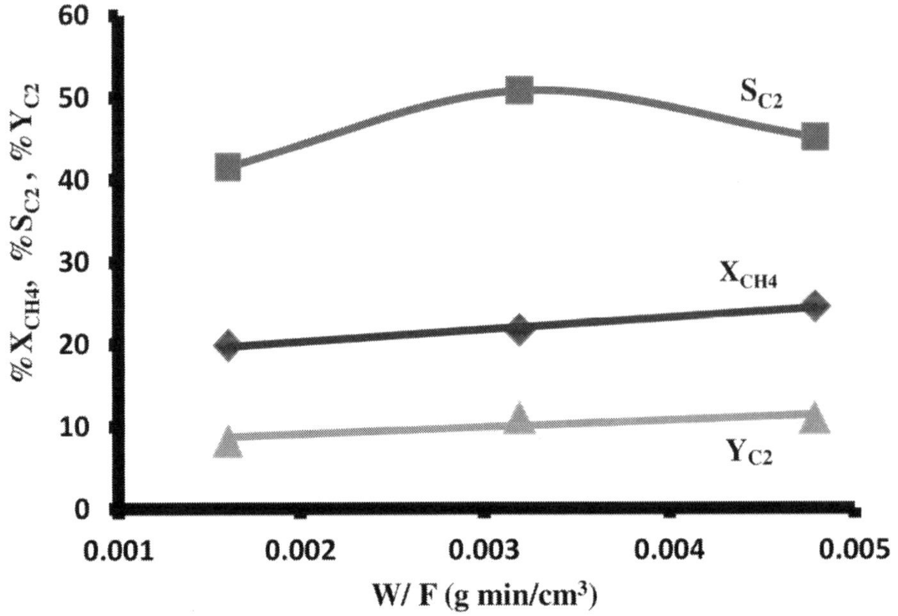

Figure 1. Dependence of CH_4 conversion, C_2 selectivity, and C_2 yield on space time at T = 800^0C and a CH_4:O_2 ratio of 3:1.

393128 The Future of the Lecture

Abstract:

The university is one of the oldest social institutions in our society, rivaled only by the church. However, many universities now are under considerable stress because of reduced public commitment to education. This stress has sparked a review of how we teach. The review has focused particularly on the traditional lecture, suggesting alternatives like MOOC's (Massive Open Online Courses). These alternatives have produced brilliant successes and significant failures. However, the alternatives have also identified ways in which we can all deliver more effective lectures.

382165 A Numerical Model for the Motion of Large Non-Spherical Solid in Fluidized Bed

Abstract:

In a number of practical applications of fluidized bed such as coal/wooden biomass gasification and combustion, the size and shape of solid materials are not uniform and large non-spherical solids are coexisting with small emulsion solids in the bed to enhance fluidization and heat transfer. The size of these solids is largely different; fuel solids are more than ten times larger comparing to emulsion particles. Fuel solids, especially pellets, are far from spherical and the non-sphericity is expected to play an important role for contact and hydrodynamic interactions. It is difficult to treat these systems by the existing numerical models. Tsuji et al. (2014) proposed a numerical model which expresses the behavior of flows including dense solids with large size difference based on the DEM-CFD coupling method. In their model, the interaction force between fluid and large solid is estimated by assuming a large solid is consisted of dense small fictitious particles. Their study was limited to spherical solids. In the present study, on the other hand, non-sphericity of large objects is taken into account by introducing Multi-sphere model. The model was applied to the motion of a single large non-spherical object in a bubbling fluidized bed and validation study was performed through the comparison with corresponding experiment. We observed similar behaviors of the large object between calculation and experiment.

Eulerian modeling of dense particle-gas flows: effect of the specularity coefficient on the fluidization hydrodynamics of bubbling fluidized beds with varying wall surface to bed volume ratios

Akhilesh Bakshi, Christos Altantzis, Richard B. Bates, Ahmed F. Ghoniem
Department of Mechanical Engineering, Massachusetts Institute of Technology
77 Massachusetts Avenue, Cambridge, MA 02139-4307, USA

Fluidized beds, owing to their excellent mass and heat transfer characteristics, constitute the most widely employed paradigm of gas-solid flow with applications ranging from energy and chemicals production to pharmaceutical and agricultural processing. The importance of these devices across many industrial fields renders the optimization of their operation a worthy modeling challenge. The ever increasing progress of computing power and the development of efficient numerical methods, has contributed to establishing Computational Fluid Dynamics (CFD) as an indispensable tool for understanding the fundamentals of gas-solid hydrodynamics and optimizing the operation of fluidized beds complementing the experiments by overcoming their limitations. One of the most widely used modeling approaches for the simulation of dense particle-gas flows is the Two-Fluid Model (TFM), which considers both gas and solid phases as interpenetrating continua. While this modeling approach balances the modeling fidelity and the computational efficiency, the closure expressions describing the gas-solid interactions and the particle-particle collisions, increase the complexity of the mathematical model and require rigorous validation.

In the framework of the Eulerian description of each phase, it is important to choose the proper boundary conditions for the velocities. For the vast majority of applications, a no-slip wall boundary condition is used for the gas phase, while a partial-slip condition, based on the formulation of Johnson and Jackson [1], is considered the most suitable for the solid phase. This model, in order to define the slip velocity of the solids along the wall, considers the momentum transfer to the wall in the tangential direction both through the collisions of the particles characterized by the specularity coefficient, ϕ, and the sliding of the particles characterized by the Coulomb friction. The influence of the boundary conditions on the fluidization hydrodynamics is an area of active research enriched by the growing number of studies on the proper description of the wall boundary conditions. Unfortunately, due to the limitations of measurement techniques, the literature on experimental investigation of the particle wall collisions is narrow, thus validation of parametric analyses using CFD is needed for the investigation of the influence of ϕ.

The current study aims at investigating the influence of the specularity coefficient on the fluidization behavior of fluidized beds with varying wall surface to bed volume ratio by means of 3D numerical simulation. The highest surface to volume ratio case investigated is a thin rectangular bed (also called pseudo-2D bed in literature) with dimensions: x=50cm, y=100cm and z=0.5cm. The front and back walls, being so close in a thin bed, influence the solids motion and the 2D assumption is strongly inaccurate. The sensitivity of the solids circulation flux to variations of the specularity coefficient is studied together with detailed analysis of the differences in bubbling dynamics. Finally the appropriate values of ϕ for different conditions extracted from the simulations are validated with the experimental results presented in [2]. It is shown that the effects of the front and the back walls are significant, influencing the sensitivity of the fluidization hydrodynamics to the third dimension whose consideration is thus mandated. For increasing values of ϕ, the fluidization behavior tends towards the bubbling regime and the increased momentum losses to the wall result in lower circulation fluxes and subsequently to larger circulation times.

The geometries with lower surface to volume ratios studied are cylindrical beds with diameters d_1=14.5cm and d_2=30cm. Detailed statistics of the bubble motion are extracted and the numerical results are compared with the experimentally measured bubble statistics presented in [3]. The sensitivity of the bed turnover time on the wall effects is studied for the 3D cylindrical cases as well. It is shown that the influence of the wall boundary conditions becomes less significant as the bed diameter increases, (decreasing surface to volume ratio).

The numerical tool used to solve the governing equations is the Multiphase Flow with Interphase eXchanges (MFIX) code developed by the U.S. Department of Energy (DOE) National Energy Technology Laboratory (NETL).

[1] P.C. Johnson and R. Jackson. **Frictional-collisional constitutive relations for granular materials, with application to plane shearing.** *Journal of Fluid Mechanics.* 176:67-93, 1987.

[2] S. Sánchez-Delgado, C. Marugán-Cruz, A. Soria-Verdugo and D. Santana. **Estimation and experimental validation of the circulation time in a 2D gas-solid fluidized beds.** *Powder Technology.* 235:669-676, 2013.

[3] M. Rüdisüli, T.J. Schildhauer, S.M.A. Biollaz, A. Wokaun and J.R. van Ommen. **Comparison of bubble growth obtained from pressure fluctuation measurements to optical probing and literature correlations.** 74:266-275, 2012.

Drag force correlations for non-spherical particles in gas solid flows

Y. Chen[1] , J. R. Third[1] , C. R. Müller[1]

[1]Department of Mechanical and Process Engineering, ETH Zürich, Leonhardstrasse 27, 8092 Zürich, Switzerland

Gas-solid systems are of significant industrial importance and are the basis of various reactor concepts, e.g. gas-fluidized and packed beds or rotary kilns. However, our fundamental understanding of these systems is still poor which is, at least partially, due to the very complex particle-particle, particle-fluid and particle-wall interactions. Thus, a good understanding of particle-fluid interactions is essential to predict more accurately the dynamics of these systems.

Due to a lack of drag force correlations for assemblies of non-spherical particles, numerical simulations of gas-fluidized beds have been restricted mostly to beds containing spherical particles. Indeed, even for *single* non-spherical particles comparatively little work has been reported (Tran-Cong et al., 2004; Hölzer et al., 2007, 2008). However, these correlations cannot be applied directly to simulate gas-fluidized beds since they do not account for the influence of the solids volume fraction on the drag coefficient. Here, using the Lattice-Boltzmann method we develop a drag force correlation for non-spherical particle assemblies. The particles used in the simulations are represented in two different ways: (1) Constructed from identical spheres, unbreakable and undeformable; (2) Continuous function representation, i.e. the super-quadric equation. The effect of particle shape and Reynolds number on the drag force is studied in detailed. Two different drag force correlations are derived based on the data of Lattice-Boltzmann simulations and the differences between these two correlations are highlighted. The new drag force correlations proposed here are suitable for Euler-Euler and Euler-Lagrangian simulations of gas-fluidized beds containing cubic particles.

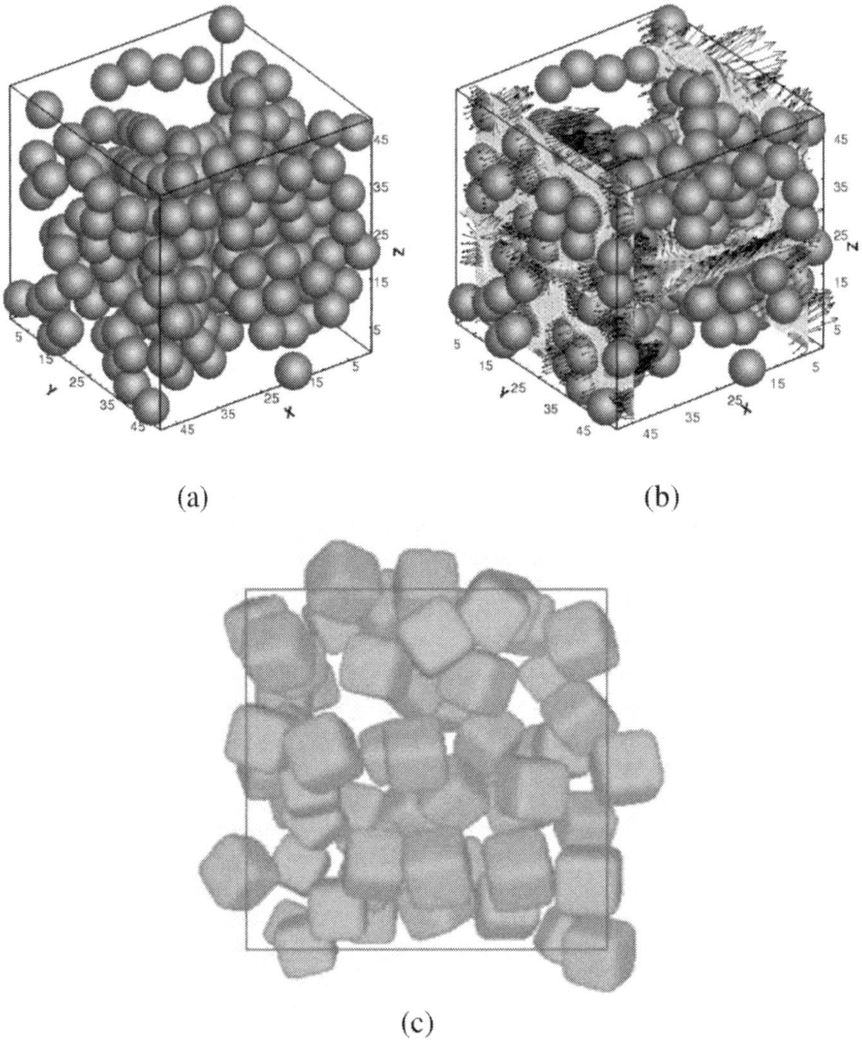

Fig. 1 (a) Random array of cubes constructed from 8 identical spheres; (b) Calculated flow profile around the array of cubes constructed from 8 spheres. The color represents the magnitude of the velocity in the x-direction; (c) Random array of super-quadric cubes.

References

Tran-Cong, S., Gay, M., Michaelides, E. E. (2004). Drag coefficients of irregularly shaped particles. Powder Technology, 139(1), 21-32.

Hölzer, A., Sommerfeld, M. (2008). New simple correlation formula for the drag coefficient of non-spherical particles. Powder Technology, 184(3), 361-365.

Hölzer, A., Sommerfeld, M. (2009). Lattice Boltzmann simulations to determine drag, lift and torque acting on non-spherical particles. Computers & Fluids, 38(3), 572-589.

Although there have been many experimental investigations into the dynamics of gas-solid fluidized beds, most experiments have focused on measuring the solid phase. Experimental studies of the dynamics of the gas phase are lacking because it is very difficult to study gas motion when the gas is obscured by opaque particles. Thus, most experimental studies have been limited to the freeboard of fluidized beds [1–7], often in pseudo 2-D fluidized beds.

Gas dynamics in the bed of particles of fluidized beds are extremely important, as they will have a strong effect on the heat and mass transfer between the solid and gas phase, and ultimately on chemical reaction kinetics in fluidized bed reactors. Despite this importance, studies of gas flow in the bed of particles has largely been restricted to theoretical predictions, starting with the theory for gas flow around bubbles based on the two-phase theory for fluidization [8]. This theory led to theoretical development of the volume averaged fluid equations for the fluid and particulate phases [9], which has led to the development of two fluid models (TFMs) [10] and discrete element models with computational fluid dynamics (DEM-CFD) [11] to simulate computationally gas and particulate dynamics in fluidized beds. Despite the current wide use of TFMs and DEM-CFD to simulate fluidized beds, there has been very little validation of the ability of these models to accurately simulate gas dynamics, due to lack of experimental data. Thus, these models have largely been validated against particle dynamics and pressure drops, with the assumption that if these properties are modelled accurately, then the gas dynamics must be accurate as well.

In this presentation, we will discuss recent spatially-resolved magnetic resonance imaging (MRI) measurements conducted of time-averaged gas velocity both in the bed of particles and the freeboard of fluidized beds. Pressurized SF_6 gas was used, such that MR signal could be obtained from the ^{19}F nuclei, and measurements could be made on the gas itself, rather than tracer particles. The bed of particles was 52 mm in diameter and filled with poppy seeds 1.2 mm in diameter to various settled bed heights. The poppy seeds contained oil with a ^{1}H MR signal, so that their velocity could be measured in separate experiments. Measurements in bubbling and slugging beds showed faster gas velocity through the centre of the bed than the sides, due to rapid passage of gas through bubbles traveling through the centre of the bed. The measurements are currently being investigated to validate gas dynamics predicted by DEM-CFD and TFM computational simulations.

References:

[1] S. T. Pemberton and J. F. Davidson, Chem. Eng. Sci. **39**, 829 (1984).
[2] Y. Levy and F. C. Lockwood, AIChE J. **29**, 889 (1983).

[3] G. Hartung, C. R. Müller, J. Hult, J. S. Dennis, and C. F. Kaminski, Ind. Eng. Chem. Res. **47**, 5686 (2008).
[4] C. R. Müller, G. Hartung, J. Hult, J. S. Dennis, and C. F. Kaminski, AIChE J. **55**, 1369 (2009).
[5] R. Solimene, A. Marzocchella, R. Ragucci, and P. Salatino, Chem. Eng. Sci. **62**, 94 (2007).
[6] G. Duursma, D. Glass, S. J. Rix, and M. Yorquez-Ramirez, Powder Technol. **120**, 2 (2001).
[7] M. Yórquez-Ramírez and G. Duursma, Powder Technol. **116**, 76 (2001).
[8] J. F. Davidson and D. Harrison, *Fluidised Particles* (Cambridge University Press, Cambridge, 1963).
[9] T. B. Anderson and R. Jackson, Ind. Eng. Chem. Fundam. **6**, 527 (1967).
[10] J. Ding and D. Gidaspow, AIChE J. **36**, 523 (1990).
[11] Y. Tsuji, T. Kawaguchi, and T. Tanaka, Powder Technol. **77**, 79 (1993).

Review of Cluster Characteristics in Gas-Solid Fluidized Bed Systems

Andy Cahyadi[1], Jia Wei Chew[1], Christine M. Hrenya[2] and Ray A. Cocco[3]

[1] *Nanyang Technological University (School of Chemical and Biomedical Engineering, Singapore)*
[2] *University of Colorado at Boulder (Chemical and Biological Engineering, Boulder, Colorado, U.S.A.)*
[3]*Particulate Solid Research Incorporated (Chicago, Illinois, U.S.A.)*

The phenomenon of particle clustering in gas-solid fluidization system has been recognized in 1948 [1], but only received more attention from researchers in the last two decades [2]. Clusters, which are thermodynamically unstable groups of loosely agglomerated particles, result from hydrodynamic effects [2, 3]. The presence of clusters in a fluidized bed system can alter the flow patterns in the fluidization column and affect system performance (namely, reaction rate, heat transfer, elutriation rate, etc), which hence warrants the need for better understanding of the clustering phenomenon [4]. Unfortunately, efforts targeted at enhanced understanding of clusters are hampered by apparent contradictions among experimental results. Therefore, this work aims at a comprehensive review of the available experimental data on clusters to provide a more complete landscape of the clustering phenomenon, thereby shedding light on the dominant underlying physics.

Accordingly, analysis of the cluster frequency trends available of Geldart Group B particles was carried out. Several interesting results are worth highlighting here: (1) Monodisperse and polydisperse (i.e., continuous particle size distributions) systems exhibit higher cluster frequency compared to binary (i.e., two particle species of different size and/or density) systems. (2) Cluster frequency increases with particle size and particle density, with the latter having a more prominent impact. (3) Higher superficial gas velocity and/or lower particle loading increases cluster frequency only for denser systems but not for dilute ones.

Keywords: Cluster; Cluster Frequency; Dense CFB Riser; Dilute CFB riser; Geldart Group B

References

1. Wilhelm, R.H. and M. Kwauk, *Fluidization of Solid Particles.* Chemical Engineering Progress, 1948. **44**(3): p. 201-218.
2. Horio, M. and R. Clift, *A Note on Terminology - Clusters and Agglomerates.* Powder Technology, 1992. **70**(3): p. 196-196.
3. Fan, L.S. and C. Zhu, *Principles of Gas Solid Flows.* 1998, New York: Cambridge University Press.
4. Chew, J.W., et al., *Cluster characteristics of continuous size distributions and binary mixtures of Group B particles in dilute riser flow.* Chemical Engineering Journal, 2011. **178**: p. 348-358.

387785 Avalanching, Bed Collapse and Cohesion: Characteristic Behavior of Geldart A, B and C Powders

Abstract:

The Geldart powder classification diagram is a useful predictor of powder fluidization behavior based on particle density and surface-volume mean particle diameter. It also provides useful identification of and insights to the flow behavior and avalanching of powders in rotating systems in the context of powder handling and process engineering. In this work, avalanching, bed collapse, and cohesion of selected Geldart Groups A, B, and C powders were measured and evaluated. Avalanche behavior was measured with a novel tumbling device equipped with a cylindrical drum, ~200 mm in diameter and ~300 mm long, and a load cell that captured changes in the centre of gravity of the tumbling powder mass; the standard deviation of the load cell signal provided an indication of avalanche activity. Powder de-aeration was measured by fluidization and the double-drainage bed collapse technique. Cohesion at zero consolidation was estimated by shear testing at low consolidation stresses. The standard deviation of the load cell signal is presented as a function of drum rotating speed, 5-30 revolutions per minute (rpm); the standard deviation profile for each powder is different and the data are reproducible. For the Group C powder used, avalanching increases with drum speed; a change of slope occurs at ~15 rpm. In comparison, the avalanche activity of the Group B powder used increased linearly up to 20 rpm before leveling off; the standard deviation values are lower than for the Group C powder. The avalanching profile of a Group A/B powder shows a characteristic similar to that of the Group B powder, but even lower standard deviation values. When two Group A powders of similar surface-volume mean particle diameter, ~40 μm, and span of size distribution, ~1.4-1.6, but different measured values of standardized bed collapse rate and cohesion are compared, the powder with a lower bed collapse rate and higher cohesion shows lower avalanche activity. Another Group A powder of ~60 μm and span of ~2.3 seem to give relatively constant avalanching behavior regardless of drum speed.

366013 Bioprocess Development of Microbial Enzymes with Outstanding Properties: An Egyptian Extremophiles Platform

Abstract:

Over two decades, researchers of Bioprocess Development Dept., Genetic Engineering and Biotechnology Research Institute have mastered the bioprocess development for the production of microbial enzymes with special preferences to extremophiles with the aim to establish an extremophiles' platform.

Within this context, we were able to define production medium to maximize the production of thermostable lipase and esterase by *Geobacillus thermoleovorans* YN with the aid of statistical-mathematical approaches that led to increase the productivity by 3 folds. The responsible gene for the production of this enzyme was identified and cloned into *E.coli* overexpression host and the enzyme was purified and well characterized. Both enzymes were most active at pH 9.5 and remarkably stable at pH 5 and 10.5. Temperature optima and stabilities (up to 70 °C) of both enzymes as well as their reaction kinetics and substrate spectra were determined.

In other work, we succeeded to optimize the production of alkaline protease from *Bacillus pseudofirmus* Mn6 in 15 L pilot plant bioreactor that lead to increase the productivity more than 10 folds initial activity under controlled PO2 and pH.

The production and purification of thermostable amylase from *Bacillus licheniformis* AI20 was also studied. Response Surface Methodology (RSM) was adopted to acquire the best process conditions among the selected variables, where a three-level Box-Behnken design was employed to create a polynomial quadratic model correlating the relationship between the three variables and -amylase activity. The predicted optimum -amylase activity was 384 U/mL/min, which is two folds more than the basal medium conditions. On enzyme level, the estimated molecular mass was 55 kDa and the -amylase had an optimal temperature and pH of 60–80 Cand 6–7.5, respectively. The enzyme showed great stability against different solvents.

360329 Perfusion Cultivation Processes for Monoclonal Antibody Production in Animal Cell Culture Technology: Cell Separation

Abstract:

Mammalian cell cultures are the major source for a large number of very high-value recombinant therapeutic proteins, monoclonal antibodies (MAbs), viral vaccines and hormones. By 2014, it is predicted that five of the top 10 products will be antibodies (Avastin, Humira, Rituxan, Herceptin, and Remicade), along with two other recombinant protein products (Enbrel and Lantus). Furthermore about half of new therapeutic products under development are antibodies. Globally, monoclonal antibodies were expected to have sales exceeding $41 billion in the year 2010. However, efficient and reliable production technologies are required to cope with the tremendous demand for these products. Generally, large-scale commercial mammalian cell cultivations are performed in batch, fed-batch and perfusion processes. Perfusion systems involve cell retention devices, and they have the advantage of continuously removing the product and the exhausted medium, as well as the simultaneous withdrawal of the by-products, specifically toxic metabolites, which largely affect the specific growth rate. Therefore, cells cultured under perfusion culture mode can be maintained at optimal growth conditions. Accordingly, growth inhibition as well as problems with product stability can be reduced. This presentation will give a short idea about the latest developments achieved in the field of industrial production of monoclonal antibodies with a special attention to the different separation devices used in perfusion cultivations, focusing on one of these systems as a case study.

373918 Industrial Platform Design for Fungal Phytase Production in Semi-Industrial Scale

Abstract:

The production of microbial industrial enzymes has recently drawn attention in the biotechnology of animal feeds. Of different enzymes applied, phytase (EC 3.2.1.8) has become an important microbial enzyme, which is applied as feed additive for monogastric livestock. Phytase hydrolyses phytate (myo-inositol-1,2,3,4,5,6-hexakisphosphate), which is the main form for storing phosphorus in most of the seeds and grains used as animal feed components. This enzyme is available in plant in low concentration and thus, its necessary to supplement the feed with exogenous addition of this enzyme. Nowadays, this enzyme was isolated from different microbial strains, however, the main production process involve the cultivation of fungi belongs to *Aspergillus* group such as *A. niger* and *A. ficcum* in either solid state or submerged fermentation system. Production of fungal phytase faces two main challenges. First, controlling the growth morphology in the bioreactor to achieve high system productivity. Second, facilitating the enzyme excretion to shorten the production time. To overcome the first bottle-neck of this process, cells were cultivated in stirred tank bioreactor under certain conditions to produce micropellet (not more than 400mm diameter) to reduce mass transfer problems inside the microbial pellet. The second problem was also solved in bioreactor system by addition of surfactants and increasing the bioreactor pressure to increase enzyme excretion rate and shorten the production time. In this work, complete industrial platform for fungal phytase proposed using starch as feed-stock to reduce the production cost.

381568 Influenced of Additive in the Post-Induction Phase on the Distribution of Phytase Expression By E. coli BL21 (DE3) When Induced with Lactose

Abstract:

Phytase is a type of phosphatase enzyme that mostly found in grain and seed that catalyzes the degradation of phytic acid realizing a useable form of inorganic phosphorus. A recombinant *Escherichia coli* BL21 (DE3) has been used for harboring thermostable phytase gene from *Bacillus* sp.MD2 and export the target protein to the periplasmic space. Lactose has been used as an alternative inducer in fermentation of recombinant *E.coli*. However, low productivity of protein expression was observed when induced with lactose due to the time needed by lactose to be transported into the cell by lactose permease and process to allolactose by β-galactosidase before binding to the repressor. To overcome this problem, the first approach used was by screening different types of additive such as Na^+, Ca^{2+}, glycine and Triton X-100 which was supplemented in the cultivation broth in the post-induction phase. Supplementation of only 10 mM of $CaCl_2$ can increased excretion of phytase outside the host cells membrane up to 33.63 % with the total productivity phytase activity only 673.75 U mL^{-1} hr^{-1} after 12 hours induced with 30 mM of lactose. In the other hand, supplementation combination of 10 mM of $CaCl_2$ with 0.5 % (w v^{-1}) of glycine in the post-induction phase did not shows drastically improved of extracellular and total phytase activity. However, the productivity of the total phytase production and extracellular phytase activity was increased up to 146.15 % and 119.9 %, respectively within 6 hours of post-induction phase. No improvement of volumetric phytase production rate was observed when 10 mM of $CaCl_2$ was combined with 0.5 % (w v^{-1}) of glycine in the pre-induction phase.

Non-chromatographic purification of recombinant surfactin from *Bacillus amyloliquefaciens* (NRC-1, KC590348) using self-cleaving purification tags (Intein)

Dr. Tarek E. Mazeed
Assistant professor in chemistry of natural and microbial products department, National Research Center, Egypt

Prof. David W. Wood
Professor in Chemical and Biomolecular Engineering Department, Ohio State University, USA

Surfactants

- Surfactants are usually organic compounds that are amphiphilic, meaning they contain both hydrophobic groups (*tails*) and hydrophilic groups (*heads*).

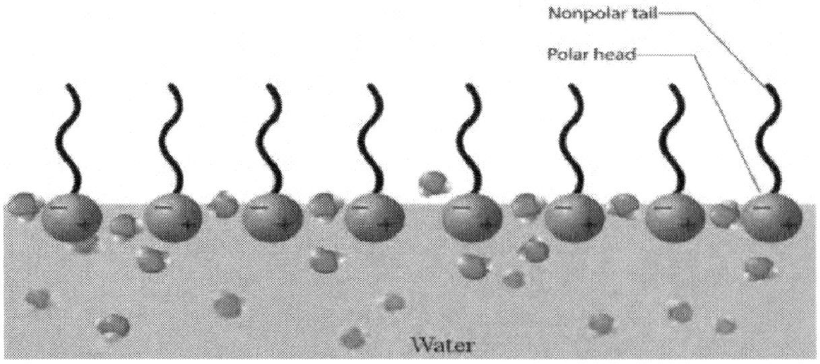

- Surfactants, are wetting agents that lowers the surface tension of a liquid, allowing easier spreading, and can also lower the interfacial tension between two liquids.

Classification of surfactants

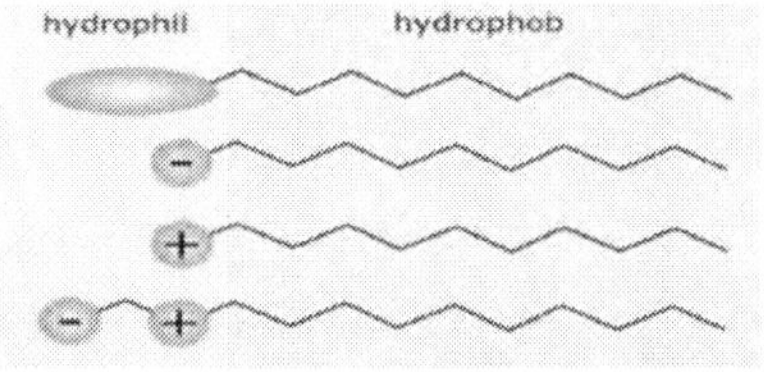

Ionic surfactant

Non-ionic surfactants

ethers of fatty alcohols

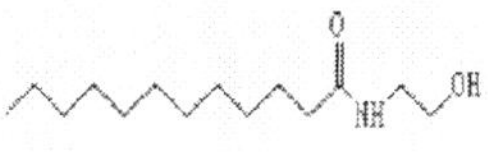

Cocamide MEA

Anionic

Sodium dodecyl sulfate (SDS)

Cationic

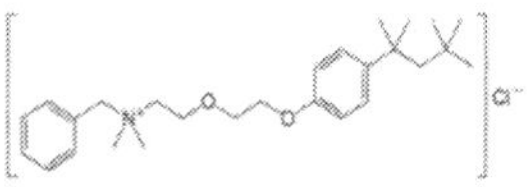

Benzethonium chloride (BZT)

Zwitterionic (amphoteric)

Alkyl Betaine

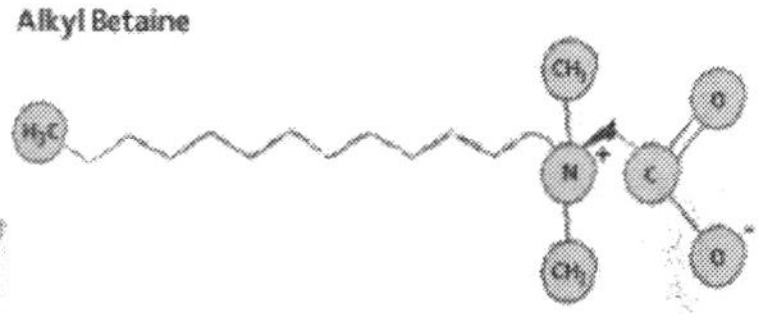

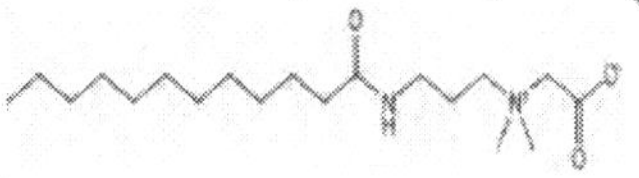

Cocamidopropyl betaine

Classification of biosurfactants

- Low molecular mass biosurfactants
  - Below the biosurfactants critical concentration → Mobilization
  - Above the biosurfactants critical concentration → Solubilization
- High molecular mass biosurfactants → Emulsification

glycolipids, rhamnolipids, trehalose lipids, sophorolipids, and lipopeptides

polymeric and particulate surfactants

Rosenberg and Ron (1999)

Commercial application of biosurfactants

Biosurfactants are promising surfactants and are useful in the fields of cosmetics, food processing, pharmaceuticals and environmental bio-remediation particularly in enhanced oil recovery (EOR) and cleaning of oil spills

- **Genetic manipulation:**

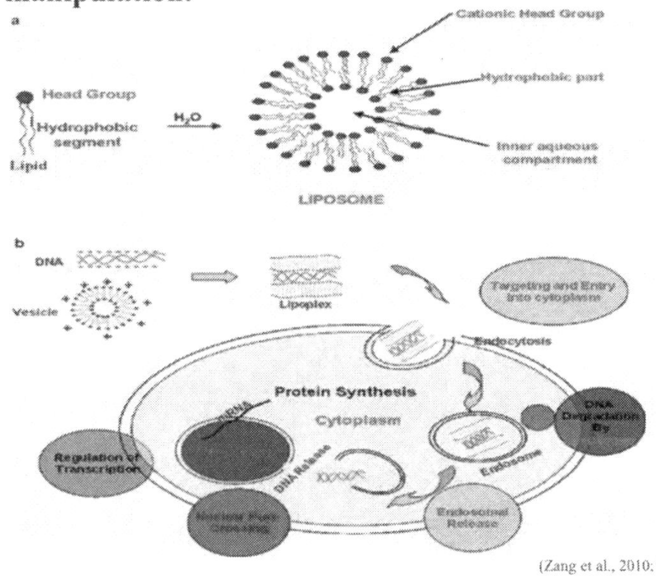

(Zang et al., 2010; Fujita et al., 2009; Liu et al., 2010)

(a) The liposome and its various loci for applications in delivery.
(b) Plausible mechanism of the lipid-based gene delivery pathway.

- **Immune modulatory action:**
  Bacterial lipopeptides constitute potent non-toxic and non-pyrogenic immunological adjuvant when mixed with conventional antigens (Gharaei-Fathabad, 2011).

- **Anti-adhesive agents**
  Adhesion of biosurfactants to solid surfaces might constitute a new and effective means of combating colonization by pathogenic micro-organisms (Rivardo et al., 2009).

- **Toxic activity against microorganisms:** (Zhoa et al., 2010).

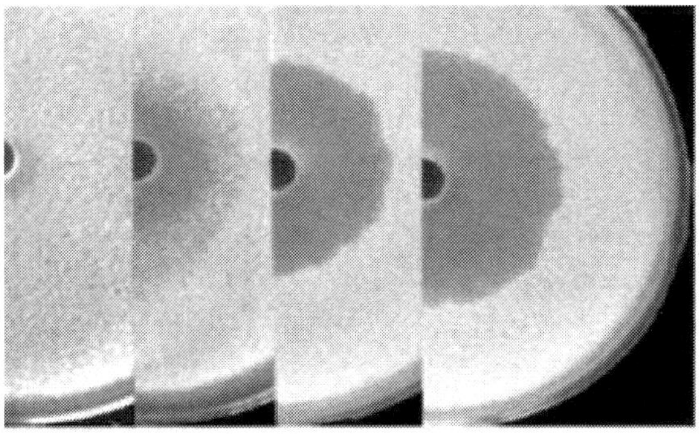

Antimicrobial activity of biosurfactants at various concentration

Microbial surfactants

Biosurfactant	Organisms
Glycolipids	
Rhamnolipids	*P.aeruginosa*
Trehalolipids	*Mycobacteria sp.*
Sophorolipids	*Nocardia sp.*
Lipopeptides and lipoproteins	
Peptide-lipid	*B. licheniformis*
Viscosin	*P. fluorescens*
Surfactin	*B. subtilis*
Gramicidins	*B. brevis*
Fatty acids, neutral lipids, and phospholipids	
Fatty acids	*C. Lepus*
Neutral lipids	*N. erythropolis*
Phospholipids	*T. thiooxidans*
Polymeric surfactants	
Emulsan	*A. calcoaceticus*

Microbial biosurfactants

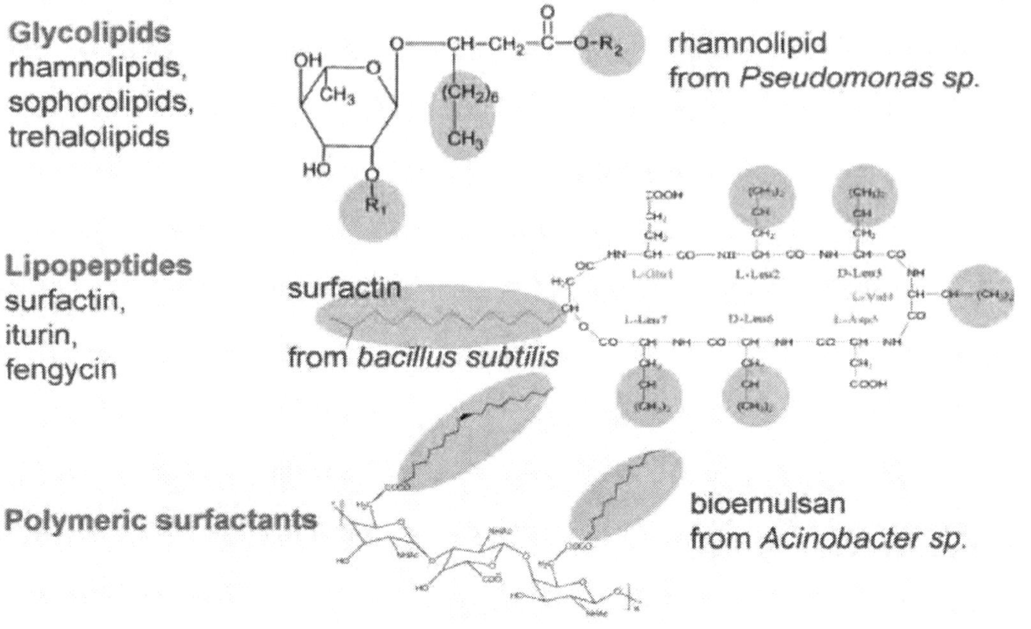

Advantage biosurfactant over chemical surfactant

Chemical surfactant	**Biosurfactant**
Toxic	Non-toxic
Not easily degraded	Biodegradable
Environmentally hazardous	Biologically safe
less sensitive in extreme environments	Effective under broad ranges of oil & reservoirs conditions (temp. ,pH…)
Required in large quantities	Required in small quantities.
Derived from petroleum	Availability of raw materials
	Specificity

Surfactin

The cyclic lipoheptapeptide surfactin is produced by different *Bacillus sp.*, and consists of a β- hydroxy fatty acid chain of variable length (12 to 15 carbon atoms).

β–hydroxy fatty acid-L_{Glu}-L_{Leu}-L_{Leu}-D_{Leu}-L_{Val}-L_{Asp}-D_{Leu}-L_{Leu}

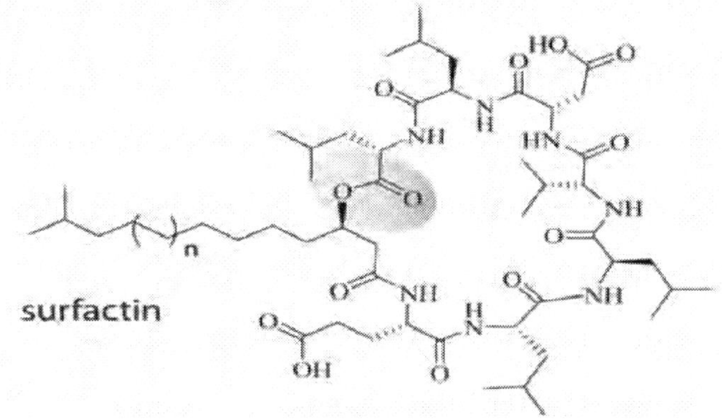

(Arima et al., 1968)

Surfactin biosynthesis

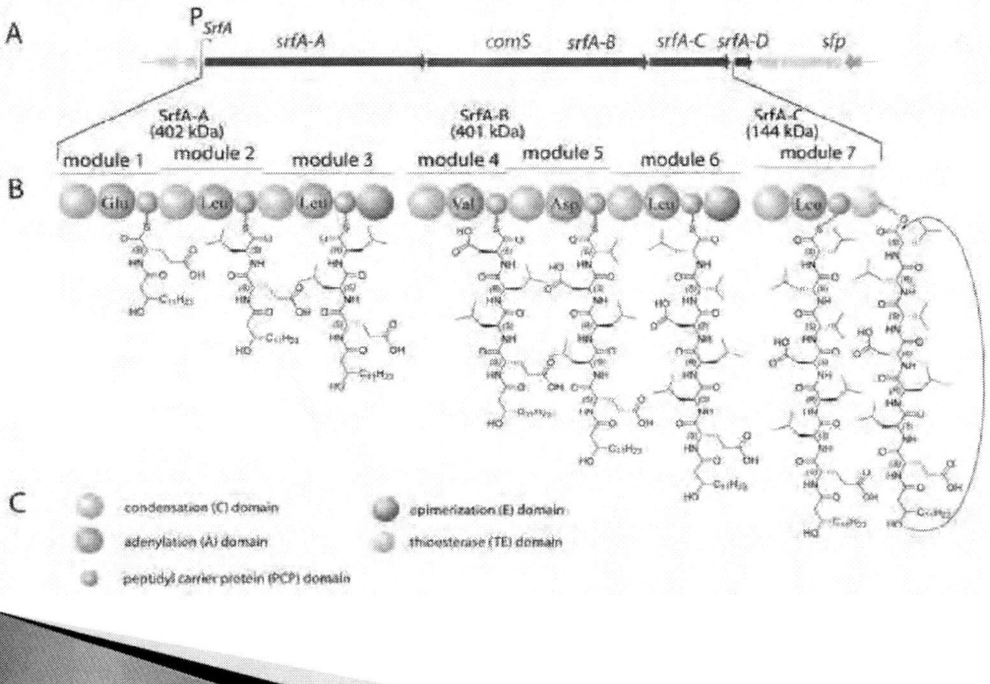

Bacillus amyloliquefaciens (NRC-1, KC590348)

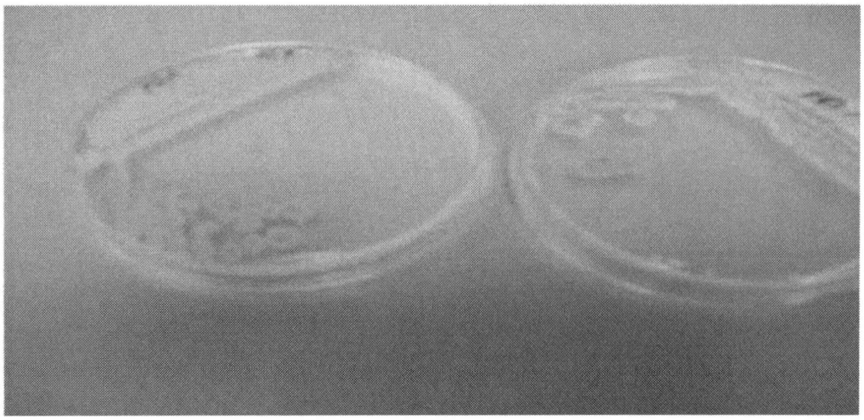

Bacillus amyloliquefaciens (NRC-1, KC590348)

- Surfactin production was optimized using a locally isolated strain identified as ***Bacillus amyloliquefaciens*** (NRC-1, KC590348), which is used in this study.
- This strain was isolated from goat skin during a search for a lipase producing bacteria in Egypt. (Mazeed, T. E., et al., 2009)

Surfactin production and foaming collection

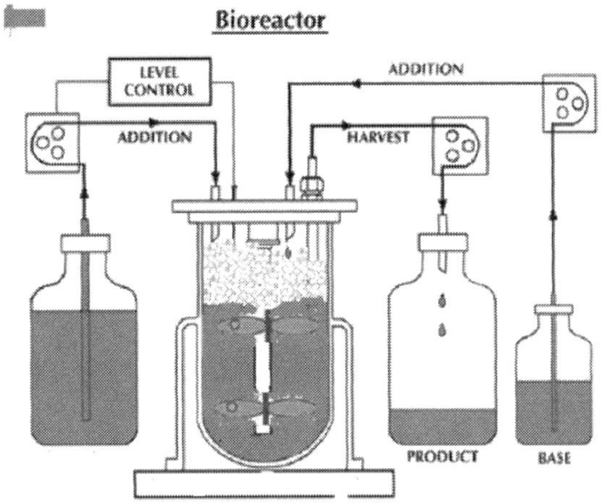

surfactin fermentation was carried out in 7 L fermentor under different combination of agitation and aeration rates to find out the optimal conditions for surfactin production

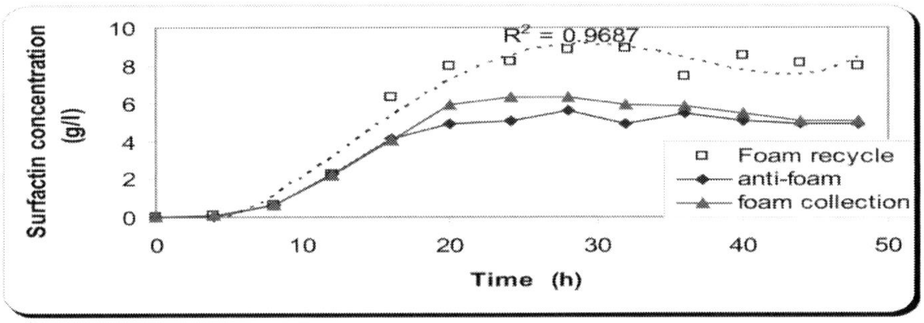

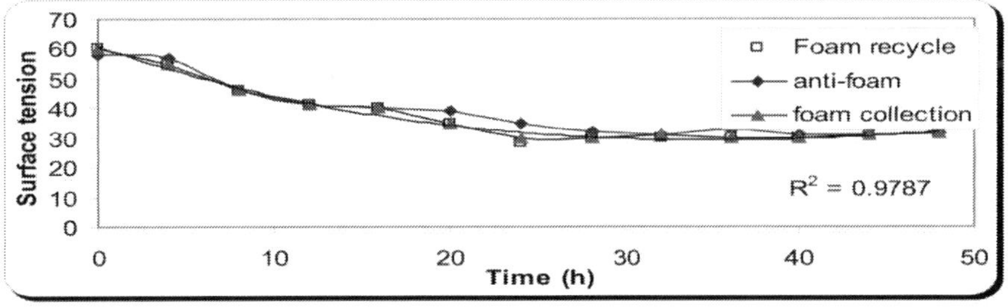

Bacillus amyloliquefaciens (NRC-1, KC590348) which is reported to produce the highest level of surfactin 8.92 g/l , and has the ability to reduce surface tension of water from 72 to 28 mN/m^2 at a concentration as low as 0.005 % . (Atwa, N. A., et al., 2013)

- Therefore, if the production cost becomes competitive with the synthetic surfactants, and as the commercial availability of surfactin will increase and its industrial use will grow tremendously in the coming decade.

- Previous studies have focused on the strain selection, mutation, or manipulation of nutritional factors for enhanced productivity of surfactin. However, only marginal increases were observed (Abdel-Mawgoud, et al. 2008, Moran, et al. 2010).

- However, there was no comprehensive report detailing the direct application of genetic manipulation or quantitative analysis of surfactin productivity and purification in any recombinant microbial hosts (Coutte, et al. 2010, Sun, et al. 2009).

Non-chromatographic purification of surfactin

- An important development in the area of recombinant protein purification has been the incorporation of self-cleaving protein elements into a variety of fusion-based purification systems. (Wood DW, et al., 1999).

- These elements are derived from naturally occurring self-splicing inteins through various protein engineering strategies, and have been combined with conventional affinity tags in a variety of configurations to yield highly effective separations methods.

- Very recently, these elements have also been combined with non-conventional purification tags to yield "self-purifying" proteins, which can deliver highly purified native products using simple mechanical means without chromatographic methods. (Banki MR, et al., 2005).

Affinity fusion technologies

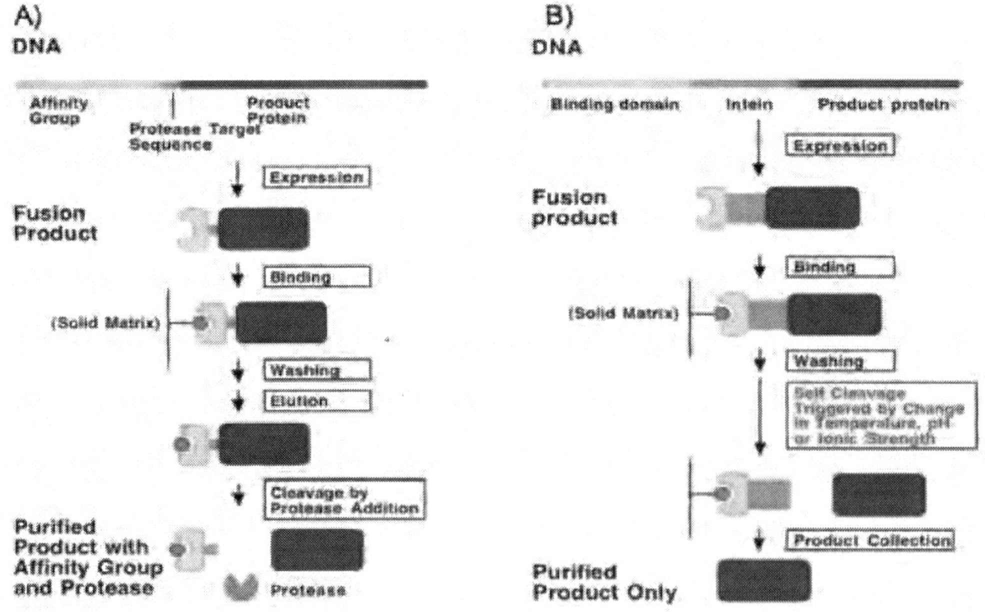

Intein = INTervening protEIN

- Are naturally occurring protein sequences capable of post-translational self-excision from a host-intein precursor protein through a process known protein splicing. Inteins are self-splicing protein elements that have the unique ability to excise themselves from within a translated host protein (Perler, F.B. et al., 1994).

- Self-splicing inteins can be mutated at N- or C-terminal splice junctions to yield self-cleaving inteins, which can then be used to mediate self-cleaving of various purification tags

- These inteins can be split into two categories:

pH-induced C-terminal cleaving activity

Thiol-induced N- and C-terminal cleaving activity

- pH- induced inteins (*e.g.* Ssp DnaB intein) are the most economical because they only require a shift in buffer pH to induce. A significant of the pH-induced inteins is premature cleaving of the tagged target during expression. (Esipov, R.S. et al., 2008, Sharma, S.S. et al., 2006 and Zhao, Z. et al., 2008)

Self purifying expression systems

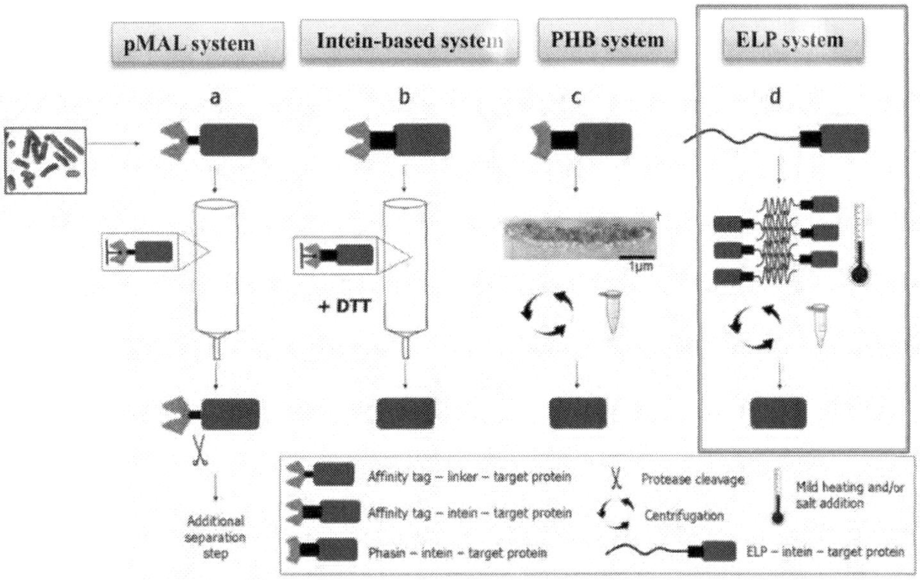

(Banki MR and Wood DW, 2005)

Elastin like polypeptide (ELP) system

Val-Pro-Gly-Xaa-Gly (Xaa is not Pro)

Target protein is tagged with reversibly-precipitating self-cleaving polypeptide (ELP system) elastin like polypeptide

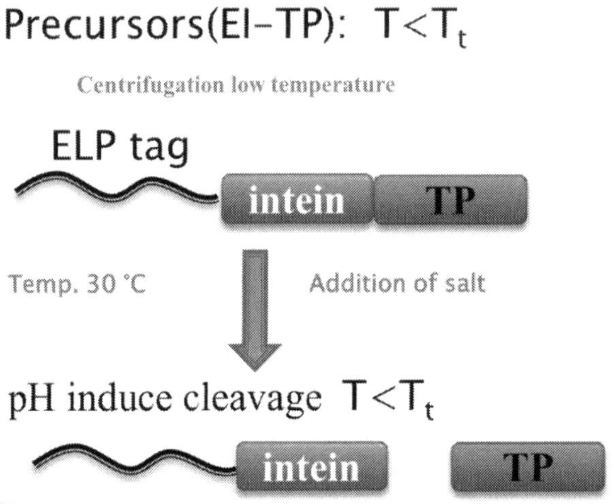

Conclusion

The genetic manipulation of the strain *Bacillus amyloliquefaciens* (NRC-1, KC590348) together with applying the self-cleaving intein purifying technique will increase the purity of the widely used biosurfactant "surfactin" with purity from 90 to 95 % .

THANK YOU

391936 Plenary Talk: Recent Advances in Computational Design of Nanomaterials for Energy Storage and Conversion

Abstract:

A variety of novel composite materials have been used/tested for next-generation energy storage and conversion devices. However, in many cases little is known about their properties and performance, although such fundamental understanding is essential for further advances in energy storage and conversion technologies. Experiments may yield many clues to the behavior of those materials, but the interpretations are often controversial due largely to the difficulty of direct characterization. Under such circumstances, first principles-based computational approaches have emerged as one of the most powerful tools for design and development of new energy materials. This talk will focus on introducing our ongoing efforts in first principles modeling of energy storage and conversion materials. In the first part of my talk, I will discuss the properties and performance of silicon-based nanomaterials for lithium-ion rechargeable batteries, particularly the impacts of surfaces, interfaces and alloying elements. From first principles-based atomistic modeling, we have found that the lithiation behavior of Si nanostructures and composites is considerably different from the case of bulk Si because of surface, interface, and/or alloying effects. The improved understanding can contribute to the rational design of Si-based anode materials to maximize capacity retention and rate performance. In the second part, I will present recent progress in our collaborative theoretical and experimental efforts to explore photocatalysts with the requisite band gaps, stability, costs, and abundance for solar-powered hydrogen production. In particular, this talk will highlight the effects of crystal structure and doping on the photocatalytic performance of $BiVO_4$ that has recently garnered considerable attention due to its high catalytic activity for photoelectrochemical water splitting.

391937 Young Faculty Talk (1): Engineering Cellular Processes through Synthetic Control

Abstract:

The past decade has witnessed the tremendous power of synthetic biology in the creation of genetic parts, devices, and systems, which helps understand complex biological systems. However, its potential for real-world applications has not been fully exploited. One of its promising applications is the construction of programmable cells that are able to integrate multiple environmental signals and to implement synthetic control over biological processes. My research interests are focused on developing microbes that are able to process multiple input signals and to generate user-defined outputs. Specifically, I aim to build genetic programs in order to control various bacterial processes such as gene expression, chemical reactions, and evolution. I will present examples of my research projects to discuss the potential and challenges of synthetic biology for practical applications.

391938 Young Faculty Talk (2): Suppressing Optical Absorption in Nanostructured Metal Electrodes in Photovoltaics

Abstract:

Electrodes are ubiquitously used in optoelectronic devices such as solar cells, infrared detectors, and light emitting diodes. Typically, metals exhibit good electrical conductivity due to their free electrons and are suitable for electrodes in these devices. However, metal electrodes cause inefficiency in the devices by absorbing light as their free electrons suffer from collisions. To circumvent this problem, nanostructured metals are being explored to realize low optical losses while maintaining large electrical conductivity. Here, we study optical absorption in helical metal nanocoils and find that absorption can be dramatically decreased when the metal is suitably nanostructured. Theoretical modeling showed that this effect is due to the increase in the effective mass of free electrons in nanostructured metals. Heavy electrons suffer a decreased rate of collision and emulate dielectric materials where optical losses are negligible. Calculations showed that, when metal electrodes are used in semiconductors devices, suitably chosen nanostructures can increase the fractional absorption in semiconductors more than 60 times by reducing losses in the metal.

392696 Panel Discussion: Career Paths in Academia, Industry and National labs

Abstract:

Moderator - Daeyeon Lee (University of Pennsylvania, Vice President of KIChE-US Chapter)

Panelists - Jaehun Chun (Pacific Northwest National Laboratory); Seung Soon Jang (Georgia Institute of Technology); Ah-Hyung (Alissa) Park (Columbia University)

Abstract:

The panel discussion will focus on choosing career paths for Chemical Engineers in academia, industry and national labs. The panel discussion is designed for students and postdoctoral fellows who are considering different career paths as chemical engineers.

391939 KIChE President Award Talk: Smartphone Integratable High Density Chemical Sensor Arrays

Abstract:

Electronic detection of molecules is rapidly emerging as an alternative to the traditional optical and electrochemical methods because of the small size, low-power consumption, improved sensing performance and most of all possibility of developing high density arrays for simultaneous analyses of multiple species in small sample volumes. Recently, one-dimensional nanostructures (e.g., carbon nanotubes (CNTs), inorganic, and organic nanowires) as conduction channels of field effect transistors (FETs) have been developed for detection of a variety of gaseous molecules with excellent low detection limit, sensitivity, and selectivity. These features are a consequence of dramatic decrease in characteristic length and increase in the ratio of surface to volume atoms, allowing for rapid diffusion into the bulk and for a more significant fraction of the atoms to participate in surface processes such as chemical binding interactions. One-dimensional geometries also enhance response times by virtue of their two-dimensional mass transfer profile. Furthermore, nanowires are heralded for device miniaturization and sensor arrays, enabling duplicate elements to reduce false positives/negatives and pattern recognition systems termed electronic noses/tongues where each sensor in the array has a unique response to every analyte creating a fingerprint type response that increases sensitivity and selectivity. Finally, sensors are also attractive for their proven commercial viability, as this approach uses a single material behaving as both the sensitive layer and transducer to directly covert chemical information into an electronic signal without the need for labels, allowing for real-time, continuous monitoring. In this presentation, synthesis, functionalization, and assembly of various nanoengineered materials including CNTs and conducting polymer nanowires will be discussed to create "true" high density gaseous sensor arrays with superior sensing performance in cost-effective manner. In addition, Android based smartphone integratable sensor is demonstrated.

367672 Removal of Cadmium Ions from Aqueous Solution Using a New Low-Cost Biosorbent

Abstract:

In recent years, heavy metal pollution has increased as a result of industrial growth, especially in industries such as electroplating, mining, metal finishing, battery production, and welding. Among the heavy metals, one of the most hazardous is cadmium, which accumulates in the human body and causes cancer along with cardiovascular, neurological, and respiratory problems. Hence, it is crucial to sequester cadmium ions from industrial effluent before the contaminated effluent is discharged into the sewage system. There are several traditional treatment methods for removing heavy metals from aqueous effluents, including chemical precipitation, ion exchange, and membrane filtration. These techniques have high operational costs and require the continuous input of chemicals, so they are neither economical nor ecologically sustainable. Therefore, biosorption has gained significant attention as an effective alternative for sequestering toxic heavy metals. In this work, *Sorghum X drummondii* (Sudangrass), an inexpensive agricultural product, is used as an adsorbent for the removal of cadmium ions under a range of conditions, including variations in initial pH, temperature, contact time, adsorbent dose and initial cadmium ion concentration. Fourier Transform Infrared (FTIR) is used to characterize the surface of the biosorbent and also to investigate the functional groups involved in the process. Scanning Electron Microscopy (SEM) is used to observe the morphology of the biosorbent before and after the adsorption process. Well-known isotherm models are used to describe the equilibrium data, from which the parameters are calculated and reported. Two kinetic stages of adsorption are identified: a fast period, followed by a relatively slow period leading to equilibrium. Thermodynamic parameters, such as the change of free energy, enthalpy, and entropy, were also evaluated for the adsorption of cadmium ions onto Sudangrass.

386521 Environmental Remediation of Dense Non-Aqueous-Phase Liquids Using Multifunctional Iron-Carbon Nanocomposites

Abstract:

Dense non-aqueous phase liquids, such as trichloroethylene (TCE) and carbon tetrachloride (CCl_4), are widespread groundwater and soil contaminants which cause long-term environmental pollution. Extensive efforts have been carried out to develop the optimal remediation strategy. Here, we will introduce a composite with carbon submicrospheres serving as the supports and embedded nanoscale zero valent iron (NZVI) being the reactive agent to treat TCE and dechlorinate it to non-toxic products. This Fe-C composite unites the current prominent method of using NZVI with the hydrophobic nature of carbon, which enables the composites to target organic contaminants and facilitates the dehalogenation reaction by increasing the local concentration of TCE in the vicinity of iron particles through rapid TCE adsorption. Besides, hydrophobically modified chitosan (HMC) can be easily coated around the carbon supports to enhance the suspension stability of these submicrospheres in groundwater.

This composite has been proven to be effective and has several conspicuous advantages: (a) the synthesize process is a simple one-step process with inexpensive precursors and can be scaled up to manufacturing level (b) the NZVI particles are well dispersed on the carbon support, preventing them from aggregation, which eliminates the major challenge of using NZVI (c) composite particles are of the optimal size for transport through sediments (d) toxic intermediates such as dichloroethane ($C_2H_2Cl_2$) and vinyl chloride (C_2H_3Cl) can be sequestered within the carbon supports until they reacted to final light gas products.

The presentation will focus on the detailed materials synthesizing and characterization, and the significant aspects of adsorption, transport and reaction of this composite in remediation of groundwater contaminants will be discussed.

Adsorption and Oxidative Degradation of Bisphenol A on Surface Modified Iron-Amended Activated Carbon: Effects of temperature on adsorption and Fenton oxidation

Jihyun R. Kim and Eunsung Kan. Department of Molecular Bioscience and Bioengineering, University of Hawaii at Manoa, Honolulu, HI 96822, USA

For the past decade, BPA as an endocrine disrupting chemical has gained great attention, and become a public concern as they can cause adverse effects in aquatic organisms and human. The present study focuses on effective adsorption and oxidation of bisphenol A (BPA) in wastewater by using newly synthesized nano iron catalyst-deposited activated carbon.

The novel synthesis of nano iron/iron oxide-deposited granular activated carbon (Fe-GAC) was developed for adsorption and oxidation of BPA by amending ferrous iron onto the acid-pretreated GAC at the pH of iron solution higher than the pH at point of zero charge (pH, pzc) of the Fe-GAC. Batch experiments for the adsorption isotherm of BPA onto the Fe-GAC at 20 °C and pH 6 followed Langmuir isotherm-type adsorption, yielding maximum monolayer adsorption capacity (q_m) of 588.2 mg BPA/g GAC which was higher than the virgin GAC with q_m of 471.2 mg BPA/g GAC. The adsorption rate of BPA was increased by factor of 3.4 when temperature increased from 20°C to 80°C. It was exactly proportional to enhancement of diffusivity of BPA at increasing temperature indicating diffusion of BPA into the pores of the GAC would be the major limiting step in the adsorption process of BPA.

Since the heterogeneous Fenton oxidation of BPA in the Fe-GAC clearly showed strong dependence on H_2O_2 concentration, 9 to 108 mol H_2O_2/mol BPA led to enhance the BPA oxidation efficiency in the GAC from 62% to 99%. The Fenton oxidation of BPA at increasing temperature (20-80°C) also resulted in enhancing the oxidation efficiency and rate of BPA while increasing the diffusivity, desorption rate and reaction rate of BPA in the GAC. The pre-optimized conditions for the heterogeneous Fenton oxidation of BPA exhibited negligible accumulation of the major intermediates and oxidative products of BPA in the aqueous and GAC phases. The mathematical model based on diffusion, adsorption and oxidative reaction and Thiele-Modulus analysis indicated that the intraparticle diffusion limitation was the major limiting step among the Fenton oxidation of BPA in the GAC. Further works will include detailed analysis of diffusion, adsorption and oxidation in the GAC using the mathematical model and engineering analysis as well as the optimization of the operating conditions over the repeated cycles of adsorption and oxidation.

365690 Removal of Metronidazole and Dimetridazole from Aqueous Solution By Adsorption on Multiwalled and Singlewalled Carbon Nanotubes

Abstract:

Antibiotics are widely prescribed in human and veterinary medicine to prevent and cure bacterial infections. Antibiotics may not be biodegraded by conventional treatments, and they are being discharged in environmental media in their parent form or as transformation products. It is very well documented that the presence of antibiotics in surface water poses serious risks to human health. Nitroimidazoles are among the most widely produced and prescribed antibiotics in many countries. Metronidazole (MNZ) and dimetridazole (DTZ) are very commonly used nitroimidazoles.

The adsorption of the antibiotics MNZ) and DTZ on multiwalled (MWCNT) and single walled (SWCNT) carbon nanotubes was investigated in this work. The effect of the adsorbent-adsorbate interactions as well as the operating conditions (ionic strength, solution pH, temperature, chemical modification of the adsorbents, presence of other antibiotic) on the adsorption capacity were analyzed to substantiate the adsorption mechanism. The MWCNT were modified using a HNO_3 solution and the modified adsorbent was denoted as MWCNT-Ox. The adsorption capacity towards MNZ and DTZ markedly varied as function of the carbon material, decreasing in the following order: SWCNT > MWCNT > MWCNT-Ox, and depended not only on their surface area and pore size distribution, but also on their chemical nature. The adsorption of MNZ and DTZ was strongly influenced by the solution pH, but was not significantly affected by the ionic strength and temperature. Desorption equilibrium data of MNZ and DTZ on all carbon materials demonstrated that the adsorption was reversible corroborating the weakness of the adsorbent-adsorbate interactions.

The binary adsorption of MNZ and DTZ on SWCNT was studied in this work. The experimental binary adsorption data were interpreted using the following multicomponent adsorption isotherms: non-modified, extended and modified Langmuir, non-modified and modified Redlich-Peterson, extended Freundlich, and Sheindorf-Rebuhn-Sheintuch. The latter model best fitted the adsorption data of MNZ-DTZ on SWCNT. The competitive adsorption of MNZ-DTZ on SWCNT revealed that both antibiotics presented very similar antagonism toward the adsorption of the other antibiotic.

Synthesis of zirconium nanoparticles doped activated carbon fiber for optimized adsorption for arsenic removal

Arsenic contamination in ground water and industrial effluent has been a great concern globally for years. Arsenic is one of the highly harmful inorganic contaminants originating from either anthropogenic activities as well as natural erosion of arsenic containing mineral substances. Arsenic exposure may cause many diseases. It may be transported to different organs of body through blood circulation and causes a variety of adverse health effects. The maximum contaminant level (MCL) of arsenic in drinking water set by USEPA had been 50 $\mu g/L$. However, a new standard for arsenic in drinking water has been set at 10 $\mu g/L$ since early 2006, which creates a great challenge to the water industry.

To enhance the removal of arsenate and prevent the secondary contamination of nanoparticles, an activated carbon fiber (ACF) was synthesized by doping of zirconium nanoparticles onto the surface of ACF with chitosan as the coating support. In recent years, nano-sized materials become more attractive for water treatment. Compared to the traditional sorbents that have relatively larger particle sizes, nano-sized particles have such advantages as higher surface to volume ratio and better physicochemical properties for adsorption. Zirconium based nanoparticle sorbent has a significantly high adsorption affinity towards the inorganic arsenic species. However, the nanoparticles may cause secondary contamination because of its dispersion in the water. It is rather difficult to remove the nanoparticles after the adsorption. A better media such as activated carbons can be used to support the nanoparticles in order that the nanoparticle can still stay on the media instead of dispersing in treated water.

Activated carbon fiber (ACF) is extensively used in the field of air and water purification because it provides an extremely high absorption capacities compared with powdered and granular activated carbons. It has great mechanical and physical properties, and is rather stable in different chemical and physical conditions (e.g. high temperature. As a result, ACF can serve as a good supporting matrix for zirconium particles to remove arsenic from water. Chitosan (CTS) is a liner polysaccharide and made by treating shrimp and other crustacean shells. CTS manifest a good film-forming property, and it was made as a membrane which has both filtration and adsorption ability. Polyethylene glycol (PEG) is a highly water-soluble polymer and it can generate pores on the chitosan film to minimize the mass transfer and diffusion resistance in adsorption. Thus, the PEG can be used as a pore generator on the CTS film. The preparation for Zr-doped ACF was optimized by an orthogonal array design matrix. Mixtures of chitosan solution, PEG solution, and zirconium nanoparticle suspension of various component ratios were prepared according to a $L_9\ (3)^4$ orthogonal array design matrix.

Through the orthogonal experiments, these four factors were proved to have determined the determined the adsorption efficiencies to some extent. The optimal composition for ACF modification was 0.3% chitosan with low molecular weight, 0.2% PEG and the mass of zirconium nanoparticles was 0.6 g. It should be noted that the particles were not dispersed uniformly once the amount of zirconium nanoparticles exceeded 0.6 g in the impregnation solution. As a result, the maximum mass of zirconium nanoparticles was selected to be 0.6 g in 50-mL impregnation solution.

The optimized sorbent was characterized by scanning electron microscope. Bath adsorption experiments were conducted for arsenic removal by this material, including pH effect experiments, adsorption kinetics and isotherms experiments; natural organic matters and coexisting anions affect experiments. From the results, we can get the highest adsorption was achieved at the optimized pH of 3.0. The kinetics study showed that the adsorption equilibrium was achieved at around 30 h. The Langmuir equation well described the adsorption isotherm, and the maximum adsorption capacity of the sorbent was 20.5 mg/g at the optimized pH. The presence of coexisting fluoride and carbonate ions did not significantly affect the adsorption. In contrast, the co-existence of silicate, phosphate ions and humic acid retarded the uptake. The X-ray photoelectron spectroscopy analysis indicated that zirconium nanoparticles were immobilized on the modified ACF. $HAsO_4^{2-}$ as the dominant species of As (V) was adsorbed onto the sorbent. This Zr-doped ACF has good adsorption ability for arsenic removal.

One-pot conversion of cellobiose into polyols and the possible reaction pathways over extremely low phosphotungstic acid with Ru/C catalyst

Kai Zhang[a], Shubin Wu [a], Xiaohong Zhang[b]*
[a]State Key Laboratory of Pulp and Paper Engineering, South China University of Technology, Guangzhou, China
[b] Guangzhou Institute of Energy Conversion, Chinese Academy of Science, Guangzhou, China
Corresponding author. Tel./fax: +86 2022236808
E-mail address: shubinwu@scut.edu.cn

Abstract

Cellobiose is a glucose dimer connected by a glycosidic bond and represents the basic repeating unit of cellulose, and it could be efficiently converted into polyols in the presence of hydrogen and precious-metal catalyst of Ru/C over extremely low phosphotungstic acid which is environmentally friendly and reduces the requirement for equipments. Then, processing parameters and the reaction performance were studied in detail, the yields of ethylene glycol (EG) and 1,2-propylene glycol (1,2-PG) were obtained up to 33.4% and 12.2% respectively at 240 ℃ for 1 h, which had a definite guidance function for catalytic conversion of cellulose.Besides, the catalytic process using cellobiose was studied to propose the possible reaction pathways of producing polyols .

Keywords Catalytic conversion; Cellobiose; Polyols; UPLC-MS; Pathway;

Introduction

In order to alleviate the shortage of fossil fuels and reduce the CO_2 emissions, the renewable biomass is a good choice for producing liquid fuels and value-added chemicals[1-2]. As the most abundant source of biomass, cellulose is currently thought as a promising alternative for fossil fuels because of its direct conversion into useful organic compounds without imposing a negative impact on food supplies [3-4]. However, unlike other water soluble polysaccharides, the low reactivity of cellulose has prevented its use in chemical industry except for the paper manufacturing[5]. In order to utilize cellulose, the key factor is to destory the glycosidic bonds connecting glucose residues to promote its dissolution in solvents. Fortunately, Luo et al[6] presented a green approach to efficient conversion of cellulose into polyols in hot water (about 473K) which could genenrate H^+ ions to catalyze the hydrolysis reactions of cellulose. Besides, solid catalyst is proved to be an environmentally benign and milder reaction in hydrolytic conversion of cellulose[7-8]. Then, Zhang Tao and coworkers[9] reported a one-pot conversion of cellulose to EG with 61% yield over a nickel-promoted tungsten carbide catalyst in hot water, which has attracted much attention from both academic and industrial societies since 2008.

It is reported that EG is an important chemical product, widely used in the synthesis of cosmetics, polyester fibers, antifreeze, etc, and its global consumption reached more than 21 million tons in 2011[10]. With the increasing demand for EG in the plastics and automotive industry, its direct production from cellulose will creat a renewable way for reducing dependence on petroleum. Therefore, more and more attentions are going to be paid to this field, meanwhile, a lot of research results have been attained. For example, combined with Ru-based or Ni-based catalyst, tungsten-based catalysts such as W, WO_3, H_2WO_4, heteropoly acids (PW, SiW), etc were found to have a higher selectivity to ethylene glycol (EG) on conversion of cellulose or glucose in presence of hydrogen[11-13]. Thereinto, Zhao Guanhong and coworkers[11] found that in combination with 4wt% Ru/AC, tungstophosphoric acid(PTA) showed a better performance on selectively producing 1,2-PG(12.7%) in addition to EG(30.1%) from the conversion of glucose. 1,2-PG is also an important bulk chemical widely used as materials of plasticizer, surfactant, resin, emulsifier and anti-emulsifier, heat carrier, antifreeze and so on[14].

However, the concentration of PTA containing 1000ppm tungsten is high in the above reaction, which leads to problems such as acid corrosion and hard disposal, and cellulose or cellobiose could used to directly produce polyols in presence of PTA over Ru/C catalyst[15]. Therefore, low concentration of PTA should be used in the reaction and the performance on selectively producing polyols from the conversion of cellulose or cellobiose should be studied.

Herein, we report a one-pot catalytic process using cellobiose as material over 5wt% Ru/C catalyst in combination with extremely low PTA which is environmentally friendly and reduces the requirement for equipments. Then, processing parameters and the reaction performance are studied in detail to have a definite guidance function for catalytic conversion cellulose. While, cellulose is a polymer with a complex molecular structure, using simple model molecules of cellulose to study the reaction process presents a promising way, and cellobiose is a glucose dimer connected by a glycosidic bond and represents the basic repeating unit of cellulose[16-17]. Therefore, according to the results from catalytic process using cellobiose and research progress from others, the possible reaction pathways of producing polyols are proposed.

Experimental Section

Materials:

Ru/C (precious-metal loading:5 wt.%) was purchased from Aladdin reagent limited company. Tungstophosphoric acid and cellobiose were obtained from Tianjin Kemiou Chemical Reagent Co. Ltd and Shanghai bioscience and technology limited company respectively.

Catalytic Reaction:

Catalytic conversion of cellobiose was typically carried out in a high pressure stainless-steel autoclave (Parr Instrument Co, 5500, 100mL). Catalysts and tungstophosphoric acid solution, cellobiose was preloaded in the reactor. Then, the reactor was filled with H_2 after expelling air, the reactor was heated to the reaction temperature from room temperature(RT), and the reaction temperature was kept a certain times. After reaction, the reactor was cooled down to RT rapidly with ice water. The liquid products were collected by filtration with 0.22 μm membrane, then they were stored in the refrigerator.

Product Analysis:

After the reaction, the liquid products were analyzed with liquid chromatography (Waters 2695 HPLC equipped with Aminex HPX-87H 300×7.8mm column, refractive index and ultraviolet detector). Besides, in order to further verify the reaction pathways of producing polyols, Ultra Performance Liquid Chromatography tandem mass spectrometry(Agilent 1290 UPLC equipped with mass spectrometry 6540 and zorbax carbohydrate 4.6×150mm column) was used.

Yield(%)=(weight of polyol in the products)/(weight of cellulose put into the reactor)×100%

Results and Discussion

1 The effect of processing parameters on the conversion of cellobiose

The effect of processing parameters such as H_2 pressure, temperature, time, and etc on the yield of polyols were investigated over extremely low phosphotungstic acid with Ru/C catalyst. When a parameter was studied, other processing parameters were kept consistent. Thereinto, the processing parameters were chosen as follows: the reaction temperature at 513K and reaction time for 1 h, 5Mpa of the H_2 pressure and 1% content of cellobiose in the reaction system, 0.1g of 5wt% Ru/C, 600rpm and 30mL extremely low phosphotungstic acid in water(0.01wt%).

1.1 The effect of H_2 pressure on the yield of polyols from cellobiose conversion

Fig 1 shows that with the increase of the H_2 pressure from 3Mpa to 5Mpa, the yield of EG reaches to the highest value of 33.4% when the pressure of H_2 is 4Mpa, and the yield of sorbitol and 1,2-PG have different trends, one gradually increases, another comes down little by little. The yiled of other polyols are too low to accurately analyze the trend. According to the yield of EG, the subsequent experiment adopted the H_2 pressure of 4Mpa.

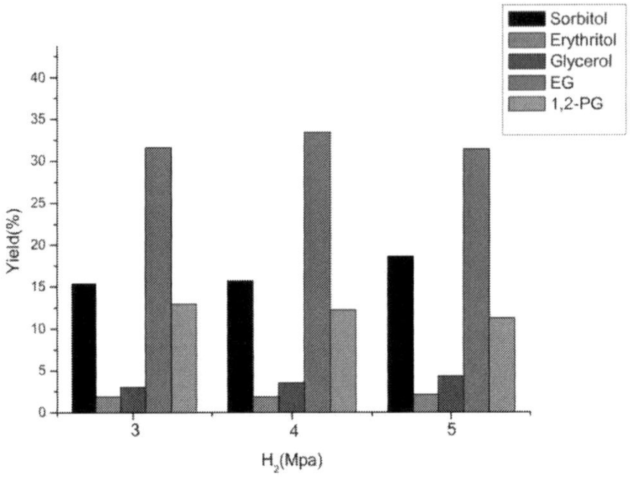

Fig 1 The effect of H_2 pressure on the yield of polyols from cellobiose conversion

1.2 The effect of temperature on the yield of polyols from cellobiose conversion

It was reported that the reaction temperature had great effect both on kind and yield of products[18]. As seen in Fig. 2, with the increase of temperature in the reaction from 220℃ to 240℃, the yield of EG and 1,2-PG have a increasing trend, when the reaction temperature is 240℃, the yield of EG and 1,2-PG are 33.4% and 12.2% respectively. However, the yield of sorbitol are greatly reducing with the increase of temperature. The subsequent experiment adopted the reaction temperature of 240℃.

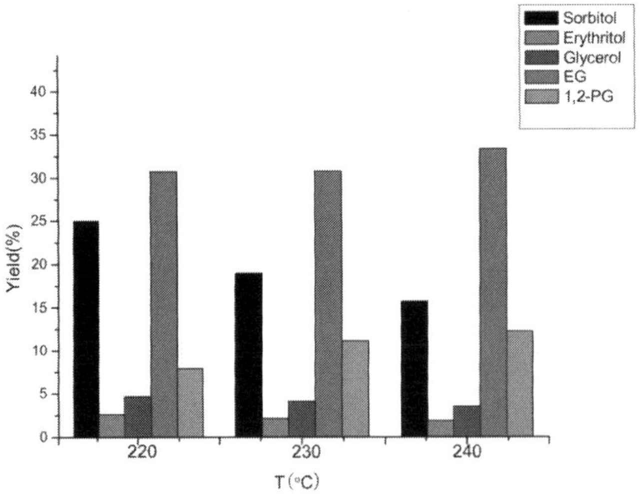

Fig 2 The effect of reaction temperature on the yield of polyols from cellobiose conversion

1.3.The effect of reaction time on the main yield of polyols from cellobiose conversion

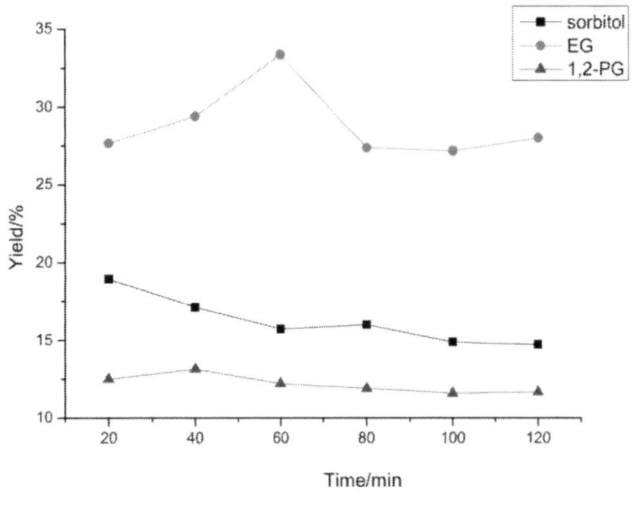

Fig 3 The effect of reaction time on the yield of the main polyols from cellobiose conversion

Six batch of experiments were done with differnet reaction time. As shown in Fig 3, the effect of time on the main yield of polyols from cellobiose conversion are obvious, from 20 min to 120 min, the yield of EG, 1,2-PG, and sorbitol present very different trends. With the increase of reaction time, the highest yield of EG achieved at 60 min, the production of sorbitol is on the decline, while the yield of 1,2-PG is about 12% all the time.

2 The cleavage of bonds and the products in the reaction

During the whole catalytic hydrogenation, the cleavage of different bonds and the production of various chemicals are involved. Broadly speaking, the catalytic hydrogenation of cellobiose involves the breakage of both C-O and C-C bonds. It is reported that the C-O cleavage proceeds through hydrolysis or dehydration or hydrogenation, whereas the C-C breaking occurs via the retro-aldol reaction[19-20], various chemicals are produced through catalytic hydrogenation.

As seen in Table 1, depending on the reaction temperature, the product varies from monose to polyol. With the increase of temperature from 120℃ to 240℃, the conversion of cellobiose reaches up to 99.0% from 27.9%, glucose and fructose undergo the catalytic hydrogenation with almost a 100% conversion at 240℃. When the reaction temperature below 150℃, glucose is producted from hydrolysis of cellobiose over an acid catalyst(phosphotungstic acid) needing a long reaction time, and trace of acetol, fructose, sorbitol, EG, 1,2-PG, erythritol and glycerol are produced, it is reported that the existence of fructose can attribute to the isomerization of glucose especially in presence of activated carbon, which promotes the isomerization of glucose into fructose by its surface basicity[21] and 1,2-PG forms via the complex pathway from fructose as the precursor[22]. Then, with the increase of reaction temperature from 150℃ to 180℃, sorbitol is produced from hydrolytic hydrogenation of cellulose under the presence of pressurized H_2 and hydrogenation catalyst(Ru/C), sorbitol is quite stable and not converted into EG under the present reaction conditions[11]. Besides, when the reaction proceeds at temperature above 210℃, EG and 1,2-PG become the major product in the presence of multifunctional catalysts[14], other polyols such as glycerol, 1,2-butanediol, erythritol, mannitol, etc are produced as by-products. Meanwhile, when cellobiose is replaced by glucose or fructose as feedstock, at the temperature above 210℃, the higher yields of EG and 1,2-PG are formed from the conversion of glucose and fructose, respectively. This part of reaction is very complex. Hence, in order to clearly understand the conversion of cellulose or hemicellulose into polyols, the possible pathways of producing polyols in the reaction are needed.

Table 1 The conversion of cellobiose, glucose and fructose at different temperatures

T/℃	Con %	Glu %	Fru %	Sor %	EG %	1,2-PG %	Acetol %	Ery %	Gly %
120[1]	27.9	trace	0	0	0	0	0	0	0
150[1]	44.4	9.0	trace	trace	trace	trace	trace	trace	trace
180[1]	69.8	11.6	4.3	6.4	9.2	2.0	2.3	trace	4.4
210[1]	98.1	trace	trace	11.0	18.7	3.0	4.0	1.4	3.8
240[1]	99.0	trace	trace	9.8	22.3	11.3	trace	2.2	1.7
210[2]	90.3	9.7	0	22.0	20.5	10.1	5.2	1.6	2.7
240[2]	92.9	7.1	0	12.5	16.5	13.5	trace	1.1	1.9
210[3]	100	0	0	-	2.5	19.1	4.0	0	4.0
240[3]	100	0	0	-	3.0	20.3	trace	0	3.7

Reaction conditions: 0.3g cellobiose, glucose or fructose, the same amount of catalyst(5wt%Ru/C), 30mL extremely low phosphotungstic acid(0.01wt%) in an 100mL autoclave and stirred at 4Mpa H_2, and 600rpm for the same time.

Con, Glu, Fru, Sor, EG, PG, Ery, Gly are abbreviations of conversion, glucose, fructose, sorbitol, ethylene glycol, 1,2-propylene glycol, erythritol and glycerol, respectively.

[1] Feedstock is cellobiose.

[2] Feedstock is glucose.

[3] Feedstock is fructose.

3 Possible pathways of the main products from cellobiose conversion

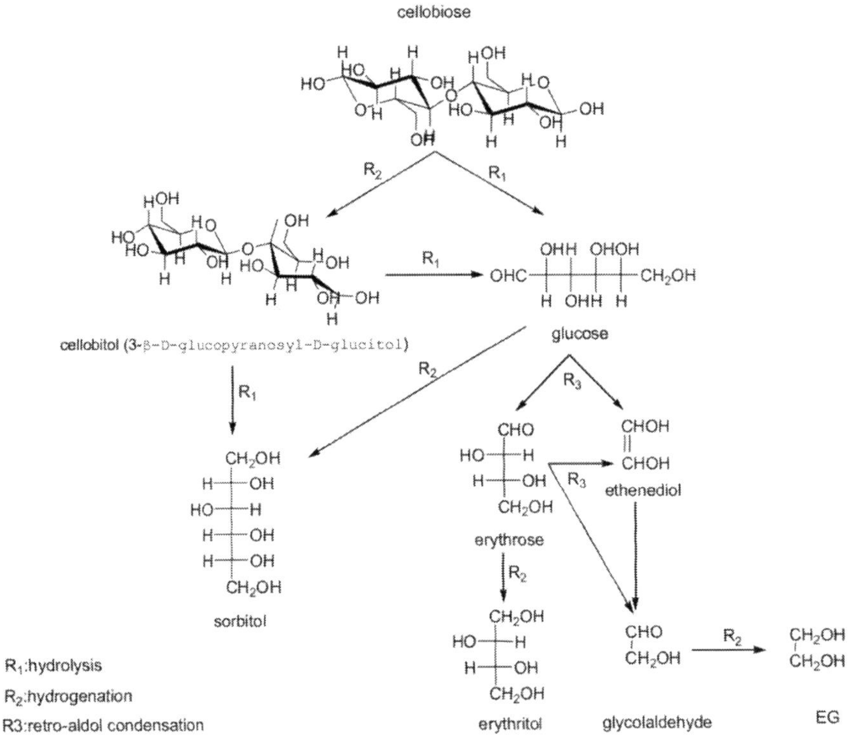

Scheme 1 The possible pathway of producing EG for cellobiose transformation over Ru/C catalyst in PTA solution

According to the results from catalytic process in presence of H_2, cellobiose undergoes hydrolysis to glucose under the catalysis of H^+ which is released from both hot water and dissolved PTA(Scheme 1). Meanwhile, on the one hand, glucose could be directly catalyzed by precious-metal catalyst (Ru/C) to produce sorbitol. On the other hand, with the increase of temperature, the C-C bond cleavage of glucose occurs via retro-aldol reaction under the dissolved tungsten from PTA, erythrose and ethenediol are produced in the reaction. For one thing, erythritol is formed by hydrogenation of erythrose in the presence of pressurized H_2 and hydrogenation catalyst (Ru/C). For another thing, the retro-aldol condensation reaction takes place, ethenediol and glycolaldehyde are produced from the C-C cleavage of erythrose. Together, ethenediol is not stable, but is easily isomerized to the glycolaldehyde which is converted to EG over hydrogenation catalyst in presence of H_2.

Besides, duo to the low concentration of H^+ from extremely low phosphotungstic acid solution, part of glycosidic bonds did not crack, cellobiose was found to undergo hydrogenation of the C-O bond on one of the glucose rings leading to cellobitol(3- β -D-glucopyranosyl-D-glucitol), which was confirmed by several researchers[17,24], and $C_{12}H_{24}O_{11}$ was also detected by UPLC-ESI-MS in our experiment (in Fig 4). In the subsequent reaction, cellobitol would be converted into sorbitol and glucose by hydrolysis. Meanwhile, cellobiose also can undergo retro-aldol reaction on one of the glucose rings to produce 2- β -D-glucopyranosyl-D-erythrose ($C_{10}H_{18}O_9$) and ethenediol. 2- β -D-glucopyranosyl-D-erythrose then was converted into 2- β -D-glucopyranosyl-D-erythritol ($C_{10}H_{20}O_9$), and $C_{10}H_{20}O_9$ was also detected by UPLC-ESI-MS (in Fig 5), the existence of 2- β -D-glucopyranosyl-D-erythritol would further confirm the possible pathway of producing EG in Scheme 1. Whether $C_{12}H_{24}O_{11}$ or $C_{10}H_{20}O_9$ should further confirm chemical structure, this part of experiment is being done.

In a word, the possible pathway of producing EG in the catalytic hydrogenation of cellobiose is proposed according to the results from our experiment and other groups' research progress[22,25]. Under the cooperation between homogeneous tungsten compound (PTA) and heterogeneous precious-metal catalyst (Ru/C), a high yield of EG can be obtained from cellobiose in one pot[12].

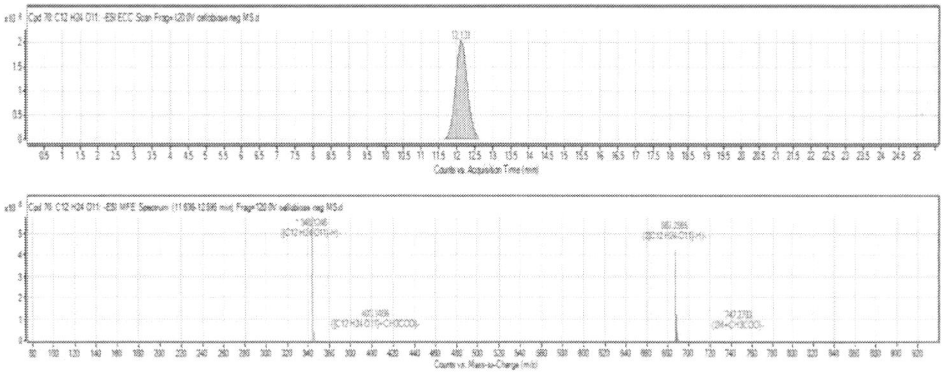

Fig 4 The spectrogram of cellobitol from UPLC-ESI-MS for cellobiose transformation over 5wt% Ru/C catalyst at 180℃, 4Mpa H_2, 60min, and 30mL 0.01wt% phosphotungstic acid solution

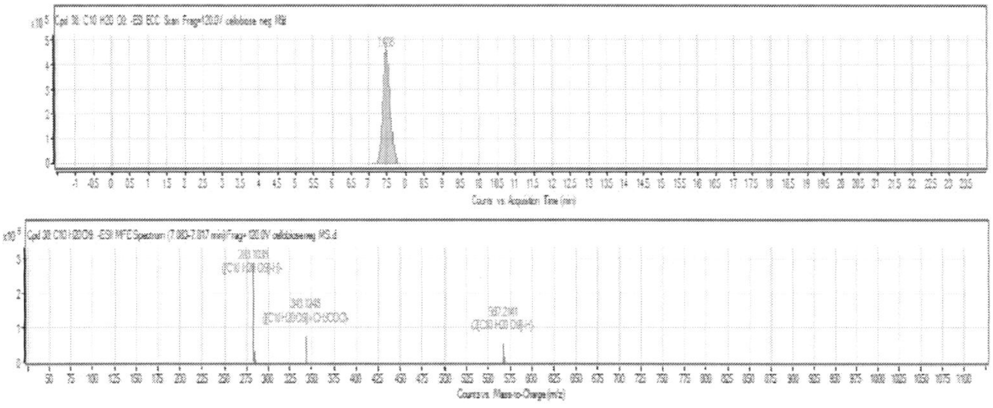

Fig 5 The spectrogram of $C_{10}H_{20}O_9$ from UPLC-ESI-MS for cellobiose transformation over 5wt% Ru/C catalyst at 180℃, 4Mpa H_2, 60min, and 30mL 0.01wt% phosphotungstic acid solution

Besides, glucose would be partly isomerized to fructose especially in presence of activated carbon, the pathway of fructose is different in the reaction. According to a paper from Zhang Tao and coworkers[22], the retro-aldol reaction of fructose produced glyceraldehyde and 1,3-dihydroxyacetone which were detected by HPLC. Then, to further find the main precursor for the formation of 1,2-PG, glyceraldehyde and 1,3-dihydroxyacetone were used as the feedstock in the reaction respectively or together, the result strongly suggested that glyceraldehyde was the main precursor for the formation of glycerol, whereas 1,3-dihydroxyacetone was responsible for the formation of 1,2-PG. While, Liu Hai-chao and coworkers[25] found that glyceraldehyde and propenetriol were produced from the retro-aldol reaction, and propenetriol was not stable, and it was easy to convert into glyceraldehyde or 1,3-dihydroxyacetone in water, but 1,3-dihydroxyacetone was not detected in the reaction. Besides, if 1,2-PG was formed from 1,3-dihydroxyacetone, the hydroxy on 1,3-dihydroxyacetone has the same chemical activity duo to the symmetrical structure, so both C_1 and C_3 on 1,3-dihydroxyacetone can become the C on methyl of 1,2-PG. While, when $^{13}C_1$- fructose and $^{13}C_6$-fructose were used as feedstock, all the C on methyl of 1,2-PG were detected to be ^{13}C after reaction. Therefore, it is glyceraldehyde rather than 1,3-dihydroxyacetone that is the main precursor for the production of 1,2-PG.

According to the conclusions above, there are disputes on the existence of 1,3-dihydroxyacetone and the main precursor for the formation of 1,2-PG in the reaction. Hence, in order to resolve disputes, we studied the possible precursor of 1,2-PG in the reaction.

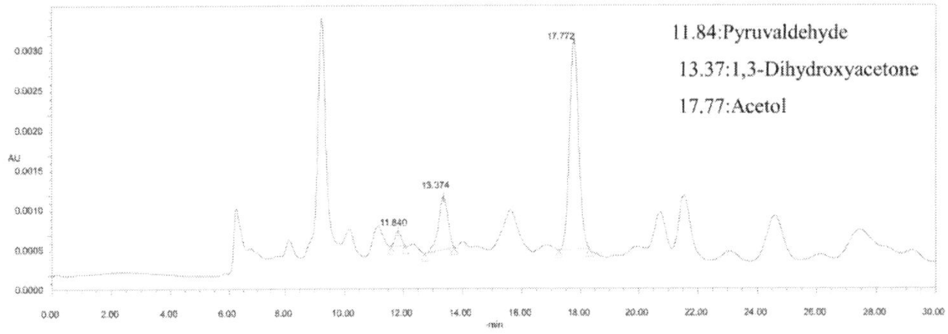

Fig 6 Liquid chromatogram by UV detector(254nm) for cellobiose transformation over 5wt% Ru/C catalyst at 180℃, 4Mpa H_2, 60min, and 30mL 0.01wt% phosphotungstic acid solution

For one part, 1,3-dihydroxyacetone was detected from the products of reaction by HPLC in Fig 6. However, it does not mean that 1,3-dihydroxyacetone is responsible for the formation of 1,2-PG. For another part, 1,3-dihydroxyacetone was used as substitution of cellobiose in the reaction at different temperatures from 120℃ to 240℃, and the results are shown in Fig 7. The highest yield of 1,2-PG is only 2.2% when 1,3-dihydroxyacetone was used for reaction at 240℃. Therefore, it is not easy for 1,3-dihydroxyacetone to produce 1,2-PG in present reaction conditions.

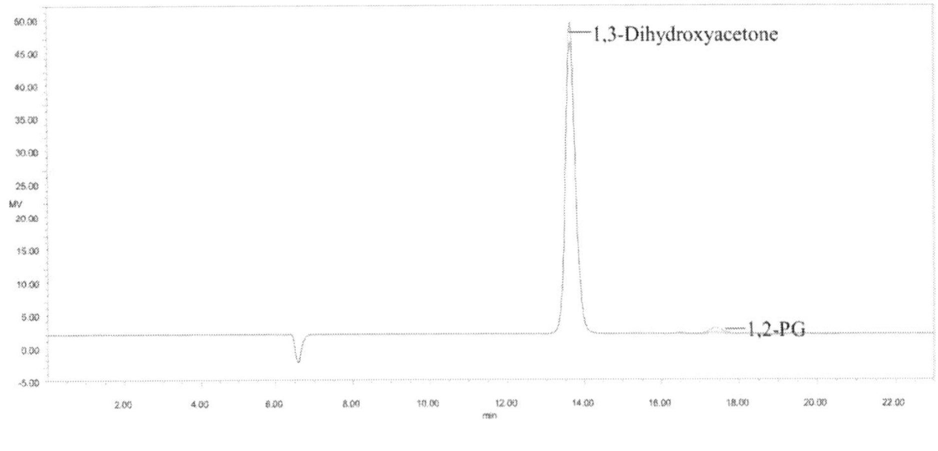

Fig 7 Liquid chromatogram by RI detector for 1,3-dihydroxyacetone transformation over 5wt% Ru/C catalyst at different temperatures from 120 to 240℃, 4Mpa H_2, 60min, and 30mL 0.01wt% phosphotungstic acid solution

Combining the research results from other groups and our own group, we can conclude that the main precursor for the formation of 1,2-PG is not 1,3-dihydroxyacetone in our reaction system. Hence, on the one hand, glyceraldehyde is responsible for the formation of glycerol through catalytic hydrogenation. On the other hand, glyceraldehyde is a precursor for the formation of 1,2-PG via cross–dehydration–hydrogenation reaction pathway (Scheme 2), pyruvaldehyde and acetol are produced as intermediates in Fig 6. However, it does not mean that the conclusions from the groups of Zhang Tao and Liu Hai-chao are not correct, there are differences from conditions to catalysts of the three reactions. Therefore, the main precursor for the formation of 1,2-PG may depend on catalysts and conditions in the reaction, and further researches on the pathways of producing 1,2-PG from cellobiose conversion are being done.

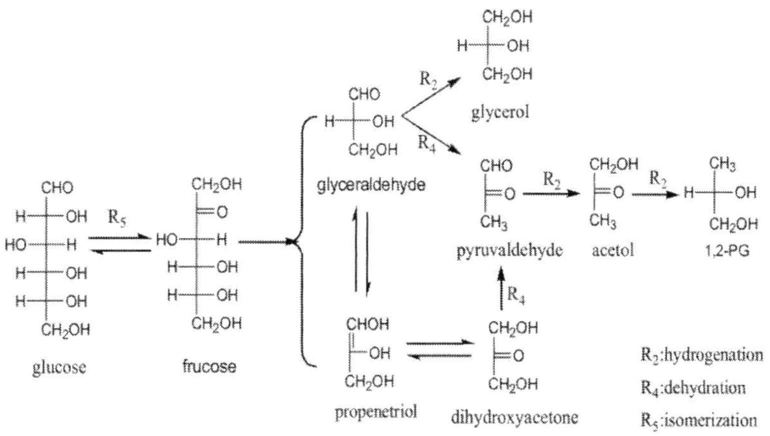

Scheme 2 The possible pathway of producing 1,2-PG for cellobiose transformation over Ru/C catalyst in PTA solution

Conclusion

Using cellobiose as feedstock to converted into polyols in the presence of hydrogen and precious-metal catalyst of Ru/C over extremely low phosphotungstic acid, the yields of ethylene glycol (EG) and 1,2-propylene glycol (1,2-PG) were obtained up to 33.4% and 12.2% respectively at 240 ℃ for 1 h, which had a definite guidance function for catalytic conversion cellulose. As the simplest and basic unit of cellulose, cellobiose is an ideal feedstock for the study of the possible reaction pathways of producing polyols. Combining the research progress from other groups, the detection of intermediates such as pyruvaldehyde, acetol and 1,3-dihydroxyacetone by HPLC and UPLC-MS verified the possible reaction pathways of producing polyols. Besides, glyceraldehyde is a mian precursor for the formation of 1,2-PG via cross–dehydration–hydrogenation reaction pathway rather than 1,3-dihydroxyacetone in the present reaction conditions.

Acknowledgments

This work was supported by the State Natural Sciences Foundation (31270635), the National High Technology Research and Development Program of China (No.2012AA101806) and the Fundamental Research Funds for the Central Universities (2014ZP14)

Reference

[1]Jiying Sun and Haichao Liu.Selective hydrogenolysis of biomass-derived xylitol to ethylene glycol and propylene glycol on supported Ru catalysts[J].Green Chem.,2011,13:135–142.

[2]D.M. Alonso, J. Q. Bond and J. A. Dumesic.Catalytic conversion of biomass to biofuels.Green Chem, 2010,12:1493–1513

[3]D. L. Klass.Biomass for Renewable Energy, Fuels, and Chemicals, Academic Press, San Diego, 1998.
[4]G.W. Huber, S. Iborra, A. Corma.Synthesis of Transportation Fuels from Biomass: Chemistry, Catalysts, and Engineering.Chem. Rev. 2006,106:4044–4098.
[5]Mizuho Yabushita,Hirokazu Kobayashi,Atsushi Fukuoka.Catalytic transformation of cellulose into platform chemicals[J].Applied Catalysis B,2014,145:1-9.
[6]C. Luo, S.Wang, C. H. Liu.Cellulose Conversion into Polyols Catalyzed by Reversibly Formed Acids and Supported Ruthenium Clusters in Hot Water[J].Angew. Chem. 2007, 119:7780–7783.
[7]Fukuoka, A. Dhepe, P. L. Catalytic Conversion of Cellulose into Sugar Alcohols[J]. Angew.Chem., Int. Ed. 2006, 45:5161–5163.
[8]Dhepe, P. L.Fukuoka, A. Cellulose Conversion under Heterogeneous Catalysis[J].ChemSusChem,2008, 1:969–975.
[9]Ji, N.Zhang, T.Zheng, M.Wang, A. Wang, H. Wang, X.Chen, J. G. Direct Catalytic Conversion of Cellulose into Ethylene Glycol Using Nickel-Promoted Tungsten Carbide Catalysts[J]. Angew. Chem., Int. Ed.2008, 47: 8510–8513.
[10]Yue, H.Zhao, Y.Ma, X.; Gong, J. Ethylene Glycol: Properties, Synthesis, and Applications.Chem. Soc. Rev. 2012, 41:4218–4244.
[11]Guanhong Zhao,Mingyuan Zheng,Junying Zhang,Aiqin Wang,and Tao Zhang.Catalytic Conversion of Concentrated Glucose to Ethylene Glycol with Semicontinuous Reaction System[J]. Ind. Eng. Chem. Res. 2013, 52:9566–9572.
[12]Zhijun Tai,Junying Zhang,Aiqin Wang,Mingyuan Zheng,Tao Zhang. Temperature -controlled phase-transfer catalysis for ethylene glycol production from cellulose[J].Chem. Commu, 2012,48:7052–7054.
[13]Zhijun Tai et al.Catalytic Conversion of Cellulose to Ethylene Glycol over a Low-Cost Binary Catalyst of Raney Ni and Tungstic Acid[J].ChemSusChem 2013,6:652–658.
[14]A.Y.Yin,X.Y.Guo,W.L.Dai,K.N.Fan. The synthesis of propylene glycol and ethylene glycol from glycerol using Raney Ni as a versatile catalyst[J].Green Chem.2009,11:1514-1516.
[15]Regina Palkovits,et al.Heteropoly acids as efficient acid catalysts in the one-step conversion of cellulose to sugar alcohols[J].Chem.Commun,2011,47:576-578.
[16]Ken-ichi Shimizu,et al.Effects of Brønsted and Lewis acidities on activity and selectivity of heteropolyacid-based catalysts for hydrolysis of cellobiose and cellulose[J]. Green Chem,2009,11:1627–1632.
[17]Leila Negahdar,Regina Palkovits,et al.Kinetic investigation of the catalytic conversion of cellobiose to sorbitol[J].Applied Catalysis B:Environmental,2014,147:677– 683.
[18]Wang Aiqin, Zhang Tao.One-Pot Conversion of Cellulose to Ethylene Glycol with Multifunctional Tungsten-Based Catalysts[J].Accounts of chemical research, 2013, 46 (7):1377-1386.
[19]B. M. Kabyemela, T. Adschiri, R. Malaluan, K. Arai.Degradation Kinetics of Dihydroxyacetone and Glyceraldehyde in Subcritical and Supercritical Water.Ind. Eng. Chem. Res.1997, 36:2025–2030.
[20]B. M. Kabyemela, T. Adschiri, R. M. Malaluan, K. Arai. Glucose and Fructose Decomposition in Subcritical and Supercritical Water: Detailed Reaction Pathway, Mechanisms, and Kinetics.Ind. Eng. Chem. Res.1999, 38:2888 –2895.
[21]Liu Y,Luo C,Liu H C.Tungsten trioxide promoted selective conversion of cellulose into propylene glycol and ethylene glycol on a ruthenium catalyst[J].Angew Chem Int Ed,2012,51(13):3249-3253.
[22]Likun Zhou,Aiqin Wang,Changzhi Li,Mingyuan Zheng,and Tao Zhang. Selective production of 1,2-propylene glycol from Jerusalem Artichoke Tuber using Ni-W2C/AC catalysts[J]. ChemSusChem,2012,5:932-938.
[23]Jifeng Pang, et al.Catalytic Conversion of Concentrated Miscanthus in Water for Ethylene Glycol Production[J]. AIChE Journal,2014,6(60):2254-2262.
[24]Ning Yan, Chen Zhao, Chen Luo, Paul J.Dyson, Haichao Liu, and Yuan Kou. One-Step Conversion of Cellobiose to C_6-Alcohols Using a Ruthenium Nanocluster Catalyst[J].J. AM. CHEM. SOC.2006,128:8714-8715
[25]Yue Liu, Haichao Liu.Mechhanism for solid acid catalysis and selective cleavage of C-Cbonds of sugars in catalytic conversion of cellulose into polyols[D].Peking University, 2013,6:101-107.

375448 Design of Process Plant for the Production of Formaldehyde and Formic Acid from Lignin

Abstract:

The production of platform chemicals from biomass is the current focus of chemical research. Platform chemicals are the precursors of many household and industrial products. These chemicals can be converted into several high-value products and materials including resins, paints, plastics, adhesives, solvents, and synthetic fibers. Most platform chemicals are currently derived from fossil fuels. One potential source to replace petroleum as a precursor for platform chemicals is plant biomass. Biomass is composed of cellulose, hemicellulose, and lignin. The composition of the component of biomass suggests that it could also be utilized in the production of platform chemicals. Most bio-refinery research has focused on converting cellulose and hemicellulose into ethanol. However, lignin, which constitutes 15-25% of plant biomass, is considered a waste product in bio-refineries. In this project, the feasibility of utilizing lignin for the production of formaldehyde and formic acid is considered. The Douglas hierarchy is utilized to develop process flowsheets for making these products from lignin. Several simulations are conducted in the process simulator ASPENPlus to study the technical as well as economic viability of these processes.

Elucidation of cellulose-lignin interaction during pyrolysis: A Py-GC-MS study

Shiliang Wu, Dekui Shen, Rui Xiao, Southeast University, NanJing, China

Abstract

Interaction between the three main components (cellulose, hemicelluloses, and lignin) of biomass is notably observed under pyrolysis condition. Method (both experimental and analytical) for estimating the extent of interactions is required in order to gain the better understanding of the mechanism of pyrolysis of whole biomass system. In this work, two kinds of mixture (direct mixture and table mixture) for cellulose-lignin and levoglucosan-lignin were pyrolyzed in Pyroprobe 5250 coupled with GC-MS under temperature range 500-700 ℃. While there is a slight difference between calculated and experimental volatile and char yield, the significant interactions on the product distribution were observed: the presence of lignin promoted the formation of light-molecular weight compounds from cellulose (such as acetone, glycolaldehyde, acetone alcohol, and butanedial) but inhibited the formation of anhydrosugars, levoglucosan and oligosaccharides. While most of lignin-derived products showed stability under co-pyrolysis with cellulose, excluding some certain compounds (p-cyresol, 2,3-dihydrobenzofuran, guaiacol and 4-vinylguaiacol). A statistical method by correlation r and p-value was adopted to evaluate the interaction strength on product distribution from two mixtures of cellulose-lignin pyrolysis, finding that interactions of table mixture were stronger than direct mixture. Moreover, interaction strength of direct mixture was decreased with increasing temperature, while table mixture achieved the strongest interaction at 600 ℃. For levoglucosan-lignin direct mixture, co-pyrolysis enhanced the levoglucosan-derived products while inhibited the lignin-derived products, revealing that hydrogen transfer and activated intermediates from cellulose and lignin maybe the dominant interaction mechanism. It is believed that the difference of interaction strength for direct and table mixture is caused by the difference of interaction phase. For table mixture, gas phase, liquid phase and even liquid-soild phase interactions are existed. But for direct mixture, only gas phase and limited liquid phase interaction is thought to be existed.

Introduction

As a CO2 neutral and sustainable energy, biomass has received more and more attention for renewable production of fuels, chemicals and energy. In United States and European Union, the renewable energy produced from biomass has surpassed the total combined contribution from hydropower, wind power, geothermal energy, and solar power [1]. In addition to energy, biomass is the only renewable energy for the production of plat chemicals; besides, it is also the only renewable source for liquid transportation fuel.

Biomass is composed of cellulose, hemicellulose and lignin, with a little amount of extractives and inorganic salts. During pyrolysis, the existence of inorganic salts influences the distribution of volatiles, accelerates the formation of low molecular weight species [2]. A number of studies on pyrolysis of biomass [3,4] and biomass components (cellulose [5,6], hemicelluloses [7] and lignin [8]) were reported in literature. But little attention was paid to the interactions between cellulose, hemicelluloses and lignin, which also played an important role during the pyrolysis of biomass.

Usually, these three components are thought to pyrolyze separately with no interactions between them. Caballero et al. investigated the pyrolysis of almond shell and its components (holocellulose and lignin) by Pyroprobe 1000 within the temperature range from 700 ℃ to 900 ℃, products from pyrolysis were analyzed by gas chromatography, they reported that the yields of gaseous product from almond shell can be obtained from the yields of holocellulose and lignin [9]. Alen et al. used pyrolysis-gas-chromatography with mass-selective detection (Py-GC/MSD) to study the thermochemical behaviour of pine wood and its three components in the temperature range 400-1000 ℃, they remarked that wood appeared to behave as the sum of its constituents [10]. Miller et al. and Rao et al. assumed that there were no interaction between cellulose, hemicellulose, and lignin, so the pyrolysis of general biomass materials could be modelled via a superposition of the kinetics of three components. They stated that experiments showed a favorable agreement with the model predictions [11,12]. Sevenson et al. carried out the pyrolysis of cellulose, birch hemicelluloses, and birch lignin in a single particle reactor at constant reactor temperature in the range 225-650 ℃, the calculated results that the three independent materials added together in the proportions of birch wood [13]. Biagini et al. applied summative law for the TG results to obtain the chemical composition of biomass, where calculated values obtained by this method are in good agreement with the experimental results [14]. Yang et al. performed pyrolysis of synthesized biomass samples and three components by a thermogravimetric analyzer (TGA), suggesting that negligible interaction among the three components was found during the pyrolysis of synthesized samples [15]. Fushimi et al. reported that there was no significant interaction for the gas evolution during pyrolysis of model biomass samples (composed of cellulose, hemicelluloses and lignin) in a continuous cross-flow moving bed type differential reactor [16]. Qu et al. claimed that the additivity law can predict reasonably the trend of product yields of biomass samples from their composition of three components, they conducted fast pyrolysis of samples in a tube furnace between 350-650 ℃ [17]. Yoon et al. validated a three-parallel reaction kinetic model which was formulated on the basis of the weight sum of reaction rates for individual components by using a combined thermogravimetry and gas chromatography [18].

In contrast, some researchers reported that the product yield and kinetics of biomass in pyrolysis cannot be explained by additivity law of the three components because of the interaction among the three components during tar production and evolution. Hosoya et al. pointed out that there were significant interactions in cellulose-lignin pyrolysis at gasification temperature (800 ℃), lignin inhibited the thermal polymerization of levoglucosan formed from cellulose and enhanced the formation of the low molecular weight products, while cellulose reduced the secondary char formation from lignin. Comparatively weak interactions were also observed in cellulose-hemicellulose pyrolysis [19]. Worasuwannarak et al. investigated pyrolysis behaviours of rice straw, rice husk, and corncob by TG-MS technique, indicating obvious interactions between cellulose and lignin during pyrolysis. The interactions lead to a decrease in tar yields but an increase in char yields [3]. Couhert et al. doubted whether additivity law is allowed to predict gas yields of biomass after rapid pyrolysis at high temperature, so experiment in an entrained flow reactor (EFR) was conducted, showing that interactions occurred both outside (i.e. in the gas phase) and inside the mixed samples [20,21]. Wang et al. tested pyrolysis characteristics of biomass by using both a thermogravimetric analyzer coupled with a Fourier-transform infrared spectrometer (TG-FTIR) and an experimental pyrolyzer. By analysing the main products by gas chromatography-mass spectrometry (GC-MS), they considered that pyrolysis characteristics of biomass cannot be predicted through its composition in the main components [22]. Liu et al. investigated the synthetic biomass samples by using the technology of thermogravimetric analyzer coupled with Fourier transform infrared spectrometer (TG-FTIR), revealing the occurrence of interactions among biomass components during pyrolysis [23].

Besides, though some researchers reported there were no interactions between three components by analysising the gas products and yields, they claimed interactions are occurred when other aspects (such as tar products, low temperatures) were analysed. Caballero et al. found that slow thermal decomposition of almond shells cannot be predicted by the sum of lignin and holocellulose [9]. Fushimi et al. observed that the interactions among cellulose, xylan and lignin suppressed the evolution of levoglucosan and significantly increase the evolution rate of 5-methylfurfural [16,24].

Finally, conclusions were different depending on the researchers and the experiment (different reactors, analytical methods and equipments, experiment conditions). Some authors claimed that there was no interaction between the three components when pyrolysised together. But others confirmed that interactions were indeed occurred under pyrolysis when products distribution was analysed. Even for the same author, different conclusions were got when different analytical methods were used. Further studies are necessary for better understanding of biomass pyrolysis. As cellulose and hemicellulose are all polysaccharides, while lignin is composed with methoxylated phenylpropane structures, besides, the interactions between cellulose-hemicellulose are smaller than cellulose-lignin [19], so the interactions between cellulose-lignin are more meaningful than cellulose-hemicellulose

In this article, pyrolyzer together with gas chromatography-mass spectrometer (Py-GC-MS) was used to confirm the interactions between cellulose and lignin in the temperature range 500-700 ℃. The influence of mixture methods for synthetic biomass samples and temperatures were checked. An evaluation method was proposed to assess the cellulose-lignin interaction. Levoglucosan-lignin mixtures were also pyrolysed to shed a light on the mechanism of the cellulose-lignin interaction. Finally, a possible cellulose-lignin interaction mechanism was put forward.

Experiment

Materials

Microcrystalline cellulose power (Sinopharm Chemical Reagent Co.) and levoglucosan (Sigma-Aldrich Co.) were used as commercial samples. Lignin was isolated and purified from black liquor, the detailed method for lignin extraction was shown elsewhere [25].

Cellulose-lignin mixture was prepared by two methods:

(i) Direct mixture: Cellulose and lignin were mixed in equal mass proportion with spoon in a container. Cellulose and lignin powder are all smaller than 100 mesh screen (<150 μm). In this case, interaction in gas phase is favoured, but the interaction between intermediate liquid form cellulose and lignin may be small, because the contact of cellulose-lignin is not very compact, air are existed in the pores of the mixture, which prevent the interaction of cellulose-lignin during pyrolysis.

(ii) Table mixture: Cellulose and lignin were firstly mixed in equal mass as direct mixture, then the miture was tabled by manual hydraulic press machine under 20 Mpa, after that, the tabled mixture was milled to smaller than 100 mesh screen (<150 μm). In this case, as air was removed from the cellulose-lignin mixture, both intermediate liquid phase and gas phase interaction will be expected.

Py-GC-MS

Py-GC-MS was used to investigate the pyrolysis characteristics of cellulose, lignin, cellulose-lignin mixture and levoglucosan-lignin mixture under flash heating mode (heating rate from 10 to 20000 K/s). Fast pyrolysis analyzer (CDS5250) is coupled with the GC-MS instrument (Trace DSQ II) to indicate the distribution of products under the fixed temperature (room temperature~1400 ℃). Only 0.5 mg samples were loaded in the pyrolysis tube before the test to avoid heat diffusion. The experiment was carried out under the temperature of 500 ℃, 600 ℃ and 700 ℃ at the heating rate of 20 ℃/ms with the residence time of 10 s. Due to the poor thermal conductivity of biomass samples and the high heating rate of pyrolyzer, its actual pyrolysis temperature was lower than the set value, reported as 100 ℃ for the difference [26]. The evolved volatiles were analyzed by the couple GC-MS, where the injector temperature was kept at 300 ℃. The chromatographic separation was performed with a TR-35MS capillary column. The oven temperature was programmed from 40 ℃ (3 min) to 200 ℃ (1 min) with 4 ℃/min heating rate, and then to 280 ℃ (1 min) with 20 ℃/min heating rate. The mass spectra were operated in EI mode at 70 eV. The mass spectra were obtained from 35 to 650 (m/z). The chromatographic peaks were identified according to the NIST MS library v2.0. Repeated tests were carried out, finding that the error for absolute peak area was no more than±10%. As calibration has not done for GC-MS, absolute peak area was used to compare the difference of prediction and experiment. Jiang reported that the linearity of peak area and mass for each standard compound was very high (the R^2 values were above 0.98) [27]. So it was receivable for the use of absolute peak area in this article.

Results and discussion

Cellulose-Lignin pyrolysis

Figure 1 shows the formation of volatiles and char of cellulose, lignin, predicted cellulose-lignin mixture without interactions, direct mixture, and table mixture. The yields of char all decreases with increasing temperature. Lignin has the maximum char yield at all conditions except 500 ℃, where char of direct mixture is a little more than lignin. Cellulose is always with the minimum char yield. As for the difference between prediction and experiment, the different is not significant at 600 ℃ and 700 ℃. But at 500 ℃, the char yield of direct mixture is higher than prediction. This may attributes to that air in the direct mixture prevents heat convection, that makes high char yield at low temperatures, when temperature increased, the hinder of air becomes insignificantly, so there is no significant difference between prediction and experiment.

Table 1 summarizes the product distribution of cellulose-lignin pyrolysis. Cellulose, lignin, direct mixed cellulose-lignin, and table mixed cellulose-lignin are pyrolysised under 500 ℃, 600 ℃, and 700 ℃, respectively. To evaluate the difference between predicted yields by additive law and experiment yields of individual products from mixtures, yield ratio γ_i is used:

$$\gamma_i = \frac{Y^i_{mix}}{x_{cel} Y^i_{cel} + x_{lig} Y^i_{lig}} \qquad (1)$$

In (1), Y_{mix} is the experimental mixture yield (for direct mixture cellulose-lignin or table mixture cellulose-lignin, they are marked withγ_D andγ_T, respectively) while Y_{cel} and Y_{lig} are the yields from single component pyrolysis of cellulose and lignin, respectively. x_{cel} and x_{lig} are the initial mass fractions, in this article, $x_{cel} = x_{lig} = 0.5$. Aγ_i value of 1 for individual product i indicates that its production is not affected by co-pyrolysis of cellulose and lignin, while value less or greater than 1 depict product inhibition or promotion, respectively.

For cellulose-derived compounds, the formation of furanone, furfural, levoglucosan, DAGP (1:4;3:6-dianhydroglucopyranose), and 5-hydroxymethylfurfural are all inhibited when lignin is mixed with cellulose, but for light weight compounds, such as acetone, glycolaldehyde, acetone alcohol, butanedial, and methyl pyruvate, are promoted. Besides, when cellulose-lignin co-pyrolysised at 500 ℃, oligosaccharides formed when cellulose pyrolysised alone can not be found. Lignin may promote the secondary reaction of cellulose so that the yields of oligosaccharides, anhydrosugars, levoglucosan and other intermediates decrease while the yields of light weight compounds increase. Piskorz et al. reported that the formation of light weight compounds (such as glycolaldehyde) and anhydrosugars (such as levoglucosan) are through a competition mechanism [28]. And it is observed that some inorganic substances and higher temperatures will reduce the yield of levoglucosan and enhance the formation of glycolaldehyde and hydroxyacetone [2]. Here, for the co-pyrolysis of cellulose and lignin, the role of co-pyrolysis is like some inorganic salts and higher temperatures, accelerating the reactions that led to the formation of low molecular weight species from cellulose as compared to those leading to anhydrosugars. This results is different from Hosoya's experiment [19,29], where both the light weight compounds and anhydrosugars are promoted.

For lignin-derived compounds, no such phenomena as cellulose are found. But phenol,4-methyl-, bezofuran,2,3-dihydro-, 2-methoxy-4-vinylphenol, and 3-tert-butyl-4-hydroxyanisole are all promoted when co-pyrolysised. Besides, phenol,

2-methoxy-4-methyl- (guaiacol) is enhanced significantly under 600 ℃. While other compounds don not show a consistent tendency for different mixture methods and temperatures, but the yield ratios for them are near 1, showing invariableness under co-pyrolysis, this may attribute to the stability of benzene ring. Hosoya et al. reported that cellulose enhanced the formation of some lignin-derived products including guaiacol, 4-methyl-guaiacol and 4-vinyl-guaiacol, while effects on other products were small[19],which is just consistent with our experiment results.

Strength of interactions evulation

Though there are some experiments have been carried on interactions of the three components, there are no efficient methods to analysis the results. It is usually by comparing the results (gas yields, product distribution of tar, TG-DTG curves) of prediction and experiment of co-pyrolysis to conclude whether interaction is occurred during co-pyrolysis or not, this method is arbitrary. That is why results are not in agreement for different researchers.

In this article, statistical method is used to evaluate the strength of interactions. For random variables X and Y, the expected values for them are E(X) and E(Y), respectively. So, covariance of X and Y could be presented as follows:

$$Cov(X,Y)=E[(X-E(X))(Y-E(Y))] \quad (2)$$

Where, Cov(X,Y) is the covariance of X and Y. If Cov(X,Y)=0, it means that random X and Y is mutual independence.

The variances of X and Y are denoted as D(X) and D(Y). Then, correlation coefficient r is defined based on variances and covariances:

$$r = Cov(X,Y)/\sqrt{D(X)D(Y)} \quad (3)$$

The value of r is between -1 to 1, if X and Y is positive correction, the value for r is 1. The bigger the absolute value of r, the correlation of X and Y is more intimate. r is used to evaluate the interactions between cellulose-lignin.

Besides, p-value is also used to check whether the difference between prediction and experiment is caused by statistical discrepancy or interactions.

Figure 2 presents the comparison of pyrolysis products between prediction and experiment. The x-axis shows the prediction values for the absolute peak area of products after co-pyrolysis. The values for the absolute peak area are took the logarithm to the base 10 for briefness. The absolute peak area is normalized with mass (so the unit for product yield is mg^{-1}). This treatment is also conducted with y-axis, which shows the experiment value. So if there is no interaction between cellulose-lignin, the data points should be located on the blue line y=x. To objectively analysis the results, r and p between prediction and experiment are calculated and presented on

Figure 2. Here, $p = 0$ means that the deviations between prediction and experiment are not caused by statistical discrepancy, maybe it is caused by interactions. And the smaller of correlation coefficient r, the stronger of the interactions.

At 500 ℃, the interactions of direct mixture are weaker than table mixture ($r_D = 0.773$ to $r_T = 0.669$), when temperature increased to 600 ℃, this difference amplifies, as $r_D = 0.841$ for direct mixture and $r_T = 0.635$ for table mixture. But at 700 ℃, this difference is very small. This may be explained by the phase of interactions happen. Under low temperature, the interactions occur at liquid phase, as table mixture is more intimate than direct mixture and there is no air in table mixture, so the interactions are stronger for table mixture. While under high temperature, the interactions are mainly happened at gas phase, so the difference is small for direct mixture and simple mixture.

When temperature increased form 500 ℃ to 700 ℃, the strength of interactions for direct mixture is becoming weak. But for table mixture, the strongest is achieved at 600 ℃, not 700 ℃. This attribute to that the interactions are mainly happened at liquid phase and gas phase, and interactions at liquid phase are stronger than gas phase, as products in gas phase are light weight compounds that reactions between them are very small. But for products in liquid phase, there are many intermediates derived from cellulose and lignin, the reactions between them (such as hydrogen transfer) are very vigorous. At 500 ℃, the interaction are mainly happened at liquid phase, but at 500 ℃, a large amount of char is also formed (Figure 1), which limits the interaction at liquid phase. But under 600 ℃, it is reported that the yield of bio-oil is largest [26], so the interactions are focused on liquid phase. When temperature increased to 700 ℃, though there are interactions at liquid phase, but the main products are changed for bio-oil to gas, so the interactions between cellulose-lignin are weakened. The reason for the trend difference of direct and table mixture with temperature is that there is air in direct mixture and the contact of direct mixture is loose, so at 600 ℃, the interactions for direct mixture may be gas interactions not liquid interactions.

So we can conclude that temperature and mixture methods all influence the interactions between cellulose and lignin. The increasing temperature weakens these interactions, and the more intimate the mixture method, the stronger the interactions.

Levoglucosan-lignin pyrolysis

During the cellulose-lignin pyrolysis, levoglucosan is greatly reduced. Levoglucosan is regarded as the intermediate of cellulose pyrolysis, which could polymerize to polysccaride or crack into low molecular products and gas [30-32]. To shed a light on the interaction mechanism of cellulose-lignin, levoglucosan and lignin are mixed by table mixture method (mass ration is also 1:1) and pyrolysised at 600 ℃, where the strongest interactions are found for table mixture.

Table 2 summarizes the products for levoglucosan-lignin pyrolysis. It is interesting to note that levoglucosan-derived products are all promoted, while lignin-derived products are all hindered. The significant promotion of the levoglucosan-derived products are mainly 3-buten-2-ol,

Acetol, 3-cyclopentene-1-acetaldehyde, 2-oxo-, and 2,5-dimethyl-4-hydroxy-3(2H)-furannone. These products could be regarded as decomposition products of levoglucosan. While anhydrosugars and polysccarides (3,4-anhydro-d-galactosan, 2,3-anhydro-d-mannosan and melezitose) are not promote significantly. Which means that lignin promotes the decomposition of levoglucosan. In the research for the secondary reaction of levoglucosan at 400 ℃, Kawamoto et al. concluded that the secondary decomposition of levoglucosan was induced by the lignin-derived products [30]. Which just verifies the results of levoglucosan-lignin pyrolysis at 600 ℃. The char for mixture (5.45%) is largely reduced compared with lignin (29.41% for lignin, though there are char formed on the wall of the quartz tube, the char are so little to be quantized by analytical balance). So the decrease of lignin-derived products may attribute to decomposition of primary products to gas, while gas can't be checked by GC-MS.

After co-pyrolysis of levoglucosan-lignin, the tar from lignin is decreased while gas is increased. This phenomenon is like the effect of increasing temperature, so the roles of interaction and increasing temperature on the lignin-derived products from levoglucosan-lignin pyrolysis are checked in Figure 3. The x-axis is the retention time of the products got from gas chromatography, the yield ration for increased temperature is product yield at 700 ℃ divided by product yield at 600 ℃. In Figure 3, the product yield ration of interaction is consistent with increased temperature. For compounds that are thermal stability, they are also stable under co-pyrolysis. This suggests that there may be a relationship between interaction and temperature. To further understand this relationship, comparison between interaction and increasing temperature is also conducted to products distribution on cellulse-lignin pyrolysis under 500 ℃ and 600 ℃. The results was show on Figure 4. Though the values of yield ration are different, the variation trends along retention time for interaction and increased time are same, except for cellulose-derived products at 600 ℃ (there is only five products can be used for comparison). This phenomenon indicates that the interaction mechanism has a relation with temperature.

Interaction mechanism of cellulose-lignin pyrolysis

There is limited research on interaction mechanism of cellulose-lignin pyrolysis. Mettler et al. investigated co-pyrolysis of levoglucosan with ^{13}C-labeled fructose or deuterated glucose, indicating that carbon atoms of the levoglucosan-derived products are mainly from levoglucosan, but there is a large portion of hydrogen atoms originated from deuterated glucose [33]. Hydrogen transfer plays an important role in co-pyrolysis. Hosoya et al. studied thermal stabilization of levoglucosan in aromatic substances, they concluded the hypothesis of the stabilizing mechanism through CH-π interaction, suggesting that the stability of levoglucosan-aromatic compound complexes is based on the π-electron densities [34]. So under temperature range 500 ℃ to 700 ℃, radicals are possible existed, hydrogen transfer between radicals and sugar-aromatic complexes formed with these radicals will all exist in intermediate pool from cellulose and lignin. A proposed interaction mechanism is presented in Figure 5. After co-pyrolysis, an intermediate pool at liquid phase is formed, where many activated complexes are formed between cellulose-derived and lignin-derived compounds, hydrogen transfer between them is also vigorous, the cellulose-derived products and lignin-derived products are formed form intermediate pool. By this process, products distribution from co-pyrolysis is different from prediction. Though gas phase intraction is also presented, but the residence time of pyrolyser is only 10s, the effect of gas phase interaction is weak than liquid phase interaction.

It is noted that there is a link between interaction and increased temperature mechanism. An increase in temperature corresponds to faster motion of molecules and more intermolecular collisions. For interaction, the reactions of radicals from co-pyrolysis are vigorous than pyrolysised alone due to the existence of hydrogen transfer and activated complexes. The effect of this is like increasing temperature makes radical reactions livelily. So the difference between direct mixture and table mixture is shown in Figure 6. For direct mixture, only gas phase interaction and limited liquid phase interaction are existed, so the intermediate pool for direct mixture is limited, that makes the whole interaction of co-pyrolysis weak. While for table mixture, liquid phase intermediates and tar surround unreacted lignin and cellulose particle, so liquid-soild phase interaction is also existed. This makes char from table mixture is much less than direct mixture at 500 ℃.

Conclusion

Cellulose-lignin pyrolysis is conducted by Py-GC-MS under temperature range 500 ℃ to 700 ℃ with two kind of mixtures (direct mixture and table mixture), indicating that lignin promotes the formation of light weight compounds and inhibits the formation of anhydrosugars, levoglucosan and oligosaccharides. While most of lignin-derived products show stability under co-pyrolysis, with an promotion of phenol,4-methyl-, bezofuran,2,3-dihydro-, 2-methoxy-4-vinylphenol, guaiacol, and 3-tert-butyl-4-hydroxyanisole.

Statistical method is introduced to evaluate the strength of interactions. Interactions between table mixture are all stronger than direct mixture. For direct mixture, the strength of direct mixture is receding with increasing temperature. For table mixture, an maximum interaction is observed at 600 ℃.

Levoglucosan-lignin pyrolysis is also investigated at 600 ℃ for table mixture. Co-pyrolysis enhances the levoglucosan-derived products while prevents the lignin-derived products.

Hydrogen transfer and activated complex in intermediate pool form cellulose-lignin pyrolysis at liquid phase is proposed to be the dominant interaction mechanism.

Reference:

1. Zakzeski J, Bruijnincx PCA, Jongerius AL, Weckhuysen BM(2010), "The catalytic valorization of lignin for the production of renewable chemicals," *Catalysts & Catalysed Reactions.* 110(6):3552-3559.
2. Patwardhan PR, Satrio JA, Brown RC, Shanks BH(2010), "Influence of inorganic salts on the primary pyrolysis products of cellulose," *Bioresource Technology.* 101(12):4646-4655.
3. Worasuwannarak N, Sonobe T, Tanthapanichakoon W(2007), "Pyrolysis behaviors of rice straw, rice husk, and corncob by TG-MS technique," *Journal of Analytical and Applied Pyrolysis.* 78(2):265-271.
4. Shafizadeh F(1982), "Introduction to pyrolysis of biomass," *Journal of Analytical and Applied Pyrolysis.* 3(4):283-305.
5. Shen D, Gu S(2009), "The mechanism for thermal decomposition of cellulose and its main products," *Bioresource Technology.* 100(24):6496-6504.
6. Shen D, Xiao R, Gu S, Luo K(2011), "The pyrolytic behavior of cellulose in lignocellulosic biomass: a

review," *RSC Adv.*;1(9):1641-1660.

7. Shen D, Gu S, Bridgwater AV(2010), "Study on the pyrolytic behaviour of xylan-based hemicellulose using TG-FTIR and Py-GC-FTIR," *Journal of Analytical and Applied Pyrolysis*.87(2):199-206.
8. Shen D, Gu S, Luo K, Wang S, Fang M(2010), "The pyrolytic degradation of wood-derived lignin from pulping process," *Bioresource Technology*.101(15):6136-6146.
9. Caballero J, Font R, Marcilla A(1996), "Comparative study of the pyrolysis of almond shells and their fractions, holocellulose and lignin. Product yields and kinetics," *Thermochimica Acta*.276:57-77.
10. Alén R, Kuoppala E, Oesch P(1996), "Formation of the main degradation compound groups from wood and its components during pyrolysis," *Journal of Analytical and Applied Pyrolysis*.36(2):137-148.
11. Miller R, Bellan J(1997), "A Generalized Biomass Pyrolysis Model Based on Superimposed Cellulose, Hemicelluloseand Liqnin Kinetics," *Combustion science and technology*.126(1-6):97-137.
12. Rao TR, Sharma A(1998), "Pyrolysis rates of biomass materials," *Energy*.23(11):973-978.
13. Svenson J, Pettersson JBC, Davidsson KO(2004), "Fast pyrolysis of the main components of birch wood," *Combustion science and technology*.176(5-6):977-990.
14. Biagini E, Barontini F, Tognotti L(2006), "Devolatilization of biomass fuels and biomass components studied by TG/FTIR technique," *Industrial & engineering chemistry research*.45(13):4486-4493.
15. Yang H, Yan R, Chen H, Zheng C, Lee DH, Liang DT(2006), "In-depth investigation of biomass pyrolysis based on three major components: hemicellulose, cellulose and lignin," *Energy & fuels*.20(1):388-393.
16. Fushimi C, Katayama S, Tsutsumi A(2009), "Elucidation of interaction among cellulose, lignin and xylan during tar and gas evolution in steam gasification," *Journal of Analytical and Applied Pyrolysis*.86(1):82-89.
17. Qu T, Guo W, Shen L, Xiao J, Zhao K(2011), "Experimental Study of Biomass Pyrolysis Based on Three Major Components: Hemicellulose, Cellulose, and Lignin," *Industrial & Engineering Chemistry Research*. 2011/09/21;50(18):10424-10433.
18. Yoon HC, Pozivil P, Steinfeld A(2011), "Thermogravimetric Pyrolysis and Gasification of Lignocellulosic Biomass and Kinetic Summative Law for Parallel Reactions with Cellulose, Xylan, and Lignin," *Energy & Fuels*. 2012/01/19;26(1):357-364.
19. Hosoya T, Kawamoto H, Saka S(2007), "Cellulose–hemicellulose and cellulose–lignin interactions in wood pyrolysis at gasification temperature," *Journal of Analytical and Applied Pyrolysis*.80(1):118-125.
20. Couhert C, Commandré JM, Salvador S(2009), "Failure of the component additivity rule to predict gas yields of biomass in flash pyrolysis at 950° C," *Biomass and Bioenergy*.33(2):316-326.
21. Couhert C, Commandre JM, Salvador S(2009), "Is it possible to predict gas yields of any biomass after rapid pyrolysis at high temperature from its composition in cellulose, hemicellulose and lignin?," *Fuel*.88(3):408-417.
22. Wang S, Guo X, Wang K, Luo Z(2011), "Influence of the interaction of components on the pyrolysis behavior of biomass," *Journal of Analytical and Applied Pyrolysis*.91(1):183-189.
23. Liu Q, Zhong Z, Wang S, Luo Z(2011), "Interactions of biomass components during pyrolysis: A TG-FTIR study," *Journal of Analytical and Applied Pyrolysis*.90(2):213-218.
24. Fushimi C, Katayama S, Tasaka K, Suzuki M, Tsutsumi A(2009), "Elucidation of the interaction among cellulose, xylan, and lignin in steam gasification of woody biomass," *Aiche J*.55(2):529-537.
25. Hu J, Xiao R, Shen D, Zhang H("Structural analysis of lignin residue from black liquor and its thermal performance in thermogravimetric- Fourier transform infrared spectroscopy," *Bioresource Technology*. (0).
26. Lu Q, Yang X, Dong C, Zhang Z, Zhang X, Zhu X(2011), "Influence of pyrolysis temperature and time on the cellulose fast pyrolysis products: Analytical Py-GC/MS study," *Journal of Analytical and Applied Pyrolysis*.92(2):430-438.
27. Jiang G, Nowakowski DJ, Bridgwater AV(2010), "Effect of the Temperature on the Composition of Lignin Pyrolysis Products," *Energy & Fuels*. 2010/08/19;24(8):4470-4475.
28. Piskorz J, Radlein D, Scott DS(1986), "On the mechanism of the rapid pyrolysis of cellulose," *Journal of Analytical and Applied Pyrolysis*.9(2):121-137.
29. Hosoya T, Kawamoto H, Saka S(2009), "Solid/liquid-and vapor-phase interactions between cellulose-and lignin-derived pyrolysis products," *Journal of Analytical and Applied Pyrolysis*.85(1-2):237-246.
30. Kawamoto H, Morisaki H, Saka S(2009), "Secondary decomposition of levoglucosan in pyrolytic production from cellulosic biomass," *Journal of Analytical and Applied Pyrolysis*.85(1-2):247-251.
31. Hosoya T, Kawamoto H, Saka S(2008), "Different pyrolytic pathways of levoglucosan in vapor-and liquid/solid-phases," *Journal of Analytical and Applied Pyrolysis*.83(1):64-70.

32. Kawamoto H, Murayama M, Saka S(2003), "Pyrolysis behavior of levoglucosan as an intermediate in cellulose pyrolysis: polymerization into polysaccharide as a key reaction to carbonized product formation," *J Wood Sci.*49(5):469-473.

33. Mettler MS, Paulsen AD, Vlachos DG, Dauenhauer PJ(2012), "Pyrolytic conversion of cellulose to fuels: levoglucosan deoxygenation via elimination and cyclization within molten biomass," *Energy & Environmental Science.*

34. Hosoya T, Kawamoto H, Saka S(2006), "Thermal stabilization of levoglucosan in aromatic substances," *Carbohydrate research.*341(13):2293-2297.

FIGURES:

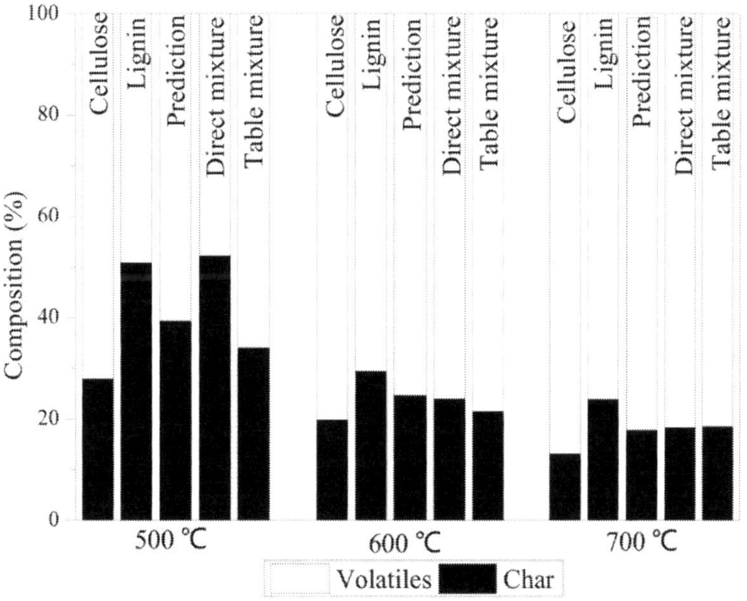

Figure 1. Volatiles and char formation of cellulose-lignin pyrolysis

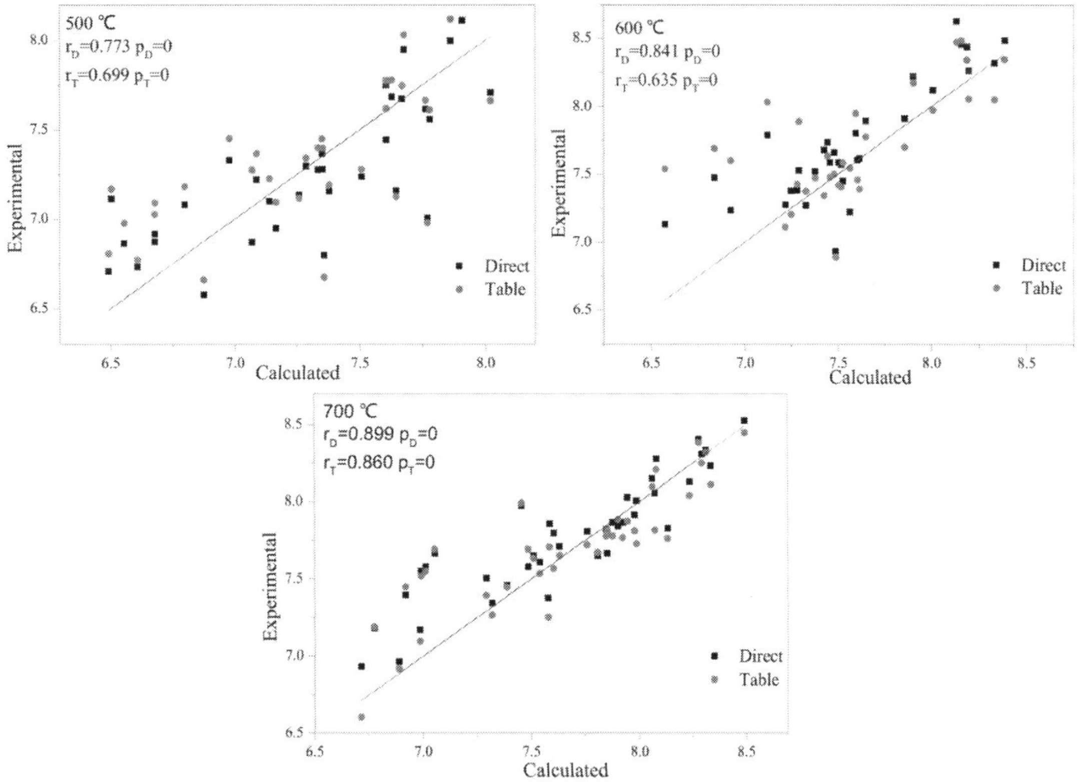

Figure 2. Comparison of pyrolysis products between calculation and experiment for different temperature (500 ℃, 600 ℃, 700 ℃) and mixture methods (direct and table).

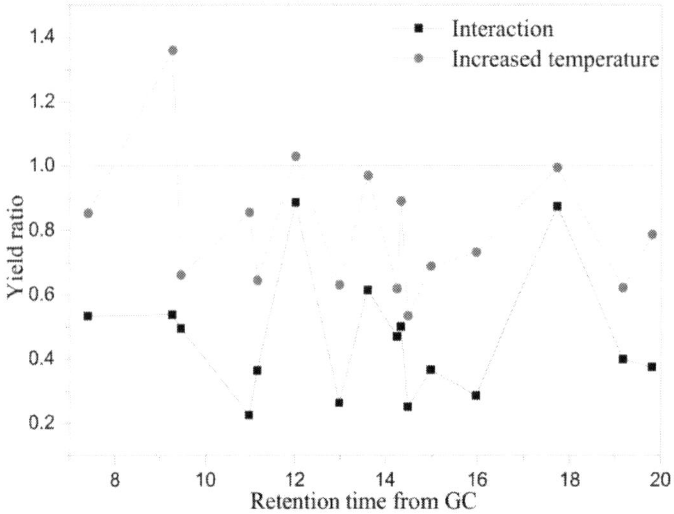

Figure 3. Effect of interaction and increased temperature[a] of 700 on lignin-derived products from levoglucosan-lignin pyrolysis

[a] the yield ration for increased temperature is product yield at 700 ℃ divided by product yield at 600 ℃

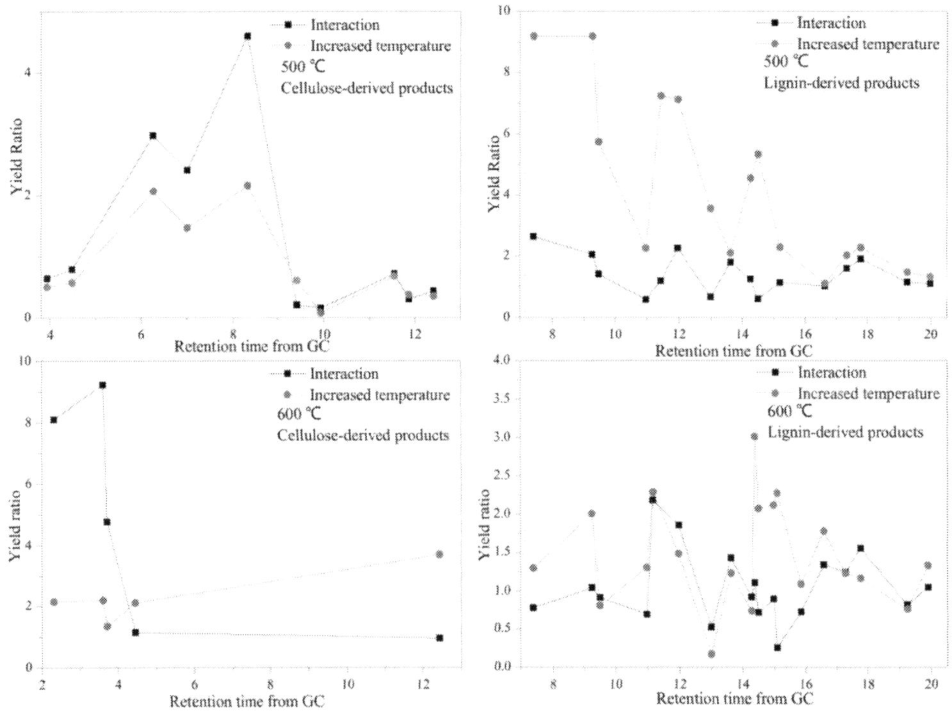

Figure 4. Effect comparison of interaction and increases time on cellulose-lignin products distribution

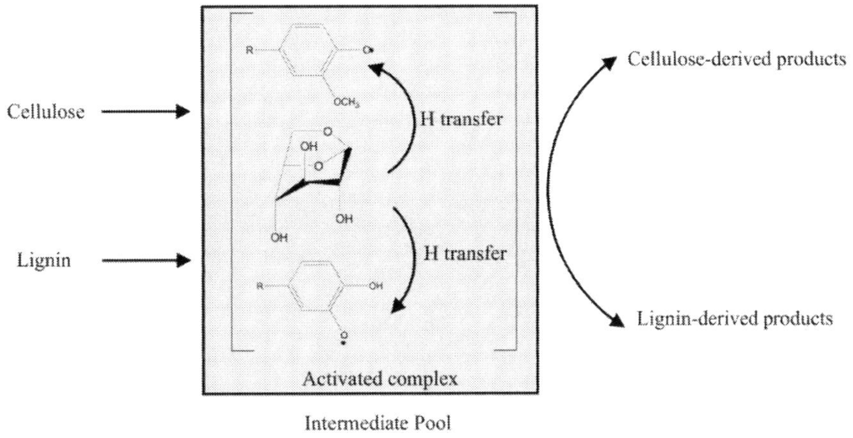

Figure 5. interaction mechanism in intermediate pool of cellulose-lignin pyrolysis

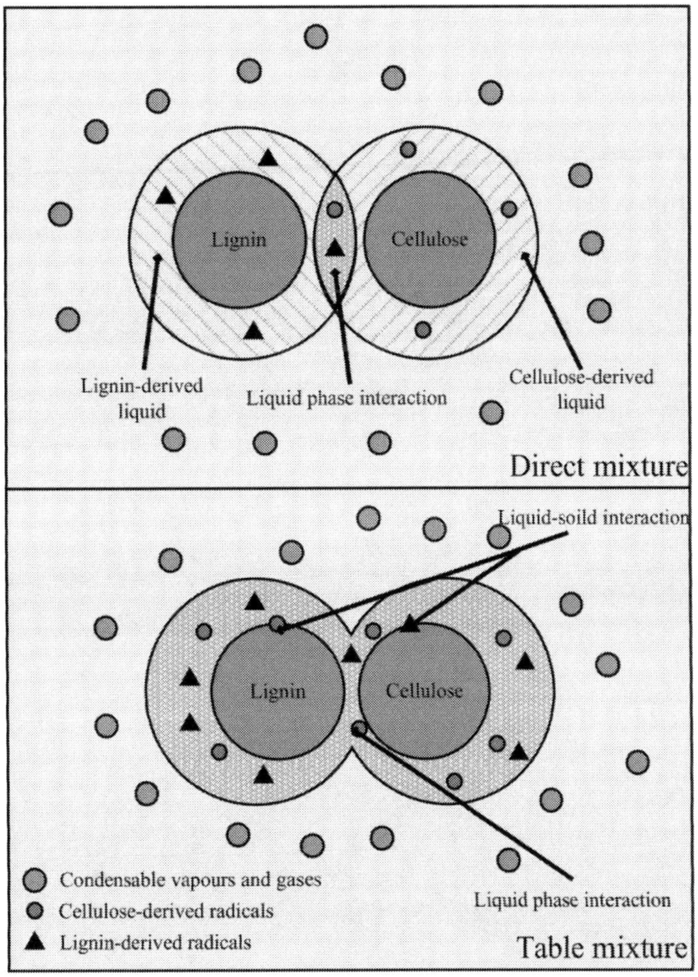

Figure 6. Different interaction phase for direct and table mixture

TABLES

Table 1. Product distribution of cellulose-lignin pyrolysis under temperature range 500-700 ℃

RT (min)	Compound	500 ℃ cellulose[a]	lignin[a]	γ_D	γ_T	600 ℃ cellulose	lignin	γ_D	γ_T	700 ℃ cellulose	lignin	γ_D	γ_T
	Only From Cellulose												
1.611	Acetone	/	/	/	/	/	/	/	/	8.73%	/	1.23	1.09
1.805	Glycolaldehyde	/	/	/	/	1.54%	/	0.99	/	2.34%	/	1.24	1.61
1.866	3-Buten-2-one	/	/	/	/	/	/	/	/	4.90%	/	0.70	0.73
2.304	Acetone alcohol	/	/	/	/	2.19%	/	4.64	8.11	2.18%	/	3.31	3.44
3.613	Butanedial	/	/	/	/	0.62%	/	3.59	9.24	0.63%	/	3.00	3.40
3.713	Methyl pyruvate	/	/	/	/	1.39%	/	2.04	4.76	0.87%	/	4.06	4.34
3.934	Furanone	3.33%	/	0.60	0.65	2.01%	/	/	/	/	/	/	/
4.452	Furfural	8.08%	/	0.71	0.79	5.54%	/	0.84	1.15	5.41%	/	0.65	0.92
6.267	2-Cyclopenten-1-one,2-hydroxy-	1.32%	/	2.25	2.98	3.23%	/	1.72	3.95	2.94%	/	1.87	1.32
6.962	Furfural,5-methyl-	0.87%	/	1.91	2.41	1.52%	/	/	/	1.59%	/	1.06	0.89
8.314	Methylcyclopentenolone	0.44%	/	4.07	4.60	1.13%	/	4.33	7.16	0.79%	/	3.69	3.45
9.939	Levoglucosan	8.21%	/	0.17	0.16	0.80%	/	/	/	/	/	/	/
11.533	DHMP[b]	1.71%	/	0.75	0.72	2.03%	/	0.00	1.82	/	/	/	/
11.866	DAGP[b]	6.14%	/	0.33	0.30	2.73%	/	/	/	3.34%	/	/	/
12.416	5-Hydroxymethylfurfural	14.56%	/	0.49	0.44	9.03%	/	0.46	0.95	7.95%	/	0.50	0.43
	Only From Lignin												
7.375	Phenol	/	0.42%	2.03	2.64	/	1.47%	1.13	0.77	/	1.39%	1.20	1.05
9.262	Phenol,4-methyl-	/	0.37%	1.64	2.06	/	1.27%	1.35	1.04	/	1.86%	1.12	0.92
9.468	Phenol,2-methoxy-	/	4.98%	1.14	1.41	/	10.81%	1.26	0.91	/	6.34%	1.04	0.91
10.613	Phenol,3,5-dimethyl	/	/	/	/	/	0.45%	/	/	/	1.12%	1.17	0.99
10.972	Phenol,4-ethyl	/	3.77%	0.54	0.59	/	3.23%	1.12	0.69	/	3.07%	0.87	0.68
11.161	Phenol,2-methoxy-4-methyl-	/	5.47%	1.01	1.20	/	15%	3.12	2.18	/	10.08%	1.08	0.90
11.831	1,2-Benzenediol	/	/	/	/	/	1.45%	1.07	0.00	/	3.14%	1.05	0.55
11.958	Benzofuran,2,3-dihydro	/	5.57%	1.87	2.26	/	3.58%	2.07	1.86	/	3.88%	1.58	1.35
12.803	1,2-Benzenediol,3-methoxy	/	/	/	/	/	2.14%	1.80	0.83	/	2.85%	1.21	0.85
12.992	Phenol,4-ethyl-2-methoxy-	/	7.09%	0.60	0.68	/	9.45%	0.97	0.52	/	6.96%	0.79	0.60
13.46	1,2-Benzenediol,4-methyl	/	/	/	/	/	/	/	/	/	3.81%	0.96	0.56

13.621	2-Methoxy-4-vinylphenol	/	8.57%	1.36	1.80	/	6.83%	1.78	1.43	/	6.15%	1.34	1.27
14.273	Phenol,2,6-dimethoxy	/	2.64%	1.04	1.26	/	4.54%	1.29	0.92	/	2.43%	0.98	0.80
14.51	1-(4-Methoxyphenyl)-1-propanol	/	0.88%	0.50	0.61	/	1.78%	1.01	0.71	/	2.70%	0.88	0.70
15.209	trans-m-Propenyl guaiacol	/	2.27%	1.02	1.14	/	1.97%	0.28	0.25	/	2.27%	0.94	0.86
15.886	cis-Isoeugenol	/	4.72%	1.40	1.48	/	6.99%	1.16	0.72	/	5.56%	0.79	0.64
16.641	4-Acetyl-2-methoxyphenol	/	4.72%	0.69	1.03	/	1.98%	1.76	1.34	/	2.57%	0.88	0.97
17.332	Guaiacylacetone	/	1.38%	0.63	1.61	/	1.06%	1.39	1.24	/	/	/	/
17.79	3-Tert-butyl-4-hydroxyanisole	/	1.43%	1.36	1.91	/	1.24%	1.96	1.55	/	1.05%	1.38	1.32
19.266	Ethyl citrate	/	2.54%	0.88	1.16	/	1.42%	1.21	0.82	/	0.79%	1.18	1.15
19.983	Methoxyeugenol	/	2.66%	0.84	1.10	/	1.34%	1.52	1.05	/	1.30%	1.56	0.92
						Else							
1.45	CO2	5.69%	4.72%	1.60	2.08	8.39%	4.14%	1.99	2.12	7.52%	3.41%	1.05	1.02

[a] The column shows the peak area percentage of the compound in volatiles from cellulose or lignin, respectively.

[b] Abbreviations: DHMP – 3,5-dihydroxy-2-methyl-4H-pyran-4-one; DAGP – 1:4;3:6-dianhydroglucopyranose.

Table 2. Product distribution of levoglucosan-lignin pyrolysis at 600 ℃

	Compound	LG	Lignin	LG-L[a]
	Only from LG			
1.625	4-Penten-1-yl acetate	7.21%	/	1.57
1.864	2,3-Butanedione	1.27%	/	2.09
1.936	3-Buten-2-ol	1.36%	/	3.93
2.301	Acetol	2.64%	/	3.01
4.493	3-cyclopentene-1-acetaldehyde,2-oxo-	0.61%	/	2.70
6.227	2-cyclopenten-1-one,2-hydroxy-	1.95%	/	1.79
8.287	2-cyclopenten-1-one,2-hydroxy-3-methyl-	1.87%	/	2.41
9.407	2,5-dimethyl-4-hydroxy-3(2H)-furanone	0.43%	/	5.35
10.083	2-cyclopenten-1-one,3-ethyl-2-hydroxy	0.81%	/	2.43
11.93	1,4:3,6-Dianhydro-d-glucopyranose	3.83%	/	1.90
12.094	3,4-anhydro-d-galactosan	35.90%	/	1.15
12.612	2,3-anhydro-d-mannosan	5.57%	/	1.28
12.73	Melezitose	1.43%	/	1.55
	Only from Lignin			
7.4	phenol	/	1.47%	0.53
9.277	phenol,3-methyl	/	1.27%	0.54
9.469	Phenol,2-methoxy-	/	10.81%	0.50
10.971	Phenol,4-ethyl-	/	3.23%	0.23
11.156	phenol,2-methoxy-4-methyl-	/	15%	0.36
12.002	Benzofuran,2,3-dihydro-	/	3.58%	0.89
12.98	phenol,4-ethyl-2-methoxy-	/	9.45%	0.26
13.611	2-methoxy-4-vinylphenol	/	6.83%	0.61
14.253	phenol,2,6-dimethoxy-	/	4.54%	0.47
14.336	phenol,2-methoxy-3-(2-propenyl)-	/	0.95%	0.50
14.495	phenol,2-methoxy-4-propyl	/	1.78%	0.25
14.981	phenol,2-methoxy-5-(1-propenyl)-	/	1.97%	0.37
15.97	phenol,2-methyoxy-4-(1-propenyl)-	/	6.99%	0.29
16.551	Acetoguaiacon	/	1.98%	0.45
17.086	5-tert-butylpyrogallol	/	1.83%	0.21
17.242	Guaiacylacetone	/	1.06%	0.36
17.719	3-tert-butyl-4-hydroxyanisole	/	1.24%	0.87
19.17	ethyl citrate	/	1.42%	0.40
19.807	Methyoxyeugenol	/	1.34%	0.38
	Others			
1.432	CO2	10.86%	4.14%	1.16

[a] LG-L is denoted for Levoglucosan-lignin

Title Mesoporous molecular sieve with strong acidity as catalysts support for pyrolysis of furfural residue to levulinic acid

Abstract Mesoporous molecular sieve with strong acidity has been prepared during a simple one step synthesis procedure. The resulting material was characterized by X-ray diffraction (XRD), transmission electronmicroscopy (TEM), N_2 adsorption/desorption, pyridine adsorbed FTIR spectroscopy and ^{27}Al MAS NMR techniques. The resulting material showed disordered worm-like mesoporous channels and strong acidity. Three TiO_2/ $SO4^{2-}$soliod catalysts were synthesized using ZSM-5□SBA-15 and self-made disordered worm-like molecular sieve as support respectively, and the catalytic performance of the resulting catalysts was examined in pyrolysis of furfural residue to levulinic acid(LA). The catalyst synthesised using self-made disordered worm-like mesoporous molecular sieve as support showed higher conversion of cellulose to levulinic acid(LA) than those of traditional microporous or mesoporous molecular sieve.

Presenter: Haiyan Xu, Energy Research Institute, Henan Academy of Sciences, and Henan Key Lab of Biomass Energy, China

Ethanol Production with Process of Simultaneous Saccharification and Fermentation at High Temperature

Xu Jingliang, Gao Yueshu, Liu Yunyun, Zhang Yu, Yuan Zhenhong, Zhuang Xinshu
Key Laboratory of Renewable Energy, Guangzhou Institute of Energy Conversion, Chinese Academy of Sciences, Guangzhou 510640, P.R. China
xjl@ms.giec.ac.cn
http://people.ucas.ac.cn/~xjl?language=en

Abstract: Simultaneous saccharification and fermentation (SSF) at high temperature is an attractive process for the cellulosic ethanol production. In the present study, different pretreatment processes, including liquid hot water (LHW) pretreatment, sodium hydroxide (NaOH) pretreatment, and their combined pretreatments, were conducted to improve the enzymatic digestibility and sugar recovery with sugarcane bagasse (SCB). LHW pretreatment could solubilize over 82.35% of xylan and 42.31% of lignin, and the NaOH pretreatment could remove 77.85% of lignin and retain most of glucan and xylan. For combined pretreatments, the sequence of two procedures had significant effects on the chemical composition, substrate characteristic and the subsequent enzymatic hydrolysis process. LHW-NaOH pretreatment could solubilize over 92.22% of xylan and remove 76.47% of lignin, and the highest total sugar recovery of 73.3% was achieved after 72 h of enzymatic hydrolysis. While NaOH-LHW pretreatment, which could remove nearly 84.33% of lignin, but only solubilize 71.15% of xylan, showed the highest enzymatic digestibility. The pretreatment efficiency order was as the following: NaOH-LHW > NaOH > LHW-NaOH > LHW. Taking the economy efficiency and simplicity into consideration, NaOH pretreatment was chosen as substrate pretreatment methods for the subsequent enzymatic hydrolysis and fermentation experiment.

To obtain high sugar yield and ethanol concentration, a fed-batch process was performed. Two processes namely simultaneous saccharification and fermentation (SSF) and separate hydrolysis and fermentation (SHF) for ethanol production had also been compared with commonly used *S. cerevisiae*. The results indicated that the ethanol concentration and yield in the SHF were higher, while the ethanol productivity (gram per unit volume and time) was lower than that of SSF. For 25% of substrate loading, the ethanol productivity and ethanol concentration could attain to 0.37 $g.L^{-1}.h^{-1}$, and 36.25 g/L for SSF in 48 hours of fermentation, respectively, while that for SHF process could attain to 0.32 $g.L^{-1}.h^{-1}$ and 47.95 g/L for SHF in 32 hours of fermentation. To alleviate the temperature incoordination during SSF process, a thermo-tolerant yeast, *Kluyveromyces marxianus* NCYC 587 was also employed, compared with *S. cerevisiae*. The results showed that the ethanol concentration remarkably increased to 42.21 g/L after 48 hours of fermentation and the ethanol productivity increased up to 0.44 $g.L^{-1}.h^{-1}$ with *Kluyveromyces marxianus*, and the maximum ethanol concentration, productivity, and yield (g ethanol /g cellulose) of SSF process with *Kluyveromyces marxianus* were all better than that of SSF process with *S. cerevisiae*. The studies also found that the ethanol concentration can improve by 30% with adding of β-glucosidase.

Key Words: Ethanol; Sugarcane bagasse; Pretreatment; Simultaneous saccharification and fermentation; Separate hydrolysis and fermentation;

Mechanism study of biomass pyrolysis

Bin Ru, Haizhou Lin, Qinjie Cai, Wuxing Sun, Shurong Wang*, Zhongyang Luo

State Key Laboratory of Clean Energy Utilization, Zhejiang University, Hangzhou310027, China

Biomass pyrolysis technology for liquid bio-fuels production has been developed for more than thirty years, while its industrial application is still restricted by the low grade of crude bio-oil. The mechanism study of biomass pyrolysis would contribute to improving the biomass pyrolysis technology, and providing guidance for bio-oil upgrading and reactor design. However, the structure of biomass is very complex, it is difficult to get insights into pyrolysis mechanism of the integral biomass. Since biomass is mainly composed of cellulose, hemicellulose and lignin, it is feasible to firstly investigate the pyrolysis mechanism of each individual components, followed by a modification of components interaction for the final integral biomass pyrolysis model. Cellulose, a polymer of glucose connected with β-(1,4)-glucosidic bond, usually has a maximum weight loss at around 350 °C. By using a high-radiation flash pyrolysis reactor, it is the first time for quantified characterization of the active cellulose, a key intermediate product of the Broido-Shafizadeh cellulose pyrolysis model. It is found that the active cellulose is a yellow, water-soluble solid material, which mainly composed of oligosaccharides. Additionally, density functional study (DFT) was introduced to investigate the possible pathway of cellulose decomposition with glucose as model compound. The Mulliken value analysis indicated that the glucose pyranoid ring could be easily opened. The followed transition state search proved that the formation of 5-hydroxymethyl-2-furaldehyde (HMF) with low energy barriers is the most favorable pathway for glucose pyrolysis. Hemicellulose, composed of many different monosaccharides, is another important carbohydrate in biomass. Xylan and other four hemicellulose monosaccharides, including two pentoses (xylose and arabinose) and two hexoses (mannose and galactose), were selected as model compounds. The TG-FTIR analysis indicated that the monosaccharides underwent similar degradation routes involving ring opening and carbon chain breakage, while xylan exhibited different decomposition behavior for the formation of acetic acid, CO_2, and CO due to the *O*-acetyl groups and 4-*O*-methylglucuronic acid units connected to the backbone. The followed DFT study indicated that furfural formation is more favorable than 1-hydroxy-2-propanone and 4-hydroxydihydrofuran-2(3H)-one during xylose pyrolysis. While hexoses preferred to form HMF because of the high energy barrier of the dissociation of hydroxymethyl group. And the breakage of O-acetyl groups for acetic acid formation could occurred at low temperature due to its lower energy barrier compared with that of the monosaccharides decomposition. As for lignin, a complex polymer of aromatic alcohols, is responsible for the phenols and pyrolytic lignin in the resulting bio-oil. FTIR, ^{1}H-NMR, ^{13}C-NMR and GPC were used to characterize the structures of lignin, it has high contents of methoxy groups and ether bond linkages. An introduced double-Gaussian distributed activation energy model indicated that the apparent activation energy for lignin pyrolysis is 148.77 kJ/mol. Moreover, it is found that the lignin from hardwood has higher content of syringyl units than softwood lignin, which results in the different distribution of pyrolytic products. And the characterization of pyrolytic lignin deduced that during the pyrolysis, the β-O-4 bonds in lignin are almost cleavage, yielding many aromatic oligomers, such as trimers and tetramers.

Acknowledgements: The authors are grateful for the financial support from the National Natural Science Foundation (51276166), the National Basic Research Program of China (2013CB228101).

387639 Where Education and Industry Meet: Lessons from a Trip to India

Abstract:

The focus of Cornell University's, NSF-funded Grass Roots GK-12 Program is to advance research and education in renewable and alternative energy. Through a two-week program in India, in cooperation with the Mumbai-based Somaiya Group, a group of Cornell chemical engineering graduate students and New York State high school teachers explored these ideas at an international level. The trip focused on educating students through inexpensive hands-on activities, visiting energy efficient schools, and learning about the Somaiya Group's use of biomass as a renewable energy source. Hands-on classroom activities involving physics (hand made centrifuge), chemistry (surface tension lab), and biology (DNA extraction) were presented to eighth and ninth grade students at four schools spread across two Indian states. The Somaiya schools and factories incorporated various renewable energy sources. For example, one rural school was installing a solar concentrator to generate heat for cooking and designing a biodigestor to produce electricity. The biorefineries utilized renewable energy at a larger scale. From sugar cane production, the company produces not only sugar, but also ethanol, specialty chemicals, wax, and electricity from burning bagasse. The processes were interconnected, with waste from sugar production serving as inputs for wax and electricity production. This talk will focus on what we learned from the trip, as well as the benefits of international experiences to education at all levels.

International educational opportunities for engineering students include, but are not limited to, internships, study abroad programs, and humanitarian programs such as Engineers Without Borders (EWB). Since 2007, a course at xxx University involving engineering students from many disciplines has focused on developing and implementing small-scale engineering projects in developing communities located in Tonga, Ghana, and Peru. The two-semester, three-credit Global Engineering Outreach (GEO) course has recently focused on projects in communities located in northern and southern Peru. Most importantly, the course focuses on integrating both engineering and social aspects that are critical for developing sustainable products. Students in the course, who speak Spanish, interact on a near-weekly basis with the communities in their native language during project development.

This work provides insights into the development of sustainable international programs in a developing community context. Highlights of this work will present student survey results associated with the social and engineering context of development projects. Recent results of implemented projects and the associated impacts will also be discussed. Particularly, recent projects involving pumps, washing machines, water filters, and ovens implemented in the Uros Islands of Southern Peru will be addressed. Feedback from those living on the islands have shown that these projects have had an impact on their health and economic conditions.

Fuel me up!

Van der Borght, K., Toch, K., Marin G.B., Thybaut, J.W.
Ghent University, Laboratory for Chemical Technology, Ghent, B-9052, Belgium.

Introduction

The training of chemical engineering students is of multi-disciplinary nature which ensures that a master of science in chemical engineering is readily employable in highly demanding industrial as well as academic environments. Numerous efforts are put in by the educational community to prepare these students at their best capabilities [1-4]. At Ghent University, Belgium (Europe), this learning trajectory comprises 5 years, i.e., 3 years bachelor of science in chemical engineering followed by 2 years master of science in chemical engineering. During the bachelor years, a fundamental basis of mathematics and sciences is laid upon which a solid engineering training can be constructed. Learning via integrated projects starts already in the 1st semester of the first bachelor year and is followed by a project in the 2nd bachelor in chemical technology and material science and culminates in a cross course project in the 3rd bachelor year focusing on chemical engineering which combines knowledge gained during the bachelor years. After obtaining the bachelor degree, students enter the master years in which chemical engineering is approached in a multi-scale ideology, see Figure 1, which corresponds to the one followed at the Laboratory for Chemical Technology [5].

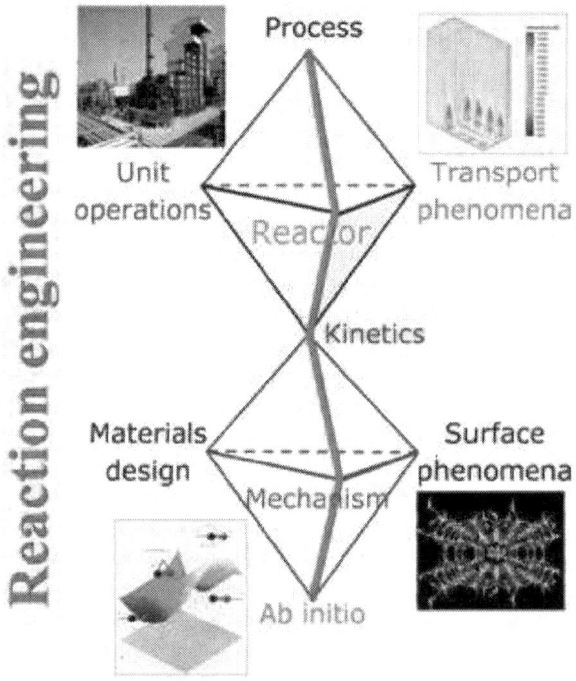

Figure 1: Multi-scale approach to chemical reaction engineering as envisioned by the Laboratory of Chemical Technology, Ghent University [5].

This work presents a case study of the cross course project for students in the 3rd bachelor year. This project already incorporates a wide variety of (chemical) engineering topics, indicated by Figure 1, among others chemical reactors, unit operations, analysis techniques, kinetic modeling and transport phenomena.

The starting point for this project was to produce a fuel for a radio controlled (RC) model car equipped with a glow engine, which is of the internal combustion type, see Figure 2. Typical RC fuels for such a two-stroke engine consist of a mixture of 10 – 40 % nitromethane mixed with methanol and additives for lubrication. In the framework of evolving towards a more sustainable society, the students' assignment was to produce a bio based fuel and have the car running. Because of the open-ended character of the assignment, an overview of possible renewable alternatives was composed and resulted in the selection of bioethanol and biodiesel for an in-depth study. Bioethanol was selected to be a possible 'green' substituent for the methanol while biodiesel would be used as a lubricant. A multi scale approach was used for a critical evaluation of the two biofuel candidates: an experimental investigation, kinetic modeling and an industrial plant simulation.

Figure 2: RC car used for the cross course project

Biofuel production: experimental investigation

For bioethanol production, a lab scale fermentation unit was implemented which started from sugar and baker's yeast, see

Figure 3, left. Special attention was given to the necessity for nutrient addition and the effect of ethanol on the performance of the yeast. The effect of different initial sugar and nutrition concentration was investigated, see

Figure 3, left. The progress of the fermentation was followed as function of time via sampling in which the unconverted sucrose and fructose is dehydrated to hydroxymethylfurfural by addition of H_2SO_4 and followed by the complexion of HMF with phenol. The formed complex can be detected by UV/VIS.

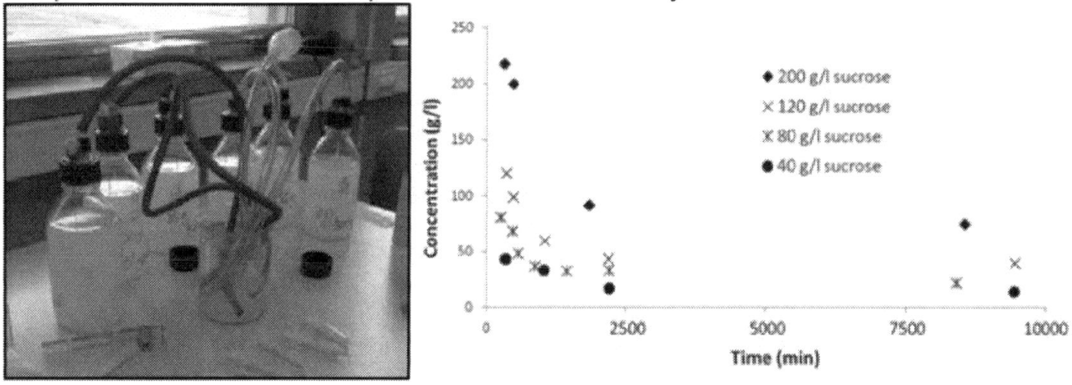

Figure 3: left: Lab scale fermentation of sugar; right: Sucrose concentration as function of initial sucrose concentrations (◆, 200 g/l sucrose; X: 120 g/l sucrose; ⁎ : 80 g/l sucrose; ●, 40 g/l sucrose

For fuel application purposes, the ethanol mixture obtained after fermentation needs to be enriched in ethanol via distillation. De distillation unit and Txy diagram is shown in Figure 4. Due to the azeotrope at 96 wt.% ethanol in water, not all water could be removed. The ultimately obtained ethanol concentration was determined using a density measurement and amounted to ± 93%.

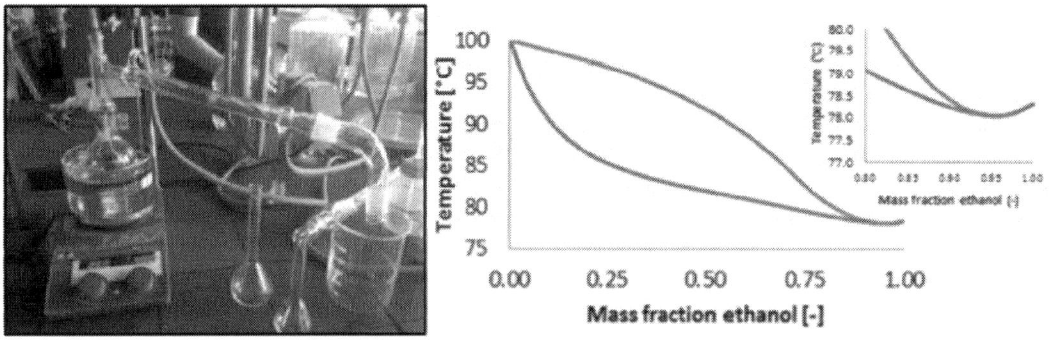

Figure 4: left: Destillation unit; right: Txy diagram of water/ethanol. The red curve is the liquid line and the blue curve is the vapor line.

The biodiesel production occurred through the transesterification of a triglyceride, e.g., sunflower oil, with an alcohol, i.e., methanol, using a homogenous catalyst, i.e., NaOH, in a lab scale experimental setup, see Figure 5, left. A hands-on approach was required to acquire reliable lab scale information on the reaction kinetics: ensuring that all reagents are dry prior to reaction, phase mixing during reaction and purification of the product by catalyst removal and separation of the organic biodiesel phase from the polar glycerol-methanol phase. The product analysis for the biodiesel experiments was performed via ATR-FTIR to follow up the yield of biodiesel as function of time, see Figure 5, right. The effect of reaction temperature (298 to 340 K), molar oil to methanol ratio (1:3 to 1:9), and catalyst concentration (0.1 to 0.6 wt%) on the reaction rates was studied.

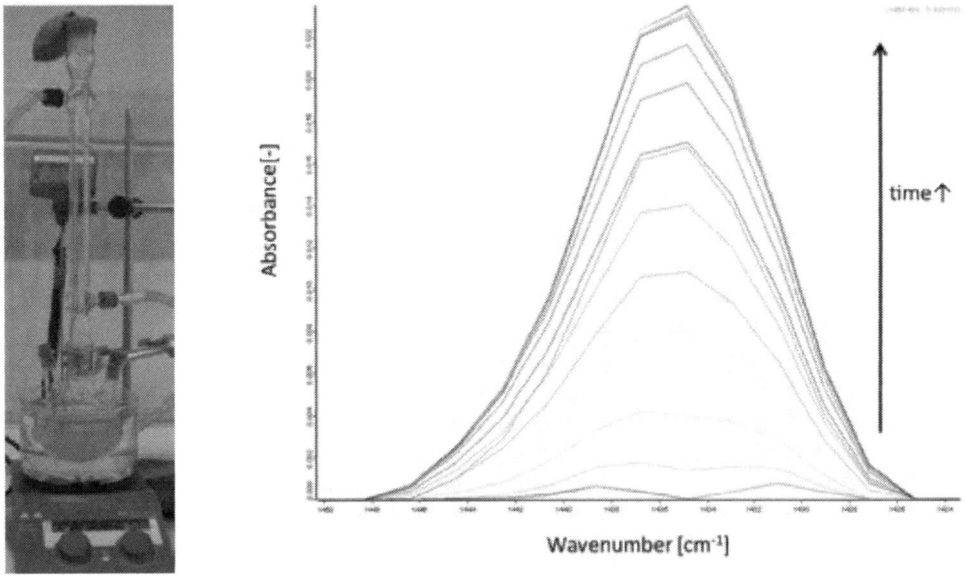

Figure 5: left: lab scale transesterification of sunflower oil, right: ATR-FTIR spectra of the product from sunflower oil transesterification as function of time

Biodiesel production: kinetic modeling

A user-friendly regression package was used for the kinetic modeling of the biodiesel production, i.e., Athena Visual Studio [6]. Based upon the experimental observations for the transesterification of sunflower oil, a kinetic model was constructed and regressed to the experimental data. Initially, a simple power law model was applied but due to some model shortcoming, it was quickly replaced by more fundamental, reaction mechanism based, kinetics. The final kinetic model was able to adequately describe the experimental data as shown in Figure 6.

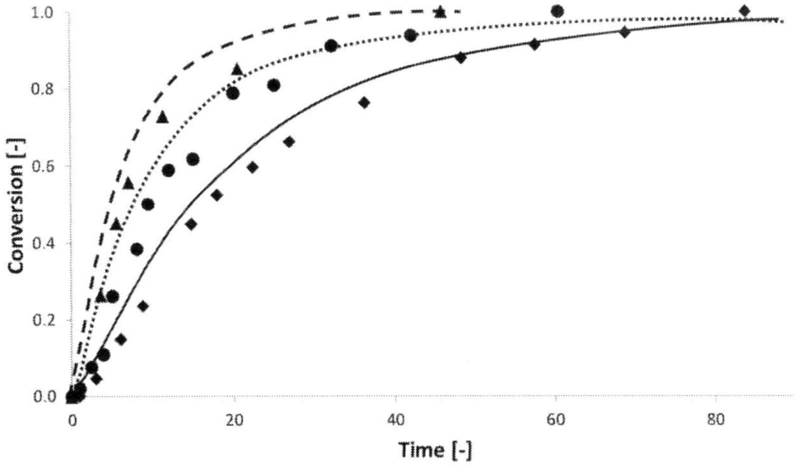

Figure 6: Sunflower oil conversion as function of time at different molar sunflower oil to methanol ratio. Symbols represent experimental data, lines represent model simulations. ◆, full line: molar ratio of sunflower oil to methanol of 1:12; ●, dotted line: : molar ratio of sunflower oil to methanol of 1:9; ▲, dashed line: molar ratio of sunflower oil to methanol of 1:6

Biofuel production: industrial plant simulation

The industrial plant simulations for the bioethanol and biodiesel production based upon typical production capacities were performed using ASPEN® software [7]. Inspired by literature and visits to bioethanol and biodiesel production plants, several plant layouts were benchmarked in ASPEN®. For the reactor, where the actual biofuel production is happening, kinetics were used based upon the experimental observations and kinetic modeling efforts. An example of such a simulation for an industrial biodiesel production plant is given in Figure 7. Similar simulations were performed for the production and purification for the production of fuel-grade ethanol where different separation methods for ethanol and water were investigated.

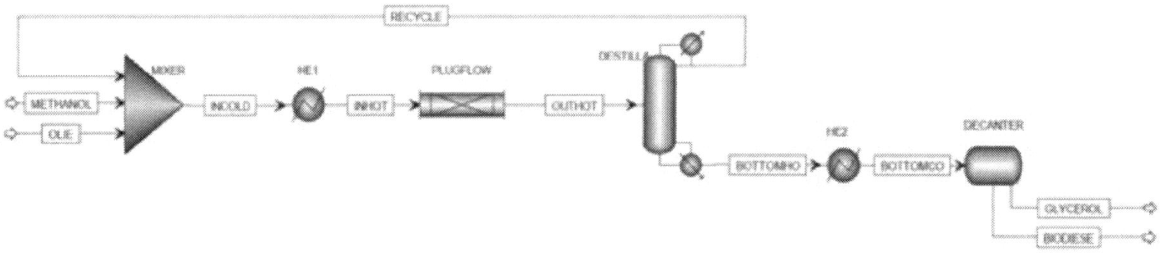

Figure 7: Example for an industrial plant simulation for biodiesel production in ASPEN®

Green RC fuel production

Finally, a fuel was composed using the produced ethanol and biodiesel, to substitute resp. the methanol and lubrication components in the original RC car fuel, and 20 wt.% of Nitromethane (Sigma Aldrich). Nitromethane still needed to be added to this mixture to ensure the proper ignition of the fuel. The fuel was successfully tested in a RC car by the students, see Figure 8. The white smoke produced during testing can be related to the remaining 7% water in the bioethanol.

Figure 8: Testing the green RC fuel

Conclusions

Overall, the project resulted in a critical evaluation of biofuels indicating both the opportunities and challenges of a bio-based society. The students had to follow a multi-disciplinary approach in which chemical reactors, unit operations, analysis techniques, kinetic modeling and transport phenomena were used. Using lab scale equipment, common reactants and easy analysis techniques, the students were able to produce and test a green alternative for RC fuel. Additionally, they gained experience in experimental investigation and kinetic modeling which were used for an industrial plant simulation.

References

[1] C.L. Dym, A.M. Agogino, O. Eris, D.D. Frey, L.J. Leifer, Engineering Design Thinking, Teaching, and Learning, Journal of Engineering Education, 94 (2005).
[2] J.G. Mackenzie, W.B. Earl, R.M. Allen, I.A. Gilmour, Amoco Computer Simulation in Chemical Engineering Education, Journal of Engineering Education, 90 (2001) 331-334.
[3] D.J. Schneck, Integrated Learning: Paradigm for a Unified Approach, Journal of Engineering Education, 90 (2001) 213-217.
[4] B.U. Sherrett, E.J. Nefcy, E.S. Gummer, M.D. Keretsky, An Expert Solution to Assess an Industrially Situated, Computer-Enabled Design Project, Journal of Engineering Education, 102 (2013) 541-576.
[5] Laboratory for Chemical Technology, Web page http://www.lct.ugent.be/.
[6] Athena Visual Studio, Web page http://www.athenavisual.com/.
[7] Aspen Plus(r) , Web page http://www.aspentech.com/products/aspen-plus.aspx.

382114 What Can We Learn from Students Using Facebook?

Abstract:

Over the past two decades, we have witnessed a remarkable growth in the internet and its usage across the globe. Furthermore, rapid advances in digital technology have allowed the internet to become more readily accessible, becoming a part of our everyday life. In recent years, social network sites (e.g. *Facebook*) have provided a new means of interaction and communication between individuals and groups. This has percolated through the education system mainly via the students. We are now starting to realise the potential for *Facebook* to be used for educational purposes; largely due to its desirable qualities which replicate other existing educational tools that allow peer feedback, possess good interaction tools, and fit into the social context of learning environments. This work investigates the extent to which *Facebook* is currently used in engineering education, the factors that motivate educators and students to use *Facebook* for study purposes, and whether *Facebook* is used as 'communities of practise' or mostly as social spaces.

The extent to which Facebook is used in the Chemical Engineering course at Monash University is presented in here. Results from a student survey carried out across different years are presented and ideas of how facebook could be further used as an education tool discussed. Similarly, limitations of using Facebook and the expectations raised by students will be presented. The outcomes of this work will serve as basis to determine if *Facebook* has a place within engineering education, and what of its features could prove to have a positive impact in the students' learning experience.

379025 Internationalizing Access to Chemical Engineering Education By a Lifelong Language Learning Environment

Abstract:

Both chemical engineering and engineering education are rapidly obtaining international dimension. The U.S. has served as home for leading chemical engineers from all over the world for the last 50 years. The challenges of finding best candidates for chemical engineering career and best chemical engineering education will shape the chemical engineering future in the USA in the 21 Century. In turn, the ability to internationalize best practices in chemical engineering (both at the education and industrial level) is necessary for the successful development of the chemical industry in the global perspective.

Attracting promising students to chemical engineering education system and their transforming into chemical engineering professionals for the US and world chemical industry requires a continuous flow of talents. Successful enrollment of talents by top world universities specialized in chemical engineering is an excellent proof of successful internationalization of chemical industry and education. These universities are, however, facing a rapidly growing challenge of dropping demand for chemical engineering programs both by domestic and international students. Chemical engineering education and engineering education in general are becoming less popular. This fact is confirmed by the statistics of the institute of International Education.

The global demand for world class chemical engineering programs taught in U.S. and other top world chemical engineering universities can be increased if we attract more potential students from Asia, Russia and post-Soviet countries. These countries are proud of talented school graduates and offer government funds programs to sponsor their education in best world chemical education centers. The problem of reduction of the influx of talented school graduates from these countries is that academic mobility in these states is closely linked to the language learning problem (in other words, the ability of involved faculty and students to speak world languages such as English at a level of fluency ensuring their integration into the global community). For example, several percent of Russian secondary school graduates and undergraduate students demonstrate TOEFL results sufficient for enrollment by chemical engineering universities in the USA. These students can be, however, excellent chemists and scientists and be motivated to obtain a good-paid position at world-class chemical industry companies requiring a degree from top universities offering chemical engineering programs.

A new approach is offered to involve more students into the U.S.-offered chemical engineering programs – a language learner focused approach. This strategy is called a "Lifelong Language Learning Environment" (a "Triple-L E") meaning that English learning in the universities from above-mentioned countries should be treated as a truly life-long process with the focus on chemical engineering linguistic aspects. Starting at the secondary school level, engineering English learning continues at a university to offer Bachelors able to pursue Master's degree in a U.S. university or a secondary school graduate with TOEFL scores allowing entering a chemical engineering program. Another component of this "Environment" is that faculty in involved in language learning together with students. An indispensible prerequisite for a successful Lifelong Language Learning Environment is involvement of U.S. universities into its development and functioning at their partnering higher education centers abroad (in other words, it requires strong international partnership that is successfully achieved by all top chemical engineering universities in the USA).

This paper also provides a case study of development and implementing of a "Triple-L E" initiative at one of Russian National Research universities specialized in chemical engineering. A peculiar feature of the Lifelong Language Learning Environment implementation in that university is that mostly technical English training is provided, English for Specific Purposes (ESP). English for Specific Purposes is aimed at developing specific skills of communicating in English on the topics of professional interest, depending on the background and career prospects. English teaching is focused on the engineering field of study with the application of computer aided language learning solutions. The core of the process if formed by professors having both engineering (major) and linguistic (minor) degrees or the linguists with 5-10 years of experience in technical English. Successful involvement of U.S. universities is achieved through the development and implementation of joint short-term and dual degree programs with the focus on the spread of disciplines taught in English from these programs to other university chemical engineering programs. This strategy allowed training of faculty, administrators and students for the internationalized chemical engineering programs and attracting English-speaking students from other countries as well as stimulating outgoing academic mobility.

393363 R&D of Green and High Efficient Cellulosic Fuel Ethanol Process Combined with Value Added By-Products By Three Main Feedstock Components

Abstract:

As the biggest developing country, China needs fuel ethanol to meet its demands of the social and economy development. Basing on the "Mid-long term plan for renewable energy" established in 2007, the annual consumption of fuel ethanol will reach 10 million tons in 2020. Cellulosic ethanol, one of the most important second generation biofuel, gains more and more attentions for its substitution of the transportation liquid, lessen the air pollution by the vehicles and promote the rural area economy. In Guangzhou Institute of Energy Conversion, a green and high efficient Cellulosic fuel ethanol process combined with value added by-products by three main feedstock components is proposed. One ton ethanol can be produced with 5.68 t sugarcane bagasse, as well as 0.96 t CO_2, 0.96 t Xylooligosaccharides and 1.59 t lignin. The lignin was used for the mushroom medium which can produce nutrient - rich *Lentinula edodes* successfully.

378318 Effect of Phospholipids on Free Lipase-Catalyzed Biodiesel Production

Abstract:

Free lipase-catalyzed biodiesel production has drawn more and more attentions in recent years because of its advantages of lower cost and faster reaction rate. Utilizing free lipase to convert low quality oils such as crude vegetable oils and microbial oils is beneficial to further reduce the cost of biodiesel production. However, these oils typically contain some amount of phospholipids. Phospholipids were found to affect the lipase-catalyzed process and further influence the enzyme's thermal stability in biodiesel production process. In this work, free lipase-catalyed biodiesel production was systematically investigated with oils containing phospholipids as feedstocks at varied conditions. It was found that increasing temperature led to a decreased fatty acid methyl esters (FAME) yield and poor reuse stability of the lipase in converting oils containing phospholipids for biodiesel production. Further study showed that this inhibitory effect was mainly attributed to the coexistence of phospholipids and methanol in the system. A novel two-step enzyme-catalyzed process was developed, with which the above-mentioned inhibitory effect was eliminated, and a FAME yield of 95.1% could be obtained with oils containing 10% phospholipids as the feedstock.

Study on syngas production by catalytic gasification of maize stalk fast pyrolysis bio-oil

Weiming Yi, Zhihe Li, Peng Fu, Xueyuan Bai, Nana Wang, Hongzhen Cai, Lihong Wang

School of Agricultural and Food Engineering, Shandong University of Technology, Zibo 255049, China
*Corresponding author. Tel.: +86 533 2786308; Fax: +86 533 2786158

Abstract

Hydrogen production via catalytic gasification of maize stalk fast pyrolysis bio-oil over the nickel/alumina supported catalysts promoted cerium was studied using a laboratory scale fixed bed coupled with Fourier transform infrared spectroscopy/thermal conductivity detection analysis (FTIR/TCD). The effects of nickel loading, reaction temperature, steam to carbon molar ratio (S/C) and bio-oil weight hourly space velocity (W_bHSV) on syngas production were investigated.

The results showed that increasing reaction temperature from 600 to 900 °C resulted in the significant increase of hydrogen yield and carbon monoxide yield. Meanwhile, the carbon conversion increased, but its increment was less pronounced than that of hydrogen yield. It was noteworthy that the carbon conversion did not reach 100% even at 950 °C. The difference related to the carbonaceous materials deposited on the catalyst bed, the carbon-containing products such as C_3H_8, C_3H_6, C_4H_{10} etc. that were difficult to be determined and the unconverted reactant itself. The hydrogen yield was significantly enhanced by increasing the S/C ratio from 1 to 3, whereas it increased slightly by further increasing S/C ratio. The CO yield always deceased with increasing the S/C ratio. The opposite trend was observed for CO_2. This was mainly because increasing the S/C ratio is conducive to the proceeding of water gas shift reaction towards hydrogen production, leading to the continuous decrease of CO yield and the increase of H_2 and CO_2 yields. The hydrogen yield decreased with the increase of W_bHSV. Meanwhile, the coke deposition amount changed little with increasing W_bHSV up to 12 h^{-1} and then it increased by 62.9% with the further increase of W_bHSV.

***Keywords*:** Maize stalk; Bio-oil; Catalytic gasification; Syngas production

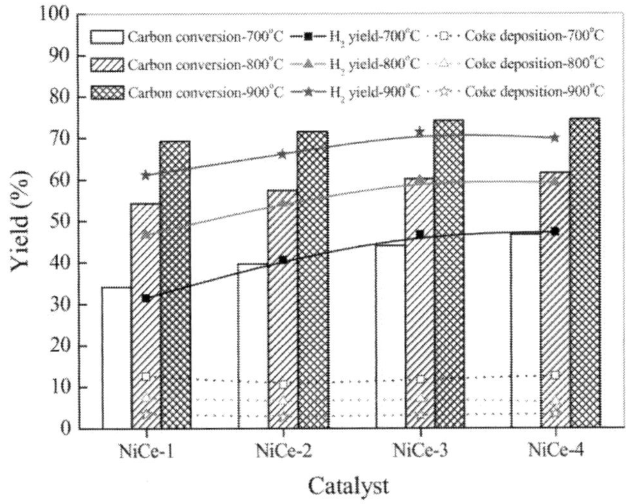

Fig. 1. Comparison of the activity of the four Ni-Ce/Al_2O_3 catalysts with different Ni loadings at the operational conditions of WCMR = 6 and W_bHSV = 12 h^{-1}.

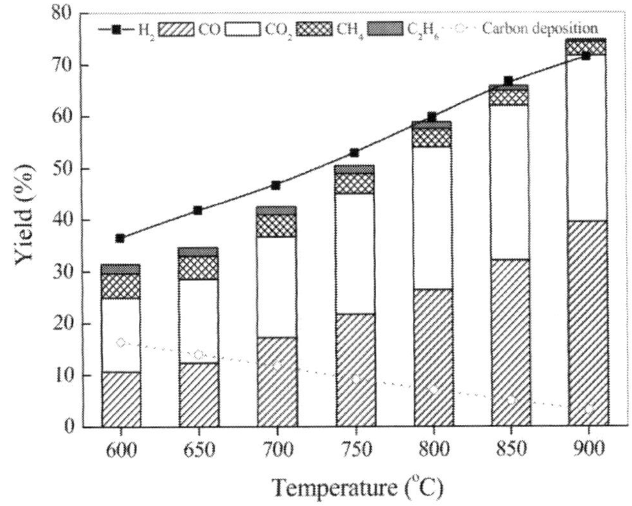

Fig. 2. Effect of reforming temperature on the yields of hydrogen and the main carbon-containing gaseous products as well as coke deposition percentage over NiCe-3 at the operational conditions of WCMR = 6 and W_bHSV = 12 h^{-1}.

Title Support effect on Fischer-Tropsch synthesis of gas from biomass gasification over a Ru-Cu-Co/Al_2O_3 catalyst

Abstract The Al_2O_3-supported Ru-Cu-Co catalysts were modified by loading with several additives (Mg Zr, and La) by incipient wetness impregnation and tested in the Fischer Tropsch reaction using a simulated syngas (CO/H_2) mixture similar to that obtained in biomass gasification. The catalysts characterization included N_2 adsorption, X-ray diffraction (XRD), temperature programmed reduction (TPR) and scanning electron microscopy (SEM) techniques. The modification of support has a great influence on the interaction between active metal and support and then the reducibility of active metal oxides.The result shows that the additives affected the catalyst activity and the C_{5+} selectivity and Zr was the best species for improving the performance.

Presenter: Shunqing LI, Energy Research Institute, Henan Academy of Sciences, and Henan Key Lab of Biomass Energy, China.

Co-cracking of bio-oil fraction and ethanol for aromatic hydrocarbon production

Qinjie Cai, Li Zhang, Junhao Chen, Bin Ru, Shurong Wang,* Zhongyang Luo

State Key Laboratory of Clean Energy Utilization, Zhejiang University, Hangzhou310027, China

Catalytic cracking using solid acidic catalyst like HZSM-5 is a promising way to transform the highly oxygenated biomass pyrolysis oil (bio-oil) into hydrocarbons, especially aromatic hydrocarbons. However, because the crude bio-oil has a high oxygen content and a high degree of unsaturation, corresponding to a low $(H/C)_{eff}$ ratio, coke easily forms during cracking, which leads to catalyst deactivation and reactor blockage. Using ethanol as the co-cracking reactant, which has a relatively high $(H/C)_{eff}$ ratio of 2, will increase the integral $(H/C)_{eff}$ ratio and thus suppress the coke formation. Moreover, some large molecular sugars and phenol oligomers in bio-oil are non-volatile and easily coke in the catalyst bed, and using molecular distillation can effective remove these compounds and enriched reactive acids and ketones in the distilled fraction (DF), which has a better cracking performance than crude bio-oil. Therefore, the bio-oil cracking technology is improved by using molecular distillation to separate bio-oil and obtain the DF first, and then adopting ethanol as the co-cracking reactant. Based on this technical route, the typical model compounds in the DF, namely hydroxypropanone, cyclopentanone, acetic acid, were selected to study their individual co-cracking performance with ethanol first. Under the condition of 400 °C and 2 MPa, with the model compound to ethanol weight ratio of 3:7, the reactants were converted completely over HZSM-5 catalyst. The yields of oil phases for co-cracking of hydroxypropanone, cyclopentanone, acetic acid with ethanol reached 31.9 wt%, 32.1 wt%, and 33.4 wt%, respectively. The aromatic hydrocarbon content in these oil phases was around 95%, and the rest were aliphatic hydrocarbons. Then, the actual bio-oil DF was used to be co-cracked with ethanol. The effect of reaction temperature, pressure, and the DF/ethanol ratio in the feed was investigated. It was found that both increasing the reaction temperature and adopting pressurized cracking benefited the yield and quality of oil phase. Using the optimum reaction temperature and pressure, co-cracking of the DF and ethanol, with different weight ratios, all generated high-quality oil phases. At 400 °C and 2 MPa, co-cracking of DF and ethanol with a weight ratio of 2:3 produced a high oil phase yield of 25.9 wt %. The aromatic hydrocarbon content in this oil phase was 94%. These generated aromatic hydrocarbons are not only important components in the commercial gasoline, but also promising chemical raw materials, especially the benzene, toluene and xylenes (BTX).

Acknowledgements: The authors are grateful for the financial support from the National Natural Science Foundation (51276166), the National Basic Research Program of China (2013CB228101).

Title Conversion of Corncob into Furfural by Acid and $FeCl_3$ Co-catalysis Process

Abstract Corncob was hydrolyzed by acid which is in steam state and $FeCl_3$ co-catalyzed process in a semi-batch tubing-bomb reactor system to produce furfural. Acid (such as acetic acid, formic acid and hydrochloric acid) and $FeCl_3$ co Catalyzed process can promote the furfural yield. A series of experiments were carried out to investigate the effects of acid, $FeCl_3$ concentrations and temperatures on furfural production. A maximum furfural yield of 72.89% (35.74% higher than that in conventional sulfuric acid-catalyzed process) was obtained at 180°C in the presence of 12g $FeCl_3$ and 3% of acid. Acid and $FeCl_3$ co-catalyzed hydrolysis was not only a high efficiency and environmental friendly technique, but also provided a possibility to utilize the furfural residue for ethanol production and other industries, such as ethanol production, levulinic acid production and so on.

Presenter: Suxia Ren, Energy Research Institute, Henan Academy of Sciences, and Henan Key Lab of Biomass Energy, China

Title Performance and emission characteristics of a diesel engine using alternative vehicle fuels formulations based on ethyl levulinate

Abstract In the paper, ethyl levulinate (EL) was taken as the key addition fuel, physical and chemical properties of different fuel formulas of EL-biodiesel-diesel were studied based on China National Standards and Codes: GB/T25199-2010 biodiesel fuel blend (B5). 13 kinds of fuel formulations were gotten and 3 of them were screened. According to theories of grey relation analysis(GRA) and analytical hierarchy process(AHP), synthesizes evaluation methods to evaluate fuel formulas was established with considering 16 kinds of indexes. The fuel formulas that satisfying the B5 standards were optimized and 4 kinds of the optimized fuel formulas were gotten. Optimized fuel formulas were selected to evaluate engine performance and emissions using an unmodification diesel engine test bench. The results show that engine power and torque with the fuel formulas were in general similar to those with diesel. Both brake specific fuel and energy consumption were lower with fuel formulas than with diesel. In general, the emissions of HC and CO and smoke opacity reduced significantly with fuel formulas compared with diesel while the emissions of NO_x and CO_2 increased.

Presenter: Zhiwei WANG, Energy Research Institute, Henan Academy of Sciences, and Henan Key Lab of Biomass Energy, China.

AUTHOR INDEX

AUTHOR INDEX

AUTHOR INDEX

AUTHOR INDEX

AUTHOR INDEX

AUTHOR INDEX

AUTHOR INDEX